CLASSICAL
THEORY OF
ELECTROMAGNETISM

Third Edition

CLASSICAL
THEORY OF
ELECTROMAGNETISM

Third Edition

BALDASSARE DI BARTOLO

Boston College, USA

 **World Scientific**

NEW JERSEY · LONDON · SINGAPORE · BEIJING · SHANGHAI · HONG KONG · TAIPEI · CHENNAI · TOKYO

Published by

World Scientific Publishing Co. Pte. Ltd.

5 Toh Tuck Link, Singapore 596224

USA office: 27 Warren Street, Suite 401-402, Hackensack, NJ 07601

UK office: 57 Shelton Street, Covent Garden, London WC2H 9HE

Library of Congress Cataloging-in-Publication Data

Names: Di Bartolo, Baldassare, author.

Title: Classical theory of electromagnetism / Baldassare Di Bartolo (Boston College, USA).

Description: Third edition. | Singapore ; Hackensack, NJ : World Scientific
    Publishing Co. Pte. Ltd., [2018] | Includes bibliographical references and index.

Identifiers: LCCN 2017047725| ISBN 9789813228191 (hardcover ; alk. paper) |
    ISBN 9813228199 (hardcover ; alk. paper) | ISBN 9789813230033 (pbk. ; alk. paper) |
    ISBN 9813230037 (pbk. ; alk. paper)

Subjects: LCSH: Electromagnetism. | Electromagnetic theory.

Classification: LCC QC760 .D49 2018 | DDC 537--dc23

LC record available at https://lccn.loc.gov/2017047725

British Library Cataloguing-in-Publication Data

A catalogue record for this book is available from the British Library.

For any available supplementary material, please visit
http://www.worldscientific.com/worldscibooks/10.1142/10660#t=suppl

Desk Editor: Christopher Teo

Typeset by Stallion Press
Email: enquiries@stallionpress.com

Printed in Singapore

To the students I have had the privilege to teach and guide

Optimus est magister,
qui docendo animos discipulorum
delectare potest.

Andrea Pozzo
Painter and Architect (1642–1709)

Contents

Foreword

Writing a new textbook on the classical theory of electromagnetism is certainly a difficult task: on the one hand, one has to compete with excellent books already in existence; on the other, the subject is very well established and consolidated, so there is not much space for original viewpoints or new considerations.

However, a different mixture and a different use of the same ingredients, for example, selection of topics versus completeness, didactical details versus erudite conciseness, use of specific examples versus abstract generalizations, and so on, can give rise to a new product that presents an original flavor and can be of great usefulness.

In *Classical Theory of Electromagnetism* by Baldassare Di Bartolo, the material is presented in a very clear way; each topic is analyzed in detail, and each new concept is illustrated by representative examples. Altogether, the text is self-contained and is clearly oriented toward the students. In fact, the book is written at a level suitable for a student graduating in physics and electrical engineering. It is concise in words, but very detailed in mathematical calculations: the equations are explicitly derived, which is of great help to students and allows them to concentrate more on physical concepts, rather than spending too much time on mathematical derivations.

The value of the presentation of the material is increased by the inclusion of a large number of exercises; in fact, it is of great importance for students to see in detail how the theoretical concepts are applied to specific problems, and the explicit examples well represent the main points.

The topics that were selected from this vast field are essentially those that a graduate student in physics and electrical engineering should be familiar with on classical electromagnetism.

There are also additional topics not included in current textbooks. In particular, I find interesting the inclusion of the Kramers–Kronig dispersion relations and their connection with the principle of causality. The introduction of the theory of special relativity is also a great challenge in teaching classical electromagnetism, and this topic is considered with particular care in Di Bartolo's textbook.

The presentation is exceptionally clear from a didactical point of view, which, combined with the wealth of mathematical detail and insistence on thoroughly comprehensive exercises, should ensure that this book will be an effective instrument for the study of electromagnetism.

Giovanni Costa
Università di Padova

Preface to the Third Edition

The appearance of the third edition of this book is due to my belief that my treatment of electromagnetic theory will continue to be well received by students and teachers and to the encouragement I have received from Mr. Christopher Teo and Ms. Yubing Zhao of World Scientific. I am truly grateful to them for their assistance.

The World Scientific Company continues to enrich the scientific literature with high level publications of great interest to scientists, teachers and students. It gives me great pleasure and pride to know that several books of mine are among those published by World Scientific.

Boston, September 2016

Preface to the Second Edition

The appearance of this second edition is due to my persistent belief that my treatment of electromagnetism will continue to be useful to students of this discipline and to the encouragement and support that I have received from Ms. Yubing Zhai, Editor at World Scientific. The preparation of this second edition has provided me with the opportunity to correct most (I hope) of the misprints or mistakes of the first edition. I want to thank the people who pointed them out to me. I want also to acknowledge the assistance received from Mr. Yeow-Hwa Quek, Production Editor at the World Scientific Publishing Company in Singapore.

In the process of reviewing the book, the matter of units came up as one deserving particular attention. The SI system was substituted for the MKS system of the previous edition, but the Gaussian system continues to be used throughout this edition. Proper connections between the two systems are established at various points in the book and at the end of the volume.

Some thoughts about units may be in order. We should not forget that fundamental truths, and scientific truths for that matter, have a life of their own, independent of the way in which they are formulated. Formulae, and the units in which they are expressed, are vehicles that we use to transfer information, and, as such, are only representations of the underlying truth. True understanding aims at the essential and is not conditioned by the particular choice of symbols or system of units. Indeed an understanding that does not transcend a particular formulation is ephemeral

The rereading of my book has brought to my attention another relevant matter: the degree of detail with which the derivations should be presented. There seem to be two schools of thought in this regard. According to one school it is better to leave to the student the task of filling up the missing steps of a derivation so that, by spending time and effort on it, the student will retain the matter better. According to the other school the student should not spend time going through long derivation, and providing these derivations allows the student to concentrate on problem solving and on the applications of the theory.

A cursory view of this book will indicate that I may have acted as a believer in the second school of thought, even if I do not discount the point of the first school. The fact is that I have felt comfortable with my style of presentation that indulges in details–galore. I have tried to write the book I would have liked to have when I was a student myself. It may be that, despite the many years of teaching, at heart I remain a student, someone who feels that his main task remains that of learning.

The experience of revising this work has been sobering: it has reminded me, with poignancy, of the fact that we may know a lot, but what we do not know is indeed much, much more than what we know. With the passing of time I have seen the line supposedly dividing me as a teacher from my students blurring and fading away altogether. Thus, I have come to the conclusion that truly "We Are All Students," and it is with this thought that I complete the preface to this second edition.

Boston, March 2002

Preface to the First Edition

Electricity is not only the powerful agent that, breaking through the atmosphere, frightens us with the flash of the lightning and the roar of the thunder, but also the life-giving agent that brings from the Sun to the Earth with light and heat the magic of colors and the breath of life, makes our eyes and heart participate in the beauty of the external world and transmits to the soul the charm of a glance and the enchantment of a smile.

Galileo Ferraris

The purpose of this textbook is to provide graduate students in physics and electrical engineering with the background in classical electricity and magnetism necessary to deal with experimental problems in electromagnetism and with the quantum mechanical treatments of radiation and matter.

This textbook is tailored to students' needs. It has evolved out of lectures that I have given for several years at Boston College and contains well tried and (I hope) clearly exposed material that can be covered in a two-term graduate course. I have not aimed at writing a "complete" book on electromagnetism; such a book will most probably never exist, considering the vastness of the field and the enormous number of subfields and possible applications. Rather, I have carefully selected a large number of topics that deal with the essential aspects of the theory of electromagnetism. I have tried to conduct the derivations in detail and to avoid the "It can be shown" method, much hated by students. I have used representative examples and reported well-tried exercises at the end of each chapter.

This book is intended for first-year graduate students in physics and in electrical engineering. The readers are assumed to have taken courses in differential equations and vector analysis and, for the part of the book starting with Chapter 6, to be familiar with classical mechanics at graduate level.

A brief description of the book follows:

Chapter 1, Mathematical Introduction, gives the mathematical foundations for the book and does not contain much physics. Chapter 2, Charges and Electrostatics, deals with electrostatics and is based on the Coulomb law. Units of measurements are discussed. Dielectric polarization and dielectrics are also treated. The concept of stress tensor is introduced. Chapter 3, Stationary Currents and Magnetostatics, deals with magnetostatics. Units of measurements are discussed again. Forces and magnetic stress tensor are examined.

Chapter 4, Induction and Quasi-stationary Phenomena, deals with quasi-stationary situations, those cases in which (physical size of the system) $\times$ frequency $\ll c =$ velocity of light. Self- and mutual-inductances are discussed. Chapter 5, General Discussion of Maxwell Equations, deals with the consequences of Maxwell equations, such as energy and momentum conservation laws. Plane waves, reflection, and refraction are also treated. In Chapter 6, Theory of Relativity: I, we begin the treatment of the theory of relativity. We expect the first term of the course to end with Chapter 6. In particular, Chapter 6 introduces the principles of relativity and the Lorentz transformation. Potential equations are expressed in Lorentz-invariant form. Plane waves and twin paradox are discussed.

A teacher of this course might begin the second term with Chapter 7, Theory of Relativity: II, and spend the first class summarizing the contents of Chapter 6. Chapter 7 contains the Lorentz transformation of the fields, the Minkowski force, Gauss's theorem in four dimensions and the electromagnetic energy–momentum tensor. It deals also with the use of Green functions for the solution of potential equations. Chapter 8, Radiation from a Moving Charge, treats the Liénard-Wiechert potentials and the radiation of a moving point charge, small periodic oscillations, and synchrotron radiation. Chapter 9, Radiation Damping and Electromagnetic Mass, introduces the related concepts of radiation damping and self-force and applies these concepts to the problems of energy loss by radiation in periodic motion and to the scattering of radiation.

Chapter 10, Radiation from Periodic Charge and Current Distributions, deals with electric and magnetic radiating multipoles and with the radiation patterns that they produce. Chapter 11, Lagrangian and Hamiltonian Formulation of Electro-dynamics, describes the Lagrangian formulation of the Maxwell equations, the field Hamiltonian, and the Poisson bracket method. The Hamiltonian of a closed system consisting of charges and radiation is treated. This chapter provides the necessary classical background for the quantum mechanical treatment of radiation. Chapter 12, Electromagnetic Properties of Matter, deals with the interaction of radiation with matter. It treats dispersion, absorption, the scattering theory of the index of refraction, relaxation, and the Kramers–Kronig relations.

Note: Since this book is directed toward the application of the theory of electromagnetism to modern physics, the Gaussian system of units is used throughout. Connections with other systems of units, such as MKS, are established.

In concluding this preface, I wish to acknowledge my great indebtedness to Professor Felix Villars, whose memorable lectures on electromagnetic theory at M. I. T. (which I had the fortune of attending while a graduate student there) have strongly influenced my way of treating this subject; to Mr. Jeremiah J. Lyons of W. H. Freeman and Company for graciously granting me the permission to adapt some of the problems of the book *Spacetime Physics* by Edwin F. Taylor and John A. Wheeler and report them in Chapters 6 and 7; to Professor Edwin F. Taylor for allowing me to do the same with some of the problems that appear in the second edition of *Spacetime Physics*; to Ms. Holly Hodder of Prentice Hall for encouragement and support and to Mrs. Mary De Luca, also of Prentice Hall, for her kind assistance; to many colleagues who contributed with helpful suggestions, and criticisms, especially Professors Giovanni Costa of the University of Padua, Daniel Schechter of the California State University, Long Beach, Silverio Almeida of Virginia Tech, T. A. K. Pillai of the University of Wisconsin, and Michael Lieber of the University of Arkansas;

and to the students who have attended my courses on electromagnetic theory at Boston College and have contributed to this book with their valuable comments.

It has been said that no book is ever "finished"; what really happens is that at a certain point the author decides to simply let it go. The experience of trying your best and never being satisfied with the results of your efforts is truly humbling. Therefore, I wish to ask the readers to point out to me mistakes, omissions, and obscurities they may find in the book. Perhaps by the time we will come out with the fiftieth edition, the book will be close to perfection! Close to, but never absolutely perfect, since, as we know well, "Nemo perfectus est, qui perfectior esse non appetit."

Baldassare Di Bartolo
Chestnut Hill, Massachusetts

1

Mathematical Introduction

1.1. Vector Notation

We shall indicate a vector by a bold letter such as $\mathbf{a}$ or by means of its three Cartesian components

$$\mathbf{a} \equiv (a_1, a_2, a_3) \quad \text{or} \quad (a_x, a_y, a_z) \tag{1.1.1}$$

The *scalar product* of two vectors is given by

$$\mathbf{a} \cdot \mathbf{b} = \sum_{i=1}^{3} a_i b_i \tag{1.1.2}$$

and the *cross product* by

$$\mathbf{a} \times \mathbf{b} = \mathbf{c} \tag{1.1.3}$$

where

$$c_1 = a_2 b_3 - a_3 b_2$$
$$c_2 = a_3 b_1 - a_1 b_3 \qquad (1.1.4)$$
$$c_3 = a_1 b_2 - a_2 b_1$$

Note also that

$$\mathbf{a} \cdot (\mathbf{b} \times \mathbf{c}) = \det \begin{vmatrix} a_1 & a_2 & a_3 \\ b_1 & b_2 & b_3 \\ c_1 & c_2 & c_3 \end{vmatrix}$$
$$= \mathbf{b} \cdot (\mathbf{c} \times \mathbf{a}) = -\mathbf{b} \cdot (\mathbf{a} \times \mathbf{c})$$
$$= \mathbf{c} \cdot (\mathbf{a} \times \mathbf{b}) = -\mathbf{c} \cdot (\mathbf{b} \times \mathbf{a}) \qquad (1.1.5)$$

and that

$$\mathbf{a} \cdot (\mathbf{b} \times \mathbf{c}) = (\mathbf{a} \times \mathbf{b}) \cdot \mathbf{c} \qquad (1.1.6)$$
$$\mathbf{a} \times (\mathbf{b} \times \mathbf{c}) = \mathbf{b}(\mathbf{a} \cdot \mathbf{c}) - \mathbf{c}(\mathbf{a} \cdot \mathbf{b}) \qquad (1.1.7)$$
$$(\mathbf{a} \times \mathbf{b}) \times \mathbf{c} = \mathbf{b}(\mathbf{a} \cdot \mathbf{c}) - \mathbf{a}(\mathbf{b} \cdot \mathbf{c}) \qquad (1.1.8)$$

Given a Cartesian coordinate system, a *position vector* identifies the position of a point in space

$$\mathbf{x} \equiv (x_1, x_2, x_3) \qquad (1.1.9)$$

In a new coordinate system obtained by performing a rotation and/or a reflection on the previous system, while keeping the origin fixed, the position vector of a point is given by

$$\mathbf{x}' \equiv (x_1', x_2', x_3') \qquad (1.1.10)$$

where

$$x_k' = \sum_{m=1}^{3} R_{mk} x_m \qquad (1.1.11)$$

On the other hand,

$$x_m = \sum_{n=1}^{3} Q_{nm} x_n' \qquad (1.1.12)$$

We must have

$$x'_k = \sum_{m=1}^{3} R_{mk} x_m = \sum_{m=1}^{3} R_{mk} \sum_{n=1}^{3} Q_{nm} x'_n$$

$$= \sum_{n=1}^{3} \left(\sum_{m=1}^{3} R_{mk} Q_{nm} \right) x'_n \qquad (1.1.13)$$

or

$$\sum_{m=1}^{3} R_{mk} Q_{nm} = \delta_{kn} \qquad (1.1.14)$$

The matrix $\mathcal{Q}$ of the coefficients Q_{nm} is the inverse of the matrix $\mathcal{R}$ of the coefficients R_{mk}.

Also,

$$\sum_{m=1}^{3} x_m^2 = \sum_{m=1}^{3} \left(\sum_{k=1}^{3} Q_{km} x'_k \right) \left(\sum_{n=1}^{3} Q_{nm} x'_n \right)$$

$$= \sum_{k=1}^{3} \sum_{n=1}^{3} \left(\sum_{m=1}^{3} Q_{km} Q_{nm} \right) x'_k x'_n = \sum_{k=1}^{3} x'^2_k \qquad (1.1.15)$$

and

$$\sum_{m=1}^{3} Q_{km} Q_{nm} = \delta_{kn} \qquad (1.1.16)$$

Comparing Eq. (1.1.16) with Eq. (1.1.14), we obtain the relation between the coefficients of $\mathcal{Q}$ and $\mathcal{R}$:

$$Q_{km} = R_{mk} \qquad (1.1.17)$$

This means that the inverse of the matrix $\mathcal{R}$ is its transpose:

$$\mathcal{R}\tilde{\mathcal{R}} = 1 \qquad (1.1.18)$$

Such a matrix is called *real orthogonal*. We can then write

$$x'_k = \sum_{m=1}^{3} R_{mk} x_m$$

$$x_m = \sum_{n=1}^{3} R_{mn} x'_n \qquad (1.1.19)$$

Other vectors that we shall encounter are the following:

$$\text{Difference of positions:} \quad \Delta\mathbf{x} = \mathbf{x}_p - \mathbf{x}_q \qquad (1.1.20)$$

$$\text{Velocity:} \quad \mathbf{v} = \frac{d\mathbf{x}}{dt} \qquad (1.1.21)$$

$$\text{Acceleration:} \quad \mathbf{a} = \frac{d^2\mathbf{x}}{dt^2} \qquad (1.1.22)$$

1.2. Fields

A *scalar field* is a function defined in a certain region of space. A change in the coordinate system does not change its value:

$$\phi(\mathbf{x}_p) = \phi'(\mathbf{x}_p') \qquad (1.2.1)$$

A *vector field* is a vector defined in a certain region of space:

$$\mathbf{A}(\mathbf{x}) = (A_1, A_2, A_3) \qquad (1.2.2)$$

where

$$\begin{aligned}
A_1 &= A_1(x_1, x_2, x_3) \\
A_2 &= A_2(x_1, x_2, x_3) \\
A_3 &= A_3(x_1, x_2, x_3)
\end{aligned} \qquad (1.2.3)$$

and

$$A_i(\mathbf{x}) = \sum_{k=1}^{3} R_{ik} A_k'(\mathbf{x}') \qquad (1.2.4)$$

A *tensor field* is a second-rank tensor and is identified by nine components that are defined in a region of space. These components transform as follows:

$$T_{ik}(\mathbf{x}) = \sum_{jl} R_{ij} R_{kl} T_{jl}'(\mathbf{x}'), \quad i, k = 1, 2, 3; \quad j, l = 1, 2, 3 \qquad (1.2.5)$$

1.3. Vector Differential Operator

We define the *gradient operator* as follows:

$$\boldsymbol{\nabla} \equiv \left(\frac{\partial}{\partial x}, \frac{\partial}{\partial y}, \frac{\partial}{\partial z} \right) \qquad (1.3.1)$$

∇ changes its components if we change the system of coordinates:

$$\frac{\partial}{\partial x_i} = \sum_k \frac{\partial x'_k}{\partial x_i} \frac{\partial}{\partial x'_k} = \sum_k R_{ik} \frac{\partial}{\partial x'_k} \qquad (1.3.2)$$

The following applications of ∇ are relevant. The *gradient* of a function $\phi(\mathbf{x})$ is a vector given by

$$\nabla \phi(\mathbf{x}) = \text{grad } \phi(\mathbf{x}) \equiv \left(\frac{\partial \phi}{\partial x_1}, \frac{\partial \phi}{\partial x_2}, \frac{\partial \phi}{\partial x_3} \right) \qquad (1.3.3)$$

The *divergence* of a vector $\mathbf{A}(\mathbf{x})$ is a scalar given by

$$\nabla \cdot \mathbf{A}(\mathbf{x}) = \text{div } \mathbf{A} = \frac{\partial A_1}{\partial x_1} + \frac{\partial A_2}{\partial x_2} + \frac{\partial A_3}{\partial x_3} \qquad (1.3.4)$$

The *curl* of a vector $\mathbf{A}(\mathbf{x})$ is a vector given by

$$\nabla \times \mathbf{A}(\mathbf{x}) = \text{curl } \mathbf{A} = \begin{vmatrix} \mathbf{i} & \mathbf{j} & \mathbf{k} \\ \dfrac{\partial}{\partial x_1} & \dfrac{\partial}{\partial x_2} & \dfrac{\partial}{\partial x_3} \\ A_1 & A_2 & A_3 \end{vmatrix}$$

$$\equiv \left\{ \left(\frac{\partial A_3}{\partial x_2} - \frac{\partial A_2}{\partial x_3} \right), \left(\frac{\partial A_1}{\partial x_3} - \frac{\partial A_3}{\partial x_1} \right), \left(\frac{\partial A_2}{\partial x_1} - \frac{\partial A_1}{\partial x_2} \right) \right\}$$

$$(1.3.5)$$

where $\mathbf{i}$, $\mathbf{j}$ and $\mathbf{k}$ are the unit vectors in the x, y, and z directions, respectively.

The *Laplacian operator* is a scalar operator:

$$\nabla \cdot \nabla = \nabla^2 = \frac{\partial^2}{\partial x_1^2} + \frac{\partial^2}{\partial x_2^2} + \frac{\partial^2}{\partial x_3^2} \qquad (1.3.6)$$

If $\phi(\mathbf{x})$ is a scalar, $\nabla^2 \phi(\mathbf{x})$ is obviously also a scalar. The Laplacian operator is invariant under linear coordinate transformations:

$$\nabla'^2 = \sum_i \frac{\partial^2}{\partial x_i'^2} = \sum_i \left(\sum_j R_{ji} \frac{\partial^2}{\partial x_j} \sum_k R_{ki} \frac{\partial}{\partial x_k} \right)$$

$$= \sum_j \sum_k \left(\sum_i R_{ji} R_{ki} \right) \frac{\partial}{\partial x_j} \frac{\partial}{\partial x_k}$$

$$= \sum_j \sum_k \delta_{jk} \frac{\partial}{\partial x_j} \frac{\partial}{\partial x_k} = \sum_k \frac{\partial^2}{\partial x_k^2} = \nabla^2 \qquad (1.3.7)$$

We know that for any vector **a**

$$\mathbf{a} \times \mathbf{a} = 0$$

Likewise,

$$\boldsymbol{\nabla} \times \boldsymbol{\nabla} = \begin{vmatrix} \mathbf{i} & \mathbf{j} & \mathbf{k} \\ \dfrac{\partial}{\partial x_1} & \dfrac{\partial}{\partial x_2} & \dfrac{\partial}{\partial x_3} \\ \dfrac{\partial}{\partial x_1} & \dfrac{\partial}{\partial x_2} & \dfrac{\partial}{\partial x_3} \end{vmatrix} = 0 \qquad (1.3.8)$$

because the order of the derivation operation is irrelevant. Therefore,

$$\boldsymbol{\nabla} \times \boldsymbol{\nabla}\phi(\mathbf{x}) = \text{curl grad } \phi(\mathbf{x}) = 0 \qquad (1.3.9)$$

Consider the vector $\boldsymbol{\nabla} \times \mathbf{A}$; we obtain

$$\boldsymbol{\nabla} \cdot (\boldsymbol{\nabla} \times \mathbf{A}) = (\boldsymbol{\nabla} \times \boldsymbol{\nabla}) \cdot \mathbf{A} \qquad (1.3.10)$$

and, because of Eq. (1.3.8),

$$\boldsymbol{\nabla} \cdot (\boldsymbol{\nabla} \times \mathbf{A}) = \text{div curl } \mathbf{A} = 0 \qquad (1.3.11)$$

We derive the property

$$\begin{aligned} \boldsymbol{\nabla} \times (\boldsymbol{\nabla} \times \mathbf{A}) &= \boldsymbol{\nabla}(\boldsymbol{\nabla} \cdot \mathbf{A}) - (\boldsymbol{\nabla} \cdot \boldsymbol{\nabla})\mathbf{A} \\ &= \boldsymbol{\nabla}(\boldsymbol{\nabla} \cdot \mathbf{A}) - \nabla^2 \mathbf{A} \\ &= \text{grad div } \mathbf{A} - \nabla^2 \mathbf{A} \end{aligned} \qquad (1.3.12)$$

$\nabla^2 \mathbf{A}$ is a vector whose components are

$$\frac{\partial^2 A_1}{\partial x_1^2} + \frac{\partial^2 A_1}{\partial x_2^2} + \frac{\partial^2 A_1}{\partial x_3^2}, \quad \frac{\partial^2 A_2}{\partial x_1^2} + \frac{\partial^2 A_2}{\partial x_2^2} + \frac{\partial^2 A_2}{\partial x_3^2},$$

$$\frac{\partial^2 A_3}{\partial x_1^2} + \frac{\partial^2 A_3}{\partial x_2^2} + \frac{\partial^2 A_3}{\partial x_3^2} \qquad (1.3.13)$$

In this book we shall use the following notation for differential operators. If one of them, say $\boldsymbol{\nabla}$, is operating on a function of more than one variable, such as $\phi(\mathbf{x}, \mathbf{x}_p)$, then we may attach a subscript to it, as in $\boldsymbol{\nabla}_x$, to indicate that the differentiation has to be carried out with respect to the variable **x**. If the differentiation operation has to be performed on a function of *one variable*, then no subscript will be affixed to the operator.

1.4. Gauss's Theorem and Related Theorems

Gauss's theorem is a relation between a volume integral and a surface integral. Consider the function $f(\mathbf{x})$, which may be a scalar or a component of a vector defined in a certain region of space. Let V be a volume inside this region and S a surface surrounding V; let dS be an infinitesimal surface element of S, and $\mathbf{n}$ a unit vector perpendicular to the surface element dS. Assume that $f(\mathbf{x})$ and its partial derivatives are continuous in V and on S. With reference to Fig. 1.1, we have at $z = z_2$

$$dx\, dy = dS\, n_z \tag{1.4.1}$$

and at $z = z_1$

$$dx\, dy = -dS\, n_z \tag{1.4.2}$$

Therefore, we can write

$$\int_V dx\, dy\, dz \frac{\partial f}{\partial z} = \int dx\, dy [f(x,y,z_2) - f(x,y,z_1)]$$

$$= \int_S dS\, n_z f(x,y,z)$$

or

$$\int_V dx\, dy\, dz \frac{\partial f}{\partial z} = \int_S dS\, n_z f(x,y,z) \tag{1.4.3}$$

Let us now examine various implications of this theorem.

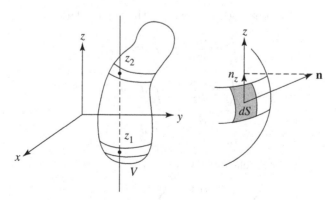

FIGURE 1.1

1.4.1. *The Gradient Theorem*

$$f = \text{scalar} = f(x, y, z) \tag{1.4.4}$$

We have

$$\int_V dx\, dy\, dz \frac{\partial f}{\partial z} = \int_S dS\, n_z f(x, y, z)$$

$$\int_V dx\, dy\, dz \frac{\partial f}{\partial y} = \int_S dS\, n_y f(x, y, z)$$

$$\int_V dx\, dy\, dz \frac{\partial f}{\partial z} = \int_S dS\, n_x f(x, y, z)$$

and, in general,

$$\int_V d\tau \boldsymbol{\nabla} f(x, y, z) = \int_S dS\, \mathbf{n} f(x, y, z) \tag{1.4.5}$$

where $d\tau = dx\, dy\, dz$. Relation (1.4.5) expresses the *gradient theorem*.

1.4.2. *The Divergence Theorem*

$$f = \text{component of a vector} = A_i$$

We have

$$\int_V d\tau \frac{\partial A_i(\mathbf{x})}{\partial x_k} = \int_S dS\, n_k A_i(\mathbf{x}) \tag{1.4.6}$$

where $i, k = 1, 2, 3$. Letting $i = k$, we get the three relations

$$\int_V d\tau \frac{\partial A_1}{\partial x_1} = \int_S dS\, n_1 A_1$$

$$\int_V d\tau \frac{\partial A_2}{\partial x_2} = \int_S dS\, n_2 A_2$$

$$\int_V d\tau \frac{\partial A_3}{\partial x_3} = \int_S dS\, n_3 A_3$$

If we sum these three relations, we obtain

$$\sum_i \int_V d\tau \frac{\partial A_i}{\partial x_i} = \sum_i \int_S dS\, n_i A_i \tag{1.4.7}$$

or

$$\int_V d\tau [\boldsymbol{\nabla} \cdot \mathbf{A}(\mathbf{x})] = \int_S dS [\mathbf{n} \cdot \mathbf{A}(\mathbf{x})] \qquad (1.4.8)$$

Relation (1.4.8) expresses the *divergence theorem*.

1.4.3. *The Curl Theorem*

In general,

$$\int_V d\tau \frac{\partial A_i}{\partial x_k} = \int_S dS \; n_k A_i \qquad (1.4.9)$$

where $i, k = 1, 2, 3$. We consider now two relations, one obtained from Eq. (1.4.9) by taking $k = 2$ and $i = 3$ and another by taking $i = 2$ and $k = 3$. If we subtract the latter from the former relation, we obtain

$$\int_V d\tau \left(\frac{\partial A_3}{\partial x_2} - \frac{\partial A_2}{\partial x_3} \right) = \int_S dS (n_2 A_3 - n_3 A_2) \qquad (1.4.10)$$

Similarly, for $k = 3$ and $i = 1$,

$$\int_V d\tau \left(\frac{\partial A_1}{\partial x_3} - \frac{\partial A_3}{\partial x_1} \right) = \int_S dS (n_3 A_1 - n_1 A_3) \qquad (1.4.11)$$

and for $k = 1$ and $i = 2$

$$\int_V d\tau \left(\frac{\partial A_2}{\partial x_1} - \frac{\partial A_1}{\partial x_2} \right) = \int_S dS (n_1 A_2 - n_2 A_1) \qquad (1.4.12)$$

The last three equations give

$$\int_V d\tau (\boldsymbol{\nabla} \times \mathbf{A}) = \int_S dS (\mathbf{n} \times \mathbf{A}) \qquad (1.4.13)$$

Relation (1.4.13) expresses the *curl theorem*.

1.4.4. *Green's Theorem*

Consider the vector

$$\mathbf{A} = \psi \boldsymbol{\nabla} \varphi \qquad (1.4.14)$$

We have

$$\text{div } \mathbf{A} = \boldsymbol{\nabla} \cdot \mathbf{A} = \boldsymbol{\nabla} \psi \cdot \boldsymbol{\nabla} \varphi + \psi \boldsymbol{\nabla}^2 \varphi \qquad (1.4.15)$$

On the other hand, because of the divergence theorem,

$$\int_V (\mathbf{\nabla} \cdot \mathbf{A})d\tau = \int_S dS(\mathbf{n} \cdot \mathbf{A}) \tag{1.4.16}$$

Then

$$\int_V (\mathbf{\nabla} \cdot \mathbf{A})d\tau = \int_V (\psi \nabla^2 \varphi + \mathbf{\nabla}\psi \cdot \mathbf{\nabla}\varphi)d\tau$$

$$= \int_S dS(\mathbf{u} \cdot \mathbf{A}) - \int_S dS(\mathbf{n} \cdot \psi \mathbf{\nabla}\varphi)$$

$$= \int_S \psi \frac{\partial \varphi}{\partial n} dS \tag{1.4.17}$$

where

$$\frac{\partial \varphi}{\partial n} = \mathbf{n} \cdot \mathbf{\nabla}\varphi \tag{1.4.18}$$

and we obtain the *first form of Green's theorem*:

$$\int_V (\psi \nabla^2 \varphi + \mathbf{\nabla}\varphi \cdot \mathbf{\nabla}\psi)d\tau = \int_S \left(\psi \frac{\partial \varphi}{\partial n}\right) dS \tag{1.4.19}$$

If we interchange ψ with φ in Eq. (1.4.19), we find

$$\int_V (\varphi \nabla^2 \psi + \mathbf{\nabla}\psi \cdot \mathbf{\nabla}\varphi)d\tau = \int_S \left(\varphi \frac{\partial \psi}{\partial n}\right) dS \tag{1.4.20}$$

Subtracting Eq. (1.4.20) from Eq. (1.4.19), we obtain the *second form of Green's theorem*:

$$\int_V [\psi \nabla^2 \varphi - \varphi \nabla^2 \psi]d\tau = \int_S \left(\psi \frac{\partial \varphi}{\partial n} - \varphi \frac{\partial \psi}{\partial n}\right) dS \tag{1.4.21}$$

If we put $\varphi = \psi$ in either Eq. (1.4.19) or Eq. (1.4.20), we find

$$\int_V [\varphi \nabla^2 \varphi + (\mathbf{\nabla}\varphi)^2]d\tau = \int_S \varphi \frac{\partial \varphi}{\partial n} dS \tag{1.4.22}$$

1.4.5. Stokes's Theorem

In three dimensions, we found in Eq. (1.4.3) that

$$\int_V d\tau \frac{\partial f}{\partial z} = \int_S dS n_z f \tag{1.4.23}$$

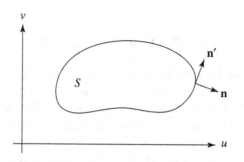

FIGURE 1.2

In two dimensions

$$\int_S \frac{\partial f}{\partial u} du\, dv = \oint_l dl\, n_u f \tag{1.4.24}$$

where $\oint_l$, indicates an integral around the contour surrounding the surface S.

Let f_u and f_v be two components in a (u,v) plane of a vector $\mathbf{f}$:

$$\iint \left(\frac{\partial f_v}{\partial u} - \frac{\partial f_u}{\partial v}\right) du\, dv = \int (f_v n_u - f_u n_v) dl \tag{1.4.25}$$

From Fig. 1.2, we find

$$n'_u = -n_v$$
$$n'_v = -n_u \tag{1.4.26}$$

Then

$$\iint \left(\frac{\partial f_v}{\partial u} - \frac{\partial f_u}{\partial v}\right) du\, dv = \int (f_v n_u - f_u n_v) dl$$

$$= \int (f_v n'_v + f_u n'_u) dl$$

$$= \oint (\mathbf{f} \cdot \mathbf{n}') dl = \oint \mathbf{f} \cdot d\mathbf{l} \tag{1.4.27}$$

But we know that

$$\frac{\partial f_v}{\partial u} - \frac{\partial f_u}{\partial v} = (\text{curl } \mathbf{f})_w = (\boldsymbol{\nabla} \times \mathbf{f})_w$$

where w is the direction perpendicular to the plane (u,v). Then we have

$$\iint (\text{curl } \mathbf{f})_w du\, dv = \oint (\mathbf{f} \cdot \mathbf{n}') dl \tag{1.4.28}$$

or

$$\iint (\boldsymbol{\nabla} \times \mathbf{f})_w \, du \, dw = \oint \mathbf{f} \cdot d\mathbf{l} \tag{1.4.29}$$

This result is valid for a plane surface, whatever its contour.

We want now to extend this result to a generic surface S surrounded by a boundary curve C; this curve in general will not be in a plane [see Fig. 1.3(a)]. We can bridge across the curve C with a path A as in Fig. 1.3(b). The sum of the line integrals of the vector $\mathbf{f}$ along the two paths $C_1 A$ and $C_2 A$ in the directions indicated is equal to the line integral

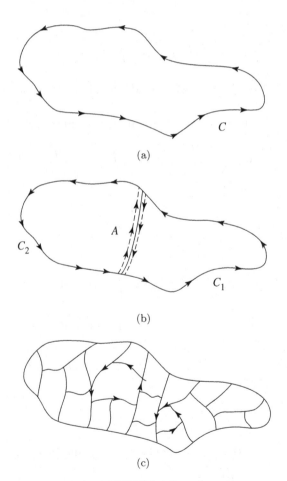

(a)

(b)

(c)

FIGURE 1.3

around C, because the integral along the path A is taken twice in opposite directions and hence cancels out. We can further subdivide the surface S into smaller and smaller surface areas as in Fig. 1.3(c), so that at some point each tiny area is planar. When this is the case, we can apply to every such area relation (1.4.29). If we then sum all these relations, the result will be that, for any surface S, we can write

$$\int_S (\nabla \times \mathbf{f}) \cdot d\mathbf{S} = \oint \mathbf{f} \cdot d\mathbf{l} \tag{1.4.30}$$

Relation (1.4.30) expresses *Stokes's theorem.*

CHAPTER 1 EXERCISES

1.1. Let $\mathbf{c}$ be a constant vector and $\mathbf{r}$ the position vector. Calculate the gradient of $\mathbf{c} \cdot \mathbf{r}$ and the divergence and the curl of $\mathbf{r}$ and $\mathbf{c} \times \mathbf{r}$.

1.2. Prove that, if a fluid experiences a pure rotation, the divergence of its velocity is zero and the curl of its velocity is twice the angular velocity.

1.3. Given a scalar $f(\mathbf{x})$, two vectors $\mathbf{A}(\mathbf{x})$ and $\mathbf{B}(\mathbf{x})$, and a volume V surrounded by a surface S, prove the following relations:

(a) $\int_V \nabla f d\tau = \int_S f\mathbf{n} \, dS$

(b) $\int_S \mathbf{A}(\mathbf{B} \cdot \mathbf{n})dS = \int_V \mathbf{A}(\nabla \cdot \mathbf{B})d\tau + \int_V (\mathbf{B} \cdot \nabla)\mathbf{A}d\tau$

Assume that you have already proved the divergence theorem.

1.4. The points of a plane rotate about a fixed point with an angular velocity $\omega = \omega(r)$, r being the distance from the fixed point. What function must $\omega(r)$ be in order for the velocity $\mathbf{v}$ to have $\nabla \times \mathbf{v} = 0$?

1.5. Prove the following relation:

$$\int_S (\nabla\phi \times \nabla\psi) \cdot d\mathbf{S} = \oint_l \phi d\psi$$

where S is a surface of contour l.

1.6. The Cartesian components of a vector are

$$a_x = y\frac{\partial\phi}{\partial z} - z\frac{\partial\phi}{\partial y}$$

$$a_y = z\frac{\partial\phi}{\partial x} - x\frac{\partial\phi}{\partial z}$$

$$a_z = x\frac{\partial\phi}{\partial y} - y\frac{\partial\phi}{\partial x}$$

where $\phi = \phi(x, y, z)$. Express **a** as the cross product of two vectors and show that

$$\mathbf{a} \cdot \mathbf{r} = 0$$

$$\mathbf{a} \cdot \nabla\phi = 0$$

1.7. Let **u** be a unit vector and let **a** be a generic vector. Prove that

$$\mathbf{a} = \mathbf{u}(\mathbf{a} \cdot \mathbf{u}) + \mathbf{u} \times (\mathbf{a} \times \mathbf{u})$$

1.8. In a certain region of space, the temperature at any point $P(x, y, z)$ is given by

$$T = T_0(x^2 + y^2 + z^2)$$

What is the rate of change of the temperature at P with respect to distance in a direction specified by the direction cosines (l, m, n)?

1.9. A "central" vector field is given by

$$\mathbf{F} = \mathbf{r}f(r)$$

Determine $f(r)$ in such a way that $\nabla \cdot \mathbf{F} = \nabla \times \mathbf{F} = 0$.

1.10. Show that, if a vector $\mathbf{A}(\mathbf{x})$ has zero divergence and zero curl in a certain region of space, then $\mathbf{A}(\mathbf{x})$ is the gradient of a solution of the Laplace equation.

1.11. Consider a vector field

$$\mathbf{A} = a\frac{\mathbf{r}}{r}$$

with $a = $ constant.

(a) Evaluate the line integral

$$\int_{r_1}^{r_2} \mathbf{A} \cdot d\mathbf{r}$$

(b) Evaluate the surface integral

$$\int_S \mathbf{A} \cdot \mathbf{n}\, dS$$

where S is a spherical surface around the origin with radius r_0 and $\mathbf{n}$ is the unit vector perpendicular to S.

1.12. A vector field $\mathbf{A}(\mathbf{x})$ is specified on the surface S that surrounds a certain region of space V. Show that, if the divergence and curl of $\mathbf{A}(\mathbf{x})$ in V are given, $\mathbf{A}(\mathbf{x})$ is uniquely determined in V.

1.13. A vector $\mathbf{A}(\mathbf{x})$ has zero value on the surface S that surrounds a certain region of space V. The divergence of $\mathbf{A}(\mathbf{x})$ and its curl are zero in V. Show that $\mathbf{A}(\mathbf{x}) = 0$ in V.

1.14. (a) A planar surface S with area A and boundary C lies in the x, y plane. Consider a coplanar vector

$$\mathbf{v} = -y\mathbf{i} + x\mathbf{j}$$

and prove that

$$\int_c \mathbf{v} \cdot d\mathbf{l} = \int_c \mathbf{v} \cdot (dx\,\mathbf{i} + dy\,\mathbf{j}) = 2A$$

(b) Use the equation of the ellipse

$$\frac{x^2}{a^2} + \frac{y^2}{b^2} = 1$$

and the result of Part (a) to find the area of the ellipse.

1.15. Let a vector field be given by $\mathbf{v} = \boldsymbol{\omega} \times \mathbf{r}$ with $\boldsymbol{\omega} = $ constant. Find the value of the line integral of such a vector around a circle C of radius a lying in a plane perpendicular to $\boldsymbol{\omega}$.

1.16. Given a scalar $f(\mathbf{x})$, a vector $\mathbf{a}(\mathbf{x})$, and a surface S of contour l, prove the following relations:

(a) $\oint d\mathbf{l} f(\mathbf{x}) = \int_S dS[\mathbf{p} \times \boldsymbol{\nabla} f(\mathbf{x})]$

(b) $\oint d\mathbf{l} \times \mathbf{a}(\mathbf{x}) = \int_S dS[(\mathbf{p} \times \boldsymbol{\nabla}) \times \mathbf{a}(\mathbf{x})]$

where $\mathbf{p} = $ unit vector perpendicular to S.

1.17. Consider a vector field $\mathbf{A}(\mathbf{x}_p)$ defined by the relation

$$\mathbf{A}(\mathbf{x}_p) = \int d\mathbf{l} \times \boldsymbol{\nabla}_x \frac{1}{r}$$

where $\mathbf{r} = \mathbf{x} - \mathbf{x}_p$ (see Fig. Pl.17). Prove by means of relation (b) of Exercise 1.16 that

$$\mathbf{A}(\mathbf{x}_p) = -\boldsymbol{\nabla}_{x_p} \Phi(\mathbf{x}_p)$$

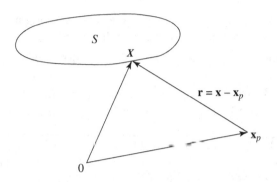

FIGURE P1.17

where

$$\Phi(\mathbf{x}_p) = \int dS(\mathbf{p} \cdot \mathbf{\nabla}_x)\frac{1}{r}$$

for all $\mathbf{x}_p$ not lying on the surface S.

1.18. (a) A vector field is defined by

$$\mathbf{E} = -\mathbf{\nabla}\phi$$

with

$$\phi = \mathbf{d} \cdot \frac{\mathbf{r}}{r^3} \quad (\mathbf{d} = \text{constant})$$

Derive an expression for $\mathbf{E}$.

(b) Prove that $\mathbf{E}$ in Part (a) can be written as follows

$$\mathbf{E} = \mathbf{\nabla} \times \mathbf{A}$$

with

$$\mathbf{A} = \mathbf{d} \times \frac{\mathbf{r}}{r^3}$$

(c) Calculate $\mathbf{\nabla} \cdot \mathbf{E}$ and $\mathbf{\nabla} \times \mathbf{E}$.

1.19. Given a vector field $\mathbf{A}$ with continuous first derivatives in a volume V, prove that

$$\int_V \mathbf{\nabla} \times \mathbf{A}\, d\tau = -\int_S (\mathbf{A} \times \mathbf{n})dS$$

where S is the surface enclosing V.

2

Charges and Electrostatics

2.1. Basic Phenomena

Let a stationary source charge q be located at a point $\mathbf{x}_q$. We measure a field of force around $\mathbf{x}_q$ by means of a test charge q'. Keeping the source charge at $\mathbf{x}_q$ and putting q' at different points in space, we see different forces acting on q'; it is by this procedure that we define a *field*. If the source charge is reduced (increased) by a certain factor, the force acting on q' is found to decrease (increase) by the same factor.

The force acting on the charge q', divided by this charge, gives a field, which we call an *electric field*:

$$\frac{\mathbf{F}(\mathbf{x})}{q'} = \mathbf{E}(\mathbf{x}) \qquad (2.1.1)$$

$\mathbf{E}(\mathbf{x})$ is a property of the source charge and is independent of the test charge q'; it is given by *Coulomb's law*, which, in the Gaussian system of units,

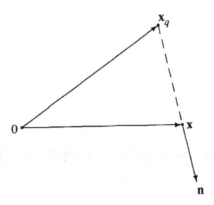

FIGURE 2.1

can be expressed as follows:

$$\mathbf{E}(\mathbf{x}) = -\nabla_x \frac{q}{|\mathbf{x} - \mathbf{x}_q|} = \frac{q\mathbf{n}}{|\mathbf{x} - \mathbf{x}_q|^2} \tag{2.1.2}$$

where (see Fig. 2.1)

$$\mathbf{n} = \frac{\mathbf{x} - \mathbf{x}_q}{|\mathbf{x} - \mathbf{x}_q|} \tag{2.1.3}$$

We can write, in general

$$\mathbf{E}(\mathbf{x}) = -\nabla \phi(\mathbf{x}) \tag{2.1.4}$$

where $\phi(\mathbf{x})$ is called *potential* and is given by

$$\phi(\mathbf{x}) = \frac{q}{|\mathbf{x} - \mathbf{x}_q|} \tag{2.1.5}$$

We now examine two important principles:

2.1.1. *The Superposition Principle*

Given two charges q_1 and q_2 at two different positions (see Fig. 2.2), the potential at point $\mathbf{x}$ is given by

$$\phi(\mathbf{x}) = \phi_1(\mathbf{x}) + \phi_2(\mathbf{x}) \tag{2.1.6}$$

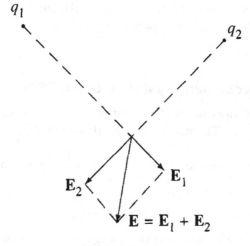

FIGURE 2.2

FIGURE 2.3

where ϕ_1 and ϕ_2 are the potentials due to the charges q_1 and q_2, respectively. Therefore,

$$\mathbf{E}(\mathbf{x}) = -\boldsymbol{\nabla}\phi(\mathbf{x}) = -\boldsymbol{\nabla}\phi_1(\mathbf{x}) - \boldsymbol{\nabla}\phi_2(\mathbf{x}) = \mathbf{E}_1(\mathbf{x}) + \mathbf{E}_2(\mathbf{x}) \quad (2.1.7)$$

2.1.2. *The Symmetry Principle*

We define the field at $\mathbf{x}$ by means of a charge q'. There is, however, no difference in concept between charge q and charge q' (see Fig. 2.3). We can consider q as the test charge and q' as the source charge. In accordance with Newton's third law,

$$\mathbf{F}(\text{on } q' \text{ at } \mathbf{x}) = -\mathbf{F}(\text{on } q \text{ at } \mathbf{x}_q) \quad (2.1.8)$$

or

$$\mathbf{E}(\mathbf{x})q' = -\mathbf{E}(\mathbf{x}_q)q \quad (2.1.9)$$

We should not forget that the experimental quantities that we measure are the forces; the concept of fields derives from these principal parameters.

2.2. Units

2.2.1. *The Electrostatic System of Units (ESU)*

The mechanical units in the ESU system are the cgs units (centimeter, gram, and second) The unit of charge, the *statcoulomb*, is derived from Coulomb's law.

We write the expression of the force acting between two charges q_1 and q_2 as follows:

$$F = |\mathbf{F}| = K\frac{q_1 q_2}{r^2} = \frac{q_1 q_2}{r^2} \tag{2.2.1}$$

where $r =$ distance between the two charges. In expression (2.2.1), we set the proportionality constant $K = 1$, and the unit charge is defined in such a way that two similar unit charges separated by 1 cm in vacuum repel each other with the force of 1 dyne. Dimensionally,

$$[q^2] = [Fr^2] = ML^3T^{-2}$$
$$[q] = LT^{-1}\sqrt{ML} \tag{2.2.2}$$

The field and the potential of a point charge in empty space are given by

$$E = \frac{q}{r^2}$$

and

$$\phi = \frac{q}{r}$$

respectively. Therefore, the dimensions of E and ϕ are given by

$$[E] = \frac{1}{T}\sqrt{\frac{M}{L}}$$
$$[\phi] = \frac{1}{T}\sqrt{ML} \tag{2.2.3}$$

The charge of the electron is -4.80286×10^{-10} statcoulomb.

2.2.2. *The International System of Units (SI)*

Before introducing the SI system of units, we shall make some observations on the system of mechanical units. In such a system, three basic units are chosen, a unit of mass, a unit of length, and a unit of time. This is one of many possible choices, and the number of units chosen (three) does not represent the minimum number of units needed. In fact it is possible to set up a mechanical system of units in which all mechanical quantities can be expressed in terms of length and time.[1]

A similar situation is present with Coulomb's law. If we set the proportionality constant equal to the pure number 1, then charges can be expressed dimensionally in terms of M, L, and T. This is the case for the electrostatic system of units. On the other hand, in the SI system, in addition to the three basic units of length (meter, m), mass (kilogram, kg), and time (second, s), a fourth basic unit is used, a unit of current called *ampere* (A). This unit is defined as follows: "That constant current which, if maintained in two straight parallel conductors of infinite length, of negligible cross section, and placed one meter apart in vacuum, would produce between these conductors a force equal to 2×10^{-7} newton per meter of length.[2]" In the SI system the *coulomb* is a derived unit defined as *ampere* × *second*. In the SI units we express Coulomb's law as follows:

$$F = \frac{1}{4\pi\varepsilon_0} \frac{q_1 q_2}{r^2} \tag{2.2.4}$$

where $1/4\pi\varepsilon_0$ is a proportionality constant and the charges are expressed in coulombs. The dimensions of ε_0 are easily derived from (2.2.4); the result of measurements gives us

$$\varepsilon_0 = 8.85 \times 10^{-12} \frac{\text{C}^2 \text{ sec}^2}{\text{kg m}^3} \tag{2.2.5}$$

We can also work with some other quantities. Since the electric field is the ratio of a force to a charge

$$[E] = \frac{\text{newton}}{\text{coulomb}} = \frac{\text{newton-meter}}{\text{coulomb-meter}} = \frac{\text{joule}}{\text{coulomb-meter}} = \frac{\text{volt}}{\text{meter}} \tag{2.2.6}$$

where we define the unit *volt* (V) as follows:

$$\text{volt} = \frac{\text{joule}}{\text{coulomb}} \tag{2.2.7}$$

---

[1] N. H. Frank, *Introduction to Electricity and Optics*, McGraw-Hill Book Company, Inc., New York, 1950, p. 40.
[2] *The International System of Units (SI)*, National Bureau of Standards Special Publication 330, 1972 edition.

On the other hand,

$$E = \frac{q}{4\pi\varepsilon_0 r^2}$$

and

$$[\varepsilon_0] = \frac{\text{coulomb}}{(\text{volt/meter})\text{meter}^2} = \frac{\text{coulomb}}{\text{volt-meter}} = \frac{\text{farad}}{\text{meter}} \qquad (2.2.8)$$

where we have introduced the new unit, the *farad*:

$$\text{farad} = \frac{\text{coulomb}}{\text{volt}} \qquad (2.2.9)$$

Since in the electrostatic system of units the proportionality constant in the Coulomb's law is a pure number, it is easy to see that the *statfarad*, the corresponding unit in the electrostatic system, has the dimension of a length and is expressed in centimeters.

Let us look for the relation between coulomb and statcoulomb. We claim that

$$1 \text{ coulomb} = 3 \times 10^9 \text{ statcoulomb} \qquad (2.2.10)$$

This can be easily shown to be true. Take two equal charges of 1 C each at a distance of 1 m: these two charges will repel each other with a force that, in the electrostatic system of units, is given by

$$F(\text{esu}) = \frac{3 \times 10^9 \times 3 \times 10^9}{(100)^2} = 9 \times 10^{14} \text{ dynes}$$

The electric field due to either charge at the location of the other charge is given by

$$E(\text{esu}) = \frac{9 \times 10^{14}}{3 \times 10^5} \text{ esu units} = 3 \times 10^5 \text{ esu units}$$

$$= 3 \times 10^5 \frac{\text{statvolt}}{\text{cm}}$$

On the other hand, in SI units,

$$F(\text{SI}) = \frac{q_1 q_2}{4\pi\varepsilon_0 r^2} = \frac{1}{4 \times \pi \times 8.85 \times 10^{-12}} \text{ N}$$

$$= 9 \times 10^9 \text{ N} = 9 \times 10^{14} \text{ dyne}$$

and

$$E(\text{SI}) = 9 \times 10^9 \frac{\text{V}}{\text{m}}$$

Therefore,

$$9 \times 10^9 \ \frac{\text{V}}{\text{m}} = 3 \times 10^5 \ \frac{\text{statvolts}}{\text{cm}}$$

or

$$1 \text{ statvolt} = 300 \, V \tag{2.2.11}$$

2.3. The Gauss Flux Theorem and the First Maxwell Equation

If we have a cloud of charge, it is convenient to consider the charge distribution continuous and introduce a *density of charge*

$$\rho(\mathbf{x}) = \frac{\left(\sum_i q_i\right) \ \text{in} \ \delta V}{\delta V} \tag{2.3.1}$$

where δV is a small volume around $\mathbf{x}$. To derive a field due to a charge density, we first consider the potential, which, because of the principle of superposition, is given by

$$\phi(\mathbf{x}) = \int d\tau' \frac{\rho(\mathbf{x}')}{|\mathbf{x} - \mathbf{x}'|} \tag{2.3.2}$$

where the integration is extended to all space in which the charge is present.

Consider now a charge q at $\mathbf{x}_q$ surrounded by a surface S. Let dS be a surface element of S at distance $\mathbf{r}$ from $\mathbf{x}_q$ (see Fig. 2.4). Because of Coulomb's law, we have at a distance $\mathbf{r}$

$$\mathbf{E} = \frac{q}{r^3}\mathbf{r} \tag{2.3.3}$$

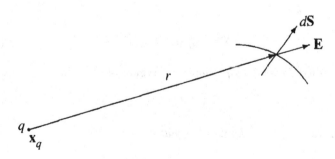

FIGURE 2.4

and

$$\mathbf{E} \cdot d\mathbf{S} = q\frac{\mathbf{r}}{r} \cdot \frac{d\mathbf{S}}{r^2} = qd\Omega \tag{2.3.4}$$

where $d\mathbf{S} = \mathbf{n}dS$ and $d\Omega$ is the solid angle through which dS is viewed from $\mathbf{x}_q$.

The integral of $d\Omega$ over a closed surface that includes the point $\mathbf{x}_q$ is 4π:

$$\int_S \mathbf{E} \cdot d\mathbf{S} = 4\pi q \tag{2.3.5}$$

If the surface encloses a number of charges $q_1, q_2, \ldots, q_n$, each charge gives rise to a flux $4\pi q_i$, and the total flux across the surface is

$$\int_S \mathbf{E} \cdot d\mathbf{S} = 4\pi q = 4\pi \sum_i q_i = 4\pi \int_V \rho \, d\tau \tag{2.3.6}$$

Relation (2.3.6) expresses the *Gauss flux theorem*.

If we make use of the divergence theorem,

$$\int_S \mathbf{E} \cdot d\mathbf{S} = \int_V \mathbf{\nabla} \cdot \mathbf{E} \, d\tau = 4\pi \int \rho \, d\tau \tag{2.3.7}$$

We shall see in the next section that

$$\mathbf{\nabla} \cdot \mathbf{E}(\mathbf{x}) = 4\pi\rho(\mathbf{x}) \tag{2.3.8}$$

Equation (2.3.8) is the *first Maxwell Equation*.

The substitution

$$\mathbf{E}(\mathbf{x}) = -\mathbf{\nabla}\phi(\mathbf{x}) \tag{2.3.9}$$

in Eq. (2.3.8) gives

$$\nabla^2\phi(\mathbf{x}) = -4\pi\rho(\mathbf{x}) \tag{2.3.10}$$

which is called *Poisson's equation*. In a region of zero charge, we have

$$\nabla^2\phi(\mathbf{x}) = 0 \tag{2.3.11}$$

a relation that is called *Laplace's equation*.

2.4. Singular and General Charge Distributions

The Dirac δ function in one dimension is defined by the functional properties

$$\delta(x - x_0) = 0, \quad \text{for all } x \neq x_0$$

$$\int_a^b \delta(x - x_0)dx = 1 \tag{2.4.1}$$

where $a < x_0 < b$ and by

$$\int_a^b \delta(x - x_0)f(x)dx = f(x_0) \tag{2.4.2}$$

where $f(x)$ is any continuous function of x. The δ function is not an analytical function of x; rather, it represents a notation that is defined by its functional properties and is always used in accordance with these properties.

In three dimensions

$$\delta^{(3)}(\mathbf{x} - \mathbf{x}_0) = 0, \quad \text{for all } \mathbf{x} \neq \mathbf{x}_0$$

$$\int d\tau \, \delta^{(3)}(\mathbf{x} - \mathbf{x}_0) = 1 \tag{2.4.3}$$

$$\iiint dx\,dy\,dz \, f(x, y, z) \, \delta^{(3)}(x - x_0, \, y - y_0, \, z - z_0)$$
$$= f(x_0, y_0, z_0) \tag{2.4.4}$$

or, more concisely,

$$\int d\tau f(\mathbf{x})\delta^{(3)}(\mathbf{x} - \mathbf{x}_0) = f(\mathbf{x}_0) \tag{2.4.5}$$

where the volume of integration includes the point $\mathbf{x}_0$.

We can now write for a point charge

$$\rho(\mathbf{x}) = q\delta^{(3)}(\mathbf{x} - \mathbf{x}_0) \tag{2.4.6}$$

That is

$$\rho(\mathbf{x}) = 0, \quad \text{if } \mathbf{x} \neq \mathbf{x}_0$$

$$\int \rho(\mathbf{x})d\tau = \int q\delta^{(3)}(\mathbf{x} - \mathbf{x}_0)d\tau = q \tag{2.4.7}$$

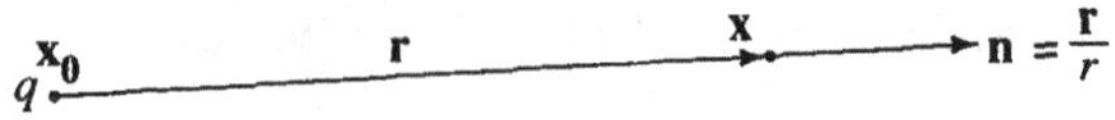

FIGURE 2.5

We know that, if a charge q is at $\mathbf{x}_0$ (see Fig. 2.5),

$$\mathbf{E}(\mathbf{x}) = -\nabla_x \frac{q}{|\mathbf{x} - \mathbf{x}_0|} = \frac{q}{|\mathbf{x} - \mathbf{x}_0|^2} \frac{\mathbf{r}}{r} \qquad (2.4.8)$$

where

$$r = \mathbf{x} - \mathbf{x}_0 \qquad (2.4.9)$$

Therefore,

$$\mathbf{E}(\mathbf{x}) = q \frac{\mathbf{r}}{r^3} = -\nabla \phi(\mathbf{x}) \qquad (2.4.10)$$

where

$$\phi(\mathbf{x}) = \frac{q}{|\mathbf{x} - \mathbf{x}_0|} = \frac{q}{r} \qquad (2.4.11)$$

This expression for $\phi(\mathbf{x})$ can also be obtained by using the integral (2.3.2) and the expression (2.4.6) for the charge distribution:

$$\phi(\mathbf{x}) = \int \frac{\rho(\mathbf{x}')}{|\mathbf{x} - \mathbf{x}'|} d\tau'$$

$$= \int \frac{q}{|\mathbf{x} - \mathbf{x}'|} \delta^{(3)}(\mathbf{x}' - \mathbf{x}_0) d\tau' = \frac{q}{|\mathbf{x} - \mathbf{x}_0|} \qquad (2.4.12)$$

Taking the divergence of $\mathbf{E}(\mathbf{x})$, we find

$$\nabla \cdot \mathbf{E}(\mathbf{x}) = -\nabla^2 \phi(\mathbf{x}) = \nabla \cdot q \frac{\mathbf{r}}{r^3} = q \nabla \cdot \frac{\mathbf{r}}{r^3}$$

$\nabla \cdot (\mathbf{r}/r^3)$ is zero for all values of $\mathbf{r}$ except $\mathbf{r} = 0$. Consider a small volume around the point $\mathbf{x}_0 (r = 0)$; the integral of $\nabla \cdot \mathbf{E}$ over this volume gives, because of the Gauss flux theorem,

$$\int d\tau [\nabla \cdot \mathbf{E}(\mathbf{x})] = \int \mathbf{n} \cdot \mathbf{E}(\mathbf{x}) dS = -\int dS \frac{\partial \phi}{\partial n} = 4\pi q \qquad (2.4.13)$$

Therefore, we can write

$$\nabla \cdot \mathbf{E}(\mathbf{x}) = 4\pi q \delta(\mathbf{x} - \mathbf{x}_0) = 4\pi \rho(\mathbf{x}) \qquad (2.4.14)$$

which expresses the *first Maxwell equation for point charges*. In general, we can express Coulomb's law as follows:

$$\mathbf{E}(\mathbf{x}) = -\nabla_x \int d\tau' \frac{\rho(\mathbf{x}')}{|\mathbf{x} - \mathbf{x}'|} \qquad (2.4.15)$$

where $\rho(\mathbf{x}')$ may include singular and/or continuous charges. Taking the divergence of $\mathbf{E}(\mathbf{x})$, we write

$$\nabla \cdot \mathbf{E}(\mathbf{x}) = -\nabla_x^2 \int d\tau' \frac{\rho(\mathbf{x}')}{|\mathbf{x} - \mathbf{x}'|}$$

$$= -\int d\tau' \rho(\mathbf{x}') \nabla_x^2 \frac{1}{|\mathbf{x} - \mathbf{x}'|} \qquad (2.4.16)$$

CLAIM[3]

$$\nabla_x^2 \frac{1}{|\mathbf{x} - \mathbf{x}'|} = -4\pi \delta(\mathbf{x} - \mathbf{x}') \qquad (2.4.17)$$

PROOF Call

$$f(\mathbf{x}) = \frac{1}{|\mathbf{x} - \mathbf{x}'|} = \frac{1}{r} \qquad (2.4.18)$$

Applying the divergence theorem, we obtain

$$\int d\tau \nabla^2 f(\mathbf{x}) = \int d\tau \nabla \cdot \nabla f(\mathbf{x}) = \int_S dS \, \mathbf{n} \cdot \nabla f(\mathbf{x}) \qquad (2.4.19)$$

Consider a small sphere of radius R around $\mathbf{x}'$ as surface S (see Fig. 2.6):

$$\mathbf{n} \cdot \nabla f = \frac{\partial}{\partial r} \frac{1}{r} = -\frac{1}{r^2}$$

and

$$\int dS \, \mathbf{n} \cdot \nabla f = -\int dS \frac{1}{r^2} = -\frac{4\pi R^2}{R^2} = -4\pi \qquad (2.4.20)$$

We can then write

$$\nabla^2 f(\mathbf{x}) = -4\pi \delta(\mathbf{x} - \mathbf{x}') \qquad (2.4.21)$$

because

$$\int \nabla^2 f(\mathbf{x}) d\tau = \int dS \, \mathbf{n} \cdot \nabla f(\mathbf{x}) = -4\pi \int d\tau \, \delta(\mathbf{x} - \mathbf{x}') = -4\pi \qquad (2.4.22)$$

in agreement with Eq. (2.4.20).

---

[3] As we move toward more complex problems, we find it convenient to introduce the following style, where we first state a *claim* and then present the *proof*.

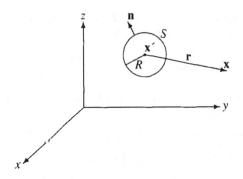

FIGURE 2.6

Having shown the validity of Eq. (2.4.22), we go back to Eq. (2.4.21) and obtain

$$\boldsymbol{\nabla} \cdot \mathbf{E}(\mathbf{x}) = - \int d\tau' \rho(\mathbf{x}') \nabla^2 \frac{1}{|\mathbf{x} - \mathbf{x}'|}$$

$$= \int d\tau' \rho(\mathbf{x}) \, 4\pi\delta(\mathbf{x} - \mathbf{x}') \, 4\pi\rho(\mathbf{x})$$

or

$$\boldsymbol{\nabla} \cdot \mathbf{E}(\mathbf{x}) = 4\pi\rho(\mathbf{x}) \tag{2.4.23}$$

which gives us the most general expression for the first Maxwell equation.

It is clear that $\phi(\mathbf{x})$ as given by

$$\phi(\mathbf{x}) = \int d\tau' \frac{\rho(\mathbf{x}')}{|\mathbf{x} - \mathbf{x}'|} \tag{2.3.2}$$

is a solution of Poisson's equation

$$\nabla^2 \phi(\mathbf{x}) = -4\pi\rho(\mathbf{x}) \tag{2.3.10}$$

The question now arises: Is the solution provided by (2.3.2) unique? We shall return to this point later.

2.5. Some Potential Theory

We have seen that the equation

$$\nabla^2 \phi(\mathbf{x}) = -4\pi\rho(\mathbf{x}) \tag{2.5.1}$$

has the particular solution

$$\phi(\mathbf{x}) = \int d\tau \, \frac{\rho(\mathbf{x}')}{|\mathbf{x} - \mathbf{x}'|} \tag{2.5.2}$$

To find other solutions, we may add to the expression above any solution of the *Laplace equation*

$$\nabla^2 \phi(\mathbf{x}) = 0 \tag{2.5.3}$$

Examples of solutions of the Laplace equation are given next.

(1) Expressions of the type

$$\sin k_1 x \, \sin k_2 y \, \sinh k_3 z \tag{2.5.4}$$

which are solutions, provided

$$-k_1^2 - k_2^2 + k_3^2 = 0 \tag{2.5.5}$$

(2) Polynomial solutions

$$\phi^l(x, y, z) = \sum_{n_1 + n_2 + n_3 = l} C_{n_1 n_2 n_3} x^{n_1} y^{n_2} z^{n_3} \tag{2.5.6}$$

These solutions are homogeneous polynomials of rank l. There are $(1/2)(l^2 + 3l + 2)$ possible $x^{n_1} y^{n_2} z^{n_3}$ products with $n_1 + n_2 + n_3 = l$ (three for $l = 1$, six for $l = 2$, ten for $l = 3$, and so on). The conditions on these polynomials are

$$\nabla^2 \phi^l = 0 \tag{2.5.7}$$

$\nabla^2 \phi^l$ are homogeneous polynomials of rank $l - 2$; there are

$$\frac{1}{2}[(l - 2)^2 + 3(l - 2) + 2] \tag{2.5.8}$$

such polynomials, all equal to zero. The difference

$$\frac{1}{2}(l^2 + 3l + 2) - \frac{1}{2}[(l - 2)^2 + 3(l - 2) + 2] = 2l + 1 \tag{2.5.9}$$

gives the number of independent homogeneous polynomials of rank l. To classify them, let us introduce spherical coordinates

$$x = r \sin \theta \cos \varphi$$
$$y = r \sin \theta \sin \varphi \tag{2.5.10}$$
$$z = r \cos \theta$$

Then
$$x + iy = r \sin \theta e^{i\varphi}$$
$$x - iy = r \sin \theta e^{-i\varphi}$$

and

$$\phi^l(x, y, z) = \sum_{n_1' + n_2' + n_3' = l} C_{n_1' n_2' n_3}(x + iy)^{n_1'}(x - iy)^{n_2'} z^{n_3'}$$

$$= \sum_{m=-l}^{l} C_{lm} r^l Y_{lm}(\theta, \varphi) \qquad (2.5.11)$$

The functions $Y_{lm}(\theta, \varphi)$ are *generalized spherical harmonics* with

$$-l \leq m \leq l \qquad (2.5.12)$$

The Laplacian operator in spherical coordinates is given by

$$\nabla^2 = \frac{1}{r}\frac{\partial^2}{\partial r^2} r + \frac{1}{r^2}\left(\frac{1}{\sin\theta}\frac{\partial}{\partial\theta}\sin\theta\frac{\partial}{\partial\theta} + \frac{1}{\sin^2\theta}\frac{\partial^2}{\partial\varphi^2} \right)$$

$$= \frac{1}{r}\frac{\partial^2}{\partial r^2} r + \frac{\Omega}{r^2} \qquad (2.5.13)$$

where

$$\Omega = \frac{1}{\sin\theta}\frac{\partial}{\partial\theta}\left(\sin\theta\frac{\partial}{\partial\theta} \right) + \frac{1}{\sin^2\theta}\frac{\partial^2}{\partial\varphi^2} \qquad (2.5.14)$$

Then Eq. (2.5.7) implies

$$\left(\frac{1}{r}\frac{\partial^2}{\partial r^2}r + \frac{\Omega}{r^2} \right) r^l Y_{lm} = \frac{1}{r}\frac{\partial^2}{\partial r^2}(r^{l+1}Y_{lm}) + \Omega r^{l-2}Y_{lm}$$

$$= \frac{1}{r}l(l+1)r^{l-1}Y_{lm} + r^{l-2}\Omega Y_{lm}$$

$$= l(l+1)r^{l-2}Y_{lm} + r^{l-2}\Omega Y_{lm} = 0$$

This gives

$$\Omega Y_{lm}(\theta, \varphi) + l(l+1)Y_{lm}(\theta, \varphi) = 0 \qquad (2.5.15)$$

which expresses the *Legendre differential equation*, an equation satisfied by spherical harmonics.[4]

---

[4]For a more extensive treatment of this subject, the reader is referred to P. M. Morse and H. Feshbach, *Methods of Theoretical Physics*, McGraw Hill, New York, 1953.

(3) Multipole potentials, such as

$$\frac{1}{r^{l+1}}Y_{lm}(\theta, \varphi) \qquad (2.5.16)$$

For $l = 0$, we have

$$\frac{1}{r}Y_{0,0} = \frac{1}{\sqrt{4\pi}}\frac{1}{r}$$

For $l = 1$,

$$\frac{1}{r^2}Y_{1,1}, \quad \frac{1}{r^2}Y_{1,-1}, \quad \frac{1}{r^2}Y_{1,0}$$

Since

$$\nabla^2(\text{multipole potential}) = \nabla^2\left(\frac{1}{r^{l+1}}Y_{lm}\right)$$

$$= \left(\frac{1}{r}\frac{\partial^2}{\partial r^2}r + \frac{\Omega}{r^2}\right)\frac{1}{r^{l+1}}Y_{lm}$$

$$= \frac{1}{r}\frac{\partial^2}{\partial r^2}\frac{1}{r^l}Y_{lm} + \frac{\Omega}{r^{l+3}}Y_{lm}$$

$$= \frac{1}{r^{l+3}}l(l+1)Y_{lm}\frac{\Omega}{r^{l+3}}Y_{lm}$$

multipole potentials are solutions of

$$\frac{1}{r^{l+3}}l(l+1)Y_{lm}(\theta, \varphi) + \Omega Y_{lm}(\theta, \varphi) = 0 \qquad (2.5.17)$$

These potentials present a singularity at $r = 0$, but not at $r = \infty$. Therefore, if we include the origin, these multipoles are not acceptable solutions of the Laplace equation. They are solutions of Poisson's equation for very special types of sources.

2.6. Properties of Spherical Harmonics

2.6.1. *Normalization*

This property is expressed by the relation

$$\int_0^{2\pi} d\varphi \int_0^\pi \sin\theta \, d\theta \, |Y_{lm}(\theta, \varphi)|^2 = 1 \qquad (2.6.1)$$

2.6.2. Orthogonality

Orthogonality is expressed by

$$\int_0^{2\pi} d\varphi \int_0^{\pi} \sin\theta \, d\theta Y_{lm}^*(\theta,\varphi) \, Y_{l'm'}(\theta,\varphi) = \delta_{ll'}\delta_{mm'} \qquad (2.6.2)$$

This property derives from the fact that spherical harmonics are solutions of a particular type of differential equation. The orthogonality over m is easily verified:

$$\int_0^{2\pi} e^{i(m-m')\varphi} d\varphi = \delta_{mm'} \qquad (2.6.3)$$

2.6.3. Symmetry

We can write

$$Y_{lm}(\theta,\varphi) = Y_{lm}\left(\frac{\mathbf{r}}{r}\right) = Y_{lm}(\mathbf{n}) \qquad (2.6.4)$$

The unit vector $-\mathbf{n}$ has the following angles:

$$\pi - \theta, \quad \pi + \varphi$$

Then

$$Y_{lm}(-\mathbf{n}) = (-1)^l Y_{lm}(\mathbf{n}) \qquad (2.6.5)$$

as can be seen in the following examples.

2.6.4. Examples

$$Y_{0,0} = \frac{1}{\sqrt{4\pi}}$$

$$Y_{1,0} = \sqrt{\frac{3}{4\pi}} \cos\theta$$

$$Y_{1,\pm 1} = \mp\sqrt{\frac{3}{8\pi}} \sin\theta \, e^{\pm i\varphi}$$

$$Y_{2,0} = \sqrt{\frac{5}{16\pi}}(3\cos^2\theta - 1)$$

$$Y_{2,\pm 1} = \mp\sqrt{\frac{15}{8\pi}} \sin\theta\cos\theta e^{\pm i\varphi}$$

$$Y_{2,\pm 2} = \sqrt{\frac{15}{32\pi}} \sin^2\theta \, e^{\pm 2i\varphi}$$

$$(2.6.6)$$

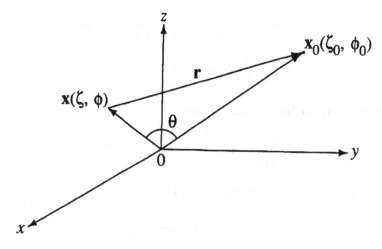

FIGURE 2.7

2.6.5. *Legendre Polynomials, $P_l(\cos\theta)$*

Legendre polynomials are, apart from a constant factor, spherical harmonics with $m = 0$:

$$P_l(\cos\theta) = \sqrt{\frac{4\pi}{2l+1}} Y_{l,0}(\theta) \qquad (2.6.7)$$

There are other definitions of the Legendre polynomials. Consider Fig. 2.7:

$$r = |\mathbf{x} - \mathbf{x}_0| = \sqrt{|\mathbf{x}|^2 + |\mathbf{x}_0|^2 - 2|\mathbf{x}||\mathbf{x}_0|\cos\theta}$$

If $|\mathbf{x}_0|/|\mathbf{x}| < 1$,

$$r = |\mathbf{x}|\sqrt{1 + (|\mathbf{x}_0|^2/|\mathbf{x}|^2) - 2(|\mathbf{x}_0|/|\mathbf{x}|)\cos\theta}$$

$$\frac{1}{r} = \frac{1}{|\mathbf{x}|\sqrt{1 + (|\mathbf{x}_0|^2/|\mathbf{x}|^2) - 2(|\mathbf{x}_0|/|\mathbf{x}|)\cos\theta}}$$

$$= \sum_l \frac{|\mathbf{x}_0|^l}{|\mathbf{x}|^{l+1}} P_l(\cos\theta) \qquad (2.6.8)$$

If $|\mathbf{x}|/|\mathbf{x}_0| < 1$,

$$r = |\mathbf{x}_0|\sqrt{1 + (|\mathbf{x}|^2/|\mathbf{x}_0|^2) - 2(|\mathbf{x}|/|\mathbf{x}_0|)\cos\theta}$$

$$\frac{1}{r} = \frac{1}{|\mathbf{x}_0|\sqrt{1 + (|\mathbf{x}|^2/|\mathbf{x}_0|^2) - 2(|\mathbf{x}|/|\mathbf{x}_0|)\cos\theta}}$$

$$= \sum_l \frac{|\mathbf{x}|^l}{|\mathbf{x}_0|^{l+1}} P_l(\cos\theta) \tag{2.6.9}$$

Examples of Legendre polynomials are

$$P_0(x) - 1$$
$$P_1(x) = x$$
$$P_2(x) = \frac{1}{2}(3x^2 - 1)$$
$$P_3(x) = \frac{1}{2}(5x^3 - 3x) \tag{2.6.10}$$
$$P_4(x) = \frac{1}{8}(35x^4 - 30x^2 + 3)$$

$1/r$ is a solution of the Laplace equation (if $r \neq 0$); therefore, if, say, $|\mathbf{x}| < |\mathbf{x}_0|$, we can write

$$\nabla_x^2[|\mathbf{x}|^l P_l(\cos\theta)] = 0 \tag{2.6.11}$$

and $P_l(\cos\theta)$ must be a sum of spherical harmonics. The situation is completely symmetrical in $\mathbf{x}$ and $\mathbf{x}_0$; we want to refer everything to x, y, z. Then

$$\mathbf{x} = \mathbf{x}(\zeta, \varphi)$$
$$\mathbf{x}_0 = \mathbf{x}_0(\zeta_0, \varphi_0) \tag{2.6.12}$$

and we have

$$\cos\theta = \cos\zeta\cos\zeta_0 + \sin\zeta\sin\zeta_0\cos(\varphi - \varphi_0) \tag{2.6.13}$$

$$P_l\cos\theta = P_l[\cos\zeta\cos\zeta_0 + \sin\zeta\sin\zeta_0\cos(\varphi - \varphi_0)] \tag{2.6.14}$$

$P_l(\cos\theta)$ must be a sum of spherical harmonics in ζ, φ:

$$P_l(\cos\theta) = \sum_{m=-l}^{l} C_m Y_{lm}(\zeta, \varphi) \tag{2.6.15}$$

In what follows the sums over m or m' will always extend from $-l$ to l.

Because of the presence of ζ_0 and φ_0, the C_m's must contain spherical harmonics in ζ_0, φ_0:

$$P_l(\cos\theta) = \sum_m \sum_{m'} \alpha_{mm'} Y_{lm'}(\zeta_0, \varphi_0) Y_{lm}(\zeta, \varphi)$$

The φ dependence must be such that

$$e^{im'\varphi_0} e^{im\varphi} = e^{i(m'\varphi_0 + m\varphi)} = e^{i(\varphi - \varphi_0)m}$$

That is, we must have $m' = -m$:

$$P_l(\cos\theta) = \sum_m \alpha_m Y_{l,-m}(\zeta_0, \varphi_0) Y_{lm}(\zeta, \phi)$$

We may verify that

$$Y_{lm}^*(\zeta, \varphi) = (-1)^m Y_{l,-m}(\zeta, \varphi) \tag{2.6.16}$$

Then

$$P_l(\cos\theta) = \sum_m (-1)^m \alpha_m Y_{lm}^*(\zeta_0, \phi_0) Y_{lm}(\zeta, \phi) \tag{2.6.17}$$

Let us take now $\theta = 0$:

$$\cos\theta = 1, \quad \mathbf{x} = \mathbf{x}_0$$

We get

$$P_l(1) = 1 = \sum_m (-1)^m \alpha_m |Y_{lm}(\zeta, \varphi)|^2$$

And let us integrate both sides of this relation over angles:

$$4\pi = \sum_{m=-l}^{l} \alpha_m (-1)^m \tag{2.6.18}$$

Let us square expression (2.6.17):

$$P_l^2(\cos\theta) = \sum_m (-1)^m \alpha_m Y_{lm}^*(\zeta_0, \varphi_0) Y_{lm}(\zeta, \varphi)$$

$$\times \sum_{m'} (-1)^{m'} \alpha_{m'} Y_{lm'}^*(\zeta_0, \varphi_0) Y_{lm'}(\zeta, \varphi)$$

$$= \sum_m (-1)^m \alpha_m Y_{lm}^*(\zeta_0, \varphi_0) Y_{lm}(\zeta, \varphi)$$

$$\times \sum_{m'} (-1)^{m'} \alpha_{m'}^* Y_{lm'}(\zeta_0, \varphi_0) Y_{lm'}^*(\zeta, \varphi)$$

and integrate over the angles (ζ, φ) to obtain

$$\int P_l^2(\cos\theta)\, d\Omega = \sum_m \sum_{m'} (-1)^{m+m'} \alpha_m \alpha_{m'}^* Y_{lm}^*(\zeta_0, \varphi_0) Y_{lm'}(\zeta_0, \varphi_0)$$

$$\times \int Y_{lm}(\zeta, \varphi) Y_{lm'}^*(\zeta, \varphi)\, d\Omega$$

and, according to Eqs. (2.6.2) and (2.6.7),

$$\frac{4\pi}{2l+1} = \sum_m |\alpha_m|^2 |Y_{lm}(\zeta_0, \varphi_0)|^2$$

We then integrate over the angles (ζ_0, φ_0) to obtain

$$4\pi \frac{4\pi}{2l+1} = \sum_{m=-1}^{l} |\alpha_m|^2 \tag{2.6.19}$$

We then have the following two equations:

$$4\pi = \sum_{m=-1}^{l} (-1)^m \alpha_m$$

$$4\pi \frac{4\pi}{2l+1} = \sum_{m=-1}^{l} |\alpha_m|^2 \tag{2.6.20}$$

which result in

$$\alpha_m = \frac{4\pi(-1)^m}{2l+1} \tag{2.6.21}$$

Now we can go back to Eq. (2.6.17) and write

$$P_l(\cos\theta) = \sum_{m=-1}^{l} (-1)^m \frac{4\pi}{2l+1} (-1)^m Y_{lm}^*(\zeta_0, \varphi_0) Y_{lm}(\zeta, \varphi)$$

$$= \frac{4\pi}{2l+1} \sum_{m=-l}^{l} Y_{lm}^*(\zeta_0, \varphi_0) Y_{lm}(\zeta, \varphi) \tag{2.6.22}$$

Relation (2.6.22) expresses the *addition theorem*.

2.7. The Mean Value Theorem

The second form of Green's theorem [see Eq. (1.4.20)] gives us

$$\int_V d\tau (\psi \nabla^2 \phi - \phi \nabla^2 \psi) = \int_S dS \left(\psi \frac{\partial \phi}{\partial n} - \phi \frac{\partial \psi}{\partial n} \right) \tag{2.7.1}$$

for two functions ψ and ϕ. If we assume that

$$\nabla^2 \phi(\mathbf{x}) = 0$$
$$\psi = \text{const} \tag{2.7.2}$$

relation (2.7.1) gives

$$\int_S dS \frac{\partial \phi}{\partial n} = 0 \tag{2.7.3}$$

If we assume that

$$\nabla^2 \phi = 0$$
$$\psi = \frac{1}{r} = \frac{1}{|\mathbf{x} - \mathbf{x}_0|} \tag{2.7.4}$$

relation (2.7.1) gives, for a spherical volume of radius R around $\mathbf{x}_0$,

$$\int_V d\tau \left(\frac{1}{r} \nabla^2 \phi - \phi \nabla^2 \frac{1}{r} \right) = \int_S dS \left(\frac{1}{R} \frac{\partial \phi}{\partial n} + \phi \frac{1}{R^2} \right)$$
$$\int_V d\tau \, 4\pi \delta(\mathbf{x} - \mathbf{x}_0) \phi(\mathbf{x}) = \int_S \frac{dS}{R^2} \phi(S)$$

or

$$\phi(\mathbf{x}_0) = \frac{1}{4\pi R^2} \int_S dS \, \phi(S) \tag{2.7.5}$$

This relation expresses the *mean value theorem*, which can also be stated as follows: *for charge-free space the value of the electrostatic potential at any point is equal to the average of the potential over a sphere centered on that point.* Several conclusions can be derived from this theorem:

(1) If $\nabla^2 \phi(\mathbf{x}) = 0$ in a spherical volume V, the maximum value of $\phi(\mathbf{x})$ is on the boundaries of V. If the maximum value were in the center, all the values on the surface would be smaller, and this would not be in accordance with the mean value theorem.

(2) If $\nabla^2\phi(\mathbf{x}) = 0$ in all space and $\phi(\infty) =$ finite, then $\phi =$ const. If the volume contains *all* space, then any point in space is the center and ϕ must be the same at all points in space.

(3) If $\phi(\mathbf{x})$ is not constant, then $\phi(\mathbf{x})$ must have singularities somewhere in space. These singularities must consist of at least *one* singular point; this point may be at finite distances (for example, multipole potentials) or at ∞ (for example, harmonic polynomials).

(4) If $\nabla^2\phi(\mathbf{x}) - 0$ in all space and $\phi(\infty) = 0$, then $\phi = 0$ in all space. We have thus shown the equivalence between Poisson's equation

$$\nabla^2\phi(\mathbf{x}) = -4\pi\rho(\mathbf{x}), \quad \phi(\infty) = 0 \tag{2.7.6}$$

and Coulomb's law expressed by

$$\phi(\mathbf{x}) = \int d\tau' \frac{\rho(\mathbf{x}')}{|\mathbf{x} - \mathbf{x}'|} \tag{2.7.7}$$

There is no other particular solution (different from zero) of the Laplace's equation that we can add to the expression for $\phi(\mathbf{x})$ in Eq. (2.7.7).

2.8. Conductors and Insulators

The distinction between *conductors* of electricity and *nonconductors* or *insulators* dates back to the experiments with electricity in the early eighteenth century. It was then found that electric charges, when applied to a substance of a certain type (insulator), remained localized for a very long time, whereas, when applied to a substance of a different type (conductor), spread over the whole body instantaneously.

The different behavior of these two types of substances is based on the mobility of the charge-carrying particles (such as electrons in metals, and ions in liquid solutions), that is, the property of these particles to move more or less easily under the action of electric forces. In insulators the electrons are attached to their parent atoms, and only very strong forces can pull them away, whereas in conductors the electrons are removed from their parent atoms and need only very small forces to move in certain directions. We should note that the distinction between conductors and insulators is not very sharp, in the sense that physical substances present a continuous range of possibilities, from almost perfect insulators that are able to retain their charges in localized form almost indefinitely, to almost perfect conductors over whose bodies the electric charges spread in an extremely short time.

We shall now examine the behavior of good conductors of electricity, such as metals, in the presence of electrostatic fields. We shall first consider a piece of conducting material, uncharged and electrically insulated (being supported by a piece of insulator material), and shall assume that electric charges are applied to it. The charges redistribute themselves in the conductor, and certain stationary, equilibrium configurations of charges and fields are reached:

(1) The charges transferred to the conductor, because of their mutual repulsions, move as far away from each other as possible, reaching the surface of the conductor. When equilibrium is reached, the transferred charges reside on the surface of the conductor. No component of the electric field parallel to the surface is present, and the field is then perpendicular to the surface of the conductor. The lines of the field start or end perpendicularly to the surface of the conductor and, therefore, the surface of the conductor is equipotential.

(2) In equilibrium, the electrons in the body of the conductor are retained with the same concentration present before the transfer of the additional charges and are compensated by the charges of the positive ions. No net force is acting on these electrons; the electric field inside the conductor is zero. Since no gradient of the potential is allowed, the entire body is at the same potential. On the other hand, in an insulator we may have potential gradients and electric fields without movement of charges.

If we bring an uncharged conductor in a region of space where there is already an electric field, different points of the conductor will be at different electric potential. Because of forces acting on them, the mobile charges will be displaced and will reach an equilibrium distribution compatible with an electric field perpendicular to the surface and a constant potential for the entire body. The total charge will continue to be zero, but there will be a nonuniform distribution of positive and negative charges on the surface of the conductor. The original field configuration of the electric field will be changed by the presence of the conductor (whether this body is charged or not) in such a way that the potential of the conductor's surface (and body) will be constant.

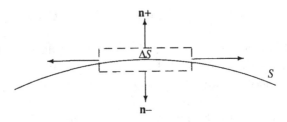

FIGURE 2.8

In any case, a relation can be found between σ, the *surface charge density*, and the electric field (which is, as noted above, perpendicular to the surface of the conductor). Considering a conductor's surface as in Fig. 2.8, we can apply the Gauss flux theorem (2.3.6) to the thin flat volume element that contains an element ΔS of the surface, as in Fig. 2.8.

$$\int dS \frac{\partial \phi}{\partial n} = \int_V \nabla^2 \phi \, d\tau = -4\pi \int \rho(\mathbf{x}) d\tau = -4\pi \sigma \Delta S \qquad (2.8.1)$$

or

$$\left(\frac{\partial \phi}{\partial n_+} + \frac{\partial \phi}{\partial n_-} \right) \Delta S = -4\pi \sigma \Delta S \qquad (2.8.2)$$

and

$$\frac{\partial \phi}{\partial n_+} + \frac{\partial \phi}{\partial n_-} = -4\pi \sigma \qquad (2.8.3)$$

Since the gradient of the potential inside the conductor is zero, we can set

$$\frac{\partial \phi}{\partial n_-} = 0 \qquad (2.8.4)$$

and obtain

$$\frac{\partial \phi}{\partial n_+} = -4\pi \sigma \qquad (2.8.5)$$

The electric field just outside the conductor is then given by

$$E = 4\pi \sigma \qquad (2.8.6)$$

2.9. General Electrostatic Problems

Let us consider the situation in Fig. 2.9. We have essentially:

A charge distribution with density $\rho(\mathbf{x})$

Charge distribution

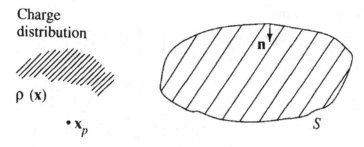

$\rho\,(\mathbf{x})$

$\bullet\,\mathbf{x}_p$

FIGURE 2.9

A volume V delimited by a surface S
A point of observation $\mathbf{x}_p$

We do not know what is inside V, but we do know the values that the potential takes on the surface surrounding V. $\phi(\mathbf{x}_p)$ is in general different from what it would be if we had only the charge distribution. Green's theorem, in its second form, gives us

$$\int_V d\tau\,(\psi\nabla^2\phi - \phi\nabla^2\psi) = \int_S dS\left(\psi\frac{\partial\phi}{\partial n} - \phi\frac{\partial\psi}{\partial n}\right) \tag{2.9.1}$$

$\phi(\mathbf{x})$ is the potential that we try to determine. We note that

$$\nabla^2\phi(\mathbf{x}) = -4\pi\rho(\mathbf{x}) \tag{2.9.2}$$

and we let

$$\psi(\mathbf{x}) = \frac{1}{|\mathbf{x} - \mathbf{x}_p|} \tag{2.9.3}$$

Obviously,

$$\nabla^2\psi(\mathbf{x}) = -4\pi\delta(\mathbf{x} - \mathbf{x}_p) \tag{2.9.4}$$

Since $\phi(\infty) = \psi(\infty) = 0$, the integral in Eq. (2.9.1) extended to all space is zero, and the integral extended to "all space minus V" is equal to minus the integral over the volume V:

$$\int_{\substack{\text{all space}\\ \text{minus } V}} d\tau(\psi\nabla^2\phi - \phi\nabla^2\psi) = \int_S dS\left(\psi\frac{\partial\phi}{\partial n} - \varphi\frac{\partial\psi}{\partial n}\right) \tag{2.9.5}$$

where vector $\mathbf{n}$ is pointing inward as in Fig. 2.9.

We can now write

$$\int d\tau (\psi \nabla^2 \phi - \phi \nabla^2 \psi) = \int d\tau \left(-\frac{4\pi \rho(\mathbf{x})}{|\mathbf{x} - \mathbf{x}_p|} \right)$$

$$- \int d\tau \phi(\mathbf{x})[-4\pi\delta(\mathbf{x} - \mathbf{x}_p)]$$

$$= -4\pi \int d\tau \frac{\rho(\mathbf{x})}{|\mathbf{x} - \mathbf{x}_p|} + 4\pi\phi(\mathbf{x}_p) \qquad (2.9.6)$$

where the integrals are extended to all space minus V.

$$\int_S dS \left(\psi \frac{\partial \phi}{\partial n} - \phi \frac{\partial \psi}{\partial n} \right)$$

$$= \int dS \left(\frac{1}{|\mathbf{x} - \mathbf{x}_p|} \frac{\partial \phi(S)}{\partial n} - \phi(S) \frac{\partial}{\partial n} \frac{1}{|\mathbf{x} - \mathbf{x}_p|} \right) \qquad (2.9.7)$$

and

$$\phi(\mathbf{x}_p) = \int_{\substack{\text{all space} \\ \text{minus } V}} d\tau \frac{\rho(\mathbf{x})}{|\mathbf{x} - \mathbf{x}_p|} + \frac{1}{4\pi} \int dS$$

$$\times \left(\frac{1}{|\mathbf{x} - \mathbf{x}_p|} \frac{\partial \phi(S)}{\partial n} - \phi(S) \frac{\partial}{\partial n} \frac{1}{|\mathbf{x} - \mathbf{x}_p|} \right) \qquad (2.9.8)$$

We shall now demonstrate the *uniqueness theorem* which can be stated as follows: to determine the value of the electric field $\mathbf{E}(\mathbf{x}_p)$, it is sufficient to know the values of ϕ on the surface S or the values of $\partial \phi / \partial n$ on the surface S.

Assume that we have two solutions $\phi_1(\mathbf{x})$ and $\phi_2(\mathbf{x})$ that satisfy Poisson's equation:

$$\nabla^2 \phi = -4\pi\rho \qquad (2.9.9)$$

Then

$$\nabla^2 \phi_1 - \nabla^2 \phi_2 = \nabla^2(\phi_1 - \phi_2) = 0 \qquad (2.9.10)$$

Consider the vector function

$$\mathbf{v} = (\phi_1 - \phi_2)\boldsymbol{\nabla}(\phi_1 - \phi_2) \qquad (2.9.11)$$

and apply Gauss's theorem to all space minus V:

$$\int_{\substack{\text{all space} \\ \text{minus } V}} d\tau (\boldsymbol{\nabla} \cdot \mathbf{v}) = \int_S dS \, \mathbf{n} \cdot \mathbf{v} = \int_S dS \, v_n \qquad (2.9.12)$$

But

$$\nabla \cdot \mathbf{v} = (\phi_1 - \phi_2)\nabla^2(\phi_1 - \phi_2) + [\nabla(\phi_1 - \phi_2)]^2$$
$$= [\nabla(\phi_1 - \phi_2)]^2 \tag{2.9.13}$$

and

$$\int_{\substack{\text{all space} \\ \text{minus } V}} d\tau [\nabla(\phi_1 - \phi_2)]^2 = \int_S dS(\phi_1 - \phi_2)\left(\frac{\partial\phi_1}{\partial n} - \frac{\partial\phi_2}{\partial n}\right) \tag{2.9.14}$$

If, on the surface S, $\phi_1 = \phi_2$, or $(\partial\phi_1/\partial n) = (\partial\phi_2/\partial n)$ then the right member is zero. The integrand of the left side, being positive definite, must then be zero for the integral to vanish; therefore, $\nabla\phi_1 = \nabla\phi_2$ outside S and $\phi_1 = \phi_2 + \text{const}$. The two potentials can then at most be different by a constant that has no effect on the electric field.

By virtue of this uniqueness theorem, expression (2.9.8) seems to require more information than needed. We have used Green's theorem in order to find the potential $\phi(\mathbf{x}_p)$, and we have used a function $\psi(\mathbf{x})$ such that

$$\nabla^2\psi(\mathbf{x}) = -4\pi\delta(\mathbf{x} - \mathbf{x}_p) \tag{2.9.4}$$

But is

$$\psi(\mathbf{x}) = \frac{1}{|\mathbf{x} - \mathbf{x}_p|} \tag{2.9.3}$$

the only solution of Eq. (2.9.4)? Since we integrate over all space minus the volume V, $\psi(\mathbf{x})$ may be in general replaced by a *Green's function*, $G(\mathbf{x}, \mathbf{x}_p)$:

$$G(\mathbf{x}, \mathbf{x}_p) = \frac{1}{|\mathbf{x} - \mathbf{x}_p|} + \chi_1(\mathbf{x}) \tag{2.9.15}$$

subject to the condition that

$$\nabla^2\chi_1(\mathbf{x}) = 0$$

in all space minus the volume V. The expression for $\phi(\mathbf{x}_p)$ in Eq. (2.9.8) becomes

$$\phi(\mathbf{x}_p) = \int_{\substack{\text{all space} \\ \text{minus } V}} d\tau\, \rho(\mathbf{x})G(\mathbf{x}, \mathbf{x}_p)$$

$$+ \frac{1}{4\pi}\int_S dS\left(G(\mathbf{x}, \mathbf{x}_p)\frac{\partial\phi(S)}{\partial n} - \phi(S)\frac{\partial G(\mathbf{x}, \mathbf{x}_p)}{\partial n}\right) \tag{2.9.16}$$

where $G(\mathbf{x}, \mathbf{x}_p)$ is given by Eq. (2.9.15).

We want now to use Green's functions that take into account the boundary conditions. We shall examine two cases.

2.9.1. *Dirichlet Boundary Value Problem*

The boundary value conditions specify the potential on a closed surface; ϕ is given on S. We choose $\chi_1(\mathbf{x})$ in such a way that

$$G(\mathbf{x}, \mathbf{x}_p) = 0 \quad \text{on } S \qquad (2.9.17)$$

Then we get

$$\phi(\mathbf{x}_p) = \int_{\substack{\text{all space} \\ \text{minus } V}} \rho(\mathbf{x}) G(\mathbf{x}, \mathbf{x}_p) d\tau$$

$$- \frac{1}{4\pi} \int_S dS \phi(S) \frac{\partial G(\mathbf{x}, \mathbf{x}_p)}{\partial n} \qquad (2.9.18)$$

2.9.2. *Neumann Boundary Value Problem*

The boundary value conditions specify the gradient of the potential (that is, the electric field) on a closed surface: $\partial \phi(\mathbf{x})/\partial n$ is given on S. We choose $\chi_1(\mathbf{x})$ in such a way that

$$\frac{\delta G(\mathbf{x}, \mathbf{x}_p)}{\partial n} = 0 \quad \text{on } S \qquad (2.9.19)$$

Then we get

$$\phi(\mathbf{x}_p) = \int_{\substack{\text{all space} \\ \text{minus } V}} \rho(\mathbf{x}) G(\mathbf{x}, \mathbf{x}_p) d\tau + \frac{1}{4\pi} \int dS \, G(S) \frac{\partial \phi(\mathbf{x})}{\partial n} \qquad (2.9.20)$$

We shall now examine by means of some examples the effect of the presence of equipotential surfaces on the electrical potential at a generic point in space.

EXAMPLE 1

A charge q is located at the point $\mathbf{x}_q$ as in Fig. 2.10. We want to calculate the potential $\phi(\mathbf{x}_p)$ at a point $\mathbf{x}_p$ in the presence of a plane, infinitely extended surface S whose electrical potential $\phi(S)$ is zero. We shall expect $\phi(\infty) = 0$.

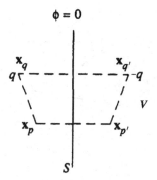

FIGURE 2.10

An appropriate Green's function is

$$G(\mathbf{x}, \mathbf{x}_p) = \frac{1}{|\mathbf{x} - \mathbf{x}_p|} + u(\mathbf{x}) \tag{2.9.21}$$

We choose u in such a way that $\nabla^2 u(\mathbf{x}) = 0$ in the region outside V and $G(S) = 0$:

$$G(\mathbf{x}, \mathbf{x}_p) = \frac{1}{|\mathbf{x} - \mathbf{x}_p|} - \frac{1}{|\mathbf{x} - \mathbf{x}_{p'}|} \tag{2.9.22}$$

$\mathbf{x}_p'$ is the mirror image point of $\mathbf{x}_p$ and

$$\nabla^2 \frac{1}{|\mathbf{x} - \mathbf{x}_p'|} = 0, \quad \text{outside } V \tag{2.9.23}$$

Now we have

$$\begin{aligned}
\phi(\mathbf{x}_p) &= \int_{\text{outside } V} d\tau \rho(\mathbf{x}) G(\mathbf{x}, \mathbf{x}_p) \\
&\quad - \frac{1}{4\pi} \int dS \phi(S) \frac{\partial G(\mathbf{x}, \mathbf{x}_p)}{\partial n} \\
&= \int d\tau \, q \delta(\mathbf{x} - \mathbf{x}_q) \left(\frac{1}{|\mathbf{x} - \mathbf{x}_p|} - \frac{1}{|\mathbf{x} - \mathbf{x}_p'|} \right) \\
&= q \left(\frac{1}{|\mathbf{x}_q - \mathbf{x}_p|} - \frac{1}{|\mathbf{x}_q - \mathbf{x}_p'|} \right) \\
&= q \left(\frac{1}{|\mathbf{x}_q - \mathbf{x}_p|} - \frac{1}{|\mathbf{x}_q' - \mathbf{x}_p|} \right) \tag{2.9.24}
\end{aligned}$$

where $\mathbf{x}_q'$ is the image point of $\mathbf{x}_q$.

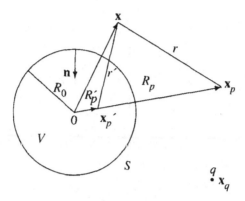

FIGURE 2.11

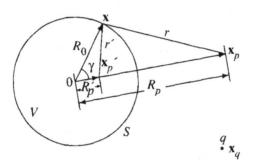

FIGURE 2.12

EXAMPLE 2

Given a spherical surface S, $\phi(S)$ known, and charge at $\mathbf{x}_q$ (see Fig. 2.11). We take the center of the sphere as the origin. We define a new point $\mathbf{x}_p'$ such that, since $R_p' = |\mathbf{x}_p'|$ and $R_p = |\mathbf{x}_p|$,

$$R_p' R_p = R_0^2 \qquad (2.9.25)$$

When $\mathbf{x}$ is on S (see Fig. 2.12),

$$r^2 = R_0^2 + R_p^2 - 2R_0 R_p \cos\gamma \qquad (2.9.26)$$

$$r'^2 = R_0^2 + R_p'^2 - 2R_0 R_p' \cos\gamma$$

$$= R_0^2 + \left(\frac{R_0^2}{R_p}\right)^2 - 2R_0\frac{R_0^2}{R_p}\cos\gamma$$

$$= R_0^2 + \frac{R_0^4}{R_p^2} - 2\frac{R_0^3}{R_p}\cos\gamma$$

$$= \frac{R_0^2}{R_p^2}(R_p^2 + R_0^2 - 2R_pR_0\cos\gamma) = \frac{R_0^2}{R_p^2}r^2 \qquad (2.9.27)$$

and

$$r' = \frac{R_0}{R_p}r \qquad (2.9.28)$$

Therefore, when $\mathbf{x}$ is on S,

$$\frac{1}{r} - \frac{R_0}{R_p}\frac{1}{r'} = \frac{1}{r} - \frac{R_0}{R_p}\frac{R_p}{R_0}\frac{1}{r} = 0 \qquad (2.9.29)$$

We note also that

$$\nabla_x^2\frac{1}{r'} = \nabla_x^2\frac{1}{|\mathbf{x} - \mathbf{x}_p'|} = 0 \qquad (2.9.30)$$

as long as $\mathbf{x}$ is outside the spherical volume V. Then we can take

$$G(\mathbf{x}, \mathbf{x}_p) = \frac{1}{|\mathbf{x} - \mathbf{x}_p|} - \frac{R_0}{R_p}\frac{1}{|\mathbf{x} - \mathbf{x}_p'|} \qquad (2.9.31)$$

as a Green's function that is zero on S. We now find

$$\phi(\mathbf{x}_p) = \int_{\text{outside the sphere}} \rho(\mathbf{x})G(\mathbf{x}, \mathbf{x}_p)d\tau$$

$$- \frac{1}{4\pi}\int \phi(S)\frac{\partial G(\mathbf{x}, \mathbf{x}_p)}{\partial n}dS$$

$$= \int_{\text{outside the sphere}} q\delta(\mathbf{x} - \mathbf{x}_q)(G(\mathbf{x}, \mathbf{x}_p))d\tau$$

$$- \frac{1}{4\pi}\int \phi(S)\frac{\partial G(\mathbf{x}, \mathbf{x}_p)}{\partial n}dS$$

$$= \frac{q}{|\mathbf{x}_p - \mathbf{x}_q|} - \frac{q}{|\mathbf{x}_p' - \mathbf{x}_q|}\frac{R_0}{R_p}$$

$$- \frac{1}{4\pi}\int \phi(S)\frac{\partial G(\mathbf{x}, \mathbf{x}_p)}{\partial n}dS \qquad (2.9.32)$$

where $\mathbf{n}$ is directed as indicated in Fig. 2.11.

We shall assume at this point that the sphere consists of a conducting material so that $\phi(S) = $ constant and that the total charge of the sphere is zero. We can write

$$\phi(\mathbf{x}_p) = \frac{q}{|\mathbf{x}_p - \mathbf{x}_q|} - \frac{q}{|\mathbf{x}'_p - \mathbf{x}_q|}\frac{R_0}{R_p} - \frac{\phi(S)}{4\pi}\int dS \frac{\partial G(\mathbf{x}, \mathbf{x}_p)}{\partial n} \quad (2.9.33)$$

We know that in general

$$\int dS \frac{\partial \chi}{\partial n} = \int dS\,(\mathbf{u}\cdot\boldsymbol{\nabla}\chi) - \int d\tau\boldsymbol{\nabla}\cdot\boldsymbol{\nabla}\chi = \int d\tau\nabla^2\chi \quad (2.9.34)$$

If $\nabla^2\chi = 0$ in a region surrounded by S, then

$$\int dS \frac{\partial \chi}{\partial n} = 0 \quad (2.9.35)$$

We have in our case

$$G(\mathbf{x}, \mathbf{x}_p) = \frac{1}{r} - \frac{R_0}{R_p}\frac{1}{r'} \quad (2.9.36)$$

and, since inside the sphere

$$\nabla^2\frac{1}{r} = 0 \quad (2.9.37)$$

$$\int dS \frac{\partial G}{\partial n} = -\frac{R_0}{R_p}\int dS \frac{\partial(1/r')}{\partial n} \quad (2.9.38)$$

Therefore,

$$\phi(\mathbf{x}_p) = \frac{q}{|\mathbf{x}_p - \mathbf{x}_q|} - \frac{q(R_0/R_p)}{|\mathbf{x}'_p - \mathbf{x}_q|} + \frac{\phi(S)}{4\pi}\frac{R_0}{R_p}\int dS \frac{\partial(1/r')}{\partial n} \quad (2.9.39)$$

But

$$\int dS \frac{\partial(1/r')}{\partial n} = -\int_V dV\,\nabla^2\frac{1}{r'}$$

$$= 4\pi\int_V \delta(\mathbf{x} - \mathbf{x}'_p)d\tau = 4\pi \quad (2.9.40)$$

Then

$$\phi(\mathbf{x}_p) = \frac{q}{|\mathbf{x}_p - \mathbf{x}_q|} - \frac{q(R_0/R_p)}{|\mathbf{x}'_p - \mathbf{x}_q|} + \phi(S)\frac{R_0}{R_p} \quad (2.9.41)$$

What is the value of the potential on the sphere? The simplest way of defining $\phi(S)$ is to see what happens when $|\mathbf{x}_p| \to \infty$. When $|\mathbf{x}_p| \to \infty$,

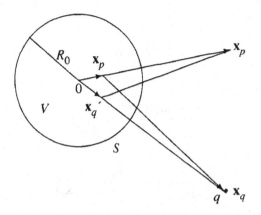

FIGURE 2.13

every charge acts as a point-like charge:

$$\phi(\mathbf{x}_p) \rightarrow \frac{q}{|\mathbf{x}_p|} \tag{2.9.42}$$

On the other hand, from Eq. (2.9.41),

$$\phi(\mathbf{x}_p) = \frac{1}{|\mathbf{x}_p|} - \frac{q(R_0/R_p)}{|\mathbf{x}_q|} + \phi(S)\frac{R_0}{|\mathbf{x}_p|} \tag{2.9.43}$$

The last two terms must cancel out:

$$\phi(S) = \frac{R_0}{|\mathbf{x}_p|} = \frac{(R_0/r_p)}{|\mathbf{x}_q|} \tag{2.9.44}$$

And

$$\phi(S) = \frac{q}{R_q} \tag{2.9.45}$$

where $R_q = |\mathbf{x}_q|$. Then

$$\phi(\mathbf{x}_p) = \frac{q}{|\mathbf{x}_p - \mathbf{x}_q|} - \frac{q(R_0/R_p)}{|\mathbf{x}_p' - \mathbf{x}_q|} + \frac{R_0}{R_p}\frac{q}{R_q} \tag{2.9.46}$$

From the following relations (see Fig. 2.13),

$$R_p' R_p = R_0^2$$
$$R_q' R_q = R_0^2 \tag{2.9.47}$$

we derive

$$\frac{R'_p}{R'_q} = \frac{R_q}{R_p} \tag{2.9.48}$$

where $R_q = |\mathbf{x}_q|$ and $R'_q = |\mathbf{x}'_q|$. We find

$$\frac{|\mathbf{x}'_p - \mathbf{x}_q|}{|\mathbf{x}'_q - \mathbf{x}_p|} = \frac{R'_p}{R'_q} = \frac{R_0^2}{R_p R'_q} \tag{2.9.49}$$

and

$$\frac{R_0}{R_p} \frac{1}{|\mathbf{x}'_p - \mathbf{x}_q|} = \frac{R'_q}{R_0} \frac{1}{|\mathbf{x}'_q - \mathbf{x}_p|}$$

We can then write

$$
\begin{aligned}
\phi(\mathbf{x}_p) &= \frac{q}{|\mathbf{x}_q - \mathbf{x}_p|} - \frac{q(R_0/R_p)}{|\mathbf{x}'_p - \mathbf{x}_q|} + \frac{R_0}{R_q} \frac{q}{R_p} \\
&= \frac{q}{|\mathbf{x}_q - \mathbf{x}_p|} - \frac{q(R'_q/R_0)}{|\mathbf{x}'_q - \mathbf{x}_p|} + q\frac{R_0/R_q}{|\mathbf{x}_p|} \\
&= \frac{q}{|\mathbf{x}_q - \mathbf{x}_p|} - q\frac{R_0}{R_q} \frac{1}{|\mathbf{x}'_q - \mathbf{x}_p|} + q\frac{R_0}{R_q} \frac{1}{|\mathbf{x}_p|} \tag{2.9.50}
\end{aligned}
$$

The potential at $\mathbf{x}_p$ is equivalent to that due to the three charges:

Charge q at $\mathbf{x}_q$
Charge $-q(R_0/R_q)$ at $\mathbf{x}'_q$
Charge $q(R_0/R_q)$ at the origin.

2.10. Forces and the Stress Tensor

Consider a charge distribution with charge density $\rho(\mathbf{x})$. A volume element $d\tau$ contains the charge

$$dq = \rho(\mathbf{x})d\tau \tag{2.10.1}$$

In the presence of an electric field, a force

$$\mathbf{E}(\mathbf{x})dq = \mathbf{E}(\mathbf{x})\rho(\mathbf{x})d\tau = d\mathbf{F} \tag{2.10.2}$$

is acting on the charges in $d\tau$. If we consider a finite region of space, a volume V, the total force acting on all the charges in V is given by

$$\mathbf{F} = \int d\tau \mathbf{E}(\mathbf{x})\rho(\mathbf{x}) = \frac{1}{4\pi} \int d\tau \mathbf{E}(\boldsymbol{\nabla} \cdot \mathbf{E}) \tag{2.10.3}$$

since

$$\mathbf{\nabla} \cdot \mathbf{E}(\mathbf{x}) = 4\pi\rho(\mathbf{x}) \tag{2.10.4}$$

It will be possible to express this force as a surface integral over the boundaries of the region considered. From Eq. (2.10.3),

$$
\begin{aligned}
F_k &= \frac{1}{4\pi} \int d\tau\, E_k \left(\sum_i \frac{\partial E_i}{\partial x_i} \right) \\
&= \frac{1}{4\pi} \int d\tau \left(\sum_i \frac{\partial}{\partial x_i}(E_i E_k) - \sum_i E_i \frac{\partial E_k}{\partial x_i} \right)
\end{aligned}
\tag{2.10.5}
$$

But

$$\mathbf{E}(\mathbf{x}) = -\mathbf{\nabla}\phi(\mathbf{x}) \tag{2.10.6}$$

and

$$\mathbf{\nabla} \times \mathbf{E}(\mathbf{x}) = 0 \tag{2.10.7}$$

which means that

$$\frac{\partial E_k}{\partial x_i} = \frac{\partial E_i}{\partial x_k} \tag{2.10.8}$$

Therefore,

$$
\begin{aligned}
F_k &= \frac{1}{4\pi} \int d\tau \left(\sum_i \frac{\partial}{\partial x_i}(E_i E_k) - \sum_i E_i \frac{\partial E_i}{\partial x_k} \right) \\
&= \frac{1}{4\pi} \int d\tau \left(\sum_i \frac{\partial}{\partial x_i}(E_i E_k) - \frac{1}{2}\frac{\partial}{\partial x_k}\left(\sum_i E_i^2 \right) \right) \\
&= \frac{1}{4\pi} \int d\tau \left(\sum_i \frac{\partial}{\partial x_i}\left[E_i E_k - \frac{1}{2}\delta_{ik}\left(\sum_l E_l^2 \right) \right] \right)
\end{aligned}
\tag{2.10.9}
$$

We can then write

$$F_k = \int_v d\tau \left(\sum_i \frac{\partial}{\partial x_i} T_{ik} \right) \tag{2.10.10}$$

where

$$T_{ik} = \frac{1}{4\pi}\left[E_i E_k - \frac{1}{2}\delta_{ik}\left(\sum_l E_l^2 \right) \right] \tag{2.10.11}$$

T_{ik} defines the *Maxwell stress tensor*. The force, as we can see, is the volume integral of a divergence. If we use the divergence theorem, we can write

$$(F_k) \text{ on charges inside } V = \int_S dS \sum_i n_i T_{ik} \qquad (2.10.12)$$

where T_{ik} is given by Eq. (2.10.11). The Maxwell stress tensor can be expressed as follows:

$$
T = \begin{pmatrix} T_{xx} & T_{xy} & T_{xz} \\ T_{yx} & T_{yy} & T_{yz} \\ T_{zx} & T_{zy} & T_{zz} \end{pmatrix} = \frac{1}{4\pi} \begin{pmatrix} E_x^2 - \frac{1}{2}E^2 & E_x E_y & E_x E_z \\ E_y E_x & E_y^2 - \frac{1}{2}E^2 & E_y E_z \\ E_z E_x & E_z E_y & E_z^2 - \frac{1}{2}E^2 \end{pmatrix}
$$

$$
= \frac{1}{4\pi} \begin{pmatrix} \frac{1}{2}(E_x^2 - E_y^2 - E_z^2) & E_x E_y & E_x E_z \\ E_x E_y & \frac{1}{2}(E_y^2 - E_x^2 - E_z^2) & E_y E_z \\ E_x E_z & E_y E_z & \frac{1}{2}(E_z^2 - E_x^2 - E_y^2) \end{pmatrix}
$$

$$(2.10.13)$$

The presence of the electric field modifies the space in such a way that it presents properties analogous to those of an elastic body. In such a body the components of a stress tensor define the forces exercised on a surface element dS, part of a boundary surface, by the material outside the volume defined by this surface.

EXAMPLE

We shall consider again the case of a charge q and a planar, infinitely extended surface with $\phi = 0$, as in Fig. 2.14. The force acting on the charge is pointing in the x direction and is given by

$$F = \frac{q^2}{(2D)^2} \qquad (2.10.14)$$

We want to calculate this force by using the formalism of the Maxwell stress tensor.

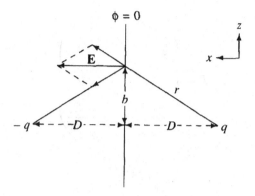

FIGURE 2.14

The electric field at the planar surface is pointing in the x direction and is given by

$$E = 2\frac{D}{r}\frac{q}{r^2} \tag{2.10.15}$$

The component of the force on the charge, as given by Eq. (2.10.12), is

$$F_k = \int dS \sum_i n_i T_{ik} = \int dS\, n_x T_{xk} = \int dS\, T_{xk}$$

where

$$T_{xx} = \frac{1}{4\pi}\left(E_x^2 - \frac{1}{2}E^2\right) = \frac{1}{4\pi}\frac{E_x^2}{2} = \frac{E_x^2}{8\pi}$$

$$T_{yy} = \frac{1}{4\pi}\left(E_y^2 - \frac{1}{2}E^2\right) = -\frac{E_x^2}{8\pi}$$

$$T_{zz} = \frac{1}{4\pi}\left(E_z^2 - \frac{1}{2}E^2\right) = -\frac{E_x^2}{8\pi}$$

$$T_{xy} = \frac{1}{4\pi}E_x E_y = 0$$

$$T_{yz} = \frac{1}{4\pi}E_y E_z = 0$$

$$T_{zx} = \frac{1}{4\pi}E_z E_x = 0$$

Therefore,

$$T = \begin{pmatrix} \dfrac{E_x^2}{8\pi} & 0 & 0 \\[2mm] 0 & -\dfrac{E_x^2}{8\pi} & 0 \\[2mm] 0 & 0 & -\dfrac{E_x^2}{8\pi} \end{pmatrix}$$

where

$$E_x = \frac{2qD}{r^3}$$

Now

$$F_x = \int dS\, T_{xx} = \int dS\, \frac{E_x^2}{8\pi} = \frac{1}{8\pi}\int dS \left(\frac{2qD}{r^3}\right)^2$$

$$= \frac{q^2 D^2}{2\pi}\int \frac{dS}{r^6}$$

But

$$dS = 2\pi b\, db = \pi\, db^2$$

$$r^6 = (D^2 + b^2)^3$$

Then

$$F_x = \frac{q^2 D^2}{2\pi}\int \frac{\pi\, db^2}{D^2 + b^2} = \frac{q^2 D^2}{2}\int_0^\infty \frac{d\lambda}{(D^2 + \lambda)^3}$$

$$= \frac{q^2}{2}\int_0^\infty \frac{D^2 d\lambda}{(D^2 + \lambda)^3} = \frac{q^2}{2}\frac{1}{2D^2} = \frac{q^2}{(2D)^2} \tag{2.10.16}$$

as expected. Also,

$$F_y = \int dS\, T_{xy} = 0$$

$$F_z = \int dS\, T_{xz} = 0$$

2.11. The Field Energy

Given an ensemble of electric charges, a displacement of these charges without external intervention corresponds to work done at the expense of the electrostatic energy of the system.

There is a force acting on each charge due to the electric field due to all other charges. Let charge q_s be at $\mathbf{x}_s$ and let $\delta\mathbf{x}_s$ be the displacement of this charge. The work done is

$$dW = \sum_s \mathbf{F}_s \cdot \delta\mathbf{x}_s \tag{2.11.1}$$

Let $\phi(\mathbf{x}_s)$ be the potential at $\mathbf{x}_s$ due to all the charges except q_s:

$$\phi(\mathbf{x}_s) = \sum_{t \neq s} \frac{q_t}{|\mathbf{x} - \mathbf{x}_t|} \tag{2.11.2}$$

Then

$$\mathbf{F}_s \cdot \delta\mathbf{x}_s = \delta\mathbf{x}_s \cdot q_s \mathbf{E}(\mathbf{x}_s) = -\delta\mathbf{x}_s \cdot \nabla_s \sum_{t \neq s} \frac{q_t q_s}{|\mathbf{x}_s - \mathbf{x}_t|} \tag{2.11.3}$$

where ∇_s = gradient operator acting on the coordinate $\mathbf{x}_s$.

Given a function

$$f = f(\mathbf{x}_1, \mathbf{x}_2, \ldots, \mathbf{x}_s, \ldots) \tag{2.11.4}$$

$$\delta f = \sum_s \delta\mathbf{x}_s \cdot \nabla_s f \tag{2.11.5}$$

Then

$$\sum_s \mathbf{F}_s \cdot \delta\mathbf{x}_s = -\delta \sum_{\substack{s,t \\ s \neq t}} \frac{1}{2} \frac{q_s q_t}{|\mathbf{x}_s - \mathbf{x}_t|} \tag{2.11.6}$$

where the factor $1/2$ avoids the double counting of each term of the sum.

EXAMPLE

Given a system of three charges: q_1 at $\mathbf{x}_1$, q_2 at $\mathbf{x}_2$, and q_3 at $\mathbf{x}_3$.

$$dW = \sum_s \mathbf{F}_s \cdot \delta\mathbf{x}_s$$

$$= -\left[q_1\, \delta\mathbf{x}_1 \cdot \nabla_1 \left(\frac{q_2}{|\mathbf{x}_1 - \mathbf{x}_2|} + \frac{q_3}{|\mathbf{x}_1 - \mathbf{x}_3|} \right) \right.$$

$$+ q_2\, \delta\mathbf{x}_2 \cdot \nabla_2 \left(\frac{q_1}{|\mathbf{x}_2 - \mathbf{x}_1|} + \frac{q_3}{|\mathbf{x}_2 - \mathbf{x}_3|} \right)$$

$$+ q_3 \, \delta\mathbf{x}_3 \cdot \boldsymbol{\nabla}_3 \left(\frac{q_1}{|\mathbf{x}_3 - \mathbf{x}_1|} + \frac{q_2}{|\mathbf{x}_3 - \mathbf{x}_2|} \right) \Bigg]$$

$$= -\delta \left(\frac{q_1 q_2}{|\mathbf{x}_1 - \mathbf{x}_2|} + \frac{q_1 q_3}{|\mathbf{x}_1 - \mathbf{x}_3|} + \frac{q_2 q_3}{|\mathbf{x}_2 - \mathbf{x}_3|} \right)$$

The work dW done by the forces and the change δU in the electrostatic energy are such that

$$dW + \delta U = 0 \tag{2.11.7}$$

Since dW is given by Eq. (2.11.6),

$$\delta U = -dW = \delta \frac{1}{2} \sum_{\substack{s,t \\ s \neq t}} \frac{q_s q_t}{|\mathbf{x}_s - \mathbf{x}_t|} \tag{2.11.8}$$

We can then express the electrostatic energy in the following way

$$U = \frac{1}{2} \sum_{\substack{s,t \\ s \neq t}} \frac{q_s q_t}{|\mathbf{x}_s - \mathbf{x}_t|} \tag{2.11.9}$$

In the case of a continuous charge distribution,

$$U = \frac{1}{2} \iint d\tau \, d\tau' \frac{\rho(\mathbf{x})\rho(\mathbf{x}')}{|\mathbf{x} - \mathbf{x}'|} \tag{2.11.10}$$

If we have continuous distributions, we forget about $s \neq t$ because the individual charges go to zero. But

$$U = \frac{1}{2} \int d\tau \rho(\mathbf{x}) \int d\tau' \frac{\rho(\mathbf{x}')}{|\mathbf{x} - \mathbf{x}'|} = \frac{1}{2} \int d\tau \rho(\mathbf{x}) \phi(\mathbf{x}) \tag{2.11.11}$$

Also,

$$\boldsymbol{\nabla} \cdot \mathbf{E}(\mathbf{x}) = 4\pi \rho(\mathbf{x})$$
$$\rho(\mathbf{x}) = \frac{1}{4\pi} \boldsymbol{\nabla} \cdot \mathbf{E}(\mathbf{x}) \tag{2.11.12}$$

Then

$$U = \frac{1}{2} \int d\tau \rho(\mathbf{x}) \phi(\mathbf{x}) = \frac{1}{8\pi} \int d\tau [\boldsymbol{\nabla} \cdot \mathbf{E}(\mathbf{x})] \phi(\mathbf{x})$$

$$= -\frac{1}{8\pi} \int d\tau \mathbf{E}(\mathbf{x}) \cdot \boldsymbol{\nabla}\phi(\mathbf{x}) = \frac{1}{8\pi} \int d\tau (\mathbf{E})^2 \tag{2.11.13}$$

where we have used partial integration over all space and taken into account the fact that, at ∞, $\mathbf{E} = 0$ and $\phi = 0$. Then

$$U = \frac{1}{8\pi} \int_{\text{all space}} d\tau (\mathbf{E})^2 \qquad (2.11.14)$$

This result is the field energy, which is equal to the potential energy of a distribution of charges.

It is interesting to compare expressions (2.11.10) and (2.11.14) for the electrostatic energy:

$$U = \frac{1}{2} \iint d\tau \, d\tau' \frac{\rho(\mathbf{x})\rho(\mathbf{x}')}{|\mathbf{x} - \mathbf{x}'|}, \qquad \text{says that the energy is} \atop \text{where the charges are}$$

$$U = \frac{1}{8\pi} \int_{\text{all space}} (\mathbf{E})^2 d\tau, \qquad \text{says that the energy is} \atop \text{where the field is}$$

The two ways of writing the expression for the energy are perfectly equivalent.

In considering the force acting on a charge $\mathbf{q}_s$, we have taken into account the field due to all the "other" charges $q_t (t \neq s)$. If we remove this provision, we run into problems. Assume that

$$\rho(\mathbf{x}) = \sum_s \mathbf{q}_s \, \delta(\mathbf{x} - \mathbf{x}_s) \qquad (2.11.15)$$

Then

$$U = \frac{1}{2} \iint d\tau \, d\tau' \frac{\rho(\mathbf{x})\rho(\mathbf{x}')}{|\mathbf{x} - \mathbf{x}'|}$$

$$= \frac{1}{2} \iint \frac{d\tau \, d\tau'}{|\mathbf{x} - \mathbf{x}'|} \sum_s \sum_t q_s \delta(\mathbf{x} - \mathbf{x}_s) q_t \delta(\mathbf{x}' - \mathbf{x}_t)$$

$$= \frac{1}{2} \sum_s \sum_t \frac{q_s q_t}{|\mathbf{x}_s - \mathbf{x}_t|} = \frac{1}{2} \sum_{s \neq t} \frac{q_s q_t}{|\mathbf{x}_s - \mathbf{x}_t|} + \frac{1}{2} \sum_s \frac{q_s^2}{0} \qquad (2.11.16)$$

If we have a point charge, the energy associated with its field, called the *self-energy* of the point charge, is ∞:

$$E = \frac{q}{r^2}$$

$$U = \frac{1}{8\pi} \int E^2 d\tau = \frac{1}{8\pi} 4\pi q^2 \int_0^\infty \frac{dr}{r^2} = \infty \qquad (2.11.17)$$

If we assume that the charge q is spread over a small sphere of radius r_0, the self-energy is given by

$$\frac{1}{8\pi} \int E^2 d\tau = \frac{1}{2} q^2 \int_{r_0}^{\infty} \frac{dr}{r^2} = \frac{q^2}{2r_0} \tag{2.11.18}$$

In any case, the self-energy is constant and is not affected by the displacement of the charges; we cannot use it to produce work.

2.12. Earnshaw's Theorem

A system of charges may have an arrangement for which every charge is in equilibrium under the action of the other charges. *Earnshaw's theorem* states that this equilibrium is unstable; that is, if we move a charge, there will be directions in which there is no force tending to push it back. We shall now prove this theorem.

The field energy can be written [see Eq. (2.11.11)]

$$U = \frac{1}{2} \sum_s q_s \phi(\mathbf{x}_s) \tag{2.12.1}$$

where

$$\phi(\mathbf{x}_s) = \sum_{t \neq s} \frac{q_t}{|\mathbf{x}_s - \mathbf{x}_t|} \tag{2.12.2}$$

Let us displace q_s by $\delta\mathbf{x}_s$

$$\delta U = \frac{1}{2} \sum_s q_s \left(\delta\mathbf{x}_s \cdot \boldsymbol{\nabla}_s \phi(\mathbf{x}_s) \right.$$
$$\left. + \frac{1}{2} \sum_i \sum_k \delta x_{si} \delta x_{sk} \frac{\partial^2 \phi(\mathbf{x}_s)}{\partial x_{si} \partial x_{sk}} \right) \tag{2.12.3}$$

where we have neglected the terms to greater than second order in the displacement. If we assume equilibrium,

$$\boldsymbol{\nabla}_s \phi(\mathbf{x}_s) = -\mathbf{E}(\mathbf{x}_s) = 0 \tag{2.12.4}$$

and

$$\delta U = \frac{1}{2} \sum_s q_s \left(\frac{1}{2} \sum_i \sum_k \delta x_{si} \, \delta x_{sk} \frac{\partial^2 \phi(\mathbf{x}_s)}{\partial x_{si} \, \partial x_{sk}} \right) \tag{2.12.5}$$

For the equilibrium to be stable we must have

$$\delta U > 0 \tag{2.12.6}$$

or

$$\sum_i \sum_k \delta x_{si} \, \delta x_{sk} \frac{\partial^2 \phi(\mathbf{x}_s)}{\partial x_{si} \, \partial x_{sk}} \gtrless 0, \quad \text{if } q_s \gtrless 0 \tag{2.12.7}$$

We set

$$a_{ik}(s) = \frac{\partial^2 \phi(\mathbf{x}_s)}{\partial x_{si} \, \partial x_{sk}}$$

Then

$$\delta U = \frac{1}{2} \sum_s q_s \left[\frac{1}{2} \sum_i \sum_k \delta x_{si} \, \delta x_{sk} \, a_{ik}(s) \right] \tag{2.12.8}$$

In stable equilibrium the quantity in brackets must have a definite positive (negative) form if q_s is positive (negative). The coefficients a_{ik} are constant, because they do not depend on the displacements $\delta \mathbf{x}$. They form a real symmetric matrix. Such a matrix may always be diagonalized by a coordinate transformation. If such diagonalization is performed,

$$\delta U = \frac{1}{2} \sum_s q_s \left[\sum_i a_{ii}(s) \delta x_{si}^2 \right] \tag{2.12.9}$$

Assuming $q_s > 0$, all these elements a_{ii} must be positive in order to have a definite positive form. But in reality

$$\sum_i a_{ii}(s) = \sum_i \frac{\partial^2 \phi(\mathbf{x}_s)}{\partial x_{si}^2} = \nabla^2 \phi(\mathbf{x}_s) = 0$$

Since $\phi(\mathbf{x}_s)$ is the potential due to all charges but q_s, it respects Laplace's equation. The situation is then such that the sum of all coefficients is zero. This means either that all these coefficients are zero or that some of them are positive and some are negative. If, starting from an equilibrium configuration, we displace a charge in a direction corresponding to a negative coefficient a_{ii}, the charge of the field energy δU will be negative and the system will not tend to go back to the original equilibrium. We can then express the Earnshaw's theorem as follows: *no stable equilibrium is possible under the influence of purely electrostatic forces.* We must be careful in interpreting this theorem. It is possible that, if we move a charge q_s along a certain path, the system will tend to go back to the original equilibrium because $[\sum_i aii(s)\delta x_{si}^2]$ is greater than zero. What the theorem says is that an equilibrium due to purely electrostatic forces is not stable with respect to the displacement of *any* charge along *any* path.

2.13. Thompson's Theorem

We can express *Thompson's theorem* as follows:

If a number of surfaces are fixed in position and a given total charge is placed on each surface, then the electrostatic energy in the region bound by the surfaces is an absolute minimum when the charges are placed so that every surface is equipotential, as happens when the surfaces enclose conducting materials.

Let us consider a number of conducting bodies, each designated by an index: $1, 2, 3, \ldots, i, \ldots, n$. Let S_i be the surface of body i and Q_i the total charge distributed over body i (see Fig. 2.15). Let $\mathbf{E}(\mathbf{x})$ and $\rho(\mathbf{x})$ be the actual field and actual charge density in equilibrium, respectively. These quantities are connected by the relation

$$\boldsymbol{\nabla} \cdot \mathbf{E}(\mathbf{x}) = 4\pi \rho(\mathbf{x}) \tag{2.13.1}$$

Let

$$\mathbf{E}' = \mathbf{E} + \delta \mathbf{E} \tag{2.13.2}$$

$$\rho' = \rho + \delta \rho \tag{2.13.3}$$

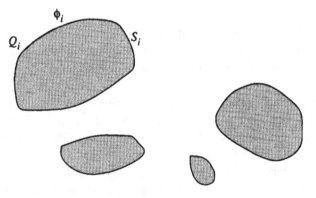

FIGURE 2.15

be a new field and a new charge density so that

$$\nabla \cdot \mathbf{E}'(\mathbf{x}) = 4\pi\rho' \tag{2.13.4}$$

and

$$\nabla \cdot \delta\mathbf{E} = 4\pi\delta\rho \tag{2.13.5}$$

Also, the change in U is

$$U' - U = \delta U = \frac{1}{8\pi} \int d\tau [(\mathbf{E} + \delta\mathbf{E})^2 - (\mathbf{E})^2]$$

$$= \frac{1}{8\pi} \int d\tau [2\mathbf{E} \cdot \delta\mathbf{E} + (\delta\mathbf{E})^2] \tag{2.13.6}$$

We know that

$$\mathbf{E} = -\nabla\phi$$

so

$$\mathbf{E} \cdot \delta\mathbf{E} = -\nabla\phi \cdot \delta\mathbf{E} = -\nabla \cdot (\phi\delta\mathbf{E}) + \phi\nabla \cdot \delta\mathbf{E}$$

and

$$U' - U = \delta U = \frac{1}{8\pi} \int d\tau [-2\nabla \cdot (\phi\delta\mathbf{E}) + 2\phi\nabla \cdot \delta\mathbf{E} + (\delta\mathbf{E})^2]$$

$$= \frac{1}{8\pi} \left(-2 \int d\mathbf{S} \cdot \phi\delta\mathbf{E} + 2 \int d\tau \, \phi 4\pi\delta\rho + \int (\delta\mathbf{E})^2 d\tau \right)$$

$$= \frac{1}{8\pi} \left(8\pi \int (\phi\delta\rho) \, d\tau + \int (\delta\mathbf{E})^2 d\tau \right) \tag{2.13.7}$$

where we have taken into account the fact that $\phi(\infty)=0$.

But, since the surfaces are equipotential,

$$\int (\phi\delta\rho)d\tau = \sum_n \phi_n \int_{S_n} (\delta\sigma)dS_n = 0 \tag{2.13.8}$$

and

$$\delta U = \frac{1}{8\pi} \int d\tau(\delta\mathbf{E})^2 > 0 \tag{2.13.9}$$

We can proceed in a different way in order to arrive at Eq. (2.13.9). We can take advantage of the knowledge that the field inside the conductors is zero. We can exclude the conductors by enclosing them in mathematical surfaces. Then

$$\delta U = \frac{1}{8\pi} \int_{\substack{\text{all space minus} \\ \text{conductors}}} d\tau [-2\boldsymbol{\nabla} \cdot (\phi \delta \mathbf{E}) + 2\phi \boldsymbol{\nabla} \cdot \delta \mathbf{E} + (\delta \mathbf{E})^2]$$

$$= \frac{1}{8\pi} \int d\tau (\delta \mathbf{E})^2 + \frac{1}{8\pi} \int d\tau \, 2\phi 4\pi \delta \rho - \frac{2}{8\pi} \int dt \boldsymbol{\nabla} \cdot (\phi \delta \mathbf{E})$$

$$= \frac{1}{8\pi} \int d\tau (\delta \mathbf{E})^2 - \frac{2}{8\pi} \sum_k \phi_k \int_{\mathcal{S}_k} dS_k (\mathbf{n} \cdot \delta \mathbf{E})$$

$$= \frac{1}{8\pi} \int d\tau (\delta \mathbf{E})^2 + \frac{2}{8\pi} \sum_k \phi_k \int_{S_k} dS_k 4\pi \delta \sigma$$

$$= \frac{1}{8\pi} \int d\tau (\delta \mathbf{E})^2 \tag{2.13.10}$$

where we have used Eq. (2.8.6).

We shall now prove that the solution of the problem is unique. Let us imagine that we have *two* possible equilibrium situations, one related to the quantities $\mathbf{E}_1$ and ρ_1 and the other to the quantities $\mathbf{E}_2$ and ρ_2, where

$$\mathbf{E}_2 = \mathbf{E}_1 + \delta \mathbf{E} \tag{2.13.11}$$

We have then

$$U(\mathbf{E}_2) = U(\mathbf{E}_1) + \frac{1}{8\pi} \int d\tau (\delta \mathbf{E})^2 \tag{2.13.12}$$

But it is also

$$\mathbf{E}_1 = \mathbf{E}_2 - \delta \mathbf{E} \tag{2.13.13}$$

and

$$U(\mathbf{E}_1) = U(\mathbf{E}_2) + \frac{1}{8\pi} \int d\tau (\delta \mathbf{E})^2$$

$$= U(\mathbf{E}_1) + \frac{2}{8\pi} \int d\tau (\delta \mathbf{E})^2 \tag{2.13.14}$$

Therefore, we must have $\delta \mathbf{E} = 0$ ($\delta \mathbf{E}$ is not an infinitesimal quantity, but a small finite quantity).

Thompson's theorem can be used to find an actual charge distribution. We shall now give an example of this application.

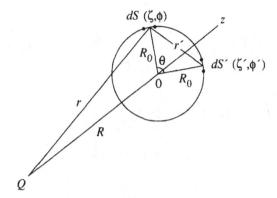

FIGURE 2.16

EXAMPLE

Let us consider the situation in Fig. 2.16 with a charge Q and a conducting uncharged sphere of radius R_0. We want to calculate the charge distribution on the sphere.

The interaction energy is given by

$$U_1 = Q \int dS \frac{\sigma(S)}{r} \tag{2.13.15}$$

where $\sigma(S)$ = surface charge density. The field energy is given by

$$U = Q \int dS \frac{\sigma(S)}{r} + \frac{1}{2} \int dS \int dS' \frac{\sigma(S)\sigma(S')}{r'} = U_1 + U_2 \tag{2.13.16}$$

Because of the symmetry of the system, the surface charge σ depends on the angle ζ and can be expanded in a series of Legendre polynomials.[5]

$$\sigma(\zeta) = \sum_l \sigma_l P_l(\cos \zeta)$$

This allows us to characterize the charge by means of the unknown coefficients σ_l.

We can write

$$\frac{1}{r} = \sum_l \frac{R_0^l}{R^{l+1}} P_l(\pi - \zeta) = \sum_l \frac{R_0^l}{R^{l+1}} (-1)^l P_l(\zeta) \tag{2.13.17}$$

---

[5]W. R. Smythe, *Static and Dynamics Electricity*, 3rd ed., McGraw-Hill Book Company, New York, 1968, p. 147.

Then

$$U_1 = Q \int dS \left[\left(\sum_l \sigma_l P_l(\cos \zeta) \right) \left(\sum_{l'} \frac{R_0^{l'}}{R^{l'+1}} (-1)^{l'} P_{l'}(\zeta) \right) \right] \qquad (2.13.18)$$

But

$$\int dS \ P_l(\zeta) P_{l'}(\zeta) = 2\pi R_0^2 \int_0^\pi \sin \zeta d\zeta P_{l'}(\zeta) P_l(\zeta)$$

$$= 2\pi R_0^2 \delta_{ll'} \frac{2}{2l+1} \qquad (2.13.19)$$

Then

$$U_1 = Q 4\pi R_0^2 \sum_l \frac{\sigma_l(-1)^l}{2l+1} \frac{R_0^l}{R^{l+1}} \qquad (2.13.20)$$

We have also

$$\frac{1}{r'} = \frac{1}{R_0} \sum_l (P_l \cos \theta)$$

$$= \frac{1}{R_0} \sum_l \frac{4\pi}{2l+1} \sum_m (-1)^m Y_{lm}(\zeta, \varphi) Y_{l,-m}(\zeta', \varphi') \qquad (2.13.21)$$

and

$$U_2 = \frac{1}{2} \int dS \int dS' \frac{\sigma(S)\sigma(S')}{r'}$$

$$= \frac{1}{2} \int dS \int dS' \left[\sum_l \sigma_l P_l(\zeta) \right]$$

$$\times \left[\sum_{l'} \sigma_{l'} P_{l'}(\zeta') \right] \frac{1}{R_0} \sum_\lambda \frac{4\pi}{2\lambda+1}$$

$$\times \sum_m (-1)^m Y_{\lambda m}(\zeta, \varphi) Y_{\lambda,-m}(\zeta', \varphi') \qquad (2.13.22)$$

But

$$\int dS \ P_l(\zeta) Y_{\lambda m}(\zeta, \varphi) = \sqrt{\frac{4\pi}{2l+1}} \int dS \ Y_{l0}(\zeta) Y_{\lambda m}(\zeta, \varphi)$$

$$= \sqrt{\frac{4\pi}{2l+1}} \int R_0^2 \sin \zeta d\zeta d\varphi \ Y_{l0}(\zeta) Y_{\lambda m}(\zeta, \varphi)$$

$$= \sqrt{\frac{4\pi}{2l+1}} R_0^2 \int \sin \zeta d\zeta d\varphi \ Y_{l0}(\zeta) Y_{\lambda m}(\zeta, \varphi)$$

$$= \sqrt{\frac{4\pi}{2l+1}} R_0^2 \delta_{\lambda l} \delta_{0m} \qquad (2.13.23)$$

and

$$\int dS' \ P_{l'}(\zeta') Y_{\lambda, -m}(\zeta', \varphi')$$

$$= \sqrt{\frac{4\pi}{2l+1}} \int dS' \ Y_{l'0}(\zeta') Y_{\lambda m}(\zeta', \varphi')$$

$$= \sqrt{\frac{4\pi}{2l+1}} R_0^2 \ \delta_{\lambda l'} \ \delta_{0m} \qquad (2.13.24)$$

Therefore,

$$U_2 = \frac{1}{2} R_0^4 \frac{1}{R_0} \sum_l \sigma_l^2 \left(\frac{4\pi}{2l+1} \right)^2 \qquad (2.13.25)$$

and

$$U = U_1 + U_2$$

$$= \sum_l \frac{4\pi}{2l+1} \left(\frac{Q(-1)^l R_0^{l+2}}{R^{l+1}} \sigma_l + \frac{1}{2} \frac{4\pi}{2l+1} R_0^3 \sigma_l^2 \right) \qquad (2.13.26)$$

The total charge on the sphere is zero:

$$\int dS \left[\sum_l P_l(\zeta) \sigma_l \right] = \sum_l \sigma_l \int dS \ P_l(\zeta) = 0 \qquad (2.13.27)$$

The integral above equal to $4\pi R^2$ if $l = 0$; otherwise, it is zero. Therefore, $\sigma_0 = 0$ and the sum in Eq. (2.13.26) starts with $l = 1$.

We must also have

$$\delta U = 0 \qquad (2.13.28)$$

The only things we can vary are the σ_l's. We have to find the σ_l's for which

$$\frac{\partial U}{\partial \sigma_l} = 0 \qquad (2.13.29)$$

This relation gives an infinite number of equations for σ_l. We have

$$\frac{\partial U}{\partial \sigma_l} = \frac{4\pi}{2l+1}\left(\frac{Q(-1)^l R_0^{l+2}}{R^{l+1}} + \frac{4\pi}{2l+1}R_0^3\sigma_l\right) = 0 \qquad (2.13.30)$$

or

$$\sigma_l = \frac{2l+1}{4\pi}Q(-1)_l\frac{R_0^{l-1}}{R^{l+1}} \qquad (2.13.31)$$

and

$$\sigma(\zeta) = \sum_{l=1}^{\infty}\sigma_l P_l(\zeta) = -\frac{Q}{4\pi}\sum_{l=1}^{\infty}(-1)^l(2l+1)\frac{R_0^{l-1}}{R^{l+1}}P_l(\zeta) \qquad (2.13.32)$$

2.14. Polarization

The introduction of the concept of dielectric derives from the fact that it is generally convenient to consider matter as a continuum characterized by certain physical parameters. If a body is subjected to the action of a homogeneous field, it is often possible to neglect the structure of matter and to introduce *macroscopic* parameters like *polarization* **P**, *displacement* **D**, and so on.

We start by considering the basic concept of the *dipole*. In certain materials, the response to the application of a homogeneous electric field is the creation of dipoles; if the field is not homogeneous, we can have the creation of multipoles. The average effect of the presence of dipoles is shown by means of the concept of polarization.

We have to distinguish between *physical dipole* and *mathematical dipole*. A physical dipole consists of two charges of opposite signs at distance **d** from each other (see Fig. 2.17). The potential produced by the two charges is given by

$$\phi(\mathbf{x}_p) = \frac{q}{|\mathbf{x}_p - \mathbf{x}|} + \frac{-q}{|\mathbf{x}_p - \mathbf{x} + \mathbf{d}|}$$

$$= \frac{q}{|\mathbf{x}_p - \mathbf{x}|} - q\left(\frac{1}{|\mathbf{x}_p - \mathbf{x}|} + \mathbf{d}\cdot\nabla_{x_p}\frac{1}{|\mathbf{x} - \mathbf{x}_p|}\right)$$

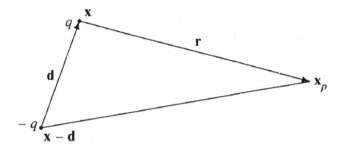

FIGURE 2.17

$+$ higher terms

$$= -q\mathbf{d} \cdot \nabla_{x_p} \frac{1}{|\mathbf{x} - \mathbf{x}_p|} + \text{higher terms} \qquad (2.14.1)$$

The first "higher term" is of the order

$$qd^2 \frac{\partial^2}{\partial r^2} \frac{1}{r} = q \frac{2d^2}{r^3} \qquad (2.14.2)$$

A mathematical dipole is simply the charge distribution that gives us the potential

$$\phi(\mathbf{x}_p) = -q\,\mathbf{d} \cdot \nabla_{x_p} \frac{1}{|\mathbf{x} - \mathbf{x}_p|} = -\boldsymbol{\mathcal{D}} \cdot \nabla_{x_p} \frac{1}{|\mathbf{x} - \mathbf{x}_p|} \qquad (2.14.3)$$

where

$$\boldsymbol{\mathcal{D}} = \text{dipole moment} = q\mathbf{d} \qquad (2.14.4)$$

The mathematical dipole may be considered a limiting case of the physical dipole; that is, in Eq. (2.14.1) we can make $q \to \infty$ and $d \to 0$ in such a way that $\boldsymbol{\mathcal{D}} = q\mathbf{d}$ remains constant. Then $qd^2 \to 0$; that is, the higher terms go to zero. We can rewrite Eq. (2.4.13) as follows:

$$\phi_{\mathcal{D}}(\mathbf{x}_p) = -\boldsymbol{\mathcal{D}} \cdot \nabla_{x_p} \frac{1}{|\mathbf{x}_p - \mathbf{x}_{\mathcal{D}}|} \qquad (2.14.5)$$

where $\mathbf{x}_{\mathcal{D}} = $ position of the dipole.

From a mathematical point of view, it is useful to introduce continuous distributions of charges. Similarly, if a macroscopically small region contains a large number of dipoles, we can talk of a *volume density of dipoles* or *polarization*, defined by

$$\mathbf{P}(\mathbf{x})d\tau = \sum_i \boldsymbol{\mathcal{D}}_i \quad \text{in } d\tau \qquad (2.14.6)$$

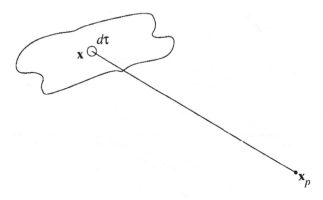

FIGURE 2.18

If all the dipoles in a volume V are identical, and if there are N dipoles in V,

$$\mathbf{P}(\mathbf{x}) = \frac{N}{V}\mathcal{D}_i \qquad (2.14.7)$$

Assume now that we have a certain polarization in a piece of matter (see Fig. 2.18). The potential at point $\mathbf{x}_p$ produced by the polarization in $d\tau$ is given by

$$d\phi(\mathbf{x}_p) = -\mathbf{P}(\mathbf{x})d\tau \cdot \boldsymbol{\nabla}_{x_p}\frac{1}{|\mathbf{x}-\mathbf{x}_p|} \qquad (2.14.8)$$

The potential produced by all the matter is

$$\phi_{\mathcal{D}}(\mathbf{x}_p) = -\int \mathbf{P}(\mathbf{x})d\tau \cdot \boldsymbol{\nabla}_{x_p}\frac{1}{|\mathbf{x}-\mathbf{x}_p|}$$

$$= \int d\tau \mathbf{P}(\mathbf{x}) \cdot \boldsymbol{\nabla}_x\frac{1}{|\mathbf{x}-\mathbf{x}_p|} = -\int d\tau \frac{\boldsymbol{\nabla}\cdot\mathbf{P}(\mathbf{x})}{|\mathbf{x}-\mathbf{x}_p|} \qquad (2.14.9)$$

where we have used integration by parts in the last step.

The potential due to a charge distribution is given by

$$\phi(\mathbf{x}_p) = \int d\tau \frac{\rho(\mathbf{x})}{|\mathbf{x}-\mathbf{x}_p|} \qquad (2.14.10)$$

Then we can write

$$\phi_{\mathcal{D}}(\mathbf{x}_p) = \int d\tau \frac{\rho_{\text{pol}}(\mathbf{x})}{|\mathbf{x}-\mathbf{x}_p|} \qquad (2.14.11)$$

where

$$\rho_{pol}(\mathbf{x}) = -\boldsymbol{\nabla} \cdot \mathbf{P}(\mathbf{x}) \tag{2.14.12}$$

We note that the polarization has the same units as the electric field.

Let us now consider a piece of substance in which the polarization is not zero, in the absence of an applied electric field. Substances of this type really occur in nature and are called *ferroelectrics*. Examples of ferroelectric substances are $BaTiO_3$ and KH_2PO_4. Let us assume that $\mathbf{P} = $ const inside and $\mathbf{P} = 0$ outside the piece of the ferroelectric material (see Fig. 2.19). Since $\mathbf{P} = $ const, inside

$$\boldsymbol{\nabla} \cdot \mathbf{P}(\mathbf{x}) = 0$$
$$\rho_{pol}(\mathbf{x}) = -\boldsymbol{\nabla} \cdot \mathbf{P}(\mathbf{x}) = 0 \tag{2.14.13}$$

If the polarization were not constant, we would have polarization charges inside the material. As things stand, the only polarization charges will be at the surfaces. In the thin flat box in Fig. 2.19,

$$\int d\tau \, \rho_{pol}(\mathbf{x}) = -\int \boldsymbol{\nabla} \cdot \mathbf{P}(\mathbf{x}) d\tau = -\int d\mathbf{S} \cdot \mathbf{P}(\mathbf{x})$$
$$= -\int dS \, \mathbf{P}(\mathbf{x}) \cdot \mathbf{n}_s$$
$$= -\int dS \, \mathbf{P}(\mathbf{x}) \cdot \mathbf{n} = \int dS \sigma_{pol} \tag{2.14.14}$$

and

$$\sigma_{pol}(\mathbf{x}) = \mathbf{n} \cdot \mathbf{P}(\mathbf{x}) \tag{2.14.15}$$

σ_{pol} has the same units as the electric field.

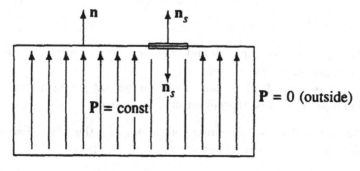

FIGURE 2.19

The introduction of $\mathbf{P}(\mathbf{x})$, a continuous function of $\mathbf{x}$, indicates that we disregard the atomic structure of matter. In electrostatics, when we deal with a field that does not vary abruptly in the region of interest, we can define an *average field*, given by

$$\langle \mathbf{E}(\mathbf{x}) \rangle = \frac{1}{\Delta V} \int_{\Delta V} d\tau \, \mathbf{E}(\mathbf{x}) \tag{2.14.16}$$

where ΔV is a small enough volume to assure that $\langle \mathbf{E}(\mathbf{x}) \rangle$ is practically a continuous function of $\mathbf{x}$.

Because of Poisson's equation,

$$\nabla \cdot \langle \mathbf{E}(\mathbf{x}) \rangle = 4\pi \langle \rho(\mathbf{x}) \rangle \tag{2.14.17}$$

where

$$\langle \rho(\mathbf{x}) \rangle = \frac{1}{\Delta V} \int_{\Delta V} d\tau \, \rho(\mathbf{x}) \tag{2.14.18}$$

We note that

$$\langle \mathbf{E}(\mathbf{x}) \rangle = -\nabla \langle \phi(\mathbf{x}) \rangle \tag{2.14.19}$$

where

$$\langle \phi(\mathbf{x}) \rangle = \frac{1}{\Delta V} \int_{\Delta V} d\tau \, \phi(\mathbf{x}) \tag{2.14.20}$$

Therefore,

$$\nabla \times \langle \mathbf{E}(\mathbf{x}) \rangle = 0 \tag{2.14.21}$$

We shall use, in what follows, $\mathbf{E}(\mathbf{x})$ for $\langle \mathbf{E}(\mathbf{x}) \rangle$ and $\rho(\mathbf{x})$ for $\langle \rho(\mathbf{x}) \rangle$. We note that

$$\rho(\mathbf{x}) = \rho_{\text{true}}(\mathbf{x}) + \rho_{\text{pol}}(\mathbf{x}) \tag{2.14.22}$$

We have two kinds of charges:

(1) Charges that we can put in and take away from a certain place, that is, charges under our control.
(2) Charges we can displace, but cannot in general put in or take away from a place, that is, charges outside our control.

The former charges are represented by $\rho_{\text{true}}(\mathbf{x})$; the latter charges are the polarization charges represented by $\rho_{pol}(\mathbf{x})$. We have

$$\nabla \cdot \mathbf{E}(\mathbf{x}) = 4\pi \rho_{\text{true}}(\mathbf{x}) + 4\pi \rho_{\text{pol}}(\mathbf{x}) \tag{2.14.23}$$

or, because of Eq. (2.14.12),

$$\nabla \cdot [\mathbf{E}(\mathbf{x}) + 4\pi\mathbf{P}(\mathbf{x})] = 4\pi\rho_{\text{true}}(\mathbf{x}) \tag{2.14.24}$$

We define the *electric displacement* $\mathbf{D}(\mathbf{x})$ as follows:

$$\mathbf{D}(\mathbf{x}) = \mathbf{E}(\mathbf{x}) + 4\pi\mathbf{P}(\mathbf{x}) \tag{2.14.25}$$

Then

$$\nabla \cdot \mathbf{D}(\mathbf{x}) = 4\pi\rho_{\text{true}}(\mathbf{x}) \tag{2.14.26}$$

This is the first Maxwell equation for systems in which $\langle \mathbf{E}(\mathbf{x}) \rangle$ and $\mathbf{P}(\mathbf{x})$ have meaning. $\mathbf{D}(\mathbf{x})$ has no meaning at all at atomic level. Equation (2.14.26) is different from its counterpart,

$$\nabla \cdot \mathbf{E}(\mathbf{x}) = 4\pi\rho(\mathbf{x}) \tag{2.14.27}$$

where $\rho(\mathbf{x})$ includes true and polarization charges as in Eq. (2.14.22).

Two important cases are defined by the relation between $\mathbf{P}$ and $\mathbf{E}$:

(1) *Dielectrics*: For these materials,

$$\mathbf{P}(\mathbf{x}) = \chi\mathbf{E}(\mathbf{x}) \tag{2.14.28}$$

where

$$\chi = electric\ susceptibility$$

For a high field, there may also be a quadratic term in the relation between $\mathbf{P}$ and $\mathbf{E}$.

(2) *Ferroelectrics*: For these materials,

$$\mathbf{P}(\mathbf{x}) \neq 0 \tag{2.14.29}$$

and is independent of $\mathbf{E}$. Ferroelectrics have Curie point-like ferromagnetics.[6]

Let us consider dielectric materials:

$$\mathbf{D}(\mathbf{x}) = \mathbf{E}(\mathbf{x}) + 4\pi\mathbf{P}(\mathbf{x}) = (1 + 4\pi\chi)\mathbf{E}(\mathbf{x}) = K\ \mathbf{E}(\mathbf{x}) \tag{2.14.30}$$

where K is called the *dielectric constant*. In general, K is a function of the spatial coordinates:

$$K = K(\mathbf{x}) \tag{2.14.31}$$

---

[6]See, for example, A. J. Dekker, *Solid State Physics*, Prentice Hall, Englewood Cliffs, N.J., 1957.

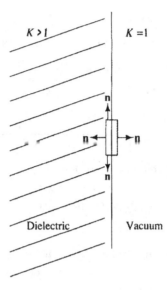

FIGURE 2.20

Therefore,

$$\nabla \times \mathbf{D}(\mathbf{x}) = \nabla \times (K\mathbf{E}) = K(\nabla \times \mathbf{E}) + (\nabla K \times \mathbf{E}) \qquad (2.14.32)$$

$\nabla \times \mathbf{E} = 0$ even in the present circumstances, but ∇K and therefore $(\nabla K \times \mathbf{E})$ may be different from zero, for example, at the surface of the dielectric. Therefore, $\mathbf{E}$ is still derivable from a potential, but in general $\mathbf{D}$ is not.

Let us examine the situation at the boundaries of a dielectric material (see Fig. 2.20). We call $\mathbf{E}^{\text{ins}}$ and $\mathbf{E}^{\text{out}}$ the electric field inside and outside the dielectric, respectively, and $\mathbf{D}^{\text{ins}}$ and $\mathbf{D}^{\text{out}}$ the electric displacement inside and outside the dielectric, respectively. We consider the thin flat volume in the figure. We know that $\nabla \cdot \mathbf{E} \neq 0$ in the volume because of the presence of polarization charges. We know also that $\nabla \times \mathbf{E} = 0$, because $\mathbf{E}$ can still be derived from a potential; therefore, applying the curl theorem (1.4.13),

$$\int_V d\tau (\nabla \times \mathbf{E}) = 0 = \int_S dS\,(\mathbf{n} \times \mathbf{E}) \qquad (2.14.33)$$

and

$$E_{\text{tang}}^{\text{ins}} = E_{\text{tang}}^{\text{out}} \qquad (2.14.34)$$

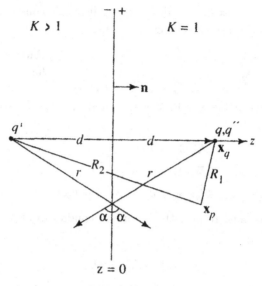

FIGURE 2.21

where the subscript indicates that we are considering the tangential components of **E**. On the other hand,

$$\boldsymbol{\nabla} \cdot \mathbf{D} = 0 \qquad (2.14.35)$$

Since there are no true charges,

$$\int_V d\tau (\boldsymbol{\nabla} \cdot \mathbf{D}) = 0 = \int_S dS \, (\mathbf{n} \cdot \mathbf{D}) \qquad (2.14.36)$$

and

$$D_{\text{normal}}^{\text{ins}} = D_{\text{normal}}^{\text{out}} \qquad (2.14.37)$$

where, obviously, the subscript indicates that we are considering the normal components of **D**. Also,

$$\mathbf{D}^{\text{ins}} = K\boldsymbol{E}^{\text{ins}}$$
$$\mathbf{D}^{\text{out}} = \boldsymbol{E}^{\text{out}} \qquad (2.14.38)$$

EXAMPLE

We shall consider the situation in Fig. 2.21. We have:

A charge q at $\mathbf{x}_q$

A half-space consisting of a dielectric of dielectric constant $K > 1$
The other half-space consisting of vacuum

We want to find the field at $\mathbf{x}_p$. The charge q polarizes the dielectric, and the field at $\mathbf{x}_p$ is due to q and the polarization charges.

We have two regions to consider, the vacuum and the dielectric, separated by the plane $z = 0$. We must find appropriate solutions to the equations

$$\nabla \cdot \mathbf{E} = 4\pi\rho, \quad z > 0$$

$$K\nabla \cdot \mathbf{E} = 0, \quad z < 0$$

$$\nabla \times \mathbf{E} = 0, \quad \text{everywhere}$$

These solutions must respect the boundary conditions at $z = 0$:

$$E_t^+ = E_t^-$$
$$D_n^+ = E_n^+ = D_n^-$$

Note the conventions: $t = $ tangential, $n = $ normal, $+ = $ outside the dielectric and $- = $ inside the dielectric. Since $\nabla \times \mathbf{E} = 0$ everywhere, $\mathbf{E}$ is derivable in the usual way from a potential.

We shall use the *method of images*, which in this case consists of replacing the dielectric by a set of charges of appropriate values and signs in adequate positions. We place an image charge q' at the symmetrical position with respect to q, so the potential at point $\mathbf{x}_p$ is given by

$$\phi(\mathbf{x}_p) = \frac{q}{R_1} + \frac{q'}{R_2} \quad (z > 0)$$

We have now to specify the potential for $z < 0$. Since there are no charges in the region $z < 0$, this potential must be a solution of Laplace's equation without singularities in the region of the dielectric. The simplest assumption is that for $z < 0$ the potential is equivalent to that of a charge q'' at the position of the actual charge q

$$\phi(\mathbf{x}_p) = \frac{q''}{R_1} \quad (z < 0)$$

We have then

$$E_t^+ = \frac{q}{r^2}\cos\alpha + \frac{q'}{r^2}\cos\alpha = (q + q')\frac{\cos\alpha}{r^2}$$

$$E_t^- = \frac{q''}{r^2}\cos\alpha$$

Since $E_t^+ = E_t^-$,

$$q + q' = q''$$

On the other hand,

$$D_n^+ = E_n^+ = \frac{q'}{r^2} \sin \alpha - \frac{q}{r^2} \sin \alpha = (q' - q) \frac{\sin \alpha}{r^2}$$

$$D_n^- = K E_n^- = -K \frac{q''}{r^2} \sin \alpha$$

Since $D_n^+ = D_n^-$,

$$q - q' = kq''$$

The two equations

$$q + q' = q''$$
$$q - q' = Kq''$$

give the solutions

$$q' = -\frac{K-1}{K+1} q$$

$$q'' = \frac{2}{K+1} q$$

Knowing these quantities, we can calculate the fields and the polarization charge densities.

2.15. Field Energy in a Dielectric with Constant K

Assume that we have a dielectric liquid in which charges can move. If a true charge Q is present in the liquid at $\mathbf{x}_Q$,

$$\nabla \cdot \mathbf{D}(\mathbf{x}) = 4\pi \rho_{\text{true}}(\mathbf{x}) = 4\pi Q \delta(\mathbf{x} - \mathbf{x}_Q) \qquad (2.15.1)$$

If the dielectric is homogeneous, that is, if the dielectric constant is not a function of the position,

$$\mathbf{D}(\mathbf{x}) = K \, \mathbf{E}(\mathbf{x}) \qquad (2.15.2)$$

and

$$\mathbf{E}(\mathbf{x}) = -\nabla_x \frac{Q}{K|\mathbf{x} - \mathbf{x}_Q|} \qquad (2.15.3)$$

The field $\mathbf{E}$ is due to *all* charges, including true charges and polarization charges.

If we consider a test charge q at $\mathbf{x}$, the field $\mathbf{E}(\mathbf{x})$ is equal to the force acting on the test charge, divided by the value of the charge. The force has the magnitude

$$|F(\mathbf{x})| = q|\mathbf{E}(\mathbf{x})| = \frac{qQ}{K|\mathbf{x} - \mathbf{x}_Q|^2} \tag{2.15.4}$$

The dielectric has the effect of reducing the force by K. Therefore, for a number of charges, the electrostatic energy is given by

$$U = \frac{1}{2} \sum_{\substack{s,t \\ s \neq t}} \frac{q_s q_t}{K|\mathbf{x}_s - \mathbf{x}_t|} \tag{2.15.5}$$

Consider now a density of true charges $\rho(\mathbf{x})$ embedded in a medium of dielectric constant K. We have in this case the following relations:

$$\mathbf{\nabla} \cdot \mathbf{D}(\mathbf{x}) = 4\pi \rho(\mathbf{x}) \tag{2.15.6}$$

$$\rho(\mathbf{x}) = \frac{1}{4\pi} \mathbf{\nabla} \cdot \mathbf{D}(\mathbf{x}) \tag{2.15.7}$$

$$\phi(\mathbf{x}) = \int \frac{\rho(\mathbf{x}')}{K|\mathbf{x} - \mathbf{x}'|} d\tau' \tag{2.15.8}$$

and

$$\begin{aligned} U &= \frac{1}{2} \iint d\tau d\tau' \frac{\rho(\mathbf{x})\rho(\mathbf{x}')}{K|\mathbf{x} - \mathbf{x}'|} \\ &= \frac{1}{2} \int d\tau \left(\frac{1}{4\pi} \mathbf{\nabla} \cdot \mathbf{D} \right) \phi(\mathbf{x}) = \frac{1}{8\pi} \int d\tau \, (\mathbf{\nabla} \cdot \mathbf{D})\phi(\mathbf{x}) \\ &= -\frac{1}{8\pi} \int d\tau \, \mathbf{D} \cdot \mathbf{\nabla}\phi = \frac{1}{8\pi} \int d\tau \, \mathbf{E} \cdot \mathbf{D} \end{aligned}$$

where we have used integration by parts. We can write

$$U = \frac{1}{8\pi} \int d\tau \, \mathbf{E} \cdot \mathbf{D} \tag{2.15.9}$$

How general is this expression? We shall see that this expression is indeed general, since it is valid even when $K = K(\mathbf{x})$.

2.16. Field Energy in a Dielectric for Which $K = K(\mathbf{x})$

We shall assume the presence of a continuous charge distribution $\rho(\mathbf{x})$. We shall set

$$\delta \mathbf{s}(\mathbf{x}) = \text{displacement of the charge element located at } \mathbf{x} \quad (2.16.1)$$

The force acting on the true charge in $d\tau$ is given by

$$\rho(\mathbf{x}) d\tau \mathbf{E}(\mathbf{x}) \quad (2.16.2)$$

The amount of work done by the forces on the charges due to the displacements δs is given by

$$\delta W = \int d\tau \rho(\mathbf{x})[\mathbf{E}(\mathbf{x}) \cdot \delta \mathbf{s}(\mathbf{x})] \quad (2.16.3)$$

The change in field energy is given by

$$\delta U = - \int d\tau \rho(\mathbf{x})[\mathbf{E}(\mathbf{x}) \cdot \delta \mathbf{s}(\mathbf{x})] \quad (2.16.4)$$

We do not change the charges, we just displace them; we have then to take into account the *continuity equation*

$$\frac{\partial \rho(\mathbf{x}, t)}{\partial t} + \boldsymbol{\nabla} \cdot [\rho(\mathbf{x}, t)\mathbf{v}(\mathbf{x})] = 0 \quad (2.16.5)$$

where we have explicitly shown the dependence of ρ on t, and $\mathbf{v}(\mathbf{x})$ is the velocity of the charge element located at $\mathbf{x}$:

$$\mathbf{v}(\mathbf{x})\delta t = \delta \mathbf{s}(\mathbf{x}) \quad (2.16.6)$$

We can then write

$$\delta \rho = \delta t \frac{\partial \rho}{\partial t} = -\boldsymbol{\nabla} \cdot (\rho \mathbf{v} \, \delta t) = -\boldsymbol{\nabla} \cdot \rho \, \delta \mathbf{s} \quad (2.16.7)$$

Then

$$\delta U = -\delta W = - \int d\tau \rho(\mathbf{x})[\mathbf{E}(\mathbf{x}) \cdot \delta \mathbf{s}(\mathbf{x})]$$

$$= \int d\tau \, \rho(\mathbf{x})\delta \mathbf{s}(\mathbf{x}) \cdot \boldsymbol{\nabla}\phi(\mathbf{x})$$

$$= - \int d\tau \{\boldsymbol{\nabla} \cdot [\rho(\mathbf{x})\delta \mathbf{s}(\mathbf{x})]\}\phi(\mathbf{x}) = \int d\tau [\delta \rho(\mathbf{x})]\phi(\mathbf{x}) \quad (2.16.8)$$

where we have used integration by parts. But

$$\nabla \cdot \mathbf{D} = 4\pi\rho \qquad (2.16.9)$$

$$4\pi\delta\rho = \nabla \cdot \delta\mathbf{D}, \quad \delta\rho = \frac{1}{4\pi}\nabla \cdot \delta\mathbf{D} \qquad (2.16.10)$$

Then

$$\delta U = \frac{1}{4\pi}\int d\tau (\nabla \cdot \delta\mathbf{D})\phi = -\frac{1}{4\pi}\int d\tau \, \delta\mathbf{D} \cdot \nabla\phi$$

$$= \frac{1}{4\pi}\int d\tau \, (\delta\mathbf{D} \cdot \mathbf{E}) \qquad (2.16.11)$$

where we have used integration by parts. But

$$\mathbf{E}(\mathbf{x}) = \frac{\mathbf{D}(\mathbf{x})}{K(\mathbf{x})} \qquad (2.16.12)$$

We note that $\delta K = 0$, because we do not act on the dielectric. Then

$$\delta U = \frac{1}{4\pi}\int d\tau \delta\mathbf{D} \cdot \mathbf{E} = \frac{1}{4\pi}\int d\tau \frac{1}{K(\mathbf{x})}\mathbf{D}(\mathbf{x}) \cdot \delta\mathbf{D}(\mathbf{x})$$

$$= \frac{1}{4\pi}\int d\tau \frac{1}{K(\mathbf{x})}\frac{1}{2}\delta(\mathbf{D})^2 = \frac{1}{8\pi}\int d\tau \frac{1}{K(\mathbf{x})}\delta(\mathbf{D})^2 \quad (2.16.13)$$

or

$$\delta U = \frac{1}{8\pi}\int d\tau \, \delta(\mathbf{D} \cdot \mathbf{E}) \qquad (2.16.14)$$

and

$$U = \frac{1}{8\pi}\int d\tau \, \mathbf{D} \cdot \mathbf{E} \qquad (2.16.15)$$

This expression, already found under more restrictive conditions in the previous section, is indeed general.

2.17. Forces on a Dielectric

We wish to calculate the force acting on a piece of dielectric due to the presence of an electric field. The variation in the electrostatic energy when the unit volume of dielectric undergoes the displacement δs is given

$$\delta U = -\int d\tau \, \delta s \cdot \mathbf{f} \qquad (2.17.1)$$

where $\mathbf{f} = force$ per unit volume acting on the dielectric. We shall assume that the true charges present in the dielectric do not move, so that only the dielectric will possibly be displaced.

The electrostatic energy is given by

$$U = \frac{1}{8\pi} \int d\tau \frac{D^2}{K} \tag{2.17.2}$$

We shall call

$$\mathbf{D}'(\mathbf{x}) = \mathbf{D}(\mathbf{x}) + \delta\mathbf{D}(\mathbf{x})$$
$$K'(\mathbf{x}) = K(\mathbf{x}) + \delta K(\mathbf{x}) \tag{2.17.3}$$

and

$$U' = \frac{1}{8\pi} \int d\tau \frac{D'^2}{K'} \tag{2.17.4}$$

But

$$\frac{(\mathbf{D}')^2}{K'} = \frac{(\mathbf{D} + \delta\mathbf{D})^2}{K + \delta K}$$

$$= \frac{D^2}{K} - \frac{D^2}{K^2}\delta K + 2\frac{\mathbf{D}}{K} \cdot \delta\mathbf{D} + \text{quadratic terms} \tag{2.17.5}$$

Then

$$U' = \frac{1}{8\pi} \int d\tau \left(\frac{D^2}{K} - \delta K \frac{D^2}{K^2} + \frac{2}{K}\delta\mathbf{D} \cdot \mathbf{D} \right) \tag{2.17.6}$$

and

$$\delta U = U' - U = \frac{1}{8\pi} \int d\tau \left(-\delta K \frac{D^2}{K^2} + \frac{2}{K}\delta\mathbf{D} \cdot \mathbf{D} \right) \tag{2.17.7}$$

Consider the term

$$\frac{1}{8\pi} \int d\tau \frac{2}{K}\delta\mathbf{D} \cdot \mathbf{D} = \frac{1}{4\pi} \int d\tau \, \delta\mathbf{D} \cdot \mathbf{E}$$

$$= -\frac{1}{4\pi} \int d\tau \, \delta\mathbf{D} \cdot \boldsymbol{\nabla}\phi$$

$$= \frac{1}{4\pi} \int d\tau \, \phi\boldsymbol{\nabla} \cdot \delta\mathbf{D} = \int d\tau \, \phi\,\delta\rho \tag{2.17.8}$$

where we have used the relation

$$\boldsymbol{\nabla} \cdot \delta\mathbf{D} = 4\pi\delta\rho \tag{2.17.9}$$

We have assumed that no displacement of charges takes place; we have then

$$\frac{1}{8\pi} \int d\tau \frac{2}{K} \delta\mathbf{D} \cdot \mathbf{D} = 0 \tag{2.17.10}$$

and

$$\delta U = -\frac{1}{8\pi} \int d\tau \, \delta K |\mathbf{E}|^2 \tag{2.17.11}$$

We can write

$$K = K(\mu) \tag{2.17.12}$$

where μ = mass density. A continuity equation for the mass density is

$$\frac{\partial\mu}{\partial t} = -\boldsymbol{\nabla} \cdot \mu\mathbf{v} \tag{2.17.13}$$

But

$$\delta\mathbf{s} = \mathbf{v}\delta t \tag{2.17.14}$$

Then

$$\delta\mu = \delta t \frac{\partial\mu}{\partial t} = \delta t(-\boldsymbol{\nabla} \cdot \mu\mathbf{v})$$

$$= -\boldsymbol{\nabla} \cdot \mu\mathbf{v} \, \delta t = -\boldsymbol{\nabla} \cdot \mu \, d\mathbf{s} \tag{2.17.15}$$

and

$$\delta K = \frac{\partial K}{\partial\mu}\delta\mu = -\frac{\partial K}{\partial\mu}\boldsymbol{\nabla} \cdot \mu \, \delta\mathbf{s}$$

$$= -\frac{\partial K}{\partial\mu}[\boldsymbol{\nabla}\mu \cdot \delta\mathbf{s} + \mu\boldsymbol{\nabla} \cdot \delta\mathbf{s}]$$

$$= -\frac{\partial K}{\partial\mu}\boldsymbol{\nabla}\mu \cdot \delta\mathbf{s} - \mu\frac{\partial K}{\partial\mu}\boldsymbol{\nabla} \cdot \delta\mathbf{s}$$

$$= -\boldsymbol{\nabla} K \cdot \delta\mathbf{s} - \mu\frac{\partial K}{\partial\mu}\boldsymbol{\nabla} \cdot \delta\mathbf{s} \tag{2.17.16}$$

$$\delta U = -\frac{1}{8\pi} \int d\tau \delta K(\mathbf{E})^2$$

$$= \frac{1}{8\pi} \int d\tau(\mathbf{E})^2 \left(\boldsymbol{\nabla} K \cdot \delta\mathbf{s} + \mu\frac{\partial K}{\partial\mu}\boldsymbol{\nabla} \cdot \delta\mathbf{s} \right)$$

$$= \frac{1}{8\pi} \int d\tau (\mathbf{E})^2 \nabla K \cdot \delta \mathbf{s} = \frac{1}{8\pi} \int d\tau (\mathbf{E})^2 \mu \frac{\partial K}{\partial \mu} \nabla \cdot \delta \mathbf{s}$$

$$= \frac{1}{8\pi} \int d\tau (\mathbf{E})^2 \nabla K \cdot \delta \mathbf{s} - \frac{1}{8\pi} \int d\tau\, \delta \mathbf{s} \cdot \nabla \left[\mu \frac{\partial K}{\partial \mu} (\mathbf{E})^2 \right]$$

$$= \frac{1}{8\pi} \int d\tau\, \delta \mathbf{s} \cdot \left[(\mathbf{E})^2 \nabla K - \nabla \left(\mu \frac{\partial K}{\partial \mu} (\mathbf{E})^2 \right) \right] \qquad (2.17.17)$$

where we have used integration by parts. Then

$$\delta U = - \int d\tau\, \delta \mathbf{s} \cdot \mathbf{f} \qquad (2.17.18)$$

where

$$\mathbf{f} = \frac{1}{8\pi} \left[-(\mathbf{E})^2 \nabla K - \nabla \left(\mu \frac{\partial K}{\partial \mu} (\mathbf{E})^2 \right) \right] \qquad (2.17.19)$$

f is the force per unit volume acting on the dielectric. The first term in the brackets represents a force that appears whenever an inhomogeneous dielectric is in an electric field. The second term, known as the *electrostriction term*, gives a volume force on a dielectric in an inhomogeneous electric field; if we integrate this term over a region delimited by an outer surface of the dielectric where the electric field is zero, it will give a total zero contribution to the force.

EXAMPLE

A U-shaped tube contains a dielectric liquid (see Fig. 2.22). One end of the tube is between the plates of a capacitor, that is, in a homogeneous field E; the other is in field-free space. We want to find the level difference between the liquid surfaces.

We have to calculate two contributions to the energy:

(1) The change in the field energy when we introduce the dielectric in the field region is given by

$$- \int d\tau\, \mathbf{f} \cdot d\mathbf{s} = -\frac{1}{8\pi} \int_0^h E^2 S |\nabla K| dh = -\frac{1}{8\pi} E^2 S (K-1) h$$

where S = cross section of the tube.

(2) The potential energy of the liquid is given by

$$Sh\rho g h = S\rho h^2 g$$

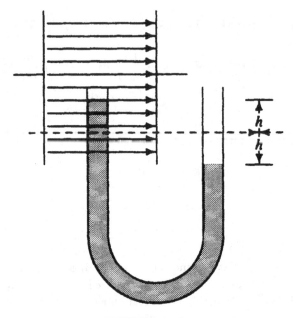

FIGURE 2.22

where ρ = density of the liquid.

Then the change in total energy is

$$\delta U_{\text{tot}} = -\frac{K-1}{8\pi}E^2 Sh + S\rho h^2 g$$

At equilibrium, h is such that

$$\frac{\partial(\delta U_{\text{tot}})}{\partial h} = -\frac{K-1}{8\pi}E^2 S + 2S\rho hg = 0$$

and

$$2h = \frac{K-1}{8\pi}\frac{E^2}{\rho g}$$

2.18. The Stress Tensor

Let us consider now a more general situation in which true charges are placed in a liquid dielectric; some conductors may be present. The force acting on a piece of dielectric, which may include some conductors, is given by

$$\mathbf{F} = \int_V d\tau [\rho_{\text{true}} \mathbf{E}(\mathbf{x}) + \mathbf{f}]$$

$$= \int_V d\tau \left\{ \frac{\boldsymbol{\nabla} \cdot \mathbf{D}}{4\pi} \mathbf{E} + \frac{1}{8\pi} \left[-E^2 \boldsymbol{\nabla} K + \boldsymbol{\nabla} \left(\mu \frac{\partial K}{\partial \mu} E^2 \right) \right] \right\} \quad (2.18.1)$$

and

$$F_k = \frac{1}{4\pi} \int d\tau \left[\sum_i \frac{\partial D_i}{\partial x_i} E_k - \frac{1}{2} E^2 \frac{\partial K}{\partial x_k} + \frac{1}{2} \frac{\partial}{\partial x_k} \left(\mu \frac{\partial K}{\partial \mu} E^2 \right) \right] \quad (2.18.2)$$

But

$$\sum_i \frac{\partial D_i}{\partial x_i} E_k = \sum_i \frac{\partial}{\partial x_i} (D_i E_k) - \sum_i D_i \frac{\partial E_k}{\partial x_i}$$

$$= \sum_i \frac{\partial}{\partial x_i} (D_i E_k) - \sum_i K E_i \frac{\partial E_i}{\partial x_k}$$

$$= \sum_i \frac{\partial}{\partial x_i} (D_i E_k) - \frac{K}{2} \frac{\partial}{\partial x_k} \left(\sum_i E_i^2 \right)$$

$$= \sum_i \frac{\partial}{\partial x_i} (D_i E_k) - \frac{K}{2} \frac{\partial E^2}{\partial x_k}$$

$$= \sum_i \frac{\partial}{\partial x_i} \left(D_i E_k - \delta_{ik} \frac{K}{2} E^2 \right) \quad (2.18.3)$$

and

$$F_k = \frac{1}{4\pi} \int d\tau \left[\sum_i \frac{\partial}{\partial x_i} (D_i E_k) - \frac{K}{2} \frac{\partial E^2}{\partial x_k} - \frac{1}{2} E^2 \frac{\partial K}{\partial x_k} \right.$$

$$\left. + \frac{1}{2} \frac{\partial}{\partial x_k} \left(\mu \frac{\partial K}{\partial \mu} E^2 \right) \right]$$

$$= \frac{1}{4\pi} \int d\tau \left[\sum_i \frac{\partial}{\partial x_i} (D_i E_k) - \frac{1}{2} \frac{\partial (K E^2)}{\partial x_k} \right.$$

$$\left. + \frac{1}{2} \frac{\partial}{\partial x_k} \left(E^2 \mu \frac{\partial K}{\partial \mu} \right) \right] \quad (2.18.4)$$

We put

$$\frac{\mu}{K}\frac{\partial K}{\partial \mu} = b \tag{2.18.5}$$

Then

$$F_k = \frac{1}{4\pi}\int d\tau \left[\sum_i \frac{\partial}{\partial x_i}(D_i E_k) - \frac{1}{2}\frac{\partial(KE^2)}{\partial x_k}\right.$$

$$\left. + \frac{1}{2}\frac{\partial}{\partial x_k}(E^2 K b)\right]$$

$$= \frac{1}{4\pi}\int d\tau \left\{\sum_i \frac{\partial}{\partial x_i}\left[D_i E_k - \frac{K}{2}(1-b)\delta_{ik}E^2\right]\right\} \tag{2.18.6}$$

or

$$F_k = \int d\tau \left(\sum_i \frac{\partial}{\partial x_i}T_{ik}\right) \tag{2.18.7}$$

where

$$T_{ik} = \frac{K}{4\pi}\left[E_i E_k - \frac{1}{2}(1-b)\delta_{ik}E^2\right] \tag{2.18.8}$$

Using the divergence theorem, we arrive at the result:

$$F_k \text{ (on everything inside volume } V) = \int_S dS \left(\sum_i n_i T_{ik}\right) \tag{2.18.9}$$

where T_{ik} is given by Eq. (2.18.8). The stress tensor T is represented by the matrix

$$T = \frac{K}{4\pi}$$

$$\begin{pmatrix} \frac{1}{2}(E_x^2 - E_y^2 - E_z^2 + bE^2) & E_x E_y & E_x E_z \\ \\ E_x E_y & \frac{1}{2}(E_y^2 - E_x^2 - E_z^2 + bE^2) & E_y E_z \\ \\ E_x E_z & E_y E_z & \frac{1}{2}(E_z^2 - E_x^2 - E_y^2 + bE^2) \end{pmatrix}$$

$$\tag{2.18.10}$$

2.19. Capacitance

A *capacitor* in its simplest form consists of two flat conducting plates of area, say, A, separated by a distance d (see Fig. 2.23). If the potentials on the plates a and b are V_a and V_b, respectively, with $V_a > V_b$, then the electric field is directed from plates a to b, perpendicularly to the plates; it is constant inside the capacitor, if we neglect the edge effects, and its value is

$$E = \frac{V_a - V_b}{d} \tag{2.19.1}$$

The surface charge density on the plate a is

$$\sigma = \frac{E}{4\pi} = \frac{V_a - V_b}{4\pi d} \tag{2.19.2}$$

and the total charge on the same plate is

$$Q = \sigma A = A \frac{V_a - V_b}{4\pi d} \tag{2.19.3}$$

The total charge on the plate b is $-Q$.

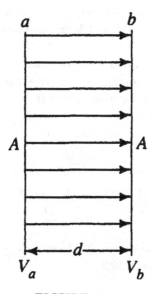

FIGURE 2.23

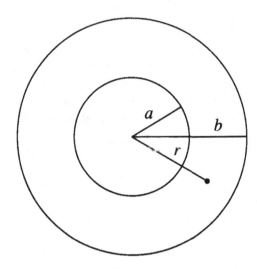

FIGURE 2.24

The ratio of the charge Q to the potential difference is called the *capacitance* of the capacitor and is designated C.

$$C = \frac{Q}{V_a - V_b} = \frac{A}{4\pi d} \qquad (2.19.4)$$

In the ESU system of units, the capacitance is measured in centimeters. A capacitor with plate area $A = 200\,\mathrm{cm}^2$ and separation $d = 1\,\mathrm{cm}$ has a capacitance $C = 15.9\,\mathrm{cm}$.

A capacitor may take other more complex geometrical forms, such as the one consisting of two conducting spherical shells of radii a and b (see Fig. 2.24). If a charge Q is distributed over the surface of the internal shell, the charge $-Q$ is distributed over the inside surface of the outer shell. The surface charge density on the internal shell is

$$\sigma = \frac{Q}{4\pi a^2} \qquad (2.19.5)$$

The electric field at the surface of the internal shell is

$$E = 4\pi\sigma = \frac{Q}{a^2} \qquad (2.19.6)$$

and is perpendicular to such surface. The electric field at the distance $r > a$ from the center of the inner shell is

$$E = \frac{Q}{r^2} \qquad (2.19.7)$$

The potential difference between the two shells is then

$$V_a - V_b = \int_a^b E dr = \int_a^b \frac{Q}{r^2} dr = Q\left(\frac{1}{a} - \frac{1}{b}\right) \qquad (2.19.8)$$

and the capacitance is given by

$$C = \frac{Q}{V_a - V_b} = \frac{1}{(1/a) - (1/b)} = \frac{ab}{b - a} \qquad (2.19.9)$$

If $b \to \infty$, $C \to a$; that is, the capacitance of an isolated conducting sphere is equal to the radius of the sphere in the ESU system of units.

Consider now a capacitor of any shape, with a potential difference

$$V_{ab} = V_a - V_b \qquad (2.19.10)$$

between the two plates and a charge Q distributed over one plate, say plate a. The work necessary to transfer the infinitesimal charge dQ from plates b to a is

$$dW = dQ(V_a - V_b) = \frac{Q}{C} dQ \qquad (2.19.11)$$

Therefore, the work necessary to charge the capacitor is

$$W = \frac{1}{C} \int_0^Q Q \, dQ = \frac{Q^2}{2C} \qquad (2.19.12)$$

But

$$Q = CV_{ab} \qquad (2.19.13)$$

Then

$$W = \frac{1}{2} CV_{ab}^2 \qquad (2.19.14)$$

and the energy stored in the capacitor is

$$U = \frac{1}{2} CV_{ab}^2 \qquad (2.19.15)$$

In the case of the parallel-plate capacitor,

$$U = \frac{1}{2} CV_{ab}^2 = \frac{1}{2} \frac{A}{4\pi d} (Ed)^2 = \frac{E^2}{8\pi} (Ad) \qquad (2.19.16)$$

This quantity is equal to the product of the electrostatic energy density $E^2/8\pi$ times the volume between the two plates.

In the case of the spherical-shell capacitor,

$$U = \frac{1}{2}CV_{ab}^2 = \frac{1}{2}\frac{ab}{b-a}Q^2 \left(\frac{1}{a} - \frac{1}{b}\right)^2$$

$$= \frac{Q^2}{2}\frac{b-a}{ab} = \frac{Q^2}{2C} \tag{2.19.17}$$

This quantity is equal to the integral of $E^2/8\pi$ over the volume between the two shells:

$$\frac{1}{8\pi}\int E^2 d\tau = \frac{1}{8\pi}\iiint \frac{Q^2}{r^4}r^2 \sin\theta \, d\theta \, d\varphi \, dr$$

$$= \frac{1}{2}Q^2 \int_a^b \frac{1}{r^2}dr = \frac{Q^2}{2}\left(\frac{1}{a} - \frac{1}{b}\right)$$

$$= \frac{Q^2}{2}\frac{b-a}{ab} \tag{2.19.18}$$

We may want to add at this point a few considerations regarding units. Since

$$\left[\frac{1}{2}CV^2\right] = \text{energy}$$

$$[C] = \frac{\text{energy}}{V^2}$$

Therefore, we have

$$\text{ESU}: C = \frac{\text{dyne cm}}{(\sqrt{\text{dyne}})^2} = \text{cm} = \text{statfarad}$$

We have also

$$\text{EMU}^7: C = \frac{s^2}{\text{cm}} = \text{abfarad}$$

$$\text{SI}: C = \text{farad} = \frac{\text{coulomb}}{\text{volt}} = \frac{3 \times 10^9 \text{ statcoulomb}}{1/300 \text{ statvolt}}$$

$$= 9 \times 10^{11} \text{ statfarad} = \frac{\frac{1}{10} \text{ abcoulomb}}{10^8 \text{ abvolt}} = 10^{-9} \text{ abfarad}$$

---

[7]The EMU system of units will be introduced in the next chapter.

CHAPTER 2 EXERCISES

2.1. **(a)** Express the Laplace equation

$$\nabla^2 \phi(\mathbf{x}) = 0$$

in spherical coordinates.

(b) Set

$$\phi(\mathbf{x}) = \phi(r, \theta, \phi) = \frac{U(r)}{r} P(\theta) Q(\phi)$$

and find the differential equations for $Q(\phi)$, $P(\theta)$, and $U(r)$.

(c) Find the forms of functions $Q(\phi)$, $P(\theta)$, and $U(r)$.

(d) Show that the functions

$$Y(\theta, \phi) = P(\theta) Q(\phi)$$

are spherical harmonics.

2.2. **(a)** Prove the relation

$$\sum_m (-1)^m Y_{lm}(\theta, \phi) Y_{l,-m}(\theta, \phi) = \frac{2l + 1}{4\pi}$$

independent of θ, ϕ.

(b) Expand the function

$$\frac{1}{d} = \frac{1}{|\mathbf{x} - \mathbf{x_0}|}$$

in Legendre polynomials for both $|\mathbf{x}| < |\mathbf{x_0}|$ and $|\mathbf{x}| > |\mathbf{x_0}|$. Set $|\mathbf{x}| = r$ and $|\mathbf{x_0}| = r_0$.

(c) Introduce polar coordinates (r, θ, ϕ) for $\mathbf{x}$ and (r_0, θ_0, ϕ_0) for $\mathbf{x_0}$, and express $1/d$ in these coordinates in the two cases given in Parts (a) and (b).

(d) Express ∇^2 in polar coordinates and show that $\nabla^2(1/d) = 0$ for $d \neq 0$.

(e) Show that

$$\delta(\mathbf{x} - \mathbf{x_0}) = \frac{1}{r_0^2} \delta(r - r_0) \sum_{l,m} Y_{lm}(\theta_0, \phi_0) Y_{lm}^*(\theta, \phi)$$

(f) The sum over lm in the equation in Part (e) is a δ function in the angles θ and ϕ. How would you write it?

2.3. In deriving the addition theorem, we obtained the following equations

$$4\pi = \sum_{m=-l}^{l} (-1)^m \alpha_m$$

$$4\pi \frac{4\pi}{2l+1} = \sum_{m=-l}^{l} |\alpha_m|^2 \tag{2.6.20}$$

Consider the case $l = 1$; the preceding two equations give

$$4\pi = -\alpha_{-1} + \alpha_0 - \alpha_1$$

$$4\pi \frac{4\pi}{3} = \alpha_{-1}^2 + \alpha_0^2 - \alpha_1^2 \tag{2.6.20'}$$

The solutions of these equations are, according to Eq. (2.6.21), given by

$$\alpha_0 = \frac{4\pi}{3}$$

$$\alpha_1 = -\frac{4\pi}{3} \tag{2.6.21'}$$

$$\alpha_{-1} = -\frac{4\pi}{3}$$

These are not the only solutions of Eq. (2.6.20'), and yet Eq. (2.6.21) is a valid relation. Show that this is so for $l = 1$.

2.4. Prove that the electric field of a point charge satisfies the equation $\nabla \times \mathbf{E} = 0$.

2.5. Assume that we have a continuous charge density symmetric with respect to a point at $x = 0$

$$\rho(\mathbf{x}) = \rho_0 e^{-\alpha|\mathbf{x}|}$$

Find the potential $\phi(\mathbf{x}_p)$ produced by this charge distribution, using the following two methods.

(a) Use Gauss's law and the fact that $\phi = \phi(|\mathbf{x}_p|) = \phi(r_p)$. Gauss's law gives you $d\phi/dr_p$, which you may integrate to get $\phi(r_p)$.

(b) Use the formal solution

$$\phi(\mathbf{x}_p) = \rho_0 \int d\tau \frac{e^{-\alpha|\mathbf{x}|}}{|\mathbf{x} - \mathbf{x}_p|}$$

and expand $1/|\mathbf{x} - \mathbf{x}_p|$ in Legendre polynomials.

2.6. Prove that in the volume V inside a hollow conductor of any shape whatsover the electric field is zero if V does not contain any charge.

2.7. A screened Coulomb potential is given by the function

$$\phi(r) = q\frac{e^{-ar}}{r}$$

(a) What is the *total* charge Q that produces this potential?
(b) What is the charge distribution $\rho(\mathbf{x})$ that produces ϕ?
(c) Check your answer to Part (a) by calculating the integral

$$Q = \int \rho(\mathbf{x})d\tau$$

Note: $\int xe^{ax} = (e^{ax}/a^2)(ax - 1)$.

2.8. A positive electric charge q is distributed uniformly inside a sphere of radius a. Inside this sphere there is a point charge $-q$. What is the force acting on this point charge as a function of its distance from the center of the sphere?

2.9. A spherical charge distribution has a volume density $\rho(r)$ that is function of r only. Find the electric field for each of the following cases, and then find the potential, subject to the condition that $\phi(\infty) = 0$:

(a)

$$\rho(r) = \frac{A}{r}, \quad 0 \leq r \leq R$$

$$\rho(r) = 0, \quad r > R$$

(b)

$$\rho = \rho_0, \quad 0 \leq r \leq R$$

$$\rho = 0, \quad r > R$$

2.10. Find the potentials in Exercise 2.9 directly from Poisson's equation. Note that

$$\nabla^2 f(r) = \frac{1}{r^2}\frac{\partial}{\partial r}\left(r^2\frac{\partial}{\partial r}\right)f(r)$$

2.11. A conducting object has a hollow cavity in its interior. If a point charge q is introduced into the cavity, what is the total charge induced on the surface of the cavity?

2.12. Given an infinitely long line charge with uniform charge density λ per unit length. Find the electric field at a distance r from the line.

2.13. A spherically symmetric charge distribution has the form

$$\rho = ar^2, \quad r \leq R$$
$$\rho = 0, \qquad r > R$$

where $a = $ const

(a) Calculate the electric field for both $r < R$ and $r > R$.
(b) Calculate the potential for both $r < R$ and $r > R$, using the condition $\varphi(\infty) = 0$.
(c) Plot both the field and the potential.

2.14. An infinitely long circular cylinder of radius a carries a charge per unit length λ, which is uniformly distributed through the volume of the cylinder.

(a) Calculate the field.
(b) Calculate the potential. Assume that the potential is zero at $r = a$.

2.15. Compute the potential of a line charge of finite length L. The charge density per unit length is λ. Take the origin to be at the center of the line charge, and align the z axis with the line charge.

2.16. A point charge q is located at distances a and b from two perpendicular conducting half-planes, both at zero potential. Calculate the force acting on the charge q.

2.17. A metallic sphere of radius R_0 and total charge zero is embedded in a homogeneous electric field $\mathbf{E}_0$. What is the value of the induced charge on the surface of the sphere. To obtain your result, use formula (2.13.32).

2.18. (a) Find the force acting on an electric dipole $\mathbf{p} = q\mathbf{d}$ in an inhomogeneous electric field.

(b) Calculate the force on a dielectric sphere of radius R and dielectric constant K in a weakly inhomogeneous field $E(\mathbf{x})$:

$$\frac{R|\nabla \cdot \mathbf{E}|}{E} \ll 1$$

Show that the sphere is pulled into the region of higher field strength.

2.19. A distribution $\rho(\mathbf{x})$ of charges is confined to a region V of linear dimension L (see Fig. P2.19). Choose an origin O inside V and write the expression for the potential $\phi(\mathbf{x})$ at a point (R, θ, ϕ) a

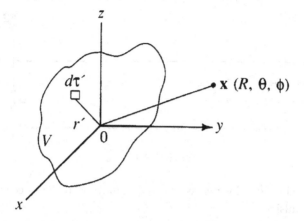

FIGURE P2.19

large distance apart $(R \gg L)$. Define multipole D_{lm} of the charge distribution:

$$D_{lm} = \sqrt{\frac{4\pi}{2l+1}} \int d\tau' \rho(x') r'^l Y_{lm}(\theta', \phi')$$

(a) Show that $\phi(\mathbf{x}) = \phi(R, \theta, \phi)$ is given by

$$\phi(R, \theta, \phi) = \sum_{l,m} \sqrt{\frac{4\pi}{2l+1}} (-1)^m D_{l,-m} \left(\frac{Y_{lm}(\theta, \phi)}{R^{l+1}} \right)$$

(b) Express the components D_{lm} of the dipole moment in terms of the Cartesian components

$$D_i = \int d\tau \rho(\mathbf{x}) x_i$$

and the components D_{2m} of the quadrupole moment in terms of

$$Q_{ik} = \int d\tau \rho(\mathbf{x}) x_i x_k$$

(c) Show that, if the total charge $Q = D_{00}$ is zero, the dipole moments D_{lm} are independent of the choice of the origin O. Can you generalize this statement?

2.20. A point charge of 0.5 statcoulomb is at a distance 1 cm from a big block limited by a plane surface. Calculate the force acting on the

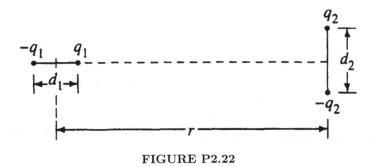

FIGURE P2.22

charge in the following two cases and indicate if the force is attractive or repulsive:

(a) The block is a conductor.

(b) The block is a dielectric with $K = 7$.

2.21. An electric dipole $\mathcal{D}$ is pointing in the z direction and is placed at the origin of the coordinate system. Working in the xz plane, find the value of the two components E_x and E_z of the electric field.

2.22. Two isolated and rigid electric dipoles lie in a plane as shown in the Fig. P2.22. The distance between the centers of the dipoles is large compared with the lengths d_1 and d_2 of the dipoles. Calculate the torques on the dipoles.

2.23. We shall take another approach to solve the problem of the metallic sphere of radius R_0 in a uniform field $\mathbf{E}_0$. Consider Example 2 of Sec. 2.9.

(a) Show that, by moving the charge q to infinite distance from the sphere and increasing its value in such a way that the electric field produced by it, $|\mathbf{E}_0| = q/R_q^2$ remains constant, the presence of the metallic sphere is equivalent to a dipole placed at the center of the sphere and given by $\mathbf{D} = \mathbf{E}_0 R_0^3$.

(b) Verify that the total field at the surface of the sphere is perpendicular to this surface.

(c) Obtain the charge distribution over the surface of the sphere.

2.24. A simple model of the hydrogen atom consists of a positively charged nucleus at the center of a sphere of radius a, which is filled with negative charge distributed with constant density. The charge of the nucleus, equal and opposite to the total charge of the sphere, is e.

Assume that, when an electric field E is applied, the sphere of negative charge is displaced with respect to the nucleus, preserving its shape and its charge density.

(a) Calculate the displacement of the nucleus from the center of the sphere in terms of a, E, and the charge e.

(b) Evaluate the *atomic polarizability* of the hydrogen atom, defined as the proportionality constant between the induced electric dipole moment and the applied electric field, and compare the value so obtained with the experimental value, which is $0.66 \times 10^{-24}\,\text{cm}^3$.

2.25. Two liquid dielectric materials of dielectric constants K_1 and K_2 are divided by an infinite plane surface. Two point charges q_1 and q_2 are placed at equal distances from and along a normal to the dividing surface. Calculate the forces acting on them and explain why they are not equal.

2.26. Calculate the force between an electron and a polarizable molecule of static polarizability α when they are many molecular diameters apart. On what power of the distance does it depend?

2.27. (a) Calculate the interaction energy of a permanent dipole $\mathbf{p}$ in an external, not necessarily uniform, field $\mathbf{E}$.

(b) Calculate the interaction energy between two dipoles p_1 and p_2 separated by a distance $\mathbf{r}_{12} = \mathbf{x}_1 - \mathbf{x}_2$.

2.28. (a) Calculate the torque on a permanent dipole $\mathbf{p}$ in a uniform external field $\mathbf{E}$.

(b) Calculate the torque on $\mathbf{p}$ in a nonuniform external field $\mathbf{E}$.

2.29. Two dipoles p_1 and p_2 lie in the (x, z) plane. p_1 is placed at the origin and is oriented in the z direction; p_2 is placed at the point (x, z) and forms an angle θ with p_1.

(a) Calculate the force on p_2 in the general case.

(b) Calculate the force on p_2 in the following cases:

$$x = a, \quad z = 0, \quad \theta = 0$$
$$x = a, \quad z = 0, \quad \theta = 90°$$
$$x = 0, \quad z = a, \quad \theta = 0$$
$$x = 0, \quad z = a, \quad \theta = 90°$$
$$x = a, \quad z = 0, \quad \theta = 45°$$

2.30. A body of dielectric constant K and conductivity σ contains some charges in its interior that are in the process of moving to the surface. Calculate the characteristic *relaxation time* with which this process takes place.

2.31. Calculate the capacitance per unit length of a coaxial cable consisting of an internal cylindrical conductor of radius a and a concentric hollow cylindrical conductor of radius b.

2.32. For a system of n conductors, each with potential ϕ_i and total charge $Q_i (i = 1, 2, \ldots, n)$, a relation exists between the charges and the potentials:

$$Q_i = \sum_{j=1}^{n} C_{ji} \Phi_j, \quad (j = 1, 2, \ldots, n)$$

where the coefficients C_{ji} are called *coefficients of capacitance*. Show that the electrostatic energy of this system can be expressed as follows:

$$U = \frac{1}{2} \sum_{i=1}^{n} \sum_{j=1}^{n} C_{ji} \Phi_i \Phi_j$$

2.33. A capacitor consists of a metallic sphere of radius a and a concentric hollow metallic sphere of radius b surrounding the first (see Fig. P2.33). The space between the two spheres is filled with two concentric layers of substances of dielectric constants K_1 and K_2. R is the radius of the sphere separating the two dielectric layers.

   (a) Calculate the capacitance C of such a capacitor.
   (b) Calculate the capacitance of an isolated metallic sphere of radius a surrounded by a spherical layer of radius R and dielectric constant K.

2.34. Two spherical conducting shells of radii r_a and r_b are concentric and are charged to potentials φ_a and φ_b, respectively. If $r_b > r_a$, find the potential for $r_a \leq r \leq r_b$ and $r > r_b$.

2.35. Three concentric conducting spherical shells, each of negligible thickness, have radii r_1, r_2, and r_3 ($r_1 < r_2 < r_3$). The innermost and the outermost shells are connected by a conducting wire. What is the capacitance of this system?

2.36. Calculate the capacitance of the earth in farads.

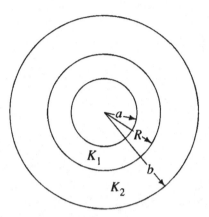

FIGURE P2.33

2.37. (a) Calculate the capacitance of a capacitor, consisting of two parallel
plates, each or area A, placed at the distance d, with the space
between the plates filled with a material of dielectric constant K.

(b) What would the effective dielectric constant K be if the material
inside the capacitor were a metal?

3

Stationary Currents and Magnetostatics

3.1. Lorentz Force and the Biot and Savart Law

Every electric charge produces an electric field $\mathbf{E}$ and feels the effect of the electric field produced by other charges. In a similar way, a moving charge produces a magnetic induction field $\mathbf{B}$ and feels the effect of the field $\mathbf{B}$ produced by other moving charges.

3.1.1. *The Force Law*

We define a vector field $\mathbf{B}(\mathbf{x})$ called *magnetic induction* as the field that describes the force acting on a charge q that moves with velocity $\mathbf{v}$:

$$\mathbf{F} = \text{const } q(\mathbf{v} \times \mathbf{B}) \qquad (3.1.1)$$

where the value of the constant will be introduced later. We can identify the field $\mathbf{B}$ by means of this force, called the *Lorentz force*.

To generalize these concepts, we introduce the notion of *current density*. If we have different charges with different velocities, the current density is given by

$$j(\mathbf{x}) = \lim_{\Delta\tau \to 0} \frac{\left(\sum_i q_i \mathbf{v}_i\right) \text{ in } \Delta\tau}{\Delta\tau} \tag{3.1.2}$$

In the special case in which all the particles in $\Delta\tau$ have the same velocities

$$\mathbf{j}(\mathbf{x}) = \mathbf{v} \lim_{\Delta\tau \to 0} \frac{\left(\sum_i q_i\right) \text{ in } \Delta\tau}{\Delta\tau} = \mathbf{v}\rho(\mathbf{x}) \tag{3.1.3}$$

where $\rho(\mathbf{x})$ = charge density.

Let us consider now a *line element* dl long with a cross section dS, and let us assume that the charges contained in this element all move the same way (see Fig. 3.1). We can write

$$\mathbf{j}(\mathbf{x})d\tau = \mathbf{j}(\mathbf{x})dl \ dS \tag{3.1.4}$$

and define a *vectorial line element* as follows:

$$d\mathbf{l} = dl \ \mathbf{n} \tag{3.1.5}$$

Then

$$\mathbf{j}(\mathbf{x})d\tau = |\mathbf{j}|dS \ d\mathbf{l} = I \ d\mathbf{l} \tag{3.1.6}$$

where I = *total current* = amount of charges that cross the plane perpendicular to $d\mathbf{l}$ in the unit time.

The force acting on a current element is given by

$$d\mathbf{F} = \text{const}[\mathbf{j}(\mathbf{x})d\tau \times \mathbf{B}(\mathbf{x})] = \text{const } I(d\mathbf{l} \times \mathbf{B}) \tag{3.1.7}$$

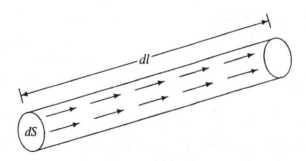

FIGURE 3.1

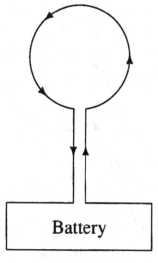

FIGURE 3.2

We have then the following relations:

$$\frac{\text{force}}{\text{volume}} = \text{force density} = \text{const}[\mathbf{j}(\mathbf{x}) \times \mathbf{B}(\mathbf{x})] \qquad (3.1.8)$$

$$\text{force on element } d\mathbf{l} \text{ of current } I = \text{const } I(d\mathbf{l} \times \mathbf{B}) \qquad (3.1.9)$$

In the present chapter, we shall deal with stationary currents. This condition of time-independence of the currents is tied to the existence of closed loops. A typical example is that of Fig. 3.2, where the effect of the battery leads is eliminated by running them close together and parallel.

Let us evaluate the total force acting on a closed-loop transversed by a current I:

$$\mathbf{F} = \text{const } I \oint d\mathbf{l} \times \mathbf{B}(\mathbf{x}) \qquad (3.1.10)$$

If the field is homogeneous ($\mathbf{B} = \text{const}$),

$$\mathbf{F} = \text{const } I \left(\oint d\mathbf{l} \right) \times \mathbf{B} = 0 \qquad (3.1.11)$$

because

$$\oint d\mathbf{l} = 0 \qquad (3.1.12)$$

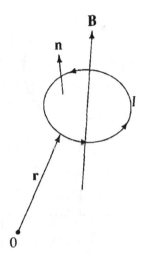

FIGURE 3.3

The force is zero, but the torque is not zero (see Fig. 3.3):

$$\mathbf{T} = \oint \mathbf{r} \times d\mathbf{F} = \text{const } I \oint \mathbf{r} \times (d\mathbf{l} \times \mathbf{B}) \qquad (3.1.13)$$

But

$$\oint \mathbf{r} \times (d\mathbf{l} \times \mathbf{B}) = \oint d\mathbf{l}(\mathbf{r} \cdot \mathbf{B}) - \mathbf{B} \oint (\mathbf{r} \cdot d\mathbf{l}) \qquad (3.1.14)$$

We know from Stokes's theorem that

$$\oint d\mathbf{l} \cdot \mathbf{r} = \int dS\, \mathbf{n} \cdot (\boldsymbol{\nabla} \times \mathbf{r}) = 0 \qquad (3.1.15)$$

A variant of the Stokes theorem (see Exercise 1.16) gives

$$\oint d\mathbf{l}\, \phi(\mathbf{x}) = \int_S dS\ \mathbf{n} \times \boldsymbol{\nabla}\phi(\mathbf{x}) \qquad (3.1.16)$$

Therefore,

$$\oint d\mathbf{l}(\mathbf{r} \cdot \mathbf{B}) = \int dS\ \mathbf{n} \times [\boldsymbol{\nabla}(\mathbf{r} \cdot \mathbf{B})]$$

$$= \int dS(\mathbf{n} \times \mathbf{B}) \qquad (3.1.17)$$

and

$$\mathbf{T} = \text{const } I \left(\int dS \, \mathbf{n} \right) \times \mathbf{B} = \mathbf{m} \times \mathbf{B} \qquad (3.1.18)$$

The quantity $\int dS \, \mathbf{n}$ is independent of the shape of the surface, as is evident from Eq. (3.1.17); we define

$$\mathbf{m} = \text{const } I \int dS \, \mathbf{n} \qquad (3.1.19)$$

the *magnetic moment* of the current loop.

3.1.2. The Biot and Savart law

We have seen that a field produces a force. Now we want to know how the field is produced by a current. The *law of Biot and Savart* allows us to determine the field produced by a current loop, as in Fig. 3.4.

$$\mathbf{B}(\mathbf{x}_p) = \text{const}' \, I \oint d\mathbf{l} \times \boldsymbol{\nabla}_x \frac{1}{|\mathbf{x} - \mathbf{x}_p|}$$

$$= \text{const}' \, I \left(\boldsymbol{\nabla}_{x_p} \times \oint \frac{d\mathbf{l}}{r} \right) \qquad (3.1.20)$$

where $r = |\mathbf{x} - \mathbf{x}_p|$ and the constant will be specified later.

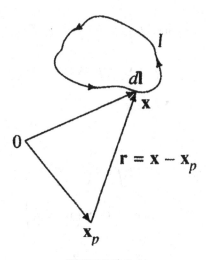

FIGURE 3.4

Let us consider now the nature of this field:

$$\mathbf{B}(\mathbf{x}_p) = \text{const}' \, I \boldsymbol{\nabla}_{x_p} \times \oint \frac{d\mathbf{l}}{r} = \text{const}' \, I \boldsymbol{\nabla}_{x_p}$$

$$\times \int dS \left(\mathbf{n} \times \boldsymbol{\nabla}_x \frac{1}{r} \right)$$

$$= \text{const}' \, I \int \left[\boldsymbol{\nabla}_{x_p} \times \left(\mathbf{n} \times \boldsymbol{\nabla}_x \frac{1}{r} \right) \right] dS \qquad (3.1.21)$$

If, in what follows, we use $\boldsymbol{\nabla}$ for $\boldsymbol{\nabla}_x$, we can write

$$\boldsymbol{\nabla}_{x_p} \times \left(\mathbf{n} \times \boldsymbol{\nabla} \frac{1}{r} \right) = \left(\boldsymbol{\nabla} \frac{1}{r} \cdot \boldsymbol{\nabla}_{x_p} \right) \mathbf{n} - (\mathbf{n} \cdot \boldsymbol{\nabla}_{x_p}) \boldsymbol{\nabla} \frac{1}{r}$$

$$+ \mathbf{n} \left(\boldsymbol{\nabla}_{x_p} \cdot \boldsymbol{\nabla} \frac{1}{r} \right) - \boldsymbol{\nabla} \frac{1}{r} (\boldsymbol{\nabla}_{x_p} \cdot \mathbf{n})$$

$$= -\mathbf{n} \boldsymbol{\nabla}_{x_p}^2 \frac{1}{r} - (\mathbf{n} \cdot \boldsymbol{\nabla}_{x_p}) \boldsymbol{\nabla} \frac{1}{r} \qquad (3.1.22)$$

where r is the distance between the point $\mathbf{x}_p$ and a point on the surface S of the loop. If $\mathbf{x}_p$ is not on S, we have $\boldsymbol{\nabla}_{x_p}^2 \frac{1}{r} = 0$. Then

$$\mathbf{B}(\mathbf{x}_p) = -\text{const}' \, I \int dS \left[(\mathbf{n} \cdot \boldsymbol{\nabla}_{x_p}) \boldsymbol{\nabla} \frac{1}{r} \right]$$

$$= \text{const}' \, I \int dS \left\{ \boldsymbol{\nabla}_{x_p} \left[(\mathbf{n} \cdot \boldsymbol{\nabla}_{x_p}) \frac{1}{r} \right] \right\}$$

$$= \text{const}' \, I \, \boldsymbol{\nabla}_{x_p} \int dS \left(-\mathbf{n} \cdot \boldsymbol{\nabla} \frac{1}{r} \right)$$

$$= -\boldsymbol{\nabla}_{x_p} \phi(\mathbf{x}_p) \qquad (3.1.23)$$

where

$$\phi(\mathbf{x}_p) = \text{const}' \, I \int dS \left(\mathbf{n} \cdot \boldsymbol{\nabla} \frac{1}{r} \right)$$

$$= -\text{const}' \, I \int dS \left(\mathbf{n} \cdot \boldsymbol{\nabla}_{x_p} \frac{1}{r} \right)$$

$$= -\text{const}' \int I \, dS \, \mathbf{n} \cdot \boldsymbol{\nabla}_{x_p} \frac{1}{|\mathbf{x} - \mathbf{x}_p|} \qquad (3.1.24)$$

The potential due to an electric dipole $\mathcal{D}$ situated at $\mathbf{x}$ is given by

$$\phi_{\mathcal{D}}(\mathbf{x}_p) = -\mathcal{D} \cdot \nabla_{x_p} \frac{1}{|\mathbf{x} - \mathbf{x}_p|} \tag{3.1.25}$$

The quantity

$$-I \, dS \, \mathbf{n} \cdot \nabla_{x_p} \frac{1}{|\mathbf{x} - \mathbf{x}_p|} \tag{3.1.26}$$

has the form of a potential due to a dipole. The loop is then equivalent to an ensemble of magnetic dipoles pointing like $\mathbf{n}$: the elementary dipole is

$$d\mathbf{m} = \text{const}' \, I \, dS \, \mathbf{n} \tag{3.1.27}$$

If the observation point $\mathbf{x}_p$ is far from the loop, we can consider the loop as a single dipole:

$$\mathbf{m} = \text{const}' \, I \int dS \, \mathbf{n}$$

3.2. Forces between Current Loops

Consider now two loops as in Fig. 3.5. Each loop will exercise a force on the other loop, the force on loop 2 due to the field set up by loop 1 is

$$\mathbf{F}_{12} = \text{const} \, I_2 \oint d\mathbf{l}_2 \times \mathbf{B}_1(\mathbf{x}_2) \tag{3.2.1}$$

where

$$\mathbf{B}_1(\mathbf{x}_2) = \text{const}' \, I_1 \left(\oint d\mathbf{l}_1 \times \nabla_1 \frac{1}{r_{12}} \right) \tag{3.2.2}$$

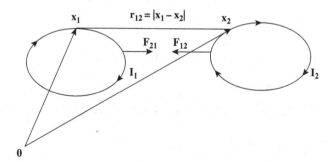

FIGURE 3.5

and

$$r_{12} = |\mathbf{x}_1 - \mathbf{x}_2| \qquad (3.2.3)$$

and $\mathbf{\nabla}_1$ is a gradient operator operating on $\mathbf{x}_1$. Then

$$\mathbf{F}_{12} = \text{const} I_2 \oint dl_2 \times \text{const}' I_1 \left(\oint dl_1 \times \mathbf{\nabla}_1 \frac{1}{r_{12}} \right)$$

$$= \text{const}'' I_1 I_2 \oint \oint dl_2 \times \left(dl_1 \times \mathbf{\nabla}_1 \frac{1}{r_{12}} \right)$$

$$= \text{const}'' I_1 I_2 \oint \oint \left[dl_1 \left(dl_2 \cdot \mathbf{\nabla}_1 \frac{1}{r_{12}} \right) \right.$$

$$\left. - (dl_1 \cdot dl_2) \mathbf{\nabla}_1 \frac{1}{r_{12}} \right] \qquad (3.2.4)$$

But

$$\oint dl_1 \oint dl_2 \cdot \mathbf{\nabla}_1 \frac{1}{r_{12}} = 0 \qquad (3.2.5)$$

since, due to Stokes's theorem

$$\oint \mathbf{\nabla} \frac{1}{r_{12}} \cdot dl = \int \left(\mathbf{\nabla} \times \mathbf{\nabla} \frac{1}{r_{12}} \right) \cdot \mathbf{n} \, dS = 0 \qquad (3.2.6)$$

We have then

$$\mathbf{F}_{12} = -\text{const}'' I_1 I_2 \oint \oint (dl_1 \cdot dl_2) \mathbf{\nabla}_1 \frac{1}{r_{12}} \qquad (3.2.7)$$

Similarly,

$$\mathbf{F}_{21} = -\text{const}'' I_1 I_2 \oint \oint (dl_1 \cdot dl_2) \mathbf{\nabla}_2 \frac{1}{r_{12}} \qquad (3.2.8)$$

Therefore,

$$\mathbf{F}_{12} = -\mathbf{F}_{21} \qquad (3.2.9)$$

In general, we can write

$$\text{force} = \text{const}'' I_1 I_2 \times \text{dimensionless factor} \qquad (3.2.10)$$

The dimensionless integral

$$\oint \oint (dl_1 \cdot dl_2) \mathbf{\nabla}_1 \frac{1}{r_{12}}$$

depends on the geometric parameters of the loops, such as shapes and relative distance.

3.3. Units

3.3.1. *The Electromagnetic System of Units* (*EMU*)

In this system of units the unit of current is defined in terms of the unit of force:

$$\mathbf{F}_{12} = \text{const}'' I_1 I_2 \left[\oint \oint (d\mathbf{l}_1 \cdot d\mathbf{l}_2) \nabla \frac{1}{r_{12}} \right] \tag{3.3.1}$$

We derive the current units from Eq. (3.3.1) in which we set $\text{const}'' = 1$:

$$[F] = [I^2] \tag{3.3.2}$$

This definition is analogous to that of the unit of charge in the electrostatic system, which is the charge that exercises a force of 1 dyne on a similar charge at a distance of 1 cm. In the present case,

$$[I] = \text{abampere} = \sqrt{\text{dyne}} \tag{3.3.3}$$

or

$$[I^2]_{\text{EMU}} = MLT^{-2} \tag{3.3.4}$$

Then

$$[q^2]_{\text{EMU}} = [I^2]T^2 = ML \tag{3.3.5}$$

On the other hand,

$$[q^2]_{\text{ESU}} = \text{dyne cm}^2 = ML^3T^{-2} \tag{3.3.6}$$

Then

$$\frac{[q^2]_{\text{ESU}}}{[q^2]_{\text{EMU}}} = \frac{ML^3T^{-2}}{ML} = L^2T^{-2} \tag{3.3.7}$$

and

$$\frac{[q]_{\text{ESU}}}{[q]_{\text{EMU}}} = LT^{-1} = \text{dimension of a velocity} \tag{3.3.8}$$

Therefore, the electromagnetic unit of charge is a constant c times larger than the electrostatic unit of charge. This constant has the dimensions of a velocity; it has been determined experimentally and found to be

$$c = 3 \times 10^{10} \text{cm/s} \qquad (3.3.9)$$

which is the speed of light.

We find

$$1 \text{ abampere} = 10 \text{ ampere}$$

$$1 \text{ abcoulomb} = 10 \text{ coulomb}$$

$$1 \text{ abvolt} = 10^{-8} \text{ volt}$$

If we have two loops such that the geometrical factor is equal to 1, and if a current of 10 A passes through each of them, they will attract or repel each other with the force of 1 dyne.

3.3.2. The SI System of Units

In the International System of Units (SI):

$$[I]_{\text{SI}} = \text{ampere} \qquad (3.3.10)$$

$$[Q]_{\text{SI}} = 1 \text{ ampere second} = \text{coulomb} \qquad (3.3.11)$$

For the magnetic induction field **B**, we rely on the relation

$$d\mathbf{F} = I d\mathbf{l} \times \mathbf{B} \qquad (3.3.12)$$

where we set the proportionality constant equal to 1. Then

$$\text{newton} = \text{ampere} \times \text{meter} \times [B] \qquad (3.3.13)$$

and

$$[B] = \frac{\text{newton}}{\text{ampere} \times \text{meter}} = \frac{\text{joule}}{\text{ampere} \times \text{meter}^2}$$

$$= \frac{\text{weber}}{\text{meter}^2} = \text{tesla} \qquad (3.3.14)$$

where

$$\text{weber} = \frac{\text{joule}}{\text{ampere}} = \text{volt second} \qquad (3.3.15)$$

The unit weber/meter^2 is called a *tesla*.

The force on a loop 2 due to a **B** field produced by a loop 1 is given, because of Eq. (3.3.12), by

$$\mathbf{F}_{12} = I_2 \oint d\mathbf{l}_2 \times \mathbf{B}_1(\mathbf{x}_2) \tag{3.3.16}$$

where

$$\mathbf{B}_1(\mathbf{x}_2) = \text{const}' I_1 \oint d\mathbf{l}_1 \times \boldsymbol{\nabla}_1 \frac{1}{r_{12}} \tag{3.3.17}$$

We shall then have

$$F_{12} = |\mathbf{F}_{12}| = \text{const}' I_1 I_2 \left[\oint d\mathbf{l}_2 \times \oint \left(d\mathbf{l}_1 \times \boldsymbol{\nabla}_1 \frac{1}{r_{12}} \right) \right]$$

$$= \frac{\mu_0}{4\pi} I_1 I_2 \times \text{dimensionless factor} \tag{3.3.18}$$

where we have taken

$$\text{const}' = \frac{\mu_0}{4\pi} \tag{3.3.19}$$

If we make the dimensionless factor equal to 1 and assume that

$$I_1 = I_2 = 1 \text{ abampere} = 10 \text{ ampere} \tag{3.3.20}$$

we obtain

$$1 \text{ dyne} = 10^{-5} \text{ newton} = \frac{\mu_0}{4\pi} (10 \text{ ampere})^2 \tag{3.3.21}$$

Then

$$\mu_0 = 4\pi \times 10^{-7} \frac{\text{newton}}{\text{ampere}^2} \tag{3.3.22}$$

3.3.3. *The Gaussian System of Units*

The Gaussian system is a mixed system, partly electrostatic and partly electromagnetic, in which

(1) The charges, currents, and **E** and **D** fields are measured in ESU units.
(2) The magnetic induction **B** is measured in EMU units.
(3) The mechanical quantities are expressed in cgs units.

Then

$$[m] = g$$

$$[l] = \text{cm}$$

$$[F] = \text{dyne} \qquad (3.3.23)$$

$$[\text{charge}] = \sqrt{\text{dyne}} \ \text{cm}$$

$$[\text{current}] = \sqrt{\text{dyne}} \ (\text{cm/s})$$

We had

$$F_{12} = \text{const} \ I_1 I_2 \times \text{dimensionless factor} \qquad (3.3.24)$$

Then

$$\text{dyne} = \text{const dyne} \ (\text{cm/s})^2 \qquad (3.3.25)$$

The constant has the dimension of $1/c^2$, where c is a *velocity*. The constant c was determined in an historic experiment designed by Weber and Kohlrausch. Two capacitors were charged and then discharged through two parallel wires. Through a ballistic experiment, the constant c was found to be equal to 3×10^{10} cm/s, the value of the velocity of light. It should be noted that thus far there is nothing in the laws of electrostatics and stationary currents that would suggest any connection of the constant c with the propagation of radiation.

In Gaussian units,

$$\mathbf{F}_{12} = \frac{I_1}{c} \frac{I_2}{c} \oint d\mathbf{l}_2 \times \oint d\mathbf{l}_1 \times \boldsymbol{\nabla}_1 \frac{1}{r_{12}} \qquad (3.3.26)$$

This formula is the resultant of the two relations

$$\mathbf{F}_{12} = \text{const} \ I_2 \oint d\mathbf{l}_2 \times \mathbf{B}_1(\mathbf{x}_2) \qquad (3.3.27)$$

$$\mathbf{B}_1(\mathbf{x}_2) = \text{const}' \ I_1 \oint d\mathbf{l}_1 \times \boldsymbol{\nabla}_1 \frac{1}{r_{12}} \qquad (3.3.28)$$

Then

$$\text{const} \times \text{const}' = \frac{1}{c^2} \qquad (3.3.29)$$

and

$$\mathbf{F}_{12} = \gamma \frac{I_2}{c} \oint d\mathbf{l}_2 \times \mathbf{B}_1(\mathbf{x}_2) \tag{3.3.30}$$

$$\mathbf{B}_1(\mathbf{x}_2) = \gamma' \frac{I_1}{c} \oint d\mathbf{l}_1 \times \boldsymbol{\nabla}_1 \frac{1}{r_{12}} \tag{3.3.31}$$

where

$$\gamma\gamma' = 1 \tag{3.3.32}$$

Also, from Eq. (3.3.30),

$$\text{dyne} = \gamma\sqrt{\text{dyne}}\frac{\text{cm}}{\text{s}}\frac{\text{s}}{\text{cm}}\text{cm}[B] = \gamma\sqrt{\text{dyne}}\ \text{cm}[B] \tag{3.3.33}$$

or

$$[B] = \frac{\sqrt{\text{dyne}}}{\gamma\ \text{cm}} \tag{3.3.34}$$

On the other hand, from Eq. (3.3.31),

$$[B] = \gamma'\sqrt{\text{dyne}}\frac{\text{cm}}{\text{s}}\frac{\text{s}}{\text{cm}}\text{cm}\frac{1}{\text{cm}^2} = \gamma'\frac{\sqrt{\text{dyne}}}{\text{cm}} \tag{3.3.35}$$

Therefore,

$$\frac{1}{\gamma}\frac{\sqrt{\text{dyne}}}{\text{cm}} = \gamma'\frac{\sqrt{\text{dyne}}}{\text{cm}} \tag{3.3.36}$$

We put now by convention

$$\gamma = \gamma' = 1 \tag{3.3.37}$$

and we then write

$$\mathbf{F}_{12} = \frac{I}{c} \oint d\mathbf{l}_2 \times \mathbf{B}_1 \tag{3.3.38}$$

$$\mathbf{B}_1 = \frac{I'}{c} \oint d\mathbf{l}_1 \times \boldsymbol{\nabla}_1 \frac{1}{r_{12}} \tag{3.3.39}$$

We define the unit of magnetic induction as follows:

$$[B] = \frac{\sqrt{\text{dyne}}}{\text{cm}} = \text{gauss} \tag{3.3.40}$$

The following should be noted:

(1) We have

$$[Q]_{\text{gauss}} = \frac{1}{3 \times 10^9}[Q]_{\text{SI}}$$

$$= \frac{1}{3 \times 10^9}\text{coulomb} \tag{3.3.41}$$

$$\text{EMU}: F_{12}(\text{dynes}) = I_1 I_2 \times \text{a dimensional factor} \tag{3.3.42}$$

$$\text{Gaussian}: F_{12}(\text{dynes}) = \frac{I_1}{c}\frac{I_2}{c} \times \text{a dimensional factor} \tag{3.3.43}$$

$$[I]_{\text{gauss}} = \frac{[I]_{\text{EMU}}}{3 \times 10^{10}} = \frac{[I]_{\text{SI}}}{3 \times 10^9} \tag{3.3.44}$$

$$1 \text{ abcoulomb} = 3 \times 10^{10} \text{ statcoulombs}$$

$$= 3 \times 10^{10}\frac{1}{3 \times 10^9}\text{coulomb}$$

$$= 10 \text{ coulombs} \tag{3.3.45}$$

(2) The advantage of the Gaussian system is the great degree of symmetry in the description of electric and magnetic phenomena:

$$[B] = \text{gauss} = \frac{\sqrt{\text{dyne}}}{\text{cm}} \tag{3.3.46}$$

$$[E] = \frac{\text{statvolt}}{\text{cm}} = \frac{\sqrt{\text{dyne}}}{\text{cm}} \tag{3.3.47}$$

(3) The force acting on a charge q moving with velocity $\mathbf{v}$ in a region where $\mathbf{E}$ and $\mathbf{B}$ fields are present is, in Gaussian units,

$$\mathbf{F} = q\left(\mathbf{E} + \frac{\mathbf{v}}{c} \times B\right) \tag{3.3.48}$$

(4) We shall see later that in electromagnetic plane waves propagating in a vacuum, in Gaussian units,

$$|\mathbf{E}| = |\mathbf{B}| \tag{3.3.49}$$

(5) The torque on a loop in a homogeneous $\mathbf{B}$ field is given by

$$\mathbf{T} = \mathbf{m} \times \mathbf{B} \tag{3.3.50}$$

where, in Gaussian units,

$$\mathbf{m} = \frac{I}{c}\oint dS \, \mathbf{n} \tag{3.3.51}$$

3.4. The Vector Potential

A stationary current I circulating in a loop (see Fig. 3.6) produces a magnetic induction field $\mathbf{B}$ given by

$$\mathbf{B}(\mathbf{x}) = \frac{I}{c} \oint d\mathbf{l} \times \nabla_{x'} \frac{1}{|\mathbf{x}' - \mathbf{x}|}$$

$$= \frac{I}{c} \nabla_x \times \oint d\mathbf{l} \frac{1}{|\mathbf{x}' - \mathbf{x}|} \tag{3.4.1}$$

The divergence of such a field is zero:

$$\nabla \cdot \mathbf{B}(\mathbf{x}) = 0 \tag{3.4.2}$$

If the divergence of a vector field is zero, the field is called *solenoidal*; this is the case for the field $\mathbf{B}$ above. If the curl of a vector field is zero, the field is called *irrotational*; this is the case for the field $\mathbf{E}$ in electrostatics.

In this regard it is worthwhile to recount a result of vector analysis: Any vector $\mathbf{a}$ that is a function of position can always be written as the sum of one part $\mathbf{a}_1$, which is solenoidal ($\nabla \cdot \mathbf{a}_1 = 0$), and one part $\mathbf{a}_2$, which is irrotational ($\nabla \times \mathbf{a}_2 = 0$).

We shall assume that *all* magnetism is produced by currents and, consequently, $\mathbf{B}$ is a solenoidal vector. The field $\mathbf{B}$ can then be expressed as the curl of a vector $\mathbf{A}$, called *vector potential*:

$$\mathbf{B}(\mathbf{x}) = \nabla \times \mathbf{A}(\mathbf{x}) \tag{3.4.3}$$

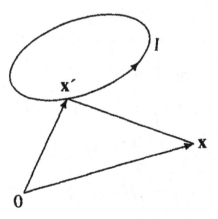

FIGURE 3.6

where

$$\mathbf{A}(\mathbf{x}) = \frac{I}{c} \oint dl \frac{1}{|\mathbf{x}' - \mathbf{x}|} + \nabla f(\mathbf{x}) \tag{3.4.4}$$

The vector potential $\mathbf{A}$ is defined apart from the gradient of an arbitrary function f; this property of $\mathbf{A}$ gives us the freedom to choose $\mathbf{A}$ by using the *gauge transformation*:

$$\mathbf{\Lambda}(\mathbf{x}) \rightarrow \mathbf{\Lambda}(\mathbf{x}) + \nabla f(\mathbf{x}) \tag{3.4.5}$$

We can generalize the expression for $\mathbf{A}(\mathbf{x})$ by replacing in Eq. (3.4.4) $I \, d\mathbf{l}$ by $\mathbf{j} \, d\tau$:

$$\mathbf{A}(\mathbf{x}) = \frac{1}{c} \int d\tau' \frac{\mathbf{j}(\mathbf{x}')}{|\mathbf{x}' - \mathbf{x}|} + \nabla f(\mathbf{x}) \tag{3.4.6}$$

Accordingly, $\mathbf{B}$ is given by

$$\mathbf{B}(\mathbf{x}) = \nabla \times \mathbf{A}(\mathbf{x}) = \nabla \times \frac{1}{c} \int d\tau' \frac{\mathbf{j}(\mathbf{x}')}{|\mathbf{x}' - \mathbf{x}|}$$

$$= \frac{1}{c} \int d\tau' \mathbf{j}(\mathbf{x}') \times \nabla_{x'} \frac{1}{|\mathbf{x}' - \mathbf{x}|} \tag{3.4.7}$$

To specify the vector potential $\mathbf{A}$, we need to give its curl and its divergence. Its curl is simply $\mathbf{B}$; its divergence is not given on account of the arbitrary function f in Eq. (3.4.4). By using the freedom allowed by the gauge transformation, we can set

$$\nabla \cdot \mathbf{A}(\mathbf{x}) = 0 \tag{3.4.8}$$

This choice is called the *Coulomb gauge* for reasons that will become clear later (see Sec. 5.10).

The condition (3.4.8) corresponds to

$$\nabla \cdot \mathbf{A}(\mathbf{x}) = \nabla \cdot \int d\tau' \frac{\mathbf{j}(\mathbf{x}')/c}{|\mathbf{x}' - \mathbf{x}|} + \nabla^2 f(\mathbf{x})$$

$$= -\int d\tau' \frac{\mathbf{j}(\mathbf{x}')}{c} \cdot \nabla_{x'} \frac{1}{|\mathbf{x}' - \mathbf{x}|} + \nabla^2 f(\mathbf{x})$$

$$= \int d\tau' \frac{1/c}{|\mathbf{x}' - \mathbf{x}|} \nabla_{x'} \cdot \mathbf{j}(\mathbf{x}') + \nabla^2 f(\mathbf{x}) = 0 \tag{3.4.9}$$

where we have used integration by parts.

If we consider the continuity equation (2.16.5) and set in it

$$\mathbf{j} = \rho\mathbf{v} \tag{3.4.10}$$

we obtain

$$\boldsymbol{\nabla} \cdot \mathbf{j} = -\left|\frac{\partial\rho}{\partial t}\right| \tag{3.4.11}$$

Equation (3.4.11) is known as the *equation of continuity of currents*.
   In the case of stationary currents,

$$\boldsymbol{\nabla} \cdot \mathbf{j} = 0 \tag{3.4.12}$$

Because of relation (3.4.12), the term containing such divergence in
Eq. (3.4.10) becomes zero, and the condition (3.4.8) implies that

$$\nabla^2 f(\mathbf{x}) = 0 \tag{3.4.13}$$

everywhere, and

$$f(\mathbf{x}) = \text{const} \tag{3.4.14}$$

everywhere, and $\boldsymbol{\nabla} f(\mathbf{x}) = 0$.
   In the framework of the Coulomb gauge, the vector potential is then
given by

$$\mathbf{A}(\mathbf{x}) = \frac{1}{c}\int d\tau'\,\frac{\mathbf{j}(\mathbf{x}')}{|\mathbf{x}' - \mathbf{x}|} \tag{3.4.15}$$

With regard to the vector field $\mathbf{B}$, we know its divergence (it is zero). We
need to determine its curl in order to specify it. We can write

$$\boldsymbol{\nabla} \times \mathbf{B}(\mathbf{x}) = \boldsymbol{\nabla} \times [\boldsymbol{\nabla} \times \mathbf{A}(\mathbf{x})]$$
$$= \boldsymbol{\nabla}[\boldsymbol{\nabla} \cdot \mathbf{A}(\mathbf{x})] - \nabla^2\mathbf{A}(\mathbf{x}) = -\nabla^2\mathbf{A}(\mathbf{x}) \tag{3.4.16}$$

But

$$\nabla^2\mathbf{A}(\mathbf{x}) = \nabla^2\frac{1}{c}\int d\tau'\,\frac{\mathbf{j}(\mathbf{x}')}{|\mathbf{x}' - \mathbf{x}|} = -\frac{4\pi}{c}\mathbf{j}(\mathbf{x}) \tag{3.4.17}$$

and therefore

$$\boldsymbol{\nabla} \times \mathbf{B}(\mathbf{x}) = \frac{4\pi}{c}\mathbf{j}(\mathbf{x}) \tag{3.4.18}$$

We can then say that the two basic differential equations of magnetostatics are given by

$$\nabla \cdot \mathbf{B}(\mathbf{x}) = 0$$

$$\nabla \times \mathbf{B}(\mathbf{x}) = \frac{4\pi}{c}\mathbf{j}(\mathbf{x})$$

(3.4.19)

From the latter equation, we can derive the relation

$$\int (\nabla \times \mathbf{B}) \cdot d\mathbf{S} = \oint \mathbf{B} \cdot d\mathbf{l} = \frac{4\pi}{c} \int \mathbf{J} \cdot d\mathbf{S}$$

(3.4.20)

an expression of *Ampere's law*, which can also be stated as follows: *The line integral of* $\mathbf{B} \cdot d\mathbf{l}$ *around any closed path is equal to* $4\pi/c$ *times the total current passing through any surface bounded by the closed path.*

3.5. Forces and the Magnetic Stress Tensor

We shall deal now with the magnetic field energy and the magnetic stress tensor. Let us assume that we have a number of current loops, as in Fig. 3.7. The force acting on all currents in the volume V is given by

$$\mathbf{F} = \int_V d\tau \frac{\mathbf{j}(\mathbf{x})}{c} \times \mathbf{B}(\mathbf{x}) = \frac{1}{4\pi} \int d\tau [(\nabla \times \mathbf{B}) \times \mathbf{B}]$$

(3.5.1)

We can write

$$(\nabla \times \mathbf{B}) \times \mathbf{B} = (\mathbf{B} \cdot \nabla)\mathbf{B} - \frac{1}{2}\nabla(B^2)$$

(3.5.2)

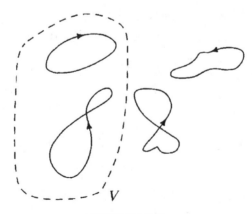

FIGURE 3.7

and

$$\mathbf{F} = \frac{1}{4\pi} \int d\tau \left[(\mathbf{B} \cdot \mathbf{\nabla})\mathbf{B} - \frac{1}{2}\mathbf{\nabla}(B^2) \right] \tag{3.5.3}$$

But

$$\mathbf{\nabla} \cdot \mathbf{B}f = (\mathbf{\nabla} \cdot \mathbf{B})f + (\mathbf{B} \cdot \mathbf{\nabla})f = (\mathbf{B} \cdot \mathbf{\nabla})f \tag{3.5.4}$$

Then

$$\begin{aligned}
F_k &= \frac{1}{4\pi} \int d\tau \left[(\mathbf{B} \cdot \mathbf{\nabla})B_k - \frac{1}{2}\frac{\partial}{\partial x_k}B^2 \right] \\
&= \frac{1}{4\pi} \int d\tau \left[\mathbf{\nabla} \cdot \mathbf{B}B_k - \frac{1}{2}\frac{\partial}{\partial x_k}B^2 \right] \\
&= \frac{1}{4\pi} \int d\tau \left[\sum_i \frac{\partial}{\partial x_i}B_i B_k - \frac{1}{2}\frac{\partial}{\partial x_k}B^2 \right] \\
&= \int d\tau \sum_i \frac{\partial T_{ik}^{(m)}}{\partial x_i} = \int dS \sum_i n_i T_{ik}^{(m)}
\end{aligned} \tag{3.5.5}$$

where the index m stands for "magnetic."

In summary,

$$F_k = \int dS \sum_i n_i T_{ik}^{(m)} \tag{3.5.6}$$

where

$$T_{ik}^{(m)} = \frac{1}{4\pi} \left(B_i B_k - \frac{1}{2}\delta_{ik}B^2 \right) \tag{3.5.7}$$

The magnetic forces do not do any work. In fact,

$$\mathbf{F} = \frac{q}{c}(\mathbf{v} \times \mathbf{B}) \tag{3.5.8}$$

and

$$\mathbf{F} \cdot d\mathbf{s} = \mathbf{F} \cdot \mathbf{v}\,dt = \frac{q}{c}dt\,\mathbf{v} \cdot (\mathbf{v} \times \mathbf{B}) = 0 \tag{3.5.9}$$

3.6. Magnetic Media

The density of magnetic dipoles in a magnetic medium is called *magnetic polarization* or *magnetization* and will be indicated by **M**. The concept of magnetization is of macroscopic nature.

The field **B** to be considered in this case is also macroscopic and is produced by the *true current density* $\mathbf{j}_{\text{true}}$ and by the *magnetization current density* $\mathbf{j}_M$. The total current density is

$$\mathbf{j}(\mathbf{x}) = \mathbf{j}_{\text{true}}(\mathbf{x}) + \mathbf{j}_M(\mathbf{x}) \tag{3.6.1}$$

In a piece of homogeneously magnetized material, on the average, the magnetization currents are zero inside the material and different from zero at the border regions. This situation is the analog of the electrostatic case involving electric polarization: in a piece of material with homogeneous electric polarization, on the average, the polarization charges are zero inside the material, but are different from zero at the border regions.

We want now to relate the magnetization **M** to the magnetization current density $\mathbf{j}_M$. Let us consider a piece of material (see Fig. 3.8) that presents a magnetization even in the absence of an applied field; such materials are called *ferromagnetic*. The field **B** is given by

$$\mathbf{B}(\mathbf{x}_p) = -\boldsymbol{\nabla}_{x_p}\phi(\mathbf{x}_p) \tag{3.6.2}$$

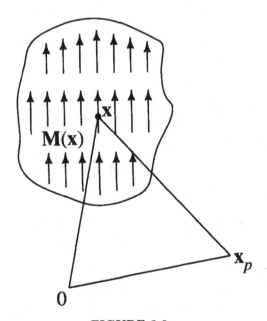

FIGURE 3.8

where, using integration by parts,

$$\phi(\mathbf{x}_p) = -\int \mathbf{M}(\mathbf{x}) \cdot \nabla_{x_p} \frac{1}{|\mathbf{x} - \mathbf{x}_p|} d\tau$$

$$= \int \mathbf{M}(\mathbf{x}) \cdot \nabla_x \frac{1}{|\mathbf{x} - \mathbf{x}_p|}$$

$$= -\int \frac{\nabla_x \cdot \mathbf{M}(\mathbf{x})}{|\mathbf{x} - \mathbf{x}_p|} d\tau \qquad (3.6.3)$$

Then

$$\mathbf{B}(\mathbf{x}_p) = \nabla_{x_p} \int d\tau \frac{\nabla_x \cdot \mathbf{M}(\mathbf{x})}{|\mathbf{x} - \mathbf{x}_p|} \qquad (3.6.4)$$

On the other hand, the field $\mathbf{B}$ due to a *magnetization current density* $\mathbf{j}_M(\mathbf{x})$ is given, according to Eq. (3.4.7), by

$$\mathbf{B}(\mathbf{x}_p) = \nabla_{x_p} \times \int d\tau \frac{(1/c)\mathbf{j}_M(\mathbf{x})}{|\mathbf{x} - \mathbf{x}_p|}$$

$$= \int d\tau \frac{(1/c)[\nabla_x \times \mathbf{j}_M(\mathbf{x})]}{|\mathbf{x} - \mathbf{x}_p|} \qquad (3.6.5)$$

We propose the relation

$$\nabla \times \mathbf{M}(\mathbf{x}) = \frac{1}{c}\mathbf{j}_M(\mathbf{x}) \qquad (3.6.6)$$

We can then write

$$\mathbf{B}(\mathbf{x}_p) = \int d\tau \frac{(1/c)[\nabla_x \times \mathbf{j}_M(\mathbf{x})]}{|\mathbf{x} - \mathbf{x}_p|}$$

$$= \int d\tau \frac{\nabla_x \times [\nabla_x \times \mathbf{M}(\mathbf{x})]}{|\mathbf{x} - \mathbf{x}_p|} \qquad (3.6.7)$$

But

$$\nabla \times (\nabla \times \mathbf{M}) = \nabla(\nabla \cdot \mathbf{M}) = \nabla^2 \mathbf{M} \qquad (3.6.8)$$

Then

$$\mathbf{B}(\mathbf{x}_p) = \int d\tau \frac{\boldsymbol{\nabla}_x[\boldsymbol{\nabla}_x \cdot \mathbf{M}(\mathbf{x})]}{|\mathbf{x} - \mathbf{x}_p|} - \int d\tau \frac{\nabla^2 \mathbf{M}(\mathbf{x})}{|\mathbf{x} - \mathbf{x}_p|}$$

$$= \int d\tau [\boldsymbol{\nabla}_x \cdot \mathbf{M}(\mathbf{x})]\boldsymbol{\nabla}_x \frac{1}{|\mathbf{x} - \mathbf{x}_p|}$$

$$- \int d\tau \mathbf{M}(\mathbf{x}) \nabla_x^2 \frac{1}{|\mathbf{x} - \mathbf{x}_p|}$$

$$= \boldsymbol{\nabla}_{x_p} \int d\tau \frac{\boldsymbol{\nabla}_x \cdot \mathbf{M}(\mathbf{x})}{|\mathbf{x} - \mathbf{x}_p|} + 4\pi \mathbf{M}(\mathbf{x}_p) \qquad (3.6.9)$$

where we have integrated by parts the first term once and the second term twice and have used the result

$$\nabla_x^2 \frac{1}{|\mathbf{x} - \mathbf{x}_p|} = -4\pi\delta(\mathbf{x} - \mathbf{x}_p) \qquad (3.6.10)$$

The expression (3.6.9) for $\mathbf{B}(\mathbf{x}_p)$ agrees with Eq. (3.6.4) if $\mathbf{x}_p$ is outside the magnetic material so that $\mathbf{M}(\mathbf{x}_p) = 0$. Therefore, the relation (3.6.6) may be considered valid.

Let us consider the general case in which we have true currents and magnetization currents. The material is magnetic, but not necessarily ferromagnetic. The $\mathbf{B}$ field inside and outside the material is given by

$$\mathbf{B}(\mathbf{x}_p) = \boldsymbol{\nabla}_{x_p} \times \int d\tau \frac{(1/c)\mathbf{j}_{\text{true}}(\mathbf{x})}{|\mathbf{x} - \mathbf{x}_p|}$$

$$+ \boldsymbol{\nabla}_{x_p} \times \int d\tau \frac{(1/c)\mathbf{j}_M(\mathbf{x})}{|\mathbf{x} - \mathbf{x}_p|}$$

$$= \boldsymbol{\nabla}_{x_p} \times \int d\tau \frac{(1/c)\mathbf{j}_{\text{true}}(\mathbf{x})}{|\mathbf{x} - \mathbf{x}_p|}$$

$$+ \boldsymbol{\nabla}_{x_p} \int d\tau \frac{\boldsymbol{\nabla}_x \cdot \mathbf{M}(\mathbf{x})}{|\mathbf{x} - \mathbf{x}_p|} + 4\pi \mathbf{M}(\mathbf{x}_p) \qquad (3.6.11)$$

We define the field

$$\mathbf{H}(\mathbf{x}_p) = \mathbf{B}(\mathbf{x}_p) - 4\pi \mathbf{M}(\mathbf{x}_p) \qquad (3.6.12)$$

In empty space

$$\mathbf{H}(\mathbf{x}_p) = \mathbf{B}(\mathbf{x}_p) \qquad (3.6.13)$$

We have then

$$\mathbf{H}(\mathbf{x}_p) = \boldsymbol{\nabla}_{x_p} \times \int d\tau \frac{(1/c)\mathbf{j}_{\text{true}}(\mathbf{x})}{|\mathbf{x} - \mathbf{x}_p|}$$

$$+ \boldsymbol{\nabla}_{x_p} \int d\tau \frac{\boldsymbol{\nabla}_x \cdot \mathbf{M}(\mathbf{x})}{|\mathbf{x} - \mathbf{x}_p|} \tag{3.6.14}$$

Therefore,

$$\boldsymbol{\nabla}_{x_p} \times \mathbf{H}(\mathbf{x}_p) = \frac{4\pi}{c} \mathbf{j}_{\text{true}}(\mathbf{x}_p)$$

$$\boldsymbol{\nabla}_{x_p} \cdot \mathbf{H}(\mathbf{x}_p) = -4\pi \boldsymbol{\nabla}_{x_p} \cdot \mathbf{M}(\mathbf{x}_p) \tag{3.6.15}$$

In deriving the above relations, we have taken into account for $\boldsymbol{\nabla}_{x_p} \times \mathbf{H}(\mathbf{x}_p)$ how Eq. (3.4.3) leads to Eq. (3.4.18) and for $\boldsymbol{\nabla}_{x_p} \cdot \mathbf{H}(\mathbf{x}_p)$ the result (3.6.10). A summary of these results is presented in Table 3.1.

The dimensions of the fields $\mathbf{B}$ and $\mathbf{H}$ are the same in the Gaussian system of units. However, $\mathbf{B}$ is expressed in *gauss* and $\mathbf{H}$ in *oersteds*; these two units are identical, but they take different names when they are used for $\mathbf{B}$ or $\mathbf{H}$. The dimensions of the magnetization $\mathbf{M}$ and of the magnetic induction field $\mathbf{B}$ are the same in the Gaussian system, but their units are different on account of the factor 4π in Eq. (3.6.12); we then designate the units of $\mathbf{M}$ as erg gauss^{-1} cm^{-3}.

In Fig. 3.9, we examine the field due to a ferromagnetic slab. In Fig. 3.10, similar results for a ferroelectric slab are reported. To derive the field patterns, we apply the divergence theorem (1.4.8) and the curl theorem (1.4.13) to a thin flat box that we position on the physical boundaries of the slab, as indicated in Table 3.2.

We have three main types of magnetic materials:

(1) *Paramagnetic materials.* The magnetization $\mathbf{M}(\mathbf{x})$ is proportional and parallel to the applied field $\mathbf{B}$.
(2) *Diamagnetic materials.* The magnetization $\mathbf{M}(\mathbf{x})$ is proportional and antiparallel to $\mathbf{B}$.
(3) *Ferromagnetic materials.* The magnetization $\mathbf{M}(\mathbf{x})$ is different from zero even if the applied field is zero.

Let us now examine the paramagnetic and diamagnetic materials.

Table 3.1.

Electrostatics	Magnetostatics
$\nabla \times \mathbf{E} = 0$	$\nabla \times \mathbf{B} = \dfrac{4\pi}{c}\mathbf{j}_{\text{true}} + \dfrac{4\pi}{c}\mathbf{j}_{M},$
	$\left(\dfrac{1}{c}\mathbf{j}_{M} = \nabla \times \mathbf{M}\right)$
$\nabla \cdot \mathbf{E} = 4\pi\rho_{\text{true}} + 4\pi\rho_{\text{pol}}$	
$(\rho_{\text{pol}} = -\nabla \cdot \mathbf{P})$	$\nabla \cdot \mathbf{B} = 0$
$\nabla \times \mathbf{D} = 4\pi\nabla \times \mathbf{P}$	$\nabla \times \mathbf{H} = \dfrac{4\pi}{c}\mathbf{j}_{\text{true}}$
$\nabla \cdot \mathbf{D} = 4\pi\rho_{\text{true}}$	$\nabla \cdot \mathbf{H} = -4\pi\nabla \cdot \mathbf{M}$
$\mathbf{D} = \mathbf{E} + 4\pi\mathbf{P}$	$\mathbf{H} = \mathbf{B} - 4\pi\mathbf{M}$

Ferroelectrics	Ferromagnetics
$\nabla \times \mathbf{E} = 0$	$\nabla \times \mathbf{B} = 4\pi\nabla \times \mathbf{M}$
$\nabla \cdot \mathbf{E} = 4\pi\rho_{\text{pol}} = -4\pi\nabla \cdot \mathbf{P}$	$\nabla \cdot \mathbf{B} = 0$
$\nabla \times \mathbf{D} = 4\pi\nabla \times \mathbf{P}$	$\nabla \times \mathbf{H} = 0$
$\nabla \cdot \mathbf{D} = 0$	$\nabla \cdot \mathbf{H} = -4\pi\nabla \cdot \mathbf{M}$

3.6.1. *Paramagnetic Materials*

In a paramagnetic material,

$$\mathbf{M}(\mathbf{x}) = \chi\mathbf{B}(\mathbf{x}) \qquad (3.6.16)$$

where $\chi = magnetic\ susceptibility > 0$. In a one-electron atom, the electron moves in an orbit of radius

$$a_0 = \frac{\hbar^2}{m_e e^2} = \text{Bohr radius} = 0.53 \times 10^{-8}\,\text{cm}$$

where $m_e = $ mass of the electron. This current gives rise to a magnetic dipole moment:

$$\mathbf{m} = -\frac{e}{2c}(\mathbf{r} \times \mathbf{v}) \qquad (3.6.17)$$

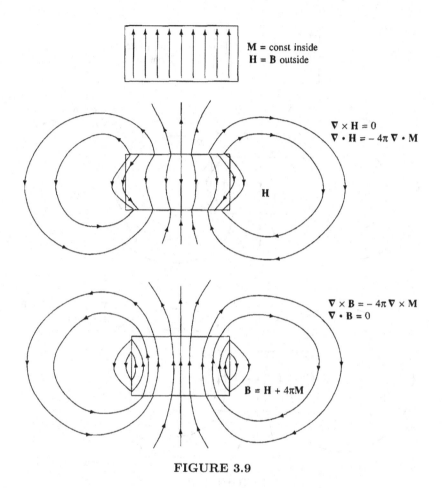

FIGURE 3.9

where e is taken to be intrinsically positive. In fact (see Fig. 3.11), we should have

$$m = |\mathbf{m}| = \frac{I}{c} \times \text{surface} = \frac{e}{T}\frac{1}{c}\pi r^2 = \frac{ev}{2\pi r}\frac{1}{c}\pi r^2 = \frac{evr}{2c} \qquad (3.6.18)$$

Then

$$\mathbf{m} = -\frac{e}{2c}(\mathbf{r} \times \mathbf{v}) = -\frac{e}{2m_e c}\mathbf{L} = -\gamma\mathbf{L} \qquad (3.6.19)$$

and

$$\gamma = \frac{|\mathbf{m}|}{|\mathbf{L}|} = gyromagnetic\ ratio = \frac{e}{2m_e c} \qquad (3.6.20)$$

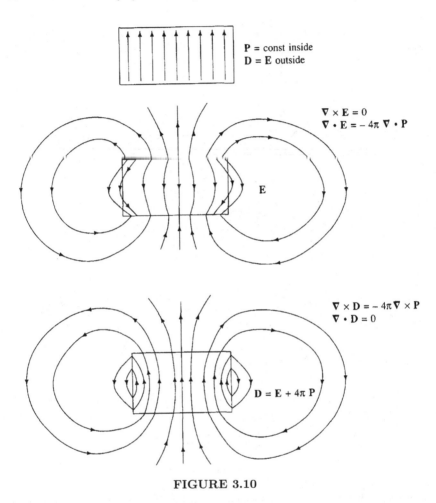

P = const inside
D = **E** outside

$\nabla \times \mathbf{E} = 0$
$\nabla \cdot \mathbf{E} = -4\pi \, \nabla \cdot \mathbf{P}$

E

$\nabla \times \mathbf{D} = -4\pi \, \nabla \times \mathbf{P}$
$\nabla \cdot \mathbf{D} = 0$

$\mathbf{D} = \mathbf{E} + 4\pi \, \mathbf{P}$

FIGURE 3.10

If we have a loop in a magnetic field, the loop will experience a torque (see Fig. 3.12):

$$\mathbf{T} = \mathbf{m} \times \mathbf{B} = \gamma(\mathbf{L} \times \mathbf{B}) = \frac{d\mathbf{L}}{dt} \qquad (3.6.21)$$

$d\mathbf{L}/dt$ is always perpendicular to **L** and to **B**. **L** precesses with constant angle and constant angular velocity about the **B** axis; this precession is called *Larmor precession*, and the frequency of precession is the *Larmor frequency*. In Table 3.3, a comparison between a precessing gyroscope and a precessing magnetic dipole is presented. As we see in the table, the Larmor

Table 3.2.

H Field

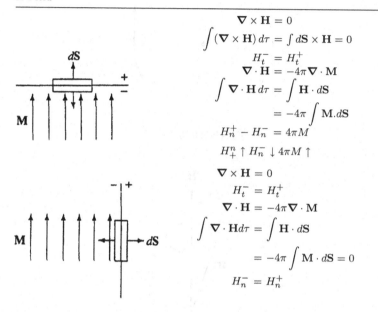

$$\boldsymbol{\nabla} \times \mathbf{H} = 0$$
$$\int (\boldsymbol{\nabla} \times \mathbf{H})\, d\tau = \int d\mathbf{S} \times \mathbf{H} = 0$$
$$H_t^- = H_t^+$$
$$\boldsymbol{\nabla} \cdot \mathbf{H} = -4\pi \boldsymbol{\nabla} \cdot \mathbf{M}$$
$$\int \boldsymbol{\nabla} \cdot \mathbf{H}\, d\tau = \int \mathbf{H} \cdot d\mathbf{S}$$
$$= -4\pi \int \mathbf{M}.d\mathbf{S}$$
$$H_n^+ - H_n^- = 4\pi M$$
$$H_+^n \uparrow H_n^- \downarrow 4\pi M \uparrow$$

$$\boldsymbol{\nabla} \times \mathbf{H} = 0$$
$$H_t^- = H_t^+$$
$$\boldsymbol{\nabla} \cdot \mathbf{H} = -4\pi \boldsymbol{\nabla} \cdot \mathbf{M}$$
$$\int \boldsymbol{\nabla} \cdot \mathbf{H} d\tau = \int \mathbf{H} \cdot d\mathbf{S}$$
$$= -4\pi \int \mathbf{M} \cdot d\mathbf{S} = 0$$
$$H_n^- = H_n^+$$

B Field

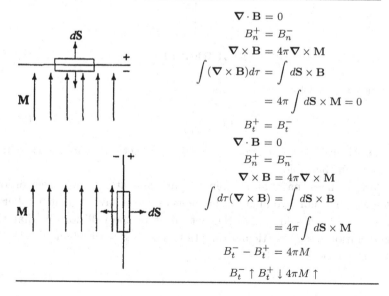

$$\boldsymbol{\nabla} \cdot \mathbf{B} = 0$$
$$B_n^+ = B_n^-$$
$$\boldsymbol{\nabla} \times \mathbf{B} = 4\pi \boldsymbol{\nabla} \times \mathbf{M}$$
$$\int (\boldsymbol{\nabla} \times \mathbf{B}) d\tau = \int d\mathbf{S} \times \mathbf{B}$$
$$= 4\pi \int d\mathbf{S} \times \mathbf{M} = 0$$
$$B_t^+ = B_t^-$$
$$\boldsymbol{\nabla} \cdot \mathbf{B} = 0$$
$$B_n^+ = B_n^-$$
$$\boldsymbol{\nabla} \times \mathbf{B} = 4\pi \boldsymbol{\nabla} \times \mathbf{M}$$
$$\int d\tau (\boldsymbol{\nabla} \times \mathbf{B}) = \int d\mathbf{S} \times \mathbf{B}$$
$$= 4\pi \int d\mathbf{S} \times \mathbf{M}$$
$$B_t^- - B_t^+ = 4\pi M$$
$$B_t^- \uparrow B_t^+ \downarrow 4\pi M \uparrow$$

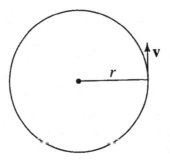

FIGURE 3.11

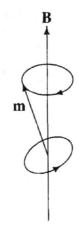

FIGURE 3.12

frequency is given by

$$\omega_L = \frac{e\mathbf{B}}{2m_e c} \tag{3.6.22}$$

A comparison between a gyroscope and a magnetic dipole in a $\mathbf{B}$ field is also given in the table.

Paramagnetic materials are generally made of atoms, molecules, or ions that have a net nonzero angular momentum. If we have a great number of atoms or molecules in a gas, the net effect of the collisions is that the magnetization in any particular direction is zero. If we apply a field $\mathbf{B}$, the energy of a dipole $\mathbf{m}$ in $\mathbf{B}$ is

$$U = -\mathbf{m} \cdot \mathbf{B} = -mB \cos\theta \tag{3.6.23}$$

Table 3.3.

Gyroscope	Magnetic dipole
Gyroscope	Magnetic dipole

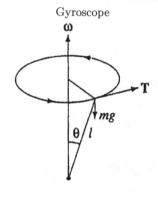

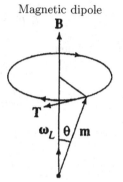

l = distance of center of mass from pivoting point

torque = $mgl \sin \theta$

$$\frac{dL}{dt} = mgl \sin \theta$$

In time dt, L changes by $mgl \sin \theta$. If we divide $mgl \sin \theta$ by $L \sin \theta$, we get the angular velocity:

$$\omega = \frac{mgl \sin \theta}{L \sin \theta} = \frac{mgl}{L}$$

torque = $|\mathbf{m} \times \mathbf{B}| = mB \sin \theta$

$$\frac{dL}{dt} = mB \sin \theta$$

In time dt, L changes by $mB \sin \theta$. If we divide $mB \sin \theta$ by $L \sin \theta$, we get the angular velocity:

$$\omega_l = \frac{mB \sin \theta}{L \sin \theta} = \frac{M}{L} B = \frac{eB}{2m_e c}$$
$$= \text{Larmor frequency}$$

Note that, since the charge of the electron is negative, $\mathbf{m}$ and $\mathbf{L}$ have opposite directions.

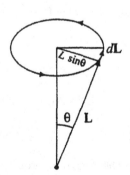

where θ is the angle between the directions of **m** and **B**. The probability that the magnetic moment lies in the range $(\theta, \theta + d\theta)$ is proportional to the Boltzmann factor and the solid angle $2\pi \sin\theta \, d\theta$:

$$P(\theta)d\theta \propto e^{\beta mB \cos\theta} \sin\theta \, d\theta \tag{3.6.24}$$

where $\beta = \frac{1}{kT}$. Then

$$\langle \cos\theta \rangle = \frac{\displaystyle\int_0^\pi e^{\beta mB \cos\theta} \sin\theta \, d\theta (\cos\theta)}{\displaystyle\int_0^\pi e^{\beta mB \cos\theta} \sin\theta \, d\theta}$$

$$= \frac{\displaystyle\int_{-1}^1 e^{\beta mB \cos\theta} \cos\theta \, d(\cos\theta)}{\displaystyle\int_{-1}^1 e^{\beta mB \cos\theta} \, d(\cos\theta)}$$

$$= \frac{1}{mB} \frac{\partial}{\partial\beta} \ln \int_{-1}^1 e^{\beta mB \cos\theta} d(\cos\theta) \tag{3.6.25}$$

On the other hand,

$$\int_{-1}^1 e^{\beta mB \cos\theta} d(\cos\theta) = \frac{2 \sinh \beta mB}{\beta mB}$$

and

$$\frac{\partial}{\partial\beta} \ln \frac{2 \sinh \beta mB}{\beta mB} = \frac{1}{\beta \sinh \beta mB}$$
$$\times (\beta mB \cosh \beta mB - \sinh \beta mB)$$

and

$$\langle \cos\theta \rangle = \frac{1}{\beta mB \sinh \beta mB}$$
$$\times (\beta mB \cosh \beta mB - \sinh \beta mB)$$
$$= \coth \beta mB - \frac{1}{\beta mB} \tag{3.6.26}$$

For $\beta mb \ll 1$, that is, $mB \ll kT$, which is in the high temperature limit,

$$\coth \beta mB \simeq \frac{1}{\beta mB} + \frac{1}{3}\beta mB \tag{3.6.27}$$

and

$$\langle \cos \theta \rangle = \frac{1}{\beta m B} + \frac{1}{3} \beta m B - \frac{1}{\beta m B} = \frac{\beta m B}{3} = \frac{m B}{3kT} \qquad (3.6.28)$$

We have then

$$M = n \langle m_z \rangle = nm \langle \cos \theta \rangle = \left(\frac{nm^2}{3kT} \right) B \qquad (3.6.29)$$

where n = density of molecules (cm^{-3}). The average of **m** in any other direction is zero. Therefore, the *paramagnetic susceptibility* is given by

$$\chi = \frac{\mathbf{M}}{\mathbf{B}} = \frac{nm^2}{3kT} \qquad (3.6.30)$$

The dependence of χ on T

$$\chi \sim \frac{\text{cont}}{T} \qquad (3.6.31)$$

expresses the *Curie's law*.

To estimate χ, we set $L \simeq \hbar$; then

$$m = \gamma L = \frac{e\hbar}{2mc} = 0.927 \times 10^{-20} \frac{\text{erg}}{\text{gauss}} \qquad (3.6.32)$$

We note that, in the Gaussian system, the magnetic dipole and the electric dipole have the same dimensions. Likewise, P = polarization and M = magnetization.

An atomic electric dipole p can be estimated by means of the formula

$$p = a_0 e = \frac{\hbar^2}{m_e e} \qquad (3.6.33)$$

We find then that

$$\frac{m}{p} = \frac{e\hbar}{2m_e c} \frac{m_e e}{\hbar^2} = \frac{1}{2} \frac{e^2}{\hbar c} = \frac{1}{2} \frac{1}{137} \qquad (3.6.34)$$

$\frac{e^2}{\hbar c} \simeq \frac{1}{137}$ is known as the *fine structure constant*. We have molecules with electric dipoles, for example, water. We may call these substances *paraelectric*; in such materials, we obtain the electric polarization by orientation of the electric dipoles in an electric field. We have

$$\frac{\chi_e}{\chi_m} = (2 \times 137)^2 \simeq (250)^2 = 6 \times^4 \qquad (3.6.35)$$

where the subscripts e and m indicate electric and magnetic susceptibilities, respectively. In the case of water $\chi_e \simeq 10$; then, in condensed matter,

$$\chi_m = \frac{10}{6 \times 10^4} \simeq 10^{-4}$$

We can also evaluate χ_m as follows:

$$\chi_m = \frac{nm^2}{3kT} = \frac{1}{3kT} nm^2 = \frac{1}{3kT} \left(\frac{1}{2a_0} \right)^3 \left(\frac{ea_0}{2 \times 137} \right)^2$$

$$= \frac{1}{32} \frac{e^2}{a_0} \frac{1}{3kT} \frac{1}{137^2} = \frac{1}{32} \frac{350}{(137)^2} \simeq 6 \times 10^{-4} \qquad (3.6.36)$$

3.6.2. *Diamagnetic Materials*

In the presence of a magnetic field, the electrons continue to move around the nucleus; but, in addition, they experience a motion of precession with angular frequency

$$\omega_L = \frac{e\mathbf{B}}{2m_e c} \qquad (3.6.37)$$

This precession produces an electric current:

$$\mathbf{I} = \text{charge} \times \text{revolutions per unit time}$$

$$= (-Ze) \frac{1}{2\pi} \frac{e\mathbf{B}}{2m_e c} = -\frac{Ze^2\mathbf{B}}{4\pi m_e c}$$

where Z = number of electrons in the atom. The magnetic moment due to this charge motion is

$$\mathbf{m} = \frac{IS}{c} = -\frac{Ze^2\mathbf{B}}{4\pi m_e c^2} \pi \langle \rho^2 \rangle = -\frac{Ze^2\mathbf{B}}{4m_e c^2} \langle \rho^2 \rangle \qquad (3.6.38)$$

where

$$\langle \rho^2 \rangle = \langle x^2 \rangle + \langle y^2 \rangle \qquad (3.6.39)$$

is the mean square radius of the electron's orbit when projected on a plane perpendicular to $\mathbf{B}$. We note that the magnetic dipole moment induced by the $\mathbf{B}$ field and related to the motion of precession of the electron's orbit is always opposite in direction to $\mathbf{B}$.

The mean square distance of the electrons from the nucleus is

$$\langle r^2 \rangle = \langle x^2 \rangle + \langle y^2 \rangle + \langle z^2 \rangle \qquad (3.6.40)$$

For spherically symmetric charge distributions,

$$\langle x^2 \rangle = \langle y^2 \rangle = \langle z^2 \rangle = \frac{1}{3}\langle r^2 \rangle \tag{3.6.41}$$

and

$$\langle \rho^2 \rangle = \langle x^2 \rangle + \langle y^2 \rangle = \frac{2}{3}\langle r^2 \rangle \tag{3.6.42}$$

Replacing this value in Eq. (3.6.38), we obtain

$$\mathbf{m} = -\frac{Ze^2}{6m_ec^2}\langle r^2 \rangle \mathbf{B} \tag{3.6.43}$$

and the *diamagnetic susceptibility* is

$$\chi = -\frac{n|\mathbf{m}|}{|\mathbf{B}|} = -\frac{nZe^2\langle r^2 \rangle}{6m_ec^2} \tag{3.6.44}$$

where n = number of atoms per cm^3. Taking $\langle r^2 \rangle = 10^{-6}\,cm^2$ and $n \simeq 5 \times 10^{22}\,cm^{-3}$, we obtain

$$\frac{ne^2\langle r^2 \rangle}{6m_ec^2} = \frac{5 \times 10^{22} \times (4.8 \times 10^{-10})^2 \times 10^{-16}}{6 \times 9.1 \times 10^{-28} \times 9 \times 10^{20}} = 2.3 \times 10^{-7}$$

and

$$\chi \simeq -10^{-7}Z \tag{3.6.45}$$

The following observations can be made:[1]

(1) The diamagnetic susceptibility is always negative.
(2) Diamagnetism is a property of *all* matter.
(3) The diamagnetic susceptibility has no explicit dependence on temperature.
(4) Small variations of χ with the temperature may be interpreted as due to a slight dependence of $\langle r^2 \rangle$ on T.

In the preceding calculations, we assumed that the radius of a certain orbit has a particular definite value. J. H. van Leeuwen[2] pointed out that in a classical model an orbit can have a continuous range of radii; this argument leads to the conclusion that, classically, the total susceptibility (diamagnetic plus paramagnetic) of matter is zero. Quantum mechanics removes these

---

[1] A. H. Morrish, *The Physical Principles of Magnetism*, John Wiley & Sons, Inc., New York, 1965, p. 39.

[2] J. H. van Leeuwen, Dissertation, Leiden, 1919; *J. de Phys.* (6) 2, 361 (1921).

objections, in that, even in its old Bohr formulation, it postulates the existence of discrete orbits.

3.7. B and H

We wish to make some additional statements concerning the *magnetic induction field* **B** and the *magnetic field* **H**. We have defined the vector field **H** by means of the relation

$$\mathbf{H}(\mathbf{x}) = \mathbf{B}(\mathbf{x}) - 4\pi\mathbf{M}(\mathbf{x}) \tag{3.7.1}$$

where $\mathbf{M}(\mathbf{x})$ = magnetization. The divergence and curl of **B** are given by

$$\boldsymbol{\nabla} \cdot \mathbf{B} = 0$$
$$\boldsymbol{\nabla} \times \mathbf{B} = \frac{4\pi}{c}\mathbf{j} \tag{3.7.2}$$

where

$$\mathbf{j} = \mathbf{j}_{\text{true}} + \mathbf{j}_M \tag{3.7.3}$$

and

$$\mathbf{j}_M = c\boldsymbol{\nabla} \times \mathbf{M} \tag{3.7.4}$$

The divergence and the curl of **H** are given by

$$\boldsymbol{\nabla} \cdot \mathbf{H} = -4\pi\boldsymbol{\nabla} \cdot \mathbf{M}$$
$$\boldsymbol{\nabla} \times \mathbf{H} = \frac{4\pi}{c}\mathbf{j}_{\text{true}} \tag{3.7.5}$$

When dealing with materials presenting an electric polarization $\mathbf{P}(\mathbf{x})$, we introduced a vector field

$$\mathbf{D}(\mathbf{x}) = \mathbf{E}(\mathbf{x}) + 4\pi\mathbf{P}(\mathbf{x}) \tag{3.7.5A}$$

The divergence and the curl of **E** are given by

$$\boldsymbol{\nabla} \cdot \mathbf{E} = 4\pi\rho$$
$$\boldsymbol{\nabla} \times \mathbf{E} = 0 \tag{3.7.6}$$

where

$$\rho = \rho_{\text{true}} + \rho_{\text{pol}} \tag{3.7.7}$$

and

$$\rho_{\text{pol}} = -\nabla \cdot \mathbf{P} \tag{3.7.8}$$

The divergence and the curl of $\mathbf{D}$ are given by

$$\nabla \cdot \mathbf{D} = 4\pi\rho_{\text{true}}$$
$$\nabla \times \mathbf{D} = 4\pi\nabla \times \mathbf{P} \tag{3.7.9}$$

There is some similarity between the vector fields $\mathbf{D}$ and $\mathbf{H}$ due to the fact that the former is determined by the true charges and the latter by the true currents. However, when treating electrical systems, little use is made of the field $\mathbf{D}$, because this physical parameter is difficult to control; we can measure and practically vary at will the potential difference between physical bodies and, therefore, the value of the electric field $\mathbf{E}$, but we cannot control easily the values of the true charges, which determine the field $\mathbf{D}$.

When treating magnetic systems, wide use is made of the vector field $\mathbf{H}$, because this field is related to true currents, entities that we can measure and control easily. On the other hand, since we cannot vary at will the magnetization $\mathbf{M}$, the field $\mathbf{B}$ is in effect out of our control.

We have defined the magnetic susceptibility χ_m by means of the relation

$$\mathbf{M} = \chi_m \mathbf{B} \tag{3.7.10}$$

where χ_m is a dimensionless quantity that is negative for diamagnetic and positive for paramagnetic materials. The relation that is traditionally used as a definition of χ_m is not, however, that given by Eq. (3.7.10), but

$$\mathbf{M} = \chi_m \mathbf{H} \tag{3.7.11}$$

For purely solid or liquid diamagnetic substances, $\chi_m \approx -10^{-6}$, and for paramagnetic substances, $\chi_m \approx 10^{-3}$. In these cases, the contribution to the field acting on the individual magnetic dipole that is due to all the other dipoles is much smaller than the applied field $\mathbf{B}$. Under these conditions $(\chi_m \ll 1)$ the difference between the two definitions of χ_m is not important, because the field acting on each dipole is essentially equal to what we would have in the absence of the magnetic sample.

For any material with a magnetization proportional to the field $\mathbf{H}$,

$$\mathbf{B} = \mathbf{H} + 4\pi\mathbf{M} = \mathbf{H} + 4\pi\chi_m\mathbf{H} = (1 + 4\pi\chi_m)\mathbf{H} = \mu\mathbf{H} \tag{3.7.12}$$

The quantity

$$\mu = l + 4\pi\chi_m \tag{3.7.13}$$

is called *magnetic permeability*.

CHAPTER 3 EXERCISES

3.1. Calculate the electric field due to an infinite flat sheet of electric charge with constant surface density σ ESU/cm^2.

3.2. A positive charge is spread over the surface of a flat disk of radius R. The surface density of this charge is constant and equal to σ ESU/cm^2.

    (a) Calculate the potential at any point on the symmetry axis of the disk.

    (b) Calculate the electric field at any point on the symmetry axis of the disk.

3.3. Find the **B** field at points on the symmetry axis of a circular loop of radius a that carries a current I.

3.4. A spherical shell of radius R carries a constant surface charge σ ESU/cm^2. Calculate the vector potential **A** and the field **B** produced when the spherical shell is set in motion with an angular frequency ω.

3.5. A small current loop of radius R lies in the xy plane (see Fig. P3.5). The current J passes through the loop.

    (a) Find the magnetic dipole moment **m** of the loop.

    (b) Find the asymptotic ($r \gg R$) magnetic field **B** from the magnetic potential due to **m**.

    (c) Write the equation of motion for a particle of mass M and charge q in the asymptotic **B** field of the loop. Show that the particle

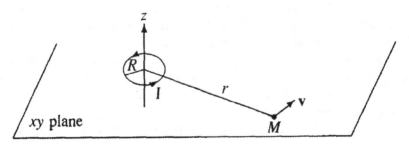

FIGURE P3.5

stays in the xy plane if its original position and velocity are in that plane.

3.6. Two long conducting coaxial cylinders are part of a device that is used for separating particles of the same charge and momentum and different mass. Between the inner cylinder of radius a and the outer cylinder of radius b, a potential difference V is maintained by a high voltage supply. A current I flows along the inner cylinder.

(a) Find the force acting on a particle of mass m and charge $+e$ traveling axially between the cylinders with (nonrelativistic) momentum p.

(b) Find the value and polarity of V that would allow a particle of mass m to pass through the system parallel to the axis without deflection.

3.7. Protons in a beam are accelerated in a discharge tube through a potential difference of 2.5×10^6 V. They then enter a region where a magnet provides a uniform magnetic field of 8000 gauss that curves their path into a circle. They follow a portion of the circular path and then leave the magnet. Consider a beam of doubly ionized helium atoms (alpha particles) that have been accelerated through a potential difference of 2×10^6 V. Calculate the magnetic field that would be required to deflect such a beam through the same path, without altering the slit system. Use nonrelativistic formulas and assume that the mass of an alpha particle is four times the mass of a proton.

3.8. A spaceship of mass M is in outer space and is propelled by ejecting a stream of singly ionized molecules. The mass of each molecule is m in atomic mass units. The ionized particles are accelerated through a potential difference of V volts, and the current of ionized molecules is I amperes.

(a) Express the resulting force acting on the ship in terms of V, m, I, and some fundamental constants.

(b) Choose reasonable values for the relevant parameters and estimate the current needed for giving the ship an acceleration of 0.1 m/s^2.

(c) Does the negative charge left on the ship pose any problem, and, if so, how would you solve it?

3.9. In a cyclotron, particles move at a constant angular velocity in a plane perpendicular to a uniform magnetic field. Assume that a cyclotron

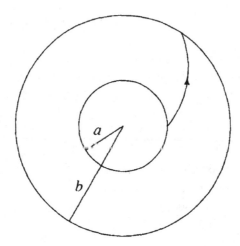

FIGURE P3.10

is used to produce deuterons with 12-MeV energy by operating with a magnetic field of 15,000 gauss.

(a) What frequency and magnetic field would be required to produce 16-MeV deuterons?

(b) How would these conditions be changed to produce 8-MeV protons?

3.10. A circular filament of radius a (see Fig. P3.10) emits electrons that reach an outer metal cylinder of radius b. The electrons leave the filament (which is grounded) with zero velocity and are accelerated by the positive voltage of V volts of the cylinder. The paths of the electrons are curved by a uniform magnetic field H applied along the axis of the cylinder. Calculate the value of the field H that, for a given voltage V, will suppress the current between the filament and the cylinder.

3.11. The magnetostatic problems can be treated in a way similar to the electrostatic ones. Show that for the calculation of the fields a material of magnetization $\mathbf{M}(\mathbf{x})$ can be replaced by a volume polarization charge density $\rho_M = -\nabla \cdot \mathbf{M}$ and by a surface polarization charge density $\sigma_M = \mathbf{n} \cdot \mathbf{M}$.

3.12. Show that, for the calculation of the fields, a material of magnetization $\mathbf{M}(\mathbf{x})$ can be replaced by a volume current density $\mathbf{j}_M = c\nabla \times \mathbf{M}$ and

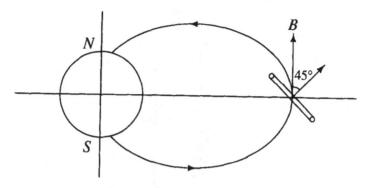

FIGURE P3.13

by a surface current density

$$\mathbf{j}_{MS} = c(\mathbf{M} \times \mathbf{n}).$$

3.13. A thin metallic ring has a radius R and mass M and moves around the earth in a circular orbit in the plane of the geomagnetic equator. It also spins about its own axis, which forms an angle of 45° with respect to the **B** field of the earth at the orbit (see Fig. P3.13). It carries an electric charge q.

(a) Derive the gyromagnetic ratio, that is, the ratio of the magnetic moment to the spin angular momentum.

(b) Derive an expression for the time it takes the spin to precess through an angle of 180°.

Assume that the only torque acting on the ring is that due to magnetic forces. Specify the system of units in which you are expressing your relations.

3.14. A cylindrical paramagnetic sample is suspended in such a way as to be able to rotate freely (see Fig. P3.14). By means of a solenoid, a magnetic field H in the z direction is turned on. This field produces a magnetization of the sample

$$\mathbf{M} = \chi\mathbf{H}$$

Explain why the cylinder will start rotating after the field H is turned on, and calculate the angular velocity of the cylinder in terms of the appropriate physical parameters. In what sense will the cylinder rotate with respect to the z axis?

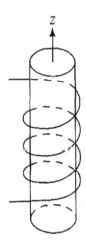

FIGURE P3.14

3.15. Show that the magnetic field just outside a medium of very high permeability is nearly perpendicular to the surface of the magnetic medium.

3.16. Calculate the force with which a very long cylindrical magnet sticks to the surface of a material of very high permeability.

3.17. Consider a cylindrical magnet of radius a and length L, and calculate the **B** and **H** fields at points along the axis of the magnet.

3.18. A bar magnet is shaped in the form of a torus and has a small gap between its ends (see Fig. P3.18). The length of the magnet is L, and the length of the gap, s. Estimate the value of the field H in the gap.

3.19. A cylindrical magnet of length L and radius a is oriented with its axis perpendicular to the surface of a material with infinite permeability (see Fig. P3.19). Calculate the **B** field on the axis of the magnet and just outside the surface of the material.

3.20. An infinite solenoid of radius a, N turns per unit length, and current I is filled with a linear magnetic material of susceptibility χ. Calculate the fields **H** and **B** inside the solenoid. Does the presence of the medium inside the solenoid enhance or reduce the **B** field?

3.21. Consider a permanent magnetic material from which a parallelopipedal hole has been cut. The hole has a square base of side L, and a thickness l, with $L > l$, as described in Fig. P3.21. Eight of the twelve edges of the hole have length L, and four have length l. There

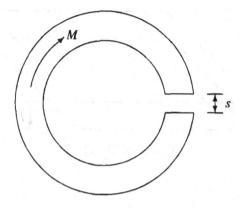

FIGURE P3.18

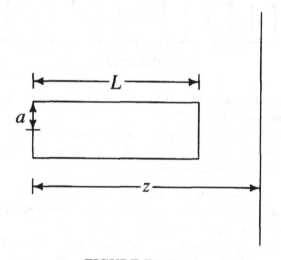

FIGURE P3.19

is a constant homogeneous magnetization **M** inside the magnetic material. Consider the following two cases:

(1) **M** is parallel to one of the eight long sides.
(2) **M** is parallel to one of the four short sides.

(a) Give an integral expression for **H** and **B** at any point.
(b) Give an explicit formula for the fields at the center of the cavity.

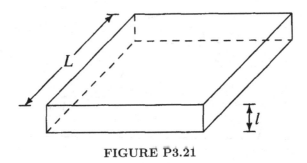

FIGURE P3.21

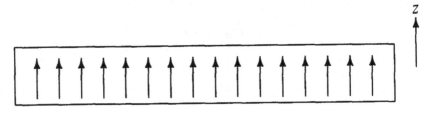

FIGURE P3.23

3.22. Calculate the fields **H** and **B** produced by a sphere of a material of uniform magnetization **M**.

3.23. A ferroelectric crystal with spontaneous electric polarization **P** (**P** = const) is cut into a thin plate (see Fig. P3.23). Consider the boundaries of the plate and determine:

   (a) The surface charge density σ.
   (b) The qualitative behavior of the z component of the **D** field, $D_z(z)$, along the z axis.
   (c) The qualitative behavior of the z component of the **E** field, $E_z(z)$, and of the potential $\phi(z)$ along the z axis.

3.24. In a C-shaped soft-iron core with permeability μ, a magnetic field is produced by means of a coil with N turns and a current J (see Fig. P3.24). The air gap has a width D, and the pole pieces have cross sections A_1 and A_2.

   (a) Calculate approximately **B** and **H** in the core and in the air gap, and, in particular, the surface values on the poles of the magnet.
   (b) A cone-shaped piece of the same soft iron is inserted into the air gap. Find the field **B** in the two air gaps and calculate the

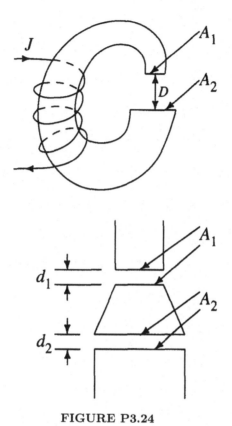

FIGURE P3.24

net force on the middle piece (you may use the magnetic stress tensor).

3.25. The particles pi mesons, also called pions, π^+ and π^- have a positive and negative charge, respectively, equal in absolute values to the charge of the electron, and have a rest mass of 139.6 MeV. This mass is in excess of the mass of the neutral pion π^0 which is 135 MeV. The mass difference is 4.6 MeV. Assume that the charged pion is a sphere in which the charge is distributed with equal density up to the radius R. Calculate R, if the mass difference above is due to the electrostatic energy of the charged particles.

4

Induction and Quasi-Stationary Phenomena

4.1. Effect of Time Variations on $\nabla \times$ B and $\nabla \times$ H

In the stationary case, the Maxwell equations are given by

$$\nabla \cdot \mathbf{E}(\mathbf{x}) = 4\pi\rho(\mathbf{x})$$
$$\nabla \times \mathbf{E}(\mathbf{x}) = 0 \rightarrow \mathbf{E}(\mathbf{x}) = -\nabla\phi(\mathbf{x})$$
$$\nabla \cdot \mathbf{B}(\mathbf{x}) = 0 \rightarrow \mathbf{B}(\mathbf{x}) = \nabla \times \mathbf{A}(\mathbf{x}) \qquad (4.1.1)$$
$$\nabla \times \mathbf{B}(\mathbf{x}) = \frac{4\pi}{c}\mathbf{j}(\mathbf{x})$$

The conservation of charge is expressed by

$$\nabla \cdot \mathbf{j}(\mathbf{x}) = 0 \qquad (4.1.2)$$

In the case of time-dependent phenomena,

$$\nabla \cdot \mathbf{j}(\mathbf{x}, t) = -\frac{\partial \rho(\mathbf{x}, t)}{\partial t} \qquad (4.1.3)$$

143

and

$$\int_V \boldsymbol{\nabla} \cdot \mathbf{j} = \int dS \ \mathbf{n} \cdot \mathbf{j} = -\frac{d}{dt} \int \rho d\tau = -\frac{d}{dt} Q \quad \text{in } V \qquad (4.1.4)$$

in accord with the *law of conservation of charges*.

If we consider the fourth equation in (4.1.1) and take the divergence of both members, we obtain

$$\boldsymbol{\nabla} \cdot \boldsymbol{\nabla} \times \mathbf{B} = 0 = \frac{4\pi}{c} \boldsymbol{\nabla} \cdot \mathbf{j} \qquad (4.1.5)$$

In the case of time-dependent phenomena, the fourth Maxwell equation needs a revision in order to make it consistent with Eq. (4.1.3). Following Maxwell, we introduce in this equation a new term, so that the equation appears as follows

$$\boldsymbol{\nabla} \times \mathbf{B} = \frac{4\pi}{c}(\mathbf{j} + \mathbf{j}_D) \qquad (4.1.6)$$

where

$$\mathbf{j}_D = \text{Maxwell's } displacement \ current$$

We must have

$$\boldsymbol{\nabla} \cdot (\mathbf{j} + \mathbf{j}_D) = 0 \qquad (4.1.7)$$

and

$$\boldsymbol{\nabla} \cdot \mathbf{j}_D = -\boldsymbol{\nabla} \cdot \mathbf{j} = \frac{\partial \rho}{\partial t} = \frac{1}{4\pi} \frac{\partial}{\partial t} \boldsymbol{\nabla} \cdot \mathbf{E} = \frac{1}{4\pi} \boldsymbol{\nabla} \cdot \frac{\partial \mathbf{E}}{\partial t} \qquad (4.1.8)$$

Maxwell, with a stroke of genius, put

$$\mathbf{j}_D = \frac{1}{4\pi} \frac{\partial \mathbf{E}}{\partial t} \qquad (4.1.9)$$

This is *not* the conclusion of a mathematical calculation, but something that was postulated and found to be valid by experimentaion.

We have now

$$\boldsymbol{\nabla} \times \mathbf{B} = -\frac{1}{c} \frac{\partial \mathbf{E}}{\partial t} = \frac{4\pi}{c} \mathbf{j} \qquad (4.1.10)$$

This equation is now consistent with Eq. (4.1.3). In fact,

$$\boldsymbol{\nabla} \cdot \boldsymbol{\nabla} \times \mathbf{B} - \frac{1}{c} \frac{\partial}{\partial t} \boldsymbol{\nabla} \cdot \mathbf{E} = \frac{4\pi}{c} \boldsymbol{\nabla} \cdot \mathbf{j} \qquad (4.1.11)$$

which agrees with Eq. (4.1.8). Equation (4.1.10) represents the fourth Maxwell equation in the case of time-dependent phenomena.

The macroscopic stationary Maxwell equations are given by

$$\nabla \cdot \mathbf{D}(\mathbf{x}) = 4\pi\rho_{\text{true}}(\mathbf{x})$$

$$\nabla \times \mathbf{E}(\mathbf{x}) = 0$$

$$\nabla \cdot \mathbf{B}(\mathbf{x}) = 0 \tag{4.1.12}$$

$$\nabla \times \mathbf{H}(\mathbf{x}) = \frac{4\pi}{c}\mathbf{j}_{\text{true}}(\mathbf{x})$$

We want now to examine the effect of time dependence on the expression for $\nabla \times \mathbf{H}$. If we use the model in which matter is a continuum, we may have, in general, time-dependent charges, polarization charges, polarization currents, true currents, and magnetization currents. We can write

$$\rho_{\text{pol}} = -\nabla \cdot \mathbf{P} \tag{4.1.13}$$

$$\nabla \cdot \mathbf{j}_{\text{pol}} = -\frac{\partial \rho_{\text{pol}}}{\partial t} = \nabla \cdot \frac{\partial \mathbf{P}}{\partial t} \tag{4.1.14}$$

and

$$\mathbf{j}_{\text{pol}} = \frac{\partial \mathbf{P}}{\partial t} \tag{4.1.15}$$

Then

$$\nabla \cdot \mathbf{E} = 4\pi(\rho_{\text{true}} + \rho_{\text{pol}}) = 4\pi\rho_{\text{true}} - 4\pi\nabla \cdot \mathbf{P} \tag{4.1.16}$$

and we again have

$$\nabla \cdot (\mathbf{E} + 4\pi\mathbf{P}) = \nabla \cdot \mathbf{D} = 4\pi\rho_{\text{true}} \tag{4.1.17}$$

On the other hand, we can write

$$\nabla \times \mathbf{B} - \frac{1}{c}\frac{\partial \mathbf{E}}{\partial t} = \frac{4\pi}{c}(\mathbf{j}_{\text{true}} + \mathbf{j}_{\text{pol}}) = \frac{4\pi}{c}\left(\mathbf{j}_{\text{true}} + \frac{\partial \mathbf{P}}{\partial t}\right) \tag{4.1.18}$$

or

$$\nabla \times \mathbf{B} - \frac{1}{c}\frac{\partial}{\partial t}(\mathbf{E} + 4\pi\mathbf{P}) = 4\pi\mathbf{j}_{\text{true}} \tag{4.1.19}$$

$$\nabla \times \mathbf{B} - \frac{1}{c}\frac{\partial \mathbf{D}}{\partial t} = \frac{4\pi}{c}\mathbf{j}_{\text{true}} \tag{4.1.20}$$

But the material may also have magnetic properties; in this case we have magnetization currents

$$\mathbf{j}_M = c\nabla \times \mathbf{M} \tag{4.1.21}$$

and the total current consists of three parts,

$$\mathbf{j} = \mathbf{j}_{\text{true}} + \mathbf{j}_{\text{pol}} + \mathbf{j}_M \tag{4.1.22}$$

and

$$\boldsymbol{\nabla} \times \mathbf{B} - \frac{1}{c}\frac{\partial \mathbf{E}}{\partial t} = \frac{4\pi}{c}(\mathbf{j}_{\text{true}} + \mathbf{j}_{\text{pol}} + \mathbf{j}_M)$$

$$= \frac{4\pi}{c}\left(\mathbf{j}_{\text{true}} + \frac{\partial \mathbf{P}}{\partial t} + c\boldsymbol{\nabla} \times \mathbf{M}\right) \tag{4.1.23}$$

or

$$\boldsymbol{\nabla} \times (\mathbf{B} - 4\pi\mathbf{M}) - \frac{1}{c}\frac{\partial}{\partial t}(\mathbf{E} + 4\pi\mathbf{P}) = \frac{4\pi}{c}\mathbf{j}_{\text{true}} \tag{4.1.24}$$

$$\boldsymbol{\nabla} \times \mathbf{H} - \frac{1}{c}\frac{\partial \mathbf{D}}{\partial t} = \frac{4\pi}{c}\mathbf{j}_{\text{true}} \tag{4.1.25}$$

Equation (4.1.25) represents the fourth Maxwell equation in the case of time-dependent phenomena.

4.2. Induction Phenomena

Consider a magnetic induction of field $\mathbf{B}$ threading a closed path as in Fig. 4.1. If the flux of $\mathbf{B}$ changes with time, an electric field is induced in the path, according to *Faraday's law of induction*:

$$\oint d\mathbf{l} \cdot \mathbf{E} = -\text{const}\frac{d}{dt}\int dS\,\mathbf{n} \cdot \mathbf{B} \tag{4.2.1}$$

If the closed path is a metallic loop, an electric current is produced in the loop by an induced *electromotive force*. The notion of electromotive

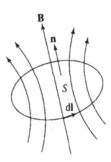

FIGURE 4.1

force, normally referred to as *emf*, includes any agent that makes a current go around a closed path.

In the Gaussian system of units, the constant in relation (4.2.1) has the dimensions of the inverse of a velocity and is found to be equal to the inverse of the velocity of light:

$$\text{const} = \frac{1}{c} \qquad (4.2.2)$$

The law of induction applies to several experimental situations:

1. Fixed loop, time-varying **B** field:

$$\frac{\partial \mathbf{B}}{\partial t} \neq 0$$

2. Loop pulled through a spatially inhomogeneous and time-independent **B** field:

$$\frac{\partial \mathbf{B}}{\partial t} = 0$$

3. Loop deformed in time; the field may be homogeneous.

All these cases are expressed by the same law. It will be our aim to express it in a differential form. This we shall do in Sec. 4.4.

4.3. Temporal Variation of a Flux through a Moving Surface Element

Let

$$\mathbf{A} = \mathbf{A}(\mathbf{x}, t) \qquad (4.3.1)$$

be a generic vector field. The flux through a surface S is given by

$$\int_S \mathbf{A} \cdot \mathbf{n} \, dS \qquad (4.3.2)$$

If the surface is at rest,

$$\frac{d}{dt} \int_S \mathbf{A} \cdot \mathbf{n} \, dS = \int_S \frac{\partial A_n}{\partial t} \, dS \qquad (4.3.3)$$

If the surface is in motion, the flux varies also because S may be brought to points where **A** is different.

Let us define a new type of derivative:

$$\frac{d}{dt} \int A_n \, dS = \int \dot{\underline{\mathbf{A}}}_n \, dS \qquad (4.3.4)$$

$\dot{\underline{\mathbf{A}}}$ is a vector whose flux through a surface in motion is equal to the temporal variation of the flux of $\mathbf{A}$ through the same surface. Let

S_1 = surface bounded by the loop at time $t - dt$
S_2 = surface bounded by the loop at time t

A velocity vector $\mathbf{u}$ is assigned to each element dS. The two surfaces S_1 and S_2 define a volume (see Fig. 4.2):

$$dt \int_{S_1} \mathbf{u} \cdot \mathbf{n} \, dS \qquad (4.3.5)$$

The variation of the flux of $\mathbf{A}$ through the surface S is then equal to (flux through S_2 at time t) — (flux through S_1 at time $t - dt$)

$$\frac{d}{dt} \int A_n \, dS \, \frac{d}{dt} = \left[\int_{S_2} A_{n,t} \, dS_2 - \int_{S_1} A_{n,t-dt} \, dS_1 \right] \qquad (4.3.6)$$

Let us apply Gauss's theorem to the box delimited by S_1 and S_2:

$$\int_S \mathbf{A} \cdot \mathbf{n} \, dS = \int_V \mathbf{\nabla} \cdot \mathbf{A} \, d\tau \qquad (4.3.7)$$

For this box, the perpendicular to S_2 is an external perpendicular, while the one to S_1 is an internal perpendicular. An element of surface on the

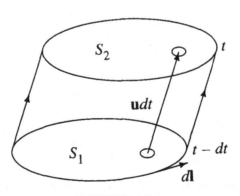

FIGURE 4.2

lateral strip is given by

$$d\mathbf{l} \times \mathbf{u}\, dt \qquad (4.3.8)$$

and

$$d\tau = dS_1 u_n\, dt \qquad (4.3.9)$$

Now, because of Gauss's theorem, we can write

$$\int_{S_2} A_{n,t}\, dS_2 + dt \oint \mathbf{A} \cdot d\mathbf{l} \times \mathbf{u} - \int_{S_1} A_{n,t}\, dS_1$$

$$= dt \int_{S_1} \boldsymbol{\nabla} \cdot \mathbf{A}\, dS_1 u_n \qquad (4.3.10)$$

and

$$\int_{S_2} A_{n,t}\, dS_2 = \int_{S_1} A_{n,t}\, dS_1 + dt \int (\boldsymbol{\nabla} \cdot \mathbf{A}) u_n\, dS_1$$

$$- dt \oint \mathbf{A} \cdot d\mathbf{l} \times \mathbf{u} \qquad (4.3.11)$$

But

$$\int_{S_1} A_{n,t-dt}\, dS_1 = \int_{S_1} A_{n,t}\, dS_1 - \int_{S_1} \left(\frac{\partial A}{\partial t}\right)_n dS_1 dt \qquad (4.3.12)$$

Therefore,

$$\int_{S_2} A_{n,t}\, dS_2 - \int_{S_1} A_{n,t-dt}\, dS_1 = dt \left[\int_{S_1} \dot{A}_n\, dS_1 \right.$$

$$\left. + \int_{S_1} (\boldsymbol{\nabla} \cdot \mathbf{A}) u_n\, dS_1 - \oint \mathbf{A} \cdot d\mathbf{l} \times \mathbf{u}\right] \qquad (4.3.13)$$

But, from Stokes's theorem,

$$\oint \mathbf{A} \cdot d\mathbf{l} \times \mathbf{u} = \oint \mathbf{u} \times \mathbf{A} \cdot d\mathbf{l} = \int \boldsymbol{\nabla} \times (\mathbf{u} \times \mathbf{A}) \cdot \mathbf{n}\, dS_1 \qquad (4.3.14)$$

Therefore,

$$\frac{d}{dt} \int A_n\, dS = \int \dot{\underline{\mathbf{A}}} \cdot \mathbf{n}\, dS$$

$$= \int \dot{A}_n dS_1 + \int (\boldsymbol{\nabla} \cdot \mathbf{A}) u_n\, dS$$

$$- \int \boldsymbol{\nabla} \times (\mathbf{u} \times \mathbf{A}) \cdot \mathbf{n}\, dS \qquad (4.3.15)$$

and

$$\underline{\dot{\mathbf{A}}} = \dot{\underline{\mathbf{A}}} + \mathbf{u}(\nabla \cdot \mathbf{A}) - \nabla \times (\mathbf{u} \times \mathbf{A}) \qquad (4.3.16)$$

Let us apply now these considerations to the $\mathbf{B}$ field:

$$\underline{\dot{\mathbf{B}}} = \dot{\underline{\mathbf{B}}} + \mathbf{u}(\nabla \cdot \mathbf{B}) - \nabla \times (\mathbf{u} \times \mathbf{B}) = \dot{\underline{\mathbf{B}}} - \nabla \times (\mathbf{u} \times \mathbf{B}) \quad (4.3.17)$$

and

$$\frac{d}{dt} \int (\mathbf{B} \cdot \mathbf{u})\, dS = \int \underline{\dot{B}}_n\, dS = \int \frac{\partial \mathbf{B}}{\partial t} \cdot \mathbf{n}\, dS$$

$$- \int [\nabla \times (\mathbf{u} \times \mathbf{B})] \cdot \mathbf{n}\, dS$$

$$= \int \frac{\partial \mathbf{B}}{\partial t} \cdot \mathbf{n}\, dS - \oint (\mathbf{u} \times \mathbf{B}) \cdot d\mathbf{l}$$

$$= \int \frac{\partial \mathbf{B}}{\partial t} \cdot \mathbf{n}\, dS - \oint \mathbf{B} \cdot (d\mathbf{l} \times \mathbf{u}) \qquad (4.3.18)$$

4.4. Differential Formulation of the Law of Induction

Let us consider a loop fixed shape and a field $\mathbf{B}$ that is time-dependent and inhomogeneous. Let us displace the loop through the field $\mathbf{B}$. We have in this case

$$\oint d\mathbf{l} \cdot \mathbf{E} = -\frac{1}{c} \frac{d}{dt} \int dS\, \mathbf{n} \cdot \mathbf{B}$$

$$= -\frac{1}{c} \left[\int dS \left(\mathbf{n} \cdot \frac{\partial \mathbf{B}}{\partial t} \right) - \oint (d\mathbf{l} \times \mathbf{v}) \cdot \mathbf{B} \right]$$

$$= -\frac{1}{c} \left[\int dS \left(\mathbf{n} \cdot \frac{\partial \mathbf{B}}{\partial t} \right) - \oint d\mathbf{l} \cdot (\mathbf{v} \times \mathbf{B}) \right] \qquad (4.4.1)$$

where we have used the relation (4.3.18). The term $\int dS[\mathbf{n} \cdot (\partial \mathbf{B}/\partial t)]$ takes into account the variation of $\mathbf{B}$ with time. The term $\oint d\mathbf{l} \cdot \mathbf{v} \times \mathbf{B}$ takes into account the variation in time of the position of S.

Relation (4.4.1) can be rewritten as follows:

$$\oint d\mathbf{l} \cdot \left(\mathbf{E} - \frac{\mathbf{v}}{c} \times \mathbf{B} \right) = -\frac{1}{c} \int dS \left(\frac{\partial \mathbf{B}}{\partial t} \cdot \mathbf{n} \right) \qquad (4.4.2)$$

Let us consider two simple situations:

(1) $\mathbf{v} = 0$; *loop at rest*: In this case

$$\oint d\mathbf{l} \cdot \mathbf{E} = \int dS\, \mathbf{n} \cdot (\mathbf{\nabla} \times \mathbf{E}) = -\frac{1}{c} \int dS \left(\frac{\partial \mathbf{B}}{\partial t} \cdot \mathbf{n} \right) \qquad (4.4.3)$$

for any kind of loop. The conclusion is that

$$\mathbf{\nabla} \times \mathbf{E} = -\frac{1}{c} \frac{\partial \mathbf{B}}{\partial t} \qquad (4.4.4)$$

This is a differential formulation of the law of induction. We shall examine now its generality.

(2) $\mathbf{B}$ *constant in time, but inhomogeneous*: In this case

$$\frac{\partial \mathbf{B}}{\partial t} = 0, \quad \mathbf{\nabla}|\mathbf{B}| \neq 0, \quad \mathbf{v} \neq 0 \qquad (4.4.5)$$

Then

$$\oint d\mathbf{l} \cdot \left(\mathbf{E} - \frac{\mathbf{v}}{c} \times \mathbf{B} \right) = 0 \qquad (4.4.6)$$

for any loop.

Let us recall the law that gives the force acting on a charge q moving with velocity $\mathbf{v}$ at time t:

$$\mathbf{F} = q \left(\mathbf{E} + \frac{\mathbf{v}}{c} \times \mathbf{B} \right) \qquad (4.4.7)$$

and let us consider two observers: observer 1, who is at rest with respect to the laboratory system, and observer 2, who moves with constant velocity $\mathbf{v}$ with respect to the laboratory system. Observer 1 measures at a certain time t a force acting on a particle of charge q and velocity $\mathbf{v}$ with respect to the laboratory given by

$$\mathbf{F}^{(1)} = q \left(\mathbf{E}^{(1)} + \frac{\mathbf{v}}{c} \times \mathbf{B}^{(1)} \right) \qquad (4.4.8)$$

Observer 2, being at rest with respect to the particle, measures, at the same time, a force

$$\mathbf{F}^{(2)} = q\mathbf{E}^{(2)} \qquad (4.4.9)$$

Since the two observes move with respect to each other with a constant velocity, according to classical (nonrelativistic) mechanics, they must measure the same force:

$$\mathbf{F}^{(1)} = \mathbf{F}^{(2)} \tag{4.4.10}$$

Therefore,

$$\mathbf{E}^{(2)} = \mathbf{E}^{(1)} + \frac{\mathbf{v}}{c} \times \mathbf{B}^{(1)} \tag{4.4.11}$$

We note that observer 2 is not aware of the presence of any $\mathbf{B}$ field. In effect, the values of the fields $\mathbf{E}$ and $\mathbf{B}$ at a certain time depend on the frame of reference of the observer.

Let us now apply these considerations to the case of a loop in motion through a time-independent and inhomogeneous $\mathbf{B}$ field. We have in this case

$$\mathbf{E}^{(1)} = 0 \tag{4.4.12}$$

$$\mathbf{E}^{(2)} = \frac{\mathbf{v}}{c} \times \mathbf{B}^{(1)} \tag{4.4.13}$$

We can see from the preceding relations that the results (4.4.6) has nothing to do with the law of induction, but is simply an expression of the law of transformation of fields.

To illustrate the generality of the differential formulation (4.4.4) of the law of induction, we shall now consider the following example.

EXAMPLE

Let us consider a loop in uniform motion with respect to a magnet and two frames of references, one (frame 1) tied to the loop and the other (frame 2) tied to the magnet (see Fig. 4.3). We shall see how relation (4.4.4) applies in both frames.

For an observer in frame 1, the field $\mathbf{B}$ produced by the magnet varies in time:

$$\frac{\partial \mathbf{B}^{(1)}}{\partial t} \neq 0 \tag{4.4.14}$$

and

$$\oint \mathbf{E}^{(1)} \cdot d\mathbf{l} = -\frac{1}{c} \int dS \frac{\partial \mathbf{B}^{(1)}}{\partial t} \cdot \mathbf{n} \tag{4.4.15}$$

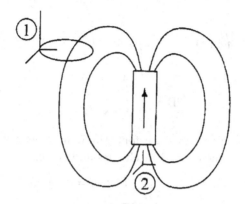

FIGURE 4.3

For an observer in frame 2, the field **B** is independent of time:

$$\frac{\partial \mathbf{B}^{(2)}}{\partial t} = 0 \tag{4.4.16}$$

$$\mathbf{E}^{(2)} = 0 \tag{4.4.17}$$

and

$$\boldsymbol{\nabla} \times \mathbf{E}^{(2)} = -\frac{1}{c}\frac{\partial \mathbf{B}^{(2)}}{\partial t} = 0 \tag{4.4.18}$$

On the other hand, if **v** is the velocity of the loop with respect to the magnet,

$$\mathbf{E}^{(1)} = \mathbf{E}^{(2)} + \frac{\mathbf{v}}{c} \times \mathbf{B}^{(2)} = \frac{\mathbf{v}}{c} \times \mathbf{B}^{(2)} \tag{4.4.19}$$

Then

$$\oint d\mathbf{l} \cdot \mathbf{E}^{(1)} = \int dS\mathbf{n} \cdot (\boldsymbol{\nabla} \times \mathbf{E}^{(1)})$$

$$= \int dS\,\mathbf{n} \cdot \boldsymbol{\nabla} \times \left(\frac{\mathbf{v}}{c} \times \mathbf{B}^{(2)}\right) \tag{4.4.20}$$

But

$$\boldsymbol{\nabla} \times (\mathbf{v} \times \mathbf{B}) = (\mathbf{B} \cdot \boldsymbol{\nabla})\mathbf{v} - (\mathbf{v} \cdot \boldsymbol{\nabla})\mathbf{B} + \mathbf{v}(\boldsymbol{\nabla} \cdot \mathbf{B}) - \mathbf{B}(\boldsymbol{\nabla} \cdot \mathbf{v})$$

$$= -(\mathbf{v} \cdot \boldsymbol{\nabla})\mathbf{B} \tag{4.4.21}$$

and therefore

$$\oint d\mathbf{l} \cdot \mathbf{E}^{(1)} = -\int dS\, \mathbf{n} \cdot \left(\frac{\mathbf{v}}{c} \cdot \boldsymbol{\nabla}\right) \mathbf{B}^{(2)} \tag{4.4.22}$$

Comparing Eq. (4.4.22) with Eq. (4.4.15), we get

$$\frac{\partial \mathbf{B}^{(1)}}{\partial t} = (\mathbf{v} \cdot \boldsymbol{\nabla})\mathbf{B}^{(2)} \tag{4.4.23}$$

We can now write down the four Maxwell equations that relate the fields $\mathbf{E}$ and $\mathbf{B}$ to time-dependent charges and currents:

$$\boldsymbol{\nabla} \cdot \mathbf{E}(\mathbf{x}, t) = 4\pi\rho(\mathbf{x}, t)$$

$$\boldsymbol{\nabla} \times \mathbf{E}(\mathbf{x}, t) + \frac{1}{c}\frac{\partial \mathbf{B}(\mathbf{x}, t)}{\partial t} = 0$$

$$\boldsymbol{\nabla} \cdot \mathbf{B}(\mathbf{x}, t) = 0 \tag{4.4.24}$$

$$\boldsymbol{\nabla} \times \mathbf{B}(\mathbf{x}, t) - \frac{1}{c}\frac{\partial \mathbf{E}(\mathbf{x}, t)}{\partial t} = \frac{4\pi}{c}\mathbf{j}(\mathbf{x}, t)$$

The equation of continuity of charges

$$\boldsymbol{\nabla} \cdot \mathbf{j}(\mathbf{x}, t) = \frac{\partial\rho(\mathbf{x}, t)}{\partial t} \tag{4.4.25}$$

follows from Eqs. (4.4.24)

For completeness, we shall write down the macroscopic version of the Maxwell equations:

$$\boldsymbol{\nabla} \cdot \mathbf{D}(\mathbf{x}, t) = 4\pi\rho_{\text{true}}(\mathbf{x}, t)$$

$$\boldsymbol{\nabla} \times \mathbf{E}(\mathbf{x}, t) + \frac{1}{c}\frac{\partial \mathbf{B}(\mathbf{x}, t)}{\partial t} = 0$$

$$\boldsymbol{\nabla} \cdot \mathbf{B}(\mathbf{x}, t) = 0 \tag{4.4.26}$$

$$\boldsymbol{\nabla} \times \mathbf{H}(\mathbf{x}, t) - \frac{1}{c}\frac{\partial \mathbf{D}(\mathbf{x}, t)}{\partial t} = \frac{4\pi}{c}\mathbf{j}_{\text{true}}(\mathbf{x}, t)$$

4.5. Quasi-Stationary Phenomena

We shall now consider the case in which the currents are *quasi-stationary*, that is, slowly varying with time. In this case the displacement current $\mathbf{j}_d$ can be neglected with respect to the conduction current $\mathbf{j}$, and $\boldsymbol{\nabla} \cdot \mathbf{j} = 0$.

As we shall see later, the introduction of the displacement current $\mathbf{j}_d$ in the Maxwell equations has the consequence of assigning a finite velocity (the velocity of light) to the electromagnetic perturbations; in its absence the results are those that could be obtained in accord with the pre-Maxwellian point of view of the *action at distance*. We can then neglect $\mathbf{j}_d$ when the time interval in which the currents undergo relevant changes is much larger than the time it takes the electromagnetic perturbations to move across the physical system under consideration.

If we assume that these changes are periodical with a frequency ν, the relevant time is the period $T = 1/\nu$, and the condition of *quasi-stationarity* is represented by the relation:

$$\text{size of the physical system} \times \text{frequency} \ll c$$
$$= \text{velocity of light} \tag{4.5.1}$$

EXAMPLE

Transformer

$$\text{size} \simeq 30\,\text{cm}$$
$$\text{frequency} = 60\,\text{s}^{-1}$$
$$\text{size} \times \text{frequency} = 60 \times 30 = 1800 \ll c$$

In a transformer we have quasi-stationary phenomena.

EXAMPLE

Waveguide

$$\text{size} = 1\,\text{cm}$$
$$\text{frequency} = 10^{10}\,\text{s}^{-1}$$
$$\text{size} \times \text{frequency} = 10^{10} \sim c$$

There are no quasi-stationary phenomena in waveguides.

4.6. Self-Inductance and Mutual Inductance

Let us consider two current loops in Fig. 4.4. Let dl_1 and dl_2 be in the same directions as the currents I_1 and I_2, and let

$$r_{12} = |\mathbf{x}_1 - \mathbf{x}_2| \tag{4.6.1}$$

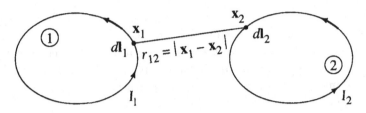

FIGURE 4.4

The current I_1 of loop 1 produces a field at point $\mathbf{x}_2$ given by

$$\mathbf{B}_1(\mathbf{x}_2) = \frac{I_1}{c} \oint d\mathbf{l}_1 \times \nabla_1 \frac{1}{r_{12}} \tag{4.6.2}$$

The flux of $\mathbf{B}_1$ through loop 2 is

$$\psi_2 = \int dS_2 \left(\mathbf{n}_2 \cdot \mathbf{B}_1 \right) = \frac{I_1}{c} \int dS_2 \, \mathbf{n}_2 \cdot \oint d\mathbf{l}_1 \times \nabla_1 \frac{1}{r_{12}}$$

$$= \frac{I_1}{c} \oint d\mathbf{l}_1 \cdot \int dS_2 \, \mathbf{n}_2 \times \nabla_2 \frac{1}{r_{12}} \tag{4.6.3}$$

Because of a variant of the Stokes theorem (see exercise 1.16) express by

$$\oint d\mathbf{l}\, \phi = \int dS\, \mathbf{n} \times \nabla \phi \tag{4.6.4}$$

we can write

$$\psi_2 = \frac{I_1}{c} \oint \oint \frac{d\mathbf{l}_1 \cdot d\mathbf{l}_2}{r_{12}} = c I_1 L_{12} \tag{4.6.5}$$

and

$$V_2 = \oint d\mathbf{l}_2 \cdot \mathbf{E}_2 = -\frac{1}{c} \frac{\partial \Psi_2}{\partial t}$$

$$= -\frac{1}{c^2} \frac{dI_1}{dt} \oint \oint \frac{d\mathbf{I}_1 \cdot d\mathbf{I}_2}{r_{12}} = -\frac{dI_1}{dt} L_{12} \tag{4.6.6}$$

where

$$L_{12} = \frac{1}{c^2} \oint \oint \frac{d\mathbf{l}_1 \cdot d\mathbf{l}_2}{r_{12}} \tag{4.6.7}$$

is called the *coefficient of mutual inductance* and depends solely on the shapes of the loops and their distance apart.

We must take into account another effect on loop 2, an effect due to variation in the time of I_2. We have to add to V_2 a term

$$-\frac{dI_2}{dt} L_{22} \tag{4.6.8}$$

where L_{22} is called the *coefficient of self-inductance*:

$$L_{22} = \frac{1}{c^2} \oint \oint \frac{d\mathbf{l}_2 \cdot d\mathbf{l}_2'}{r_{22}'} \tag{4.6.9}$$

$d\mathbf{l}_2$ and $d\mathbf{l}_2'$ are elements of the same loop 2. If we calculate L_{22} by using formula (4.6.9), we get $L_{22} = \infty$, because we do not take into account the thickness of the wire. We shall obtain later a convenient expression for L_{22}.

For the moment, let us consider a loop powered by a battery that produces a voltage V_B (see Fig. 4.5). Let L be the self-inductance of the loop and R' the *electrical resistance per unit length*, defined by

$$R' = \frac{E_{\text{tang}}}{I} \tag{4.6.10}$$

where

$$E_{\text{tang}} = \frac{\mathbf{E} \cdot d\mathbf{l}}{dl} \tag{4.6.11}$$

We can write

$$V_B = -\int_b^a \mathbf{E} \cdot d\mathbf{l} \quad \begin{array}{l}\text{(the integral is performed across}\\ \text{the battery)}\end{array} \tag{4.6.12}$$

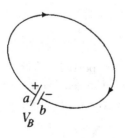

FIGURE 4.5

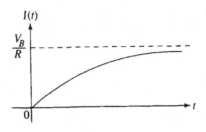

FIGURE 4 6

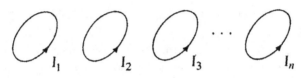

FIGURE 4.7

Then

$$\oint \mathbf{E} \cdot d\mathbf{l} = \int_a^b \mathbf{E} \cdot d\mathbf{l} \quad \text{(along the wire)}$$

$$+ \int_a^b \mathbf{E} \cdot d\mathbf{l} \quad \text{(across the battery)}$$

$$= I \int_a^b R' d\mathbf{l} - V_B = -\frac{dJ}{dt} L \qquad (4.6.13)$$

and

$$RI + L\frac{dJ}{dt} = V_B \qquad (4.6.14)$$

where $R = \int_a^b R' d\mathbf{l}$. This differential equation, if the initial condition is $I(0) = 0$, integrates as follows:

$$I(t) = \frac{V_B}{R}[1 - e^{-(R/L)t}] \qquad (4.6.15)$$

The time evolution of $I(t)$ is reported in Fig. 4.6.

We want now to examine the general case (see Fig. 4.7) of several loops $1,2,\ldots,n$ with currents $I_1, I_2, \ldots, I_n$, respectively. We have for each loop

$$R_n I_n + \sum_m L_{mn} \frac{DI_m}{dt} = V_{Bn} \qquad (4.6.16)$$

Let us multiply each term by I_n and sum over n:

$$\sum_n V_{Bn} I_n - \sum_m \sum_n \frac{1}{2} L_{mn} \frac{dI_m}{dt} I_n - \sum_n R_n I_n^2 = 0 \qquad (4.6.17)$$

or

$$\sum_n V_{Bn} I_n - \frac{d}{dt} \frac{1}{2} \sum_{m,n} L_{m,n} I_m I_n - \sum_n R_n I_n^2 = 0 \qquad (4.6.18)$$

Let us examine the different terms:

1. $\sum_n V_{Bn} I_n$ is the energy extracted from the batteries in the unit time.
2. $\sum_n R_n I_n^2$ is the work done by moving the currents against the resistance in the unit time.
3. $\frac{1}{2}(d/dt) \sum_m \sum_n L_{mn} I_n$ is the variation of the magnetic field energy in the unit time.

We can write

$$\sum_m \sum_n \frac{1}{2} I_m I_n L_{mn} = \frac{1}{2c^2} \sum_m \sum_n I_m I_n \oint \oint \frac{d\mathbf{l}_m \cdot d\mathbf{l}_n}{r_{mn}}$$

$$= \frac{1}{2c^2} \sum_n I_n \oint d\mathbf{l}_n \cdot \sum_m \frac{I_m}{c} \oint \frac{d\mathbf{l}_m}{r_{mn}} \qquad (4.6.19)$$

But

$$\mathbf{A}(\mathbf{x}) = \frac{1}{c} \int d\tau' \frac{\mathbf{j}(\mathbf{x}')}{|\mathbf{x}' - \mathbf{x}|} = \sum_m \frac{I_m}{c} \oint \frac{d\mathbf{l}_m}{|\mathbf{x}_m - \mathbf{x}|} \qquad (4.6.20)$$

Then

$$\sum_m \sum_n \frac{1}{2} I_m I_n L_{mn} = \frac{1}{2c} \sum_n I_n \oint d\mathbf{l}_n \cdot \mathbf{A}(\mathbf{x}_n)$$

$$= \frac{1}{2c} \int d\tau \, \mathbf{j}(\mathbf{x}) \cdot \mathbf{A}(\mathbf{x}) \qquad (4.6.21)$$

$\mathbf{A}(\mathbf{x_n})$ is the vector potential produced by all the currents, including the current I_n. We have, from the fourth equation in (4.4.24), disregarding the term $-(1/c)(\partial \mathbf{E}/\partial t)$

$$\frac{\mathbf{j}}{c} = \frac{1}{4\pi} \nabla \times \mathbf{B} \qquad (4.6.22)$$

Then

$$\frac{1}{2c}\int d\tau \mathbf{j}(\mathbf{x}) \cdot \mathbf{A}(\mathbf{x}) = \frac{1}{8\pi}\int d\tau (\boldsymbol{\nabla} \times \mathbf{B}) \cdot \mathbf{A}$$

$$= \frac{1}{8\pi}\int \left[\left(\frac{\partial B_z}{\partial y} - \frac{\partial B_z}{\partial x} \right) A_x + \left(\frac{\partial B_x}{\partial z} - \frac{\partial B_z}{\partial x} \right) A_y \right.$$

$$\left. + \left(\frac{\partial B_y}{\partial x} - \frac{\partial B_x}{\partial y} \right) Z_z \right] d\tau \overset{IP}{=} -\frac{1}{8\pi}\int \left[\left(B_z \frac{\partial A_x}{\partial y} - B_y \frac{\partial A_x}{\partial z} \right) \right.$$

$$\left. + \left(B_x \frac{\partial A_y}{\partial z} - B_z \frac{\partial A_y}{\partial x} \right) + \left(B_y \frac{\partial A_z}{\partial x} - B_x \frac{\partial A_z}{\partial y} \right) \right] d\tau$$

$$= \frac{1}{8\pi}\int \left[B_x \left(\frac{\partial A_z}{\partial y} - \frac{\partial A_y}{\partial z} \right) + B_y \left(\frac{\partial A_x}{\partial z} - \frac{\partial A_z}{\partial x} \right) \right.$$

$$\left. + B_z \left(\frac{\partial A_y}{\partial x} - \frac{\partial A_x}{\partial y} \right) \right] d\tau = \frac{1}{8\pi}\int [\mathbf{B} \cdot (\boldsymbol{\nabla} \times \mathbf{A})] d\tau$$

$$= \frac{1}{8\pi}\int (\mathbf{B})^2 d\tau \qquad\qquad (4.6.23)$$

Therefore,

$$\frac{1}{2}\sum_m \sum_n I_m I_n L_{mn} = \frac{1}{8\pi}\int (\mathbf{B})^2 d\tau \qquad\qquad (4.6.24)$$

represents the *magnetic field energy*. If we have a single loop, the magnetic field energy is given by

$$\frac{1}{8\pi}\int B^2 d\tau = \frac{1}{2}L I_2 \qquad\qquad (4.6.25)$$

Formula (4.6.25) provides a less ambiguous definition of the coefficient of self-inductance L.

We note that the quantity $\sum_m \sum_n I_m I_n L_{mn}$ is positive and that all the L_{mn} coefficients are positive.

4.7. About Units

Resistance. The work done by currents against the resistance in the unit time is

$$[I^2 R] = \frac{\text{energy}}{\text{time}}$$

Then

$$[R] = \frac{\text{energy}}{I^2 \text{time}}$$

$$\text{ESU} : [R] = \frac{\text{dyne cm}}{[\sqrt{\text{dyne}}(\text{cm/s})]^2 \text{s}} = \frac{\text{s}}{\text{cm}} = \text{statohm}$$

$$\text{ESU} : [R] = \frac{\text{dyne cm}}{(\sqrt{\text{dyne}})^2 \text{s}} = \frac{\text{cm}}{\text{s}} = \text{abohm}$$

$$\text{SI} : [R] = \frac{\text{volt}}{\text{ampere}} = \text{ohm} = \frac{(1/300)\text{statvolt}}{3 \times 10^9 \text{statampere}}$$

$$= \frac{1}{9 \times 10^{11}} \frac{\text{s}}{\text{cm}} (\text{statohm})$$

$$= \frac{10^8 \text{abvolt}}{(1/10)\text{abampere}} = 10^9 \frac{\text{cm}}{\text{s}} (\text{abohm})$$

Inductance

$$\left[\frac{1}{2}LI^2\right] = \text{energy}$$

$$[L] = \frac{\text{energy}}{[I^2]}$$

$$\text{ESU} : [L] = \frac{\text{dyne cm}}{[\sqrt{\text{dyne}}(\text{cm}^2/\text{s}^2)]} = \frac{\text{s}^2}{\text{cm}} = \text{stathenry}$$

$$\text{EMU} : [L] = \frac{\text{dyne cm}}{\text{dyne}} = \text{cm} = \text{abhenry}$$

$$\text{SI} : [L] = \frac{\text{joule}}{(\text{ampere})^2} = \frac{\text{volt second}}{\text{ampere}} = \text{henry}$$

$$= \frac{(1/300) \text{ statvolt second}}{3 \times 10^9 \text{ statampere}} = \frac{1}{9 \times 10^{11}} \frac{\text{s}^2}{\text{cm}} (\text{stathenry})$$

$$= \frac{10^8 \text{ abvolt}}{(1/10) \text{ abampere}} = 10^9 \text{ cm (abhenry)}$$

Law of Induction

$$\oint \mathbf{E} \cdot d\mathbf{l} = -\text{const}\frac{d}{dt}\int_S dS\, \mathbf{n} \cdot \mathbf{B}$$

$$\text{ESU}: \sqrt{\text{dyne}} = \text{const}\frac{1}{T}L^2\frac{1}{c}\text{gauss}$$

$$\sqrt{\text{dyne}} = \text{const}\frac{L^2}{Tc}\frac{\sqrt{\text{dyne}}}{\text{cm}}$$

$$\text{const} = 1$$

$$\oint \mathbf{E} \cdot d\mathbf{l} = -\frac{d}{dt}\int_S dS\, \mathbf{n} \cdot \mathbf{B}$$

$$\text{EMU}: \frac{\sqrt{\text{dyne}}}{\text{second}}\text{cm} = \text{const}\frac{1}{T}L^2\frac{\sqrt{\text{dyne}}}{\text{cm}}$$

$$\text{const} = 1$$

$$\oint y\mathbf{E} \cdot d\mathbf{l} = -\frac{d}{dt}\int_S dS\, \mathbf{n} \cdot \mathbf{B}$$

Gaussian:

$$\sqrt{\text{dyne}} = \text{const}\frac{1}{T}L^2\frac{\sqrt{\text{dyne}}}{\text{cm}}$$

$$\text{const} = \frac{1}{c}$$

$$\oint \mathbf{E} \cdot d\mathbf{l} = -\frac{1}{c}\frac{d}{dt}\int_S dS\mathbf{n} \cdot \mathbf{B}$$

$$\text{SI}: \text{volt} = \text{const}\frac{1}{T}\text{meter}^2\frac{\text{weber}}{\text{meter}^2}$$

$$\text{volt} = \text{const}\frac{1}{\text{second}}\text{meter}^2\frac{\text{volt second}}{\text{meter}^2}$$

$$\text{const} = 1$$

$$\oint \mathbf{E} \cdot d\mathbf{l} = -\frac{d}{dt}\int_S dS\, \mathbf{n} \cdot \mathbf{B}$$

CHAPTER 4 EXERCISES

4.1. A uniform material has an electrical conductivity σ and a dielectric constant K.

(a) Under what circumstances can we neglect the displacement current with respect to the conduction current?

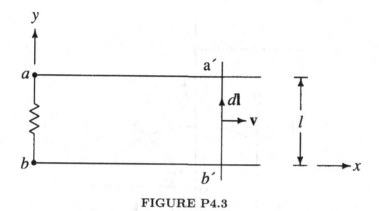

FIGURE P4.3

(b) Assuming that the displacement current is negligible, derive from Maxwell equations the equation satisfied at any point by the current density.

4.2. A straight segment of conducting wire of length l moves with velocity **v** in a uniform field **B**. Let α be the angle between **v** and **B**. Calculated the emf in the wire.

4.3. A conducting stick is made to slide with constant velocity v over a pair of metal wires as in Fig. P4.3. A uniform field **B** is oriented in the z direction, perpendicular to the plane (x, y) of the circuit. Assume that the resistance R that we way connect to the terminals a and b is the only nonnegligible resistance.

   (a) Determine the open-circuit voltage across terminals a and b.
   (b) Show that the electric power dissipated in a resistance R connected to a and b is equal to the mechanical power necessary to keep the stick in motion at constant velocity v.

4.4. Two parallel and infinitely long wires, made of a conducting material, carry currents I_1 and I_2 in the same direction. The distance between them is d. Calculate the force per unit length between the two wires.

4.5. A Faraday disk generator is a metal disk rotated about its axis with a constant angular velocity ω in a uniform field **B** parallel to the axis of rotation (see Fig. P4.5). Let R be the radius of the disk. Calculate the emf between the axis and rim.

4.6. A current I in an N-turn coil produces a flux ϕ in the magnetic circuit of an electromagnet of cross section S (see Fig. P4.6). Calculate the lifting force **F** of the electromagnet on the armature.

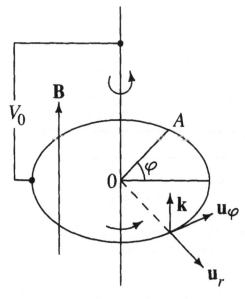

FIGURE P4.5

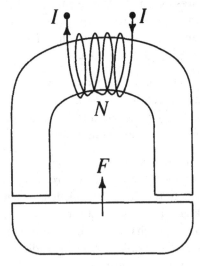

FIGURE P4.6

4.7. Calculate the inductance per unit length of a transmission line that consists of two long parallel conducting wires of radius R that carry currents in opposite directions. Assume that the axes of the wires are separated by a distance $l \gg R$.

4.8. Two circular coils have radii a_1 and a_2, N_1 and N_2 closely wound turns, and carry currents I_1 and I_2, respectively. They are separated by a distance d that is much larger than either radius and have the same symmetry axis. Calculate the force between them.

4.9. (a) Calculate the value of the field B inside a very long solenoid with an air core and n turns per unit length carrying a current I.

(b) Calculate the inductance per unit length of a very long solenoid with an air core and n turns per unit length.

4.10. A circular coil with radius a and N turns of conducting wire lies in the (x, y) plane with its center at the origin. A B field is specified by the relation

$$\mathbf{B} = \mathbf{k}B_0 \cos\left(\frac{\pi r}{2a}\right) \sin \omega t$$

where r is the distance from the z axis, which is the symmetry axis of the coil. Calculate the emf induced in the loop.

4.11. A loop, called a, lies on a horizontal plane and carries a current in the clockwise direction. Another loop, called b, lies on a horizontal plane above that of a. Determine the sense of the emf induced in b when the current in a is interrupted.

4.12. A source produces a voltage

$$V = V_0 \sin \omega t$$

which is connected across a parallel-plate capacitor, as in Fig. P4.12. Verify that the displacement current in the capacitor is equal to the conduction currents in the wires.

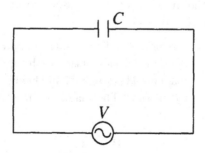

FIGURE P4.12

4.13. A coil with N_1 turns and length l_1 is wound concentrically with a coil of turns N_2 and length l_2 on an empty cylinder core of radius R. Calculate the mutual inductance between the two coils.

4.14. A parallel-plate capacitor has circular plates of radius $r = 0.1\,\mathrm{m}$ and a plate separation of $d = 1.5 \times 10^{-3}\,\mathrm{m}$. It is immersed in oil with dielectric constant $k = 9$ and a specific resistivity $\rho = 8\pi \times 106\,\Omega\text{-m}$. An AC potential of $100\,\mathrm{V}$ rms of frequency $\omega = 400\,\mathrm{s}^{-1}$ is applied across the plates. Calculate the amplitude of the magnetic field H at a point between the plates at a distance $0.02\,\mathrm{m}$ from the axis of the capacitor.

4.15. Calculate the self-inductance per unit length of a coaxial cable when the inner conductor is a solid of radius a and has a uniform current density, while the outer conductor is a shell of infinitesimal thickness at radius b. Take into account the contribution to the self-inductance due to the magnetic field *inside* the inner conductor. Use Gaussian units throughout.

4.16. A charge Q is uniformly distributed on a thin ring-shaped insulator of radius a.

   (a) What are the electric field E are the electric potential V at great distances $r \gg a$ to the lowest order of approximation?

   (b) The ring is rotated around its axis with angular velocity ω. What is the field B at great distances?

4.17. A circular coil forming a closed circuit of n turns of radius a rotates with constant angular velocity ω about a diameter perpendicular to a uniform magnetic field H.

   (a) Derive an expression for the instantaneous current in the coil in terms of ω, the resistance R, and the self-inductance L of the coil.

   (b) What is the average power necessary to maintain the rotation, neglecting the friction?

4.18. A rigid circular wire loop of radius a, mass m, and ohmic resistance R moves under the combined action of the earth's gravitational field and the field of a bar magnet. The gravitational field is uniform and points in the negative z direction. The magnetic induction $\mathbf{B}$ is expressed by the relation

$$\mathbf{B} = -C\frac{\mathbf{r}}{r^3}$$

where C is a constant. The loop moves in such a way that its center always lies on the z axis, and the plane of the loop is always perpendicular to the z axis. What is the resultant vertical force on the loop when its center is at z and its velocity is v? Neglect the self-inductance of the loop.

4.19. The betatron is an electron accelerator that is based on the following principle. Electrons move in a circular orbit of radius r perpendicular to a uniform magnetic induction field B. An additional magnetic induction field parallel to B and localized near the center of the orbit gives rise to a flux that increases at a constant rate; the electrons are accelerated by the resulting induced electromotive force. The field B increases at such a rate that r remains constant.

(a) Find the momentum increase of an electron during one revolution.

(b) Find the relation between the flux and B under the conditions given.

4.20. A charged particle moves in a plane perpendicular to a magnetic induction field $\mathbf{B}$ that is uniform, but varies slowly in time.

(a) Making use of Faraday's induction law in integral form, derive an approximate relation between the magnitude of the induced electromotive force around the orbit, the time derivative of $|\mathbf{B}|$, and the gyration radius R.

(b) Utilizing the answer to Part (a), show that $p^2/|\mathbf{B}|$, where p is the momentum of the particle, remains constant in time.

4.21. A particle carrying a charge q rotates around the origin of coordinates in the (x, y) plane. A homogeneous field B is applied along the z direction and is decreasing with time according to the law

$$\mathbf{B} = \frac{\mathbf{B}_0 t_0}{t + t_0}$$

where t_0 is a positive constant.

(a) Write the equation of motion for the particle in circular cylindrical coordinates r, θ, and z.

(b) Assume that q is negative and prove that the particle's radial position is inversely proportional to the square root of the field strength; that is, $r \propto l/(B^{1/2})$.

(c) Show that the orbital angular momentum of the particle in Part (b) is independent of the B field.

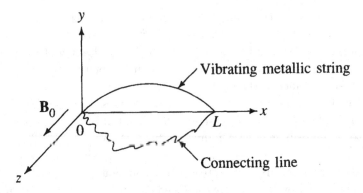

FIGURE P4.22

4.22. The two ends of a metallic string are fixed at points $x = 0$ and $x = L$ (see Fig. P4.22). The string has a linear mass density ρ and is under tension T. These ends are electrically connected by a piece of wire, and the entire circuit has a resistance R. A static constant magnetic induction field **B** is applied in the z direction, and the string is vibrating in the xy plane. Assume that the vibrations are small.

(a) Set up the equation of motion for the vibrating string.

(b) Find the damping, to the lowest order in **B**, for the fundamental mode of the string.

5

General Discussion of Maxwell Equations

5.1. Introduction

The electromagnetic field is produced by charged particles that have masses that are subject to the laws of mechanics. We have then:

(1) The system of charges
(2) The system of masses

We shall show that if we introduce the concepts of *electromagnetic energy* and *electromagnetic momentum* we can formulate conservation laws.

5.2. Field Equations, Forces Acting on Charged Matter, and Conservation Laws

Let us write down the Maxwell equations:

$$\nabla \cdot \mathbf{E} = 4\pi\rho \qquad (5.2.1a)$$

$$\mathbf{\nabla} \times \mathbf{E} + \frac{1}{c}\frac{\partial \mathbf{B}}{\partial t} = 0 \qquad (5.2.1\text{b})$$

$$\mathbf{\nabla} \cdot \mathbf{B} = 0 \qquad (5.2.1\text{c})$$

$$\mathbf{\nabla} \times \mathbf{B} - \frac{1}{c}\frac{\partial \mathbf{E}}{\partial t} = \frac{4\pi}{c}\mathbf{j} \qquad (5.2.1\text{d})$$

If we multiply Eq. (5.2.1b) by **B** and Eq. (5.2.1d) by **E**, we get

$$\mathbf{B} \cdot \mathbf{\nabla} \times \mathbf{E} + \frac{1}{c}\mathbf{B} \cdot \frac{\partial \mathbf{B}}{\partial t} = 0$$

$$\mathbf{E} \cdot \mathbf{\nabla} \times \mathbf{B} - \frac{1}{c}\mathbf{E} \cdot \frac{\partial \mathbf{E}}{\partial t} = \frac{4\pi}{c}\mathbf{j} \cdot \mathbf{E}$$

which can be rewritten

$$\mathbf{B} \cdot \mathbf{\nabla} \times \mathbf{E} + \frac{1}{2c}\frac{\partial B^2}{\partial t} = 0 \qquad (5.2.2)$$

$$\mathbf{E} \cdot \mathbf{\nabla} \times \mathbf{B} - \frac{1}{2c}\frac{\partial E^2}{\partial t} = \frac{4\pi}{c}\mathbf{j} \cdot \mathbf{E} \qquad (5.2.3)$$

Then, subtracting Eq. (5.2.3) from Eq. (5.2.2), we obtain

$$\mathbf{B} \cdot (\mathbf{\nabla} \times \mathbf{E}) - \mathbf{E} \cdot (\mathbf{\nabla} \times \mathbf{B}) + \frac{1}{2c}\frac{\partial B^2}{\partial t} + \frac{1}{2c}\frac{\partial E^2}{\partial t} = -\frac{4\pi}{c}\mathbf{j} \cdot \mathbf{E} \qquad (5.2.4)$$

Multiplying every term by $c/4\pi$, we get

$$\frac{1}{8\pi}\frac{\partial}{\partial t}(B^2 + E^2) + \frac{c}{4\pi}[\mathbf{B} \cdot (\mathbf{\nabla} \times \mathbf{E}) - \mathbf{E} \cdot (\mathbf{\nabla} \times \mathbf{B})] = -\mathbf{j} \cdot \mathbf{E} \qquad (5.2.5)$$

But,

$$\mathbf{\nabla} \cdot (\mathbf{E} \times \mathbf{B}) = \mathbf{B} \cdot (\mathbf{\nabla} \times \mathbf{E}) - \mathbf{E} \cdot (\mathbf{\nabla} \times \mathbf{B}) \qquad (5.2.6)$$

Therefore,

$$\frac{\partial}{\partial t}\left[\frac{1}{8\pi}(E^2 + B^2)\right] + \frac{c}{4\pi}[\mathbf{\nabla} \cdot (\mathbf{E} \times \mathbf{B})] = -\mathbf{j} \cdot \mathbf{E} \qquad (5.2.7)$$

We shall now assume that the quantity in the first pair of brackets in the left member of Eq. (5.2.7) represents the total electromagnetic energy density, even for fields that vary with time:

$$\frac{1}{8\pi}(E^2 + B^2) = \text{electromagnetic energy per unit volume} = U \quad (5.2.8)$$

We also note that

$$\mathbf{j} \cdot \mathbf{E} = \begin{pmatrix} \text{work done per unit time per unit volume} \\ \text{by the electric field on moving charges} \end{pmatrix} = \frac{\partial W}{\partial t} \quad (5.2.9)$$

and set

$$\frac{c}{4\pi}(\mathbf{E} \times \mathbf{B}) = \mathbf{N}, \quad \frac{\text{erg}}{\text{cm}^2\,\text{s}} \quad (5.2.10)$$

We can then write

$$\frac{\partial U}{\partial t} + \boldsymbol{\nabla} \cdot \mathbf{N} = -\frac{\partial W}{\partial t} \quad (5.2.11)$$

Integrating over a certain volume

$$\frac{\partial}{\partial t} \int d\tau (U + W) + \int d\mathbf{S} \cdot \mathbf{N} = 0 \quad (5.2.12)$$

According to relation (5.2.12), the decrease in the unit time of the electromagnetic energy contained in a volume is given by the flux of the vector **N** through the surface surrounding the volume, plus the work done by the electric field on the charges inside the volume. The vector **N** represents the energy crossing the unit surface in the unit time and is called the *Poynting vector*.

We note that, if the volume encompasses the entire (infinite) space, the integral over the surface of **N** is zero, since **N** at ∞ is zero.

Given certain distributions of charges ρ and of currents **j**, the force per unit volume acting on the charges and currents is given by

$$\mathbf{f} = \rho\mathbf{E} + \frac{\mathbf{j}}{c} \times \mathbf{B} \quad (5.2.13)$$

Let us multiply the fourth Maxwell equation [(5.2.1d)] by **B**, and the second Maxwell equation [(5.2.1b)] by $-\mathbf{E}$ in the following manner:

$$(\boldsymbol{\nabla} \times \mathbf{B}) \times \mathbf{B} - \frac{1}{c}\frac{\partial \mathbf{E}}{\partial t} \times \mathbf{B} = \frac{4\pi}{c}\mathbf{j} \times \mathbf{B} \quad (5.2.14)$$

$$-\mathbf{E} \times (\boldsymbol{\nabla} \times \mathbf{E}) - \frac{1}{c}\mathbf{E} \times \frac{\partial \mathbf{B}}{\partial t} = 0 \quad (5.2.15)$$

Summing up

$$(\boldsymbol{\nabla} \times \mathbf{B}) \times \mathbf{B} - \mathbf{E} \times (\boldsymbol{\nabla} \times \mathbf{E}) - \frac{1}{c}\frac{\partial}{\partial t}(\mathbf{E} \times \mathbf{B}) = \frac{4\pi}{c}\mathbf{j} \times \mathbf{B} \qquad (5.2.16)$$

But

$$(\boldsymbol{\nabla} \times \mathbf{B}) \times \mathbf{B} = (\mathbf{B} \cdot \boldsymbol{\nabla})\mathbf{B} - \frac{1}{2}\boldsymbol{\nabla}B^2 \qquad (5.2.17)$$

$$\mathbf{E} \times (\boldsymbol{\nabla} \times \mathbf{E}) = \frac{1}{2}\boldsymbol{\nabla}E^2 - (\mathbf{E} \cdot \boldsymbol{\nabla})\mathbf{E} \qquad (5.2.18)$$

Then

$$\frac{4\pi}{c}\mathbf{j} \times \mathbf{B} = (\mathbf{B} \cdot \boldsymbol{\nabla})\mathbf{B} - \frac{1}{2}\boldsymbol{\nabla}B^2 - \frac{1}{2}\boldsymbol{\nabla}E^2 + (\mathbf{E} \cdot \boldsymbol{\nabla})\mathbf{E}$$
$$- \frac{1}{c}\frac{\partial}{\partial t}(\mathbf{E} \times \mathbf{B}) \qquad (5.2.19)$$

But

$$\mathbf{E}(\boldsymbol{\nabla} \cdot \mathbf{E}) = 4\pi\rho\,\mathbf{E} \qquad (5.2.20)$$

and

$$\rho\mathbf{E} = \frac{1}{4\pi}\mathbf{E}(\boldsymbol{\nabla} \cdot \mathbf{E}) \qquad (5.2.21)$$

We can then write

$$\mathbf{f} = \rho\mathbf{E} + \frac{\mathbf{j}}{c} \times \mathbf{B} = \frac{1}{4\pi}\mathbf{E}(\boldsymbol{\nabla} \cdot \mathbf{E})$$
$$+ \frac{1}{4\pi}\left[(\mathbf{B} \cdot \boldsymbol{\nabla})\mathbf{B} - \frac{1}{2}\boldsymbol{\nabla}B^2 - \frac{1}{2}\boldsymbol{\nabla}E^2\right.$$
$$\left. + (\mathbf{E} \cdot \boldsymbol{\nabla})\mathbf{E} - \frac{1}{c}\frac{\partial}{\partial t}(\mathbf{E} \times \mathbf{B})\right] \qquad (5.2.22)$$

$$\mathbf{f} + \frac{1}{4\pi c}\frac{\partial}{\partial t}(\mathbf{E} \times \mathbf{B}) = \frac{1}{4\pi}\left[(\mathbf{B} \cdot \boldsymbol{\nabla})\mathbf{B} - \frac{1}{2}\boldsymbol{\nabla}B^2\right.$$
$$\left. + (\mathbf{E} \cdot \boldsymbol{\nabla})\mathbf{E} - \frac{1}{2}\boldsymbol{\nabla}E^2 + \mathbf{E}(\boldsymbol{\nabla} \cdot \mathbf{E})\right] \qquad (5.2.23)$$

or

$$\mathbf{f} + \frac{\partial \mathbf{G}}{\partial t} = \frac{1}{4\pi} \left[(\mathbf{B} \cdot \boldsymbol{\nabla}) B - \frac{1}{2} \boldsymbol{\nabla} B^2 \right.$$

$$\left. + (\mathbf{E} \cdot \boldsymbol{\nabla}) \mathbf{E} - \frac{1}{2} \boldsymbol{\nabla} E^2 + \mathbf{E}(\boldsymbol{\nabla} \cdot \mathbf{E}) \right] \tag{5.2.24}$$

where

$$\mathbf{G} = \frac{\mathbf{E} \times \mathbf{B}}{4\pi c} = electromagnetic\ momentum\ density$$

Now

$$f_k + \frac{\partial G_k}{\partial t} = \frac{1}{4\pi} \left[\sum_i B_i \frac{\partial B_k}{\partial x_i} - \frac{1}{2} \sum_i \frac{\partial B_i^2}{\partial x_k} \right.$$

$$\left. + \sum_i E_i \frac{\partial E_k}{\partial x_i} - \frac{1}{2} \sum_i \frac{\partial E_i^2}{\partial x_k} + E_k \sum_i \frac{\partial E_i}{\partial x_i} \right]$$

$$= \frac{1}{4\pi} \sum_i \left[B_i \frac{\partial B_k}{\partial x_i} - \frac{1}{2} \frac{\partial B_i^2}{\partial x_k} \right.$$

$$\left. + E_i \frac{\partial E_k}{\partial x_i} - \frac{1}{2} \frac{\partial E_i^2}{\partial x_k} + E_k \frac{\partial E_i}{\partial x_i} \right] \tag{5.2.25}$$

But

$$\sum_i B_i \frac{\partial B_k}{\partial x_i} = \sum_i \frac{\partial}{\partial x_i} (B_i B_k) \tag{5.2.26}$$

because

$$\sum_i \frac{\partial}{\partial x_i} (B_i B_k) = \left(\sum_i \frac{\partial B_i}{\partial x_i} \right) B_k + \sum_i B_i \frac{\partial B_k}{\partial x_i}$$

$$= \sum_i B_i \frac{\partial B_k}{\partial x_i} \tag{5.2.27}$$

since

$$\boldsymbol{\nabla} \cdot \mathbf{B} = \sum_i \frac{\partial B_i}{\partial x_i} = 0 \tag{5.2.28}$$

Also,

$$E_i \frac{\partial E_k}{\partial x_i} + E_k \frac{\partial E_i}{\partial x_i} = \frac{\partial}{\partial x_i} (E_i E_k) \tag{5.2.29}$$

Therefore,

$$f_k + \frac{\partial G_k}{\partial t} = \frac{1}{4\pi} \sum_i \left[\frac{\partial}{\partial x_i}(B_i B_k) - \frac{1}{2}\frac{\partial B_i^2}{\partial x_k} + \frac{\partial}{\partial x_i}(E_i E_k) - \frac{1}{2}\frac{\partial E_i^2}{\partial x_k} \right]$$

$$= \sum_i \frac{\partial}{\partial x_i} \left[\frac{1}{4\pi} \left(B_i B_k - \frac{1}{2}\delta_{ik}B^2 \right) \right.$$

$$\left. + \frac{1}{4\pi} \left(E_i E_k - \frac{1}{2}\delta_{ik}E^2 \right) \right]$$

$$= \sum_i \frac{\partial}{\partial x_i}[T_{ik}^{\text{magn}} + T_{ik}^{\text{el}}]$$

or

$$f_k + \frac{\partial G_k}{\partial t} = \sum_i \frac{\partial T_{ik}}{\partial x_i} \tag{5.2.30}$$

where

$$T_{ik}^{\text{magn}} = \frac{1}{4\pi} \left(B_i B_k - \frac{1}{2}\delta_{ik}B^2 \right) \tag{5.2.31}$$

$$T_{ik}^{\text{el}} = \frac{1}{4\pi} \left(E_i E_k - \frac{1}{2}\delta_{ik}E^2 \right) \tag{5.2.32}$$

$$T_{ik} = T_{ik}^{\text{magn}} + T_{ik}^{\text{el}} \tag{5.2.33}$$

Integrating Eq. (5.2.30), we obtain

$$\int f_k d\tau + \int \frac{\partial G_k}{\partial t}d\tau = \int \sum_i \frac{\partial T_{ik}}{\partial x_i}d\tau$$

$$= \sum_i \int_S T_{ik}n_i dS \tag{5.2.34a}$$

or

$$\frac{d}{dt}(P_{k\ \text{mech}} + P_{k\ \text{field}}) = \int_S \sum_i T_{ik}n_i dS \tag{5.2.34b}$$

where

$$\mathbf{P}_{\text{mech}} = \text{mechanical momentum in volume } V$$

$$\mathbf{P}_{\text{field}} = \text{electromagnetic momentum in volume } V$$

$$\sum_i T_{ik} n_i = k\text{th component of the force per unit area that is}$$

transmitted across the surface S into the
volume V, acting on the particles
and fields inside S

Equation (5.2.12) expresses the *conservation of energy*. Equation (5.2.34) expresses the *conservation of momentum*.

In dealing with the above, we have not taken into account the possible flow of matter into the volume V.

For completeness we want to deal with the notion of electromagnetic angular momentum. We define the *density of electromagnetic angular momentum* as the quantity

$$\mathbf{r} \times \mathbf{G} = \mathbf{r} \times \frac{1}{4\pi c}[\mathbf{E} \times \mathbf{B}] \tag{5.2.35}$$

The total electromagnetic angular momentum in a volume V is given by the integral

$$\int_V d\tau(\mathbf{r} \times \mathbf{G}) = \int_V d\tau \left[\mathbf{r} \times \frac{1}{4\pi c}(\mathbf{E} \times \mathbf{B}) \right] \tag{5.2.36}$$

In a closed system composed of particles and an electromagnetic field, the sum of the angular momentum of the particles

$$\mathbf{L} = \sum_i (\mathbf{r}_i \times \mathbf{p}_i) \tag{5.2.37}$$

and of the angular momentum of the field (5.2.36) is conserved.

5.3. Conservation Laws for the Macroscopic Case

Let us now consider the macroscopic Maxwell equations:

$$\nabla \cdot \mathbf{D} = 4\pi \rho_{\text{true}} \tag{5.3.1a}$$

$$\nabla \times \mathbf{E} + \frac{1}{c}\frac{\partial \mathbf{B}}{\partial t} = 0 \tag{5.3.1b}$$

$$\nabla \cdot \mathbf{B} = 0 \tag{5.3.1c}$$

$$\nabla \times \mathbf{H} - \frac{1}{c}\frac{\partial \mathbf{D}}{\partial t} = \frac{4\pi}{c}\mathbf{j}_{\text{true}} \tag{5.3.1d}$$

Let us assume the following:

(1) The macroscopic medium in which charges, currents, and fields reside is *linear* in its electric and magnetic properties.
(2) The macroscopic medium is also *homogeneous*; that is, the dielectric constant K and the magnetic permeability μ are constant throughout the system ($\nabla K = 0$ and $\nabla \mu = 0$)
(3) For sinusoidally varying fields, K and μ depend on the frequency of the time variation. We shall assume that either the time variations of the fields are slow enough to allow us to retain for K and μ their static values, or that such variations involve only the frequencies of a small interval in which K and μ are constant. In any case, we shall assume that at any place in the system and at any time

$$\mathbf{B} = \mu \mathbf{H} \tag{5.3.2}$$

$$\mathbf{D} = K\mathbf{E} \tag{5.3.3}$$

If we multiply Eq. (5.3.1b) by $\mathbf{H}$ and Eq. (5.3.1d) by $\mathbf{E}$, we get

$$\mathbf{H} \cdot (\nabla \times \mathbf{E}) + \frac{1}{c}\mathbf{H} \cdot \frac{\partial \mathbf{B}}{\partial t} = 0 \tag{5.3.4}$$

$$\mathbf{E} \cdot (\nabla \times \mathbf{H}) - \frac{1}{c}\mathbf{E} \cdot \frac{\partial \mathbf{D}}{\partial t} = \frac{4\pi}{c}\mathbf{j}_{\text{true}} \cdot \mathbf{E} \tag{5.3.5}$$

and

$$\mathbf{H} \cdot (\nabla \times \mathbf{E}) - \mathbf{E} \cdot (\nabla \times \mathbf{H}) + \frac{1}{c}\mathbf{H} \cdot \frac{\partial \mathbf{B}}{\partial t} + \frac{1}{c}\mathbf{E} \cdot \frac{\partial \mathbf{D}}{\partial t}$$

$$= -\frac{4\pi}{c}\mathbf{j}_{\text{true}} \cdot \mathbf{E} \tag{5.3.6}$$

But

$$\mathbf{H} \cdot \nabla \times \mathbf{E} - \mathbf{E} \cdot (\nabla \times \mathbf{H}) = \nabla \cdot (\mathbf{E} \times \mathbf{H}) \tag{5.3.7}$$

and

$$\frac{c}{4\pi}\nabla \cdot (\mathbf{E} \times \mathbf{H}) = \frac{\mu}{8\pi}\frac{\partial H^2}{\partial t} + \frac{K}{8\pi}\frac{\partial E^2}{\partial t} = -\mathbf{j}_{\text{true}} \cdot \mathbf{E} \tag{5.3.8}$$

or

$$\frac{c}{4\pi}\nabla \cdot (\mathbf{E} \times \mathbf{H}) + \frac{\partial}{\partial t}\left[\frac{1}{8\pi}(\mathbf{B} \cdot \mathbf{H} + \mathbf{E} \cdot \mathbf{D})\right] = -\mathbf{j}_{\text{true}} \cdot \mathbf{E} \tag{5.3.9}$$

Then, in accordance with the treatment of the previous section, we assume

$$\frac{1}{8\pi}(\mathbf{B} \cdot \mathbf{H} + \mathbf{E} \cdot \mathbf{D}) = \text{electromagnetic energy per}$$
$$\text{unit volume} = U \tag{5.3.10}$$

Note that

$$\mathbf{j} \cdot \mathbf{E} = \text{work done per unit time per unit volume by}$$
$$\text{the electric field on moving charges}$$

$$= \frac{dW}{dt} \tag{5.3.11}$$

and set

$$\frac{c}{4\pi}(\mathbf{E} \times \mathbf{H}) = \mathbf{N} = \text{Poynting vector}, \quad \frac{\text{erg}}{\text{cm}^2\,\text{s}} \tag{5.3.12}$$

We can then write

$$\frac{\partial U}{\partial t} + \boldsymbol{\nabla} \cdot \mathbf{N} = -\frac{\partial W}{\partial t} \tag{5.3.13}$$

which can be integrated over a volume as follows:

$$\frac{d}{dt}\int d\tau(U + W) + \int d\mathbf{S} \cdot \mathbf{N} = 0 \tag{5.3.14}$$

Formula (5.3.13) or (5.3.14) expresses the conservation of energy.

The force per unit volume is given by

$$\mathbf{f} = \rho_{\text{true}}\mathbf{E} + \frac{\mathbf{j}_{\text{true}}}{c} \times \mathbf{B} \tag{5.3.15}$$

Let us multiply the fourth Maxwell equation [(5.3.1d)] by $\mathbf{B}$ and the second Maxwell equation [(5.3.1b)] by $-\mathbf{D}$, as follows:

$$(\boldsymbol{\nabla} \times \mathbf{H}) \times \mathbf{B} - \frac{1}{c}\frac{\partial \mathbf{D}}{\partial t} \times \mathbf{B} = \frac{4\pi}{c}\mathbf{j}_{\text{true}} \times \mathbf{B} \tag{5.3.16}$$

$$-\mathbf{D} \times (\boldsymbol{\nabla} \times \mathbf{E}) - \frac{1}{c}\mathbf{D} \times \frac{\partial \mathbf{B}}{\partial t} = 0 \tag{5.3.17}$$

Summing up,

$$(\boldsymbol{\nabla} \times \mathbf{H}) \times \mathbf{B} - \mathbf{D} \times (\boldsymbol{\nabla} \times \mathbf{E}) - \frac{1}{c}\frac{\partial}{\partial t}(\mathbf{D} \times \mathbf{B})$$

$$= \frac{4\pi}{c}\mathbf{j}_{\text{true}} \times \mathbf{B} \tag{5.3.18}$$

But

$$(\boldsymbol{\nabla} \times \mathbf{B}) \times \mathbf{B} = (\mathbf{B} \cdot \boldsymbol{\nabla})\mathbf{B} - \frac{1}{2}\boldsymbol{\nabla}B^2 \qquad (5.3.19)$$

$$\mathbf{E} \times (\boldsymbol{\nabla} \times \mathbf{E}) = \frac{1}{2}\boldsymbol{\nabla}E^2 - (\mathbf{E} \cdot \boldsymbol{\nabla})\mathbf{E} \qquad (5.3.20)$$

Therefore,

$$(\boldsymbol{\nabla} \times \mathbf{H}) \times \mathbf{B} = \frac{1}{\mu}\left[(\mathbf{B} \cdot \boldsymbol{\nabla})\mathbf{B} - \frac{1}{2}\boldsymbol{\nabla}B^2\right] \qquad (5.3.21)$$

$$\mathbf{D} \times (\boldsymbol{\nabla} \times \mathbf{E}) = K\left[\frac{1}{2}\boldsymbol{\nabla}E^2 - (\mathbf{E} \cdot \boldsymbol{\nabla})\mathbf{E}\right] \qquad (5.3.22)$$

and

$$\frac{4\pi}{c}\mathbf{j}_{\text{true}} \times \mathbf{B} = \frac{1}{\mu}\left[(\mathbf{B} \cdot \boldsymbol{\nabla})\mathbf{B} - \frac{1}{2}\boldsymbol{\nabla}B^2\right]$$

$$- K\left[\frac{1}{2}\boldsymbol{\nabla}E^2 - (\mathbf{E} \cdot \boldsymbol{\nabla})\mathbf{E}\right] - \frac{1}{c}\frac{\partial}{\partial t}(\mathbf{D} \times \mathbf{B}) \qquad (5.3.23)$$

Also,

$$\mathbf{E}(\boldsymbol{\nabla} \cdot \mathbf{D}) = 4\pi\rho_{\text{true}}\,\mathbf{E} \qquad (5.3.24)$$

and

$$\rho_{\text{true}}\,\mathbf{E} = \frac{1}{4\pi}\mathbf{E}(\boldsymbol{\nabla} \cdot \mathbf{D}) \qquad (5.3.25)$$

We can then write

$$\mathbf{f} = \rho_{\text{true}}\,\mathbf{E} + \frac{\mathbf{j}_{true}}{c} \times \mathbf{B}$$

$$= \frac{1}{4\pi}\mathbf{E}(\boldsymbol{\nabla} \cdot \mathbf{D}) + \frac{1}{4\pi\mu}\left[(\mathbf{B} \cdot \boldsymbol{\nabla})\mathbf{B} - \frac{1}{2}\boldsymbol{\nabla}B^2\right]$$

$$+ \frac{K}{4\pi}\left[(\mathbf{E} \cdot \boldsymbol{\nabla})\mathbf{E} - \frac{1}{2}\boldsymbol{\nabla}E^2\right] - \frac{1}{4\pi c}\frac{\partial}{\partial t}(\mathbf{D} \times \mathbf{B}) \qquad (5.3.26)$$

or

$$\mathbf{f} + \frac{1}{4\pi c}\frac{\partial}{\partial t}(\mathbf{D} \times \mathbf{B}) = \frac{1}{4\pi\mu}\left[(\mathbf{B} \cdot \boldsymbol{\nabla})\mathbf{B} - \frac{1}{2}\boldsymbol{\nabla}B^2\right]$$

$$+ \frac{K}{4\pi}\left[(\mathbf{E} \cdot \boldsymbol{\nabla})\mathbf{E} - \frac{1}{2}\boldsymbol{\nabla}E^2 + \mathbf{E}(\boldsymbol{\nabla} \cdot \mathbf{E})\right] \qquad (5.3.27)$$

or

$$\mathbf{f} + \frac{\partial \mathbf{G}}{\partial t} = \frac{1}{4\pi\mu} \left[(\mathbf{B} \cdot \boldsymbol{\nabla})\mathbf{B} - \frac{1}{2}\boldsymbol{\nabla}B^2 \right]$$

$$+ \frac{K}{4\pi} \left[(\mathbf{E} \cdot \boldsymbol{\nabla})\mathbf{E} - \frac{1}{2}\boldsymbol{\nabla}E^2 + \mathbf{E}(\boldsymbol{\nabla} \cdot \mathbf{E}) \right] \qquad (5.3.28)$$

where

$$\mathbf{G} = \frac{\mathbf{D} \times \mathbf{B}}{4\pi c} = electromagnetic\ momentum\ density \qquad (5.3.29)$$

The kth component Eq. (5.3.28) is

$$f_k + \frac{\partial G_k}{\partial t} = \frac{1}{4\pi\mu} \left[\sum_i B_i \frac{\partial B_k}{\partial x_i} - \frac{1}{2} \sum_i \frac{\partial B_i^2}{\partial x_k} \right]$$

$$+ \frac{K}{4\pi} \left[\sum_i E_i \frac{\partial E_k}{\partial x_i} - \frac{1}{2} \sum_i \frac{\partial E_i^2}{\partial x_k} + E_k \sum_i \frac{\partial E_i}{\partial x_i} \right] \qquad (5.3.30)$$

But

$$\sum_i B_i \frac{\partial B_k}{\partial x_i} = \sum_i \frac{\partial}{\partial x_i}(B_i B_k) \qquad (5.3.31)$$

$$E_i \frac{\partial E_k}{\partial x_i} + E_k \frac{\partial E_i}{\partial x_i} = \frac{\partial}{\partial x_i}(E_i E_k) \qquad (5.3.32)$$

Therefore,

$$f_k + \frac{\partial G_k}{\partial t} = \frac{1}{4\pi\mu} \sum_i \frac{\partial}{\partial x_i} \left(B_i B_k - \frac{1}{2}\delta_{ik}B^2 \right)$$

$$+ \frac{K}{4\pi} \sum_i \frac{\partial}{\partial x_i} \left(E_i E_k - \frac{1}{2}\delta_{ik}E^2 \right)$$

$$= \sum_i \frac{\partial}{\partial x_i} \left[\frac{1}{4\pi} \left(H_i B_k - \delta_{ik}\frac{\mathbf{B} \cdot \mathbf{H}}{2} \right) \right.$$

$$\left. + \frac{1}{4\pi} \left(E_i D_k - \delta_{ik}\frac{\mathbf{E} \cdot \mathbf{D}}{2} \right) \right]$$

$$= \sum_i \frac{\partial}{\partial x_i}[T_{ik}^{\mathrm{magn}} + T_{ik}^{\mathrm{el}}] \qquad (5.3.33)$$

where

$$T_{ik}^{\mathrm{magn}} = \frac{1}{4\pi} \left(H_i B_k - \delta_{ik} \frac{\mathbf{H} \cdot \mathbf{B}}{2} \right) \tag{5.3.34}$$

$$T_{ik}^{\mathrm{el}} = \frac{1}{4\pi} \left(E_i D_k - \delta_{ik} \frac{\mathbf{E} \cdot \mathbf{D}}{2} \right) \tag{5.3.35}$$

5.4. Energy and Momentum Conservation in General

The energy conservation, already expressed by means of Eqs. (5.2.11) and (5.3.13), can be further generalized as follows:

$$\frac{\partial}{\partial t}(U^{\mathrm{el}} + U^{\mathrm{mech}} + \boldsymbol{\nabla} \cdot (\mathbf{N} + \mathbf{N}^{\mathrm{mech}}) = 0 \tag{5.4.1}$$

where

U^{el} = electromagnetic energy per unit volume

U^{mech} = mechanical energy per unit volume

N = Poynting vector

$\mathbf{N}^{\mathrm{mech}}$ = vector describing a flow of matter inside the volume

The momentum conservation, already expressed by means of Eqs. (5.2.30) and (5.3.33), can be generalized as follows:

$$\frac{\partial}{\partial t}(G_k^{\mathrm{mech}} + G_k^{\mathrm{el}}) = \sum_i \frac{\partial T_{ik}^{\mathrm{total}}}{\partial x_i} \tag{5.4.2}$$

where

$$T_{ik}^{\mathrm{total}} = T_{ik}^{\mathrm{magn}} + T_{ik}^{\mathrm{el}} + T_{ik}^{\mathrm{mech}}$$

and

$$\sum_i T_{ik}^{\mathrm{mech}} n_i = k\text{th component of the mechanical momentum}$$
$$\text{transmitted across the surface } S \text{ into the}$$
$$\text{volume } V \text{ in the unit time}$$

5.5. Complex Field

The use of a complex exponential time dependence for the fields is a convenience that may allow us to express Maxwell equations in simple forms. The physically meaningful fields are the real parts of the complex quantities. The notation

$$\mathbf{E} = \mathbf{E}_0 \, e^{i\omega t} \tag{5.5.1}$$

implies that the field is actually given by

$$\mathbf{E} = \mathbf{E}_0 \cos \omega t \tag{5.5.2}$$

The representation of physical variables by means of complex quantities is based on the understanding that, at the end of the calculations involving these variables, only their real parts will be used. This prescription applies also to the square of a physical variable, for which the meaningful quantity is the square of the real part (which is different from the real part of the square). Therefore, if a field $\mathbf{A}$ is expressed in complex form, in order to obtain the time average of its square value, we have first to find the real part of $\mathbf{A}$, square it, and then average the result over one period. In many cases, such as that of a plane, monochromatic wave, such a calculation is simple. In other circumstances the complex functions are more complicated than $e^{i\omega t}$ and the calculations require more elaborate efforts. However, some useful relations may be found that simplify the task of obtaining the time average of squared monochromatic quantities.

Given a complex vector $\mathbf{A}$ with a time dependence $e^{i\omega t}$, we can write

$$\mathrm{Re}\ \mathbf{A} = \frac{1}{2}(\mathbf{A} + \mathbf{A}^*) \tag{5.5.3}$$

and

$$(\mathrm{Re}\ \mathbf{A})^2 = \frac{1}{4}[(\mathbf{A})^2 + (\mathbf{A}^*)^2 + 2\mathbf{A} \cdot \mathbf{A}^*] \tag{5.5.4}$$

where the first and second terms in the brackets are functions of frequency 2ω and the third term is time independent. Taking the time average of Eq. (5.5.4), we obtain

$$\overline{(\mathrm{Re}\ \mathbf{A})^2} = \frac{1}{2}\mathbf{A} \cdot \mathbf{A}^* \tag{5.5.5}$$

Similarly, given two vectors $\mathbf{A}$ and $\mathbf{B}$ with the same harmonic time dependence, we can write

$$\begin{aligned}
\mathrm{Re}\ \mathbf{A} \cdot \mathrm{Re}\ \mathbf{B} &= \frac{1}{2}(\mathbf{A} + \mathbf{A}^*) \cdot \frac{1}{2}(\mathbf{B} + \mathbf{B}^*) \\
&= \frac{1}{4}(\mathbf{A} \cdot \mathbf{B} + \mathbf{A}^* \cdot \mathbf{B}^* + \mathbf{A} \cdot \mathbf{B}^* + \mathbf{A}^* \cdot \mathbf{B})
\end{aligned} \tag{5.5.6}$$

and taking the time averages,

$$\overline{\text{Re } \mathbf{A} \cdot \text{Re } \mathbf{B}} = \frac{1}{4}(\mathbf{A} \cdot \mathbf{B}^* + \mathbf{A}^* \cdot \mathbf{B}) \tag{5.5.7}$$

But

$$\mathbf{A}^* \cdot \mathbf{B} = (\mathbf{A} \cdot \mathbf{B}^*)^* \tag{5.5.8}$$

Then

$$\mathbf{A} \cdot \mathbf{B}^* + \mathbf{A}^* \cdot \mathbf{B} = \mathbf{A} \cdot \mathbf{B}^* + (\mathbf{A} \cdot \mathbf{B}^*)^*$$
$$= 2\,\text{Re}(\mathbf{A} \cdot \mathbf{B}^*) = 2\,\text{Re}(\mathbf{A}^* \cdot \mathbf{B}) \tag{5.5.9}$$

and

$$\overline{\text{Re } \mathbf{A} \cdot \text{Re } \mathbf{B}} = \frac{1}{2}\,\text{Re}(\mathbf{A} \cdot \mathbf{B}^*) = \frac{1}{2}\text{Re}(\mathbf{A}^* \cdot \mathbf{B}) \tag{5.5.10}$$

Relations (5.5.5) and (5.5.10) are very useful for calculating time averages.

5.6. Electromagnetic Waves in Vacuum and in Continuous Media

The Maxwell equations in the absence of charges and currents ($\rho = j = 0$) become homogeneous equations that may have solutions different from zero. We will not be concerned at this point with the production of electromagnetic waves, but rather with waves that can exist independently from the sources.

The macroscopic Maxwell equations, in the absence of true charges and true currents ($\rho_{\text{true}} = \mathbf{j}_{\text{true}} = 0$) can be written as follows:

$$\nabla \cdot \mathbf{D} = 0 \tag{5.6.1a}$$

$$\nabla \times \mathbf{E} + \frac{1}{c}\frac{\partial \mathbf{B}}{\partial t} = 0 \tag{5.6.1b}$$

$$\nabla \cdot \mathbf{B} = 0 \tag{5.6.1c}$$

$$\nabla \times \mathbf{H} - \frac{1}{c}\frac{\partial \mathbf{D}}{\partial t} = 0 \tag{5.6.1d}$$

We will assume that the macroscopic medium, as in Sec. 5.3, is linear in its electric and magnetic properties and homogeneous ($\nabla K = \nabla \mu = 0$). In addition, since we will be dealing with harmonic monochromatic solutions of the Maxwell equations, we will assume that at any place an at any time

$$\mathbf{B} = \mu\mathbf{H}$$
$$\mathbf{D} = K\mathbf{E}$$

(5.6.2)

Equations (5.6.1b) and (5.6.1d) can be written as

$$\mathbf{\nabla} \times \mathbf{E} = -\frac{1}{c}\frac{\partial \mathbf{B}}{\partial t}$$

(5.6.3)

$$\mathbf{\nabla} \times \mathbf{B} = \frac{\mu K}{c}\frac{\partial \mathbf{E}}{\partial t}$$

(5.6.4)

Therefore,

$$\mathbf{\nabla} \times (\mathbf{\nabla} \times \mathbf{E}) = \mathbf{\nabla}(\mathbf{\nabla} \cdot \mathbf{E}) - \nabla^2\mathbf{E} = -\frac{1}{c}\mathbf{\nabla} \times \dot{\mathbf{B}}$$

(5.6.5)

$$\mathbf{\nabla} \times (\mathbf{\nabla} \times \mathbf{B}) = \mathbf{\nabla}(\mathbf{\nabla} \cdot \mathbf{B}) - \nabla^2\mathbf{B} = -\frac{\mu K}{c}(\mathbf{\nabla} \times \dot{\mathbf{E}})$$

(5.6.6)

and, since $\mathbf{\nabla} \cdot \mathbf{E} = \mathbf{\nabla} \cdot \mathbf{B} = 0$,

$$\nabla^2\mathbf{E} = \frac{1}{c}\mathbf{\nabla} \times \mathbf{B} = \frac{\mu K}{c^2}\frac{\partial^2 \mathbf{E}}{\partial t^2}$$
$$\nabla^2\mathbf{B} = -\frac{\mu K}{c} - (\mathbf{\nabla} \times \dot{\mathbf{E}}) = -\frac{\mu K}{c^2}\frac{\partial^2 \mathbf{B}}{\partial t^2}$$

(5.6.7)

These equations have plane wave solutions of the type

$$\mathbf{E}(x,t) = \mathbf{E}_0\, e^{i\mathbf{k}\cdot\mathbf{x} - i\omega t}$$
$$\mathbf{B}(\mathbf{x},t) = \mathbf{B}_0\, e^{i\mathbf{k}\cdot\mathbf{x} - i\omega t}$$

(5.6.8)

Let us see the consequences of the Eqs. (5.6.1) on $\mathbf{E}$ and $\mathbf{B}$:
    For $\mathbf{\nabla} \cdot \mathbf{E} = 0$,

$$\mathbf{\nabla} \cdot \mathbf{E} = i\mathbf{k} \cdot \mathbf{E} = 0$$

(5.6.9)

This result means that

$$\mathbf{k} \cdot \mathbf{E}_0 = 0$$

(5.6.10)

and thus $\mathbf{E}_0$ is perpendicular to $\mathbf{k}$.

For $\mathbf{\nabla} \cdot \mathbf{B} = 0$,

$$\mathbf{k} \cdot \mathbf{B}_0 = 0 \qquad (5.6.11)$$

$\mathbf{B}_0$ is also perpendicular to $\mathbf{k}$.

For $\mathbf{\nabla} \times \mathbf{E} + \frac{1}{c}\frac{\partial \mathbf{B}}{\partial t} = 0$,

$$i\mathbf{k} \times \mathbf{E} + \frac{1}{c}(-i\omega)\mathbf{B} = 0$$

$$i\mathbf{k} \times \mathbf{E}_0 - \frac{i\omega}{c}\mathbf{B}_0 = 0 \qquad (5.6.12)$$

and

$$\mathbf{k} \times \mathbf{E}_0 = \frac{\omega}{c}\mathbf{B}_0 \qquad (5.6.13)$$

For $\mathbf{\nabla} \times \mathbf{B} - \frac{\mu K}{c}\frac{\partial \mathbf{E}}{\partial t} = 0$,

$$i\mathbf{k} \times \mathbf{B}_0 + i\frac{\mu K \omega}{c}\mathbf{E}_0 = 0 \qquad (5.6.14)$$

$$\mathbf{k} \times \mathbf{B}_0 = -\frac{\mu K \omega}{c}\mathbf{E}_0 \qquad (5.6.15)$$

In summary,

$$\mathbf{k} \cdot \mathbf{E}_0 = \mathbf{k} \cdot \mathbf{B}_0 = 0$$

$$\mathbf{k} \times \mathbf{E}_0 = \frac{\omega}{c}\mathbf{B}_0 \qquad (5.6.16)$$

$$\mathbf{k} \times \mathbf{B}_0 = -\frac{\mu K \omega}{c}\mathbf{E}_0$$

Let us introduce the unit vector in the $\mathbf{k}$ direction:

$$\mathbf{k}_0 = \frac{\mathbf{k}}{|\mathbf{k}|} \qquad (5.6.17)$$

Equations (5.6.16) can be expressed as follows:

$$\mathbf{k}_0 \cdot \mathbf{E}_0 = \mathbf{k}_0 \cdot \mathbf{B}_0 = 0$$

$$\frac{ck}{\omega}(\mathbf{k}_0 \times \mathbf{E}_0) = \mathbf{B}_0 \qquad (5.6.18)$$

$$\frac{ck}{\omega}(\mathbf{k}_0 \times \mathbf{B}_0) = -K\mu\mathbf{E}_0$$

The correct orientations of the vectors $\mathbf{B}$, $\mathbf{E}$ and $\mathbf{k}$ are given in Fig. 5.1.

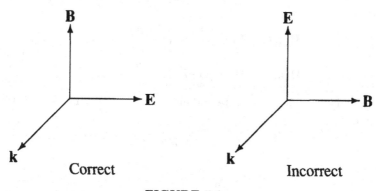

FIGURE 5.1

We set

$$K\mu = n^2 \tag{5.6.19}$$

where $n = $ *index of refraction*. Then

$$\frac{ck}{\omega}|\mathbf{E}_0| = |\mathbf{B}_0|$$

$$\frac{ck}{\omega}|\mathbf{B}_0| = n^2|\mathbf{E}_0| \tag{5.6.20}$$

and

$$\frac{|\mathbf{B}_0|}{|\mathbf{E}_0|} = n^2\frac{|\mathbf{E}_0|}{|\mathbf{B}_0|} \tag{5.6.21}$$

$$\frac{|\mathbf{B}_0|}{|\mathbf{E}_0|} = n \tag{5.6.22}$$

We have then

$$\frac{ck}{\omega} = \frac{|\mathbf{B}_0|}{|\mathbf{E}_0|} = n \tag{5.6.23}$$

or

$$k = \frac{n\omega}{c} \tag{5.6.24}$$

We can write

$$\mathbf{k} = k\mathbf{k}_0 = \frac{n\omega}{c}\mathbf{k}_0 \tag{5.6.25}$$

Then

$$\mathbf{E}(\mathbf{x}, t) = \mathbf{E}_0 \exp\left[-i\omega\left(t - \frac{n}{c}\mathbf{k}_0 \cdot \mathbf{x}\right)\right]$$

$$\mathbf{B}(\mathbf{x}, t) = \mathbf{B}_0 \exp\left[-i\omega\left(t - \frac{n}{c}\mathbf{k}_0 \cdot \mathbf{x}\right)\right]$$

(5.6.26)

These fields exhibit spatial periodicity: In fact, taking $\mathbf{k}_0$ in the x direction,

$$e^{i\mathbf{k}\cdot\mathbf{x}} = e^{ikx} = e^{(in\omega/c)x}$$

(5.6.27)

and for

$$x = n\frac{2\pi}{k}, \quad n \text{ integer}$$

(5.6.28)

we find

$$e^{ikx} = e^{ikn(2\pi/k)}e = e^{in2\pi} = 1$$

(5.6.29)

The *wavelength* of the wave is given by

$$\lambda = \frac{2\pi}{k} = \frac{2\pi c}{n\omega} = \frac{c}{n\nu}$$

(5.6.30)

and the *phase velocity* of the wave by

$$w = \frac{c}{n} = \lambda\nu$$

(5.6.31)

n is generally greater than 1; that is, the phase velocity of a wave in a medium is less than c.

If the space is empty,

$$K\mu = n^2 = 1$$

(5.6.32)

and

$$|\mathbf{B}_0| = |\mathbf{E}_0|$$

(5.6.33)

The Maxwell equations are linear. Expressions for the fields that are linear combinations of two or more solutions of the Maxwell equations are also solutions of these equations. The physical meaning of this mathematical property is that the presence of a wave does not affect the propagation of another wave. This fact is correct according to the classical electromagnetic theory, but is not exactly true in the quantum theory of the electromagnetic field.

In the quantum view, scattering of light by light can take place, in principle. The nonlinearity of the electromagnetic fields responsible for such scattering is due to the uncertainty principle that allows the annihilation of two photons accompanied by the temporary creation of a positron-electron pair, followed by the annihilation of this pair and the creation of two different photons. Nobody has ever observed the scattering of light by light; the difficulty of detecting such an effect is due to the scattering of light by the dust that is present even in the highest vacuum.

We want now to evaluate the energy of a plane monochromatic electromagnetic wave. Assume that this wave travels in the x direction and that the **E** and **H** are pointing in the y and z directions, respectively. Let

$$E_y = f(x - wt) \tag{5.6.34}$$

$$H_z = \frac{B_z}{\mu} = \frac{nE_y}{\mu} = \frac{\sqrt{K\mu}E_y}{\mu}$$

$$= \sqrt{\frac{K}{\mu}}E_y = \sqrt{\frac{K}{\mu}}f(x - wt) \tag{5.6.35}$$

The energy that crosses the unit area of a plane parallel to the yz plane in the unit time is given by the absolute value of the Poynting vector, which is parallel to the x axis:

$$N_x = \frac{c}{4\pi}E_y H_z = \frac{c}{4\pi}\sqrt{\frac{K}{\mu}}f^2 \tag{5.6.36}$$

On the other hand, the energy density of the electromagnetic field is given by

$$u = \frac{1}{8\pi}(KE^2 + \mu H^2) = \frac{1}{8\pi}\left(Kf^2 + \mu\frac{K}{\mu}f^2\right) = \frac{k}{4\pi}f^2 \tag{5.6.37}$$

The electric and magnetic fields contribute equally to the energy. The energy density u and the energy intensity N_x must obey the relation

$$uw = N_x \tag{5.6.38}$$

In fact, using expression (5.6.36) for N_x and Eq. (5.6.37) for u, Eq. (5.6.38) gives us

$$\frac{K}{4\pi}f^2 w = \frac{c}{4\pi}\sqrt{\frac{K}{\mu}}f^2 \tag{5.6.39}$$

or

$$w = \frac{c}{\sqrt{K\mu}} = \frac{c}{n} \tag{5.6.40}$$

Let us consider now the time average of the energy density in its general form:

$$\bar{u} = \frac{1}{8\pi}\overline{(\mathbf{H} \cdot \mathbf{B} + \mathbf{E} \cdot \mathbf{D})} = \frac{1}{2}\frac{1}{8\pi}\mathrm{Re}(\mathbf{H} \cdot \mathbf{B}^* + \mathbf{E} \cdot \mathbf{D}^*)$$

$$= \frac{1}{2}\frac{1}{8\pi}(H_0 B_0 + E_0 D_0) = \frac{1}{2}\left[\frac{1}{8\pi}\left(\frac{1}{\mu}B_0^2 + KE_0^2\right)\right] \quad (5.6.41)$$

$$= \frac{1}{2}\left[\frac{1}{8\pi}\left(\frac{1}{\mu}K\mu E_0^2 + KE_0^2\right)\right] = \frac{1}{8\pi}KE_0^2$$

because

$$B_0^2 = E_0^2 n^2 = E_0^2 K\mu \quad (5.6.42)$$

The time average of the absolute value of the Poynting vector is given by

$$\overline{N} = \left|\frac{c}{4\pi}\frac{1}{2}\mathrm{Re}(\mathbf{E} \times \mathbf{H}^*)\right|$$

$$= \frac{c}{4\pi}\left|\frac{1}{2}\mathrm{Re}\left(\frac{\mathbf{E} \times \mathbf{B}^*}{\mu}\right)\right| = \frac{c}{4\pi}\frac{1}{2}\frac{E_0 B_0}{\mu}$$

$$= \frac{c}{4\pi}\frac{1}{2}\frac{E_0 E_0 \sqrt{K\mu}}{\mu} = \frac{c}{4\pi}\frac{1}{2}\sqrt{\frac{K}{\mu}}E_0^2 \quad (5.6.43)$$

and the velocity of the propagation of the electromagnetic wave is again

$$\frac{\overline{N}}{\bar{u}} = \frac{(c/4\pi)(1/2)\sqrt{\frac{K}{\mu}}E_0^2}{\frac{1}{8\pi}KE_0^2} = \frac{c}{\sqrt{K\mu}} = \frac{c}{n} \quad (5.6.44)$$

5.7. Radiation Pressure

If $\mathbf{N}$ is the Poynting vector,

$$\mathbf{N} = \frac{c}{4\pi}(\mathbf{E} \times \mathbf{H}) \quad (5.7.1)$$

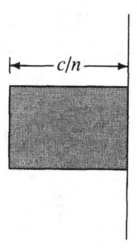

FIGURE 5.2

we can write

$$\frac{\mathbf{N}}{(c/n)^2} = \frac{n^2}{c^2}\frac{c}{4\pi}(\mathbf{E} \times \mathbf{H})$$

$$= \frac{1}{4\pi c}K\mu(\mathbf{E} \times \mathbf{H}) = \frac{1}{4\pi c}(\mathbf{D} \times \mathbf{B}) = \mathbf{G} \qquad (5.7.2)$$

where $\mathbf{G}$ = density of electromagnetic momentum.

Consider now a plane wave encountering a physical surface perpendicular to the direction of its propagation as in Fig. 5.2. If the surface absorbs the wave entirely, all the electromagnetic momentum contained in the cylinder of volume c/n, that is $\mathbf{G}c/n$, will be delivered to the surface in the unit time, and the *radiation pressure* exercised on the surface will be given by Gc/n, where $G = |\mathbf{G}|$. If the surface reflects the plane wave entirely, then the radiation pressure will be $2Gc/n$.

Relation (5.6.38) gives us

$$N = |\mathbf{N}| = u\frac{c}{n} \qquad (5.7.3)$$

Then

$$\frac{N}{(c/n)^2} = G = \frac{u(c/n)}{(c/n)^2} = \frac{u}{c/n} \qquad (5.7.4)$$

Table 5.1.

Microscopic Case	Macroscopic Case
$u = \dfrac{1}{8\pi}(E^2 + B^2)$	$u = \dfrac{1}{8\pi}(\mathbf{E} \cdot \mathbf{D} + \mathbf{B} \cdot \mathbf{H})$
$\mathbf{N} = \dfrac{c}{4\pi}(\mathbf{E} \times \mathbf{B})$	$\mathbf{N} = \dfrac{c}{4\pi}(\mathbf{E} \times \mathbf{H})$
$\mathbf{G} = \dfrac{1}{4\pi c}(\mathbf{E} \times \mathbf{B}) = \dfrac{\mathbf{N}}{c^2}$	$\mathbf{G} = \dfrac{1}{4\pi c}(\mathbf{D} \times \mathbf{B}) = \dfrac{\mathbf{N}}{(c/n)^2}$
$\bar{u} = \dfrac{1}{8\pi}E_0^2$	$\bar{u} = \dfrac{1}{8\pi}KE_0^2$
$\dfrac{\overline{N}}{\bar{u}} = c$	$\dfrac{\overline{N}}{\bar{u}} = \dfrac{c}{n}$
$\overline{G} = \dfrac{\overline{N}}{c^2} = \dfrac{\bar{u}c}{c^2} = \dfrac{\bar{u}}{c}$	$\overline{G} = \dfrac{\overline{N}}{(c/n)^2} = \dfrac{\bar{u}(c/n)^2}{(c/n)^2} = \dfrac{\bar{u}}{c/n}$
$\overline{G}c = $ radiation pressure $= \bar{u}$	Radiation pressure $= \dfrac{\overline{G}c}{n} = \bar{u}$

and the radiation pressure is give by

$$\frac{Gc}{n} = u \tag{5.7.5}$$

Table 5.1 lists some important results. Note that the upper half of the table treats the general case and the lower half refers to plane waves.

EXAMPLE: Sunlight

For this case,

$$\overline{N} = \frac{1 \text{ kw}}{\text{m}^2} = 10^{10}\frac{\text{erg/s}}{10^4 \text{cm}^2} = 10^6 \frac{\text{erg}}{\text{cm}^2\,\text{s}}$$

The pressure exercised by the radiation on the surfaces exposed to it is

$$2cG = \frac{2\overline{N}}{c} = \frac{2 \times 10^6}{3 \times 10^{10}} = \frac{2}{3} \times 10^{-4}\frac{\text{dyne}}{\text{cm}^2}$$

$$= 6.58 \times 10^{-11} \text{ atm}$$

where the extra factor of 2 is due to the fact that light is reflected.

The phenomenon of radiation pressure can also be considered in terms of photons. Consider a number of photons of the same frequency ν traveling in the same direction. All the photons in a volume c/n will strike the unit area perpendicular to their trajectory in the unit time. The number of photons in the volume c/n is given by $(uc/n)/h\nu$. Each photon carries a momentum $h\nu/(c/n)$. The total momentum delivered to the unit area in the unit time is

$$\frac{u(c/n)}{h\nu}\frac{h\nu}{c/n} = u \tag{5.7.6}$$

5.8. Reflection and Refraction of Waves

Assume that, as in Fig. 5.3, we have a wave that is incident on a plane xy in such a way that the $\mathbf{k}$ vector is in the xz plane. This plane separates the medium above the plane, which has an index of refraction n_1, dielectric constant K, and permeability μ from the medium below whose parameters are n_2, K'', and μ''. These parameters are related as follows:

$$n_1 = \sqrt{K\mu} \tag{5.8.1}$$

$$n_2 = \sqrt{K''\mu''} \tag{5.8.2}$$

We shall begin by examining the boundary conditions at the plane xy.

For $\boldsymbol{\nabla} \cdot \mathbf{D} = 0$,

$$D_n^+ = D_n^- \tag{5.8.3}$$

$+$ stands for above and $-$ for below the xy plane.

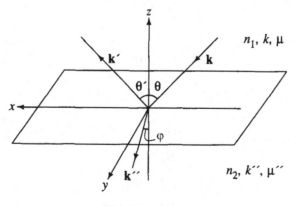

FIGURE 5.3

For $\nabla \cdot \mathbf{B} = 0$,

$$B_n^+ = B_n^- \tag{5.8.4}$$

For $\nabla \times \mathbf{E} + \frac{1}{c}\frac{\partial \mathbf{B}}{\partial t} = 0$,

$$\int (\nabla \times \mathbf{E}) \cdot d\mathbf{S} + \frac{1}{c}\frac{\partial}{\partial t} \int \mathbf{B} \cdot d\mathbf{S}$$
$$= \int (\nabla \times \mathbf{E}) \cdot d\mathbf{S} + \frac{1}{c}\frac{\partial}{\partial t} \int \nabla \cdot \mathbf{B}\, d\tau$$
$$= \int (\nabla \times \mathbf{E}) \cdot d\mathbf{S} = 0$$

Then

$$E_t^+ = E_t^- \tag{5.8.5}$$

For $\nabla \times \mathbf{H} - \frac{1}{c}\frac{\partial \mathbf{D}}{\partial t} = 0$,

$$\int (\nabla \times \mathbf{H}) \cdot d\mathbf{S} - \frac{1}{c}\frac{\partial}{\partial t} \int \mathbf{D} \cdot d\mathbf{S}$$
$$= \int (\nabla \times \mathbf{H}) \cdot d\mathbf{S} - \frac{1}{c}\frac{\partial}{\partial t} \int \nabla \cdot \mathbf{D}\, d\tau$$
$$= \int (\nabla \times \mathbf{H}) \cdot d\mathbf{S} = 0$$

Then

$$H_t^+ = H_t^- \tag{5.8.6}$$

In summary,

$$D_n^+ = D_n^- \tag{5.8.7a}$$
$$B_n^+ = B_n^- \tag{5.8.7b}$$
$$E_t^+ = E_t^- \tag{5.8.7c}$$
$$H_t^+ = H_t^- \tag{5.8.7d}$$

These boundary conditions must be valid *at any time*. We can write

$$\mathbf{E} = \mathbf{E}_0 \exp[i(k_x x + k_z z) - i\omega t]$$

$$= \text{incident wave}, \quad z \geq 0$$

$$\mathbf{E}' = \mathbf{E}_0' \exp[i(k_x' x + k_y' y + k_z' z) - i\omega' t]$$

$$= \text{reflected wave}, \quad z \geq 0 \tag{5.8.8}$$

$$\mathbf{E}'' = \mathbf{E}_0'' \exp[i(k_x'' x + k_y'' y + k_z'' z) - i\omega'' t]$$

$$= \text{refracted wave}, \quad z \leq 0$$

Dependence on Time. The fact that the boundary conditions must be respected at any time imposes

$$\omega = \omega' = \omega'' \tag{5.8.9}$$

Then relation (5.8.8) can be written as follows:

$$\mathbf{E} = \mathbf{E}_0 \exp[i(k_x x + k_z z) - i\omega t]$$

$$\mathbf{E}' = \mathbf{E}_0' = \exp[i(k_x' x + k_y' y + k_z' z) - i\omega t] \tag{5.8.10}$$

$$\mathbf{E}'' = \mathbf{E}_0'' \exp[i(k_x'' x + k_y'' y + k_z'' z) - i\omega t]$$

Dependence on y. To match incident, refracted, and reflected waves at any point along the y axis, we must have

$$a + be^{ik_y' y} = ce^{ik_y'' y} \tag{5.8.11}$$

for all y. This can be true only if

$$k_y' = k_y'' = 0 \tag{5.8.12}$$

Then

$$\mathbf{E} = \mathbf{E}_0 \exp[i(k_x x + k_z z) - i\omega t]$$

$$\mathbf{E}' = \mathbf{E}_0' \exp[i(k_x' x + k_z' z) - i\omega t] \tag{5.8.13}$$

$$\mathbf{E}'' = \mathbf{E}_0'' \exp[i(k_x'' x + k_z'' z) - i\omega t]$$

These results indicate that the reflected and the refracted waves are in the plane of incidence.

Dependence on x. Since we need the boundary conditions respected for all x, we must have

$$k_x' = k_x'' = k_x \tag{5.8.14}$$

or

$$\mathbf{E} = \mathbf{E}_0 \exp[i(k_x x + k_z z) - i\omega t]$$
$$\mathbf{E}' = \mathbf{E}_0' \exp[i(k_x x + k_z' z) - i\omega t] \qquad (5.8.15)$$
$$\mathbf{E}'' = \mathbf{E}_0'' \exp[i(k_x x + k_z'' z) - i\omega t]$$

We have

$$k^2 = k'^2 = \frac{n_1^2 \omega^2}{c^2} \qquad (5.8.16)$$

$$k''^2 = n_2^2 \frac{\omega^2}{c^2} = \left(\frac{\omega^2}{c^2} n_1^2\right) \frac{n_2^2}{n_1^2} = k^2 \frac{n_2^2}{n_1^2} \qquad (5.8.17)$$

Reflected Wave.

$$k_x^2 + k_z^2 = k_x'^2 + k_z'^2 = k_x^2 + k_x'^2 \qquad (5.8.18)$$

or

$$k_z'^2 = k_z^2 \qquad (5.8.19)$$

$$k_z' = \pm k_z \qquad (5.8.20)$$

We choose the minus sign:

$$k_z' = -k_z \qquad (5.8.21)$$

This result, together with the condition $k_x' = k_x$, implies

$$\theta = \theta' \qquad (5.8.22)$$

Refracted Wave. Taking advantage of Fig. 5.4, we can write

$$k^2 = \frac{k_x^2}{\sin^2 \theta} \qquad (5.8.23)$$

$$k''^2 = \frac{k_x''^2}{\sin^2 \varphi''} = \frac{k_x^2}{\sin^2 \varphi''} = k^2 \frac{n_2^2}{n_1^2} \qquad (5.8.24)$$

Therefore,

$$k^2 = \frac{k_x^2}{\sin^2 \theta} = \frac{k_x^2}{\sin^2 \varphi''} = \frac{n_1^2}{n_2^2} \qquad (5.8.25)$$

or

$$\frac{\sin \theta}{\sin \varphi''} = \frac{n_2}{n_1} \qquad (5.8.26)$$

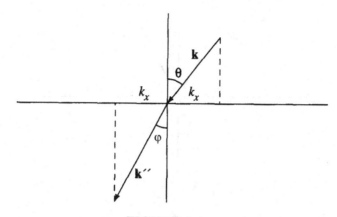

FIGURE 5.4

Let us now consider in detail the question of polarization of waves. We shall begin by considering the case in Fig. 5.5, where the incoming wave is such that

$$\mathbf{E} \parallel \text{ plane of incidence}$$

$$\mathbf{B} \parallel y \text{ axis} \tag{5.8.27}$$

Boundary condition (5.8.7c) gives

$$E_0 \cos \theta - E_0' \cos \theta = E_0'' \cos \varphi \tag{5.8.28}$$

Boundary condition (5.8.7d) gives

$$\frac{1}{\mu} B_0 + \frac{1}{\mu} B_0' = \frac{1}{\mu''} B_0''$$

$$\sqrt{\frac{K}{\mu}} (E_0 + E_0') = \sqrt{\frac{K''}{\mu''}} E_0'' \tag{5.8.29}$$

Boundary condition (5.8.7b) is automatically fulfilled because

$$B_n^+ = B_n^- = 0 \tag{5.8.30}$$

That is, no normal component of **B** is present in any of the three waves. Finally, boundary condition (5.8.7a) gives

$$D_n^+ = D_n^- \tag{5.8.31}$$

or

$$K(E_0 + E_0') \sin \theta = K''(E_0'' \sin r)$$

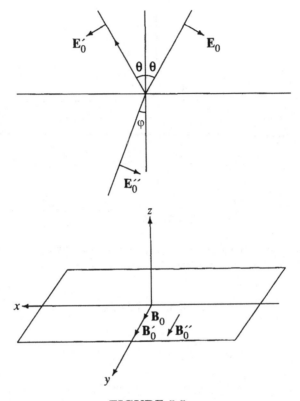

FIGURE 5.5

We obtain then

$$(E_0 + E_0')K\frac{\sin\theta}{\sin\varphi} = K''E_0''$$

$$(E_0 + E_0')K\frac{\sqrt{K''\mu''}}{\sqrt{K\mu}} = K''E_0''$$

and

$$(E_0 + E_0')\sqrt{\frac{K}{\mu}} = E_0''\sqrt{\frac{K''}{\mu''}} \qquad (5.8.32)$$

which has already been obtained in Eq. (5.8.29).

In summary, the boundary conditions have given us

$$(E_0 - E_0') \cos \theta = E_0'' \cos \varphi$$

$$\sqrt{\frac{K}{\mu}}(E_0 + E_0') = \sqrt{\frac{K''}{\mu''}} E_0'' \tag{5.8.33}$$

Then

$$E_0'' = \sqrt{\frac{K\mu''}{K''\mu}}(E_0 + E_0') \tag{5.8.34}$$

$$(E_0 - E_0') \cos \theta = \sqrt{\frac{K\mu''}{K''\mu}}(E_0 + E_0') \cos \varphi \tag{5.8.35}$$

which gives

$$\frac{E_0'}{E_0} = \frac{(\mu/\mu'') \sin 2\theta - \sin 2\varphi}{(\mu/\mu'') \sin 2\theta + \sin 2\varphi} \tag{5.8.36}$$

Also

$$E_0'' = \sqrt{\frac{K\mu''}{K''\mu}} E_0 \left(1 + \frac{E_0'}{E_0}\right)$$

$$= \sqrt{\frac{K\mu''}{K''\mu}} E_0 \frac{(2\mu/\mu'') \sin 2\theta}{(\mu/\mu'') \sin 2\theta + \sin 2\varphi}$$

$$= \sqrt{\frac{K\mu}{K''\mu''}} E_0 \frac{2 \sin 2\theta}{(\mu/\mu'') \sin 2\theta + \sin 2\varphi} \tag{5.8.37}$$

In summary, for **E** parallel to the plane of incidence we have

$$\frac{E_0'}{E_0} = \frac{(\mu/\mu'') \sin 2\theta - \sin 2\varphi}{(\mu/\mu'') \sin 2\theta + \sin 2\varphi}$$

$$\frac{E_0''}{E_0} = 2\sqrt{\frac{K\mu}{K''\mu''}} \frac{\sin 2\theta}{(\mu/\mu'') \sin 2\theta + \sin 2\varphi} \tag{5.8.38}$$

These formulas for $\mu = \mu''$ become

$$\frac{E_0'}{E_0} = \frac{\sin 2\theta - \sin 2\varphi}{\sin 2\theta + \sin 2\varphi} = \frac{\tan(\theta - \varphi)}{\tan(\theta + \varphi)}$$

$$\frac{E_0''}{E_0} = 2\sqrt{\frac{K}{K''}} \frac{\sin 2\theta}{\sin 2\theta + \sin 2\varphi} = \frac{2 \sin \varphi \cos \theta}{\sin(\theta + \varphi) \cos(\theta - \varphi)} \tag{5.8.39}$$

For normal incidence $\theta = 0$, $r = 0$, and formulas (5.8.39) become

$$\frac{E_0'}{E_0} = \frac{\sin 2\theta - \sin 2\varphi}{\sin 2\theta + \sin 2\varphi} = \frac{\sin i \cos i - \sin r \cos r}{\sin \theta \cos \theta + \sin \varphi \cos \varphi}$$

$$\rightarrow \frac{(\sin \theta / \sin \varphi) - 1}{(\sin \theta / \sin \varphi) + 1} \bigg|_{\substack{\theta \to 0 \\ \varphi \to 0}} = \frac{n_2 - n_1}{n_2 + n_1} \tag{5.8.40}$$

$$\frac{E_0''}{E_0} = 2\sqrt{\frac{K}{K''}} \frac{\sin 2\theta}{\sin 2\theta + \sin 2\varphi} = 2\frac{\sin \varphi}{\sin \theta} \frac{\sin \theta \cos \theta}{\sin \theta \cos \theta + \sin \varphi \cos \varphi}$$

$$\times 2\frac{\sin \varphi}{\sin i} \frac{\sin \theta}{\sin \theta + \sin \varphi} \bigg|_{\substack{\theta \to 0 \\ \varphi \to 0}} = 2\frac{\sin \varphi}{\sin \theta} \frac{(\sin \theta / \sin \varphi)}{(\sin \theta / \sin \varphi) + 1}$$

$$= 2\frac{n_1}{n_2} \frac{(n_2/n_1)}{(n_2/n_1) + 1} = \frac{2n_1}{n_1 + n_2} \tag{5.8.41}$$

In summary, for $\theta = 0$,

$$\frac{E_0'}{E_0} = \frac{n_2 - n_1}{n_2 + n_1}$$

$$\frac{E_0''}{E_0} = \frac{2n_1}{n_1 + n_2} \tag{5.8.42}$$

Let us consider again relations (5.8.39). If we set $i = i_B$, where

$$\theta_B + \varphi = \frac{\pi}{2} \tag{5.8.43}$$

then

$$\tan(\theta_B + \varphi) = \infty \tag{5.8.44}$$

and

$$E_0' = 0 \tag{5.8.45}$$

Thus, there is no reflected wave. Also, in this case

$$\frac{\sin \theta_B}{\sin \varphi} = \frac{\sin \theta_B}{\sin[(\pi/2) - \theta_B]} = \frac{\sin \theta_B}{\cos \theta_B} = \tan \theta_B = \frac{n_2}{n_1} \tag{5.8.46}$$

This relation defines the angle i_B, which is called the *Brewster angle*. For a typical ratio $n2/n1 = 1.5$, $i_B = 56°$.

If a plane were of mixed polarization is incident on a plane interface with an angle of incidence θ_B, the reflected radiation is completely plane polarized with **E** perpendicular to the plane of incidence. Even if the unpolarized wave is reflected at angles other than θ_B, there is a tendency for the reflected wave to be predominantly polarized perpendicular to the plane of incidence.

Let us now consider the case in Fig. 5.6, where the incoming is such that

$$\mathbf{E} \perp \text{ plane of incidence}$$
$$\mathbf{B} \parallel \text{ plane of incidence}$$

(5.8.47)

Boundary condition (5.8.7c) gives

$$E_0 + E_0' = E_0''$$

(5.8.48)

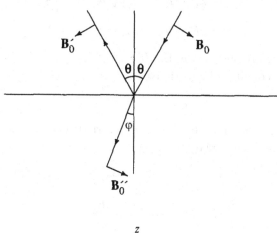

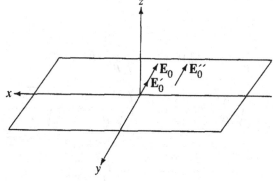

FIGURE 5.6

Boundary condition (5.8.7d) gives

$$\frac{1}{\mu}B_0 \cos\theta - \frac{1}{\mu}B_0' \cos\theta = \frac{1}{\mu''}B_0'' \cos\varphi \qquad (5.8.49)$$

But $B_0 = E_0\sqrt{K\mu}$. Then

$$\sqrt{\frac{K}{\mu}}(E_0 - E_0') - \sqrt{\frac{K''}{\mu''}}E_0' \cos\varphi = 0 \qquad (5.8.50)$$

Boundary condition (5.8.7b) gives

$$B_0 \sin\theta + B_0' \sin\theta = B_0'' \sin\varphi$$

or

$$(E_0 + E_0')\sqrt{K\mu} = E_0''\sqrt{K''\mu''}\frac{\sin\varphi}{\sin\theta}$$

$$E_0 + E_0' = E_0''$$

a condition already obtained in Eq. (5.8.48).

Boundary condition (5.8.7a) gives

$$KE_n^+ = K''E_n^- \qquad (5.8.51)$$

This condition is observed because the electric fields have no components along the z axis. In summary, we have

$$E_0 + E_0' = E_0''$$

$$\sqrt{\frac{K}{\mu}}(E_0 - E_0')\cos\theta = \sqrt{\frac{K''}{\mu''}}E_0'' \cos\varphi \qquad (5.8.52)$$

Then

$$\sqrt{\frac{K}{\mu}}(E_0 - E_0')\cos\theta = \sqrt{\frac{K''}{\mu''}}(E_0 + E_0')\cos\varphi$$

$$\left(\sqrt{\frac{K}{\mu}}\cos\theta - \sqrt{\frac{K''}{\mu''}}\cos\varphi\right)E_0 = \left(\sqrt{\frac{K}{\mu}}\cos\theta + \sqrt{\frac{K''}{\mu''}}\cos\varphi\right)E_0'$$

$$\frac{E_0'}{E_0} = \frac{\sqrt{\frac{K}{\mu}}\cos\theta - \sqrt{\frac{K''}{\mu''}}\cos\varphi}{\sqrt{\frac{K}{\mu}}\cos\theta + \sqrt{\frac{K''}{\mu''}}\cos\varphi} = \frac{1 - \dfrac{\mu}{\mu''}\dfrac{\tan\theta}{\tan\varphi}}{1 + \dfrac{\mu}{\mu''}\dfrac{\tan\theta}{\tan\varphi}} \qquad (5.8.53)$$

On the other hand,

$$E_0'' = E_0 + E_0' = E_0 \left(1 + \frac{E_0'}{E_0} \right) = E_0 \frac{2}{1 + \dfrac{\mu}{\mu''} \dfrac{\tan \theta}{\tan \varphi}} \qquad (5.8.54)$$

In summary, for **E** perpendicular to the plane of incidence,

$$\frac{E_0'}{E_0} = \frac{1 - (\mu/\mu'')(\tan \theta / \tan \varphi)}{1 + (\mu/\mu'')(\tan \theta / \tan \varphi)}$$

$$\frac{E_0''}{E_0} = \frac{2}{1 + (\mu/\mu'')(\tan \theta / \tan \varphi)} \qquad (5.8.55)$$

These formulas for $\mu = \mu''$ become

$$\frac{E_0'}{E_0} = \frac{1 - (\tan \theta / \tan \varphi)}{1 + (\tan \theta / \tan \varphi)} = -\frac{\sin(\theta - \varphi)}{\sin(\theta + \varphi)}$$

$$\frac{E_0''}{E_0} = \frac{2}{1 + (\tan \theta / \tan \varphi)} = -\frac{2 \cos \theta \sin \varphi}{\sin(\theta + \varphi)} \qquad (5.8.56)$$

For normal incidence ($i = r = 0$), formulas (5.8.56) become

$$\frac{E_0'}{E_0} = \frac{1 - \dfrac{\sin \theta \cos \varphi}{\cos \theta \sin \varphi}}{1 + \dfrac{\sin \theta \cos \varphi}{\cos \theta \sin \varphi}} \xrightarrow[\varphi \to 0]{\theta \to 0} \frac{1 - \dfrac{\sin \theta}{\sin \varphi}}{1 + \dfrac{\sin \theta}{\sin \varphi}} = \frac{1 - \sqrt{\dfrac{K''}{K}}}{1 + \sqrt{\dfrac{K''}{K}}}$$

$$= \frac{1 - (n_2/n_1)}{1 + (n_2/n_1)} = \frac{n_1 - n_2}{n_1 + n_2} \qquad (5.8.57)$$

$$\frac{E_0''}{E_0} = \frac{2}{1 + \dfrac{\sin \theta}{\sin \varphi}} = \frac{2}{1 + (n_2/n_1)} = \frac{2n_1}{n_1 + n_2} \qquad (5.8.58)$$

or

$$\frac{E_0'}{E_0} = \frac{n_1 - n_2}{n_1 + n_2}$$

$$\frac{E_0''}{E_0} = \frac{2n_1}{n_1 + n_2} \qquad (5.8.59)$$

5.9. Electromagnetic Waves in a Conducting Medium

We shall now consider the problem of the existence and propagation of electromagnetic waves in a medium that is characterized by a dielectric constant K and a magnetic permeability μ and in addition presents the property that at any point the current density is proportional to the electric field:

$$\mathbf{j}_{\text{true}} = \sigma \mathbf{E} \qquad (5.9.1)$$

This relation is known as *Ohm's law* and applies in general to solid homogeneous materials; it has also been found valid in a wide frequency range for conducting liquids and dense ionized gases. The parameter a is called *electrical conductivity* and has the dimensions s^{-1} in the Gaussian system of units.

When we take Ohm's law into account, the Maxwell equations are written as follows:

$$\nabla \cdot K\mathbf{E} = 0$$

$$\nabla \cdot \mu\mathbf{H} = 0$$

$$\nabla \times \mathbf{E} + \frac{\mu}{c}\frac{\partial \mathbf{H}}{\partial t} = 0 \qquad (5.9.2)$$

$$\nabla \times \mathbf{H} - \frac{K}{c}\frac{\partial \mathbf{E}}{\partial t} = \frac{4\pi\sigma}{c}\mathbf{E}$$

If the conductivity is zero, as it is in the case for the nonconducting dielectric materials examined in Sec. 5.6, the divergences of the timevarying fields $\mathbf{E}$ and $\mathbf{H}$ are zero and the fields are *transverse*, that is, perpendicular to the direction of the vector $\mathbf{k}$ in which the spatial variations occur.

On the other hand, our previous treatments of stationary fields indicate that in nonconducting dielectric materials the curls of such fields are zero and the fields are *longitudinal*; that is, they point in the direction of the spatial variation.

In the case we are presently considering, the electrical conductivity is not zero and the fields vary with time. If for simplicity we assume that the spatial variations of the fields occur only in correspondence to one variable, say x, then the fields can be expressed as follows:

$$\mathbf{E}(x,t) = \mathbf{E}_{\text{long}}(x,t) + \mathbf{E}_{\text{tr}}(x,t)$$
$$\mathbf{H}(x,t) = \mathbf{H}_{\text{long}}(x,t) + \mathbf{H}_{\text{tr}}(x,t) \qquad (5.9.3)$$

where the subscripts long and tr represent longitudinal and transverse fields, respectively. We can then write

$$\nabla \cdot K(\mathbf{E}_{\text{long}} + \mathbf{E}_{\text{tr}}) = 0$$

$$\nabla \cdot \mu(\mathbf{H}_{\text{long}} + \mathbf{H}_{\text{tr}}) = 0$$

$$\nabla \times (\mathbf{E}_{\text{long}} + \mathbf{E}_{\text{tr}}) + \frac{\mu}{c} \frac{\partial}{\partial t}(\mathbf{H}_{\text{long}} + \mathbf{H}_{\text{tr}}) = 0 \tag{5.9.4}$$

$$\nabla \times (\mathbf{H}_{\text{long}} + \mathbf{H}_{\text{tr}}) - \frac{K}{c} \frac{\partial}{\partial t}(\mathbf{E}_{\text{long}} + \mathbf{E}_{\text{tr}})$$

$$- \frac{4\pi\sigma}{c}(\mathbf{E}_{\text{long}} + \mathbf{E}_{\text{tr}}) = 0$$

But

$$\nabla \cdot \mathbf{E}_{\text{tr}} = 0$$

$$\nabla \cdot \mathbf{H}_{\text{tr}} = 0$$

$$\nabla \times \mathbf{E}_{\text{long}} = 0 \tag{5.9.5}$$

$$\nabla \times \mathbf{H}_{\text{long}} = 0$$

Therefore, the Maxwell equations can be written as follows:

$$\nabla \cdot \mathbf{E}_{\text{long}} = \frac{\partial \mathbf{E}_{\text{long}}}{\partial x} = 0$$

$$\nabla \cdot \mathbf{H}_{\text{long}} = \frac{\partial \mathbf{H}_{\text{long}}}{\partial x} = 0$$

$$\nabla \times \mathbf{E}_{\text{tr}} + \frac{\mu}{c} \frac{\partial}{\partial t}\mathbf{H}_{\text{tr}} + \frac{\mu}{c} \frac{\partial}{\partial t}\mathbf{H}_{\text{long}} = 0 \tag{5.9.6}$$

$$\nabla \times \mathbf{H}_{\text{tr}} - \left(\frac{K}{c} \frac{\partial}{\partial t} + \frac{4\pi\sigma}{c}\right)\mathbf{E}_{\text{tr}}$$

$$- \left(\frac{K}{c} \frac{\partial}{\partial t} + \frac{4\pi\sigma}{c}\right)\mathbf{E}_{\text{long}} = 0$$

Therefore,

$$\frac{\partial \mathbf{H}_{\text{long}}}{\partial t} = 0$$

$$\left(\frac{K}{c} \frac{\partial}{\partial t} + \frac{4\pi\sigma}{c}\right)\mathbf{E}_{\text{long}} = 0 \tag{5.9.7}$$

We have for $\mathbf{E}_{\text{long}}$

$$\frac{\partial \mathbf{E}_{\text{long}}}{\partial x} = 0$$

$$\frac{\partial \mathbf{E}_{\text{long}}}{\partial t} + \frac{4\pi\sigma}{K} \mathbf{E}_{\text{long}} = 0 \tag{5.9.8}$$

and for $\mathbf{H}_{\text{long}}$

$$\frac{\partial \mathbf{H}_{\text{long}}}{\partial x} - \frac{\partial \mathbf{H}_{\text{long}}}{\partial t} = 0 \tag{5.9.9}$$

Therefore, longitudinal magnetic fields are time independent and uniform.

The longitudinal electric field is uniform in space, but has a time variation

$$\mathbf{E}_{\text{long}}(\mathbf{x}, t) = \mathbf{E}_0 \, e^{-(4\pi\sigma/K)t} \tag{5.9.10}$$

This means that no stationary longitudinal electric field can exit if there is no applied current density. The perturbation (5.9.10) is damped in a very short time if σ is high ($\sigma \simeq 10^{17}\,\text{s}^{-1}$ for copper).

Let us consider now the transverse fields. Equations (5.9.6) give

$$\boldsymbol{\nabla} \times \mathbf{E}_{\text{tr}} + \frac{1}{c}\frac{\partial}{\partial t}\mathbf{B}_{\text{tr}} = 0$$

$$\frac{1}{\mu}\boldsymbol{\nabla} \times \mathbf{B}_{\text{tr}} - \frac{K}{c}\frac{\partial}{\partial t}\mathbf{E}_{\text{tr}} - \frac{4\pi\sigma}{c}\mathbf{E}_{\text{tr}} = 0 \tag{5.9.11}$$

Assume

$$\mathbf{E} = \mathbf{E}_0 \, e^{i(\mathbf{k}\cdot\mathbf{x} - \omega t)}$$

$$\mathbf{H} = \mathbf{H}_0 \, e^{i(\mathbf{k}\cdot\mathbf{x} - \omega t)} \tag{5.9.12}$$

where we have dropped the subscript tr. Then, using the expressions (5.9.12) in (5.9.11), we get

$$i\mathbf{k} \times \mathbf{E}_0 - \frac{i\omega}{c}\mathbf{B}_0 = 0$$

$$\frac{1}{\mu}i\mathbf{k} \times \mathbf{B}_0 + \frac{K}{c}i\omega\mathbf{E}_0 = \frac{4\pi\sigma}{c}\mathbf{E}_0 \tag{5.9.13}$$

or

$$\mathbf{k} \times \mathbf{E}_0 = \frac{\omega}{c}\mathbf{B}_0 \tag{5.9.14a}$$

$$\mathbf{k} \times \mathbf{B}_0 + \frac{\mu K}{c}\omega\mathbf{E}_0 = \frac{4\pi\sigma\mu}{ic}\mathbf{E}_0 \tag{5.9.14b}$$

From Eqs. (5.9.14a),

$$\mathbf{k} \times (\mathbf{k} \times \mathbf{E}_0) = \frac{\omega}{c} \mathbf{k} \times \mathbf{B}_0$$

or

$$\mathbf{k} \times \mathbf{B}_0 = \frac{c}{\omega}[\mathbf{k} \times (\mathbf{k} \times \mathbf{E}_0)] = -\frac{c}{\omega} k^2 \mathbf{E}_0 \qquad (5.9.15)$$

Replacing this expression in Eq. (5.9.14b), we obtain

$$-\frac{c}{\omega} k^2 \mathbf{E}_0 + \frac{\mu K}{c} \omega \mathbf{E}_0 = \frac{4\pi\sigma\mu}{ic} \mathbf{E}_0$$

which gives

$$k^2 = \mu K \frac{\omega^2}{c^2} + 4\pi i \frac{\mu\omega\sigma}{c^2} = \mu K \left(\frac{\omega}{c}\right)^2 \left(1 + i\frac{4\pi\sigma}{\omega K}\right) \qquad (5.9.16)$$

The first term in the expression for k^2 is the displacement current contribution; the second term is the conduction current contribution.

Let us put

$$k = \frac{\omega}{c}(\eta + i\chi) \qquad (5.9.17)$$

Then

$$k^2 = \left(\frac{\omega}{c}\right)^2 (\eta^2 - \chi^2 + i2\eta\chi) \qquad (5.9.18)$$

and

$$\eta^2 - \chi^2 = \mu K$$
$$\eta\chi = \frac{2\pi\sigma\mu}{\omega} \qquad (5.9.19)$$

This gives

$$\eta = \sqrt{\mu K} \left[\frac{1 + \sqrt{1 + (4\pi\sigma/\omega K)^2}}{2}\right]^{1/2} \xrightarrow[\sigma \to 0]{} \sqrt{\mu K}$$

$$\chi = \sqrt{\mu K} \left[\frac{-1 + \sqrt{1 + (4\pi\sigma/\omega K)^2}}{2}\right]^{1/2} \xrightarrow[\sigma \to 0]{} 0 \qquad (5.9.20)$$

For a poor conductor,

$$\frac{4\pi\sigma}{\omega K} \ll 1 \qquad (5.9.21)$$

and

$$\eta = \sqrt{\mu K}$$

$$\chi = \sqrt{\mu K} \left(\frac{2\pi\sigma}{\omega K} \right) = \frac{2\pi\sigma}{\omega} \sqrt{\frac{\mu}{K}} \tag{5.9.22}$$

$$k = \frac{\omega}{c}(\eta + i\chi) = \sqrt{\mu K}\frac{\omega}{c} + i\frac{2\pi\sigma}{c}\sqrt{\frac{\mu}{K}} \tag{5.9.23}$$

Because of Eq. (5.9.21), the real part of k is much greater than the imaginary part of k. The attenuation, which is related to the imaginary part of k, is dependent of frequency, apart from a possible variation of σ with frequency.

For a good conductor,

$$\frac{4\pi\sigma}{\omega K} \gg 1 \tag{5.9.24}$$

and

$$\eta = \sqrt{\mu K} \left(\frac{2\pi\sigma}{\omega K} \right)^{1/2} = \sqrt{\frac{2\pi\sigma\mu}{\omega}}$$

$$\chi = \sqrt{\frac{2\pi\sigma\mu}{\omega}} \tag{5.9.25}$$

$$k = \frac{\omega}{c}(\eta + i\chi) = (1+i)\frac{\omega}{c}\sqrt{\frac{2\pi\sigma\mu}{\omega}} = (1+i)\frac{\sqrt{2\pi\sigma\mu\omega}}{c} \tag{5.9.26}$$

The attenuation is now frequency dependent, apart from any dependent of σ on ω. Note that for a very good conductor $\sigma \gg 1$, and also $\eta \gg 1$.

Let us assume now that a plane wave, whose fields lie in a plane parallel to the xy plane, is traveling in the z direction. The exponential factor in the wave is

$$\exp[-i\omega t + ikz] = \exp\left[-i\omega \left(t - \frac{k}{\omega}z \right) \right]$$

$$= \exp\left[-i\omega \left(t - \frac{\eta + i\chi}{c}z \right) \right]$$

$$= \exp\left[-\frac{\omega\chi}{c}z \right] \exp\left[-i\omega \left(t - \frac{\eta z}{c} \right) \right] \tag{5.9.27}$$

The exponential $e^{-(\omega\chi/c)z}$ is a damping factor. The wave entering a conductor is damped to $1/e$ of its initial amplitude at a distance

$$\delta = \frac{1}{\omega\chi/c} - \frac{c}{\omega\chi} = \frac{c}{\omega\sqrt{2\pi\sigma\mu/\omega}} = \frac{c}{\sqrt{2\pi\sigma\mu\omega}} \qquad (5.9.28)$$

This distance δ is called *skin depth* or *penetration depth*. For a conductor like copper, $\delta = 0.85$ and $0.71 \times 10^{-3}\,\text{cm}$ at 60 and 100 MHz, respectively.

The wave propagates as if the metal had an index of refraction η. For a wave traveling in the z direction,

$$\mathbf{E}(z,t) = \mathbf{E}_0 \exp\left[-\frac{\omega}{c}\chi z\right] \exp\left[-i\omega\left(t - \frac{\eta}{c}z\right)\right] \qquad (5.9.29)$$

The phase velocity is given by c/η.

Let us consider now the case of a normal incidence of a plane wave on a metallic surface (see Fig. 5.7). The boundary conditions impose

$$E_0 - E_0' = E_0''$$
$$\frac{1}{\mu}B_0 + \frac{1}{\mu}B_0' = \frac{1}{\mu''}B_0'' \qquad (5.9.30)$$

Assume $\mu = \mu'' = 1$:

$$E_0 - E_0' = E_0''$$
$$B_0 + B_0' = B_0'' \qquad (5.9.31)$$

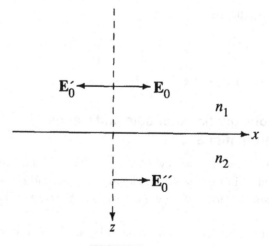

FIGURE 5.7

In empty space,

$$E_0 = B_0$$
$$E_0' = B_0'$$

(5.9.32)

In the metal [see Eq. (5.9.14a)],

$$\mathbf{B}_0 = \frac{c}{\omega}(\mathbf{k} \times \mathbf{E}_0) = \mathbf{j}\frac{c}{\omega}k_z E_x$$

(5.9.33)

if the wave travels in the z direction and is polarized in the x direction. Then

$$B'' = \frac{1}{\omega}k_z E_x = \frac{c}{\omega}kE'' = \frac{c}{\omega}\frac{\omega}{c}(\eta + i\chi)E'' = (\eta + i\chi)E''$$

(5.9.34)

There is a phase shift inside the metal between $\mathbf{B}$ and $\mathbf{E}$.

The boundary conditions (5.9.31) can now be written as

$$E_0 - E_0' = E_0''$$
$$E_0 + E_0' = (\eta + i\chi)E_0''$$

(5.9.35)

which give

$$\frac{E_0'}{E_0} = -\frac{1 - \eta - i\chi}{1 + \eta + i\chi} \xrightarrow[\sigma \to 0]{} -\frac{1 - n_2}{1 + n_2}$$

$$\frac{E_0''}{E_0} = \frac{2}{1 + \eta + i\chi} \xrightarrow[\sigma \to 0]{} \frac{2}{1 + n_2}$$

(5.9.36)

The reflection coefficient is given by

$$R = \frac{(1 - \eta)^2 + \chi^2}{(1 + \eta)^2 + \chi^2}$$

(5.9.37)

The values of η and χ are frequency dependent.

5.10. Electromagnetic Potentials and Gauge Transformations

This section begins with a review of the notions of scalar potential Φ and vector potential $\mathbf{A}$ in the case os stationary charges and currents and then moves to the consideration of these potentials in the case of time-dependent phenomena.

We shall use a device that may be didactically useful and consists in placing a line in the middle of a page so that two parallel treatments, one pertaining to the **E** field on the left side and the other pertaining to the **B** field on the right side, can be presented. We give first the stationary phenomena and then the time-dependent phenomena.

Stationary Phenomena

$$\nabla \cdot \mathbf{E} = 4\pi\rho \tag{5.10.1a}$$

$$\nabla \times \mathbf{E} = 0 \tag{5.10.1b}$$

$$\nabla \cdot \mathbf{B} = 0 \tag{5.10.1c}$$

$$\nabla \times \mathbf{B} = \frac{4\pi}{c}\mathbf{j} \tag{5.10.1d}$$

We recall

curl grad $f = 0$	div curl $\mathbf{u} = 0$
$\nabla \times \nabla f = 0$	$\nabla \cdot \nabla \times \mathbf{u} = 0$
(b) $\quad \nabla \times \mathbf{E} = 0$	(c) $\quad \nabla \cdot \mathbf{B} = 0$
$\mathbf{E} = -\nabla\phi \qquad (5.10.2)$	$\mathbf{B} = \nabla \times \mathbf{A} \qquad (5.10.3)$
(a) $\quad \nabla \cdot \mathbf{E} = 4\pi\rho$	(d) $\quad \nabla \times \mathbf{B} = \dfrac{4\pi}{c}\mathbf{j}$
$\nabla \cdot \nabla\phi = \nabla^2\phi = -4\pi\rho$	$\nabla \times (\nabla \times \mathbf{A}) = \nabla(\nabla \cdot \mathbf{A})$
	$-\nabla^2\mathbf{A} = \dfrac{4\pi}{c}\mathbf{j}$
	But (Coulomb gauge)
	$\nabla \cdot \mathbf{A} = 0$
	Then
$\nabla^2\phi = -4\pi\rho \qquad (5.10.4)$	$\nabla^2\mathbf{A} = -\dfrac{4\pi}{c}\mathbf{j} \qquad (5.10.5)$
$\phi = \displaystyle\int \frac{\rho(\mathbf{x}')d\tau'}{|\mathbf{x} - \mathbf{x}'|}$	$\mathbf{A} = \displaystyle\int \frac{\mathbf{j}(\mathbf{x}')d\tau'}{c|\mathbf{x} - \mathbf{x}'|}$
$(\phi = 0 \text{ at } \infty)$	$(\mathbf{A} = 0 \text{ at } \infty)$

Time-dependent Phenomena

$$\nabla \cdot \mathbf{E} = 4\pi\rho \tag{5.10.6a}$$

$$\nabla \times \mathbf{E} + \frac{1}{c}\frac{\partial \mathbf{B}}{\partial t} = 0 \tag{5.10.6b}$$

$$\nabla \cdot \mathbf{B} = 0 \tag{5.10.6c}$$

$$\nabla \times \mathbf{B} - \frac{1}{c}\frac{\partial \mathbf{E}}{\partial t} = \frac{4\pi}{c}\mathbf{j} \tag{5.10.6d}$$

curl grad $f = 0$

$$\nabla \times \nabla f = 0$$

(b) $\nabla \times \mathbf{E} = -\dfrac{1}{c}\dfrac{\partial \mathbf{B}}{\partial t}$

$$\nabla \times \mathbf{E} = -\dfrac{1}{c}\dfrac{\partial}{\partial t}(\nabla \times \mathbf{A})$$

$$= -\nabla \times \dfrac{1}{c}\dfrac{\partial \mathbf{A}}{\partial t}$$

$$\nabla \times \left(\mathbf{E} + \dfrac{1}{c}\dfrac{\partial \mathbf{A}}{\partial t} \right) = 0$$

$$\mathbf{E} + \dfrac{1}{c}\dfrac{\partial \mathbf{A}}{\partial t} = -\nabla \phi$$

$$\mathbf{E} = -\nabla \phi - \dfrac{1}{c}\dfrac{\partial \mathbf{A}}{\partial t} \qquad (5.10.7)$$

div curl $\mathbf{u} = 0$

$$\nabla \cdot \nabla \times \mathbf{u} = 0$$

(c) $\nabla \cdot \mathbf{B} = 0$

$$\mathbf{B} = \nabla \times \mathbf{A} \qquad (5.10.8)$$

(a) $\nabla \cdot \mathbf{E} = 4\pi \rho$

$$\nabla \cdot \left(-\nabla \phi - \dfrac{1}{c}\dfrac{\partial \mathbf{A}}{\partial t} \right) = 4\pi \rho$$

$$-\nabla^2 \phi - \dfrac{1}{c}\dfrac{\partial}{\partial t}\nabla \cdot \mathbf{A} = 4\pi \rho$$

$$\left(\nabla^2 \phi - \dfrac{1}{c^2}\dfrac{\partial^2 \phi}{\partial t^2} \right)$$

$$+ \dfrac{1}{c}\dfrac{\partial}{\partial t}\left(\nabla \cdot \mathbf{A} + \dfrac{1}{c}\dfrac{\partial \phi}{\partial t} \right) = -4\pi \rho$$

$$(5.10.9)$$

(d) $\nabla \times (\nabla \times \mathbf{A})$

$$+ \dfrac{1}{c}\dfrac{\partial}{\partial t}\left(\nabla \phi + \dfrac{1}{c}\dfrac{\partial \mathbf{A}}{\partial t} \right)$$

$$= \dfrac{4\pi}{c}\mathbf{j}$$

$$\nabla(\nabla \cdot \mathbf{A}) - \nabla^2 \mathbf{A}$$

$$+ \dfrac{1}{c}\dfrac{\partial}{\partial t}\left(\nabla \phi + \dfrac{1}{c}\dfrac{\partial \mathbf{A}}{\partial t} \right)$$

$$= \dfrac{4\pi}{c}\mathbf{j}$$

$$\left(\nabla^2 \mathbf{A} - \dfrac{1}{c^2}\dfrac{\partial^2 \mathbf{A}}{\partial t^2} \right)$$

$$- \nabla \left(\nabla \cdot \mathbf{A} + \dfrac{1}{c}\dfrac{\partial \phi}{\partial t} \right)$$

$$= -\dfrac{4\pi}{c}\mathbf{j} \qquad (5.10.10)$$

ϕ is not uniquely defined. **A** is not uniquely defined. Consider a function

$$f = f(\mathbf{x}, t)$$

and make

$$\phi' = \phi - \frac{1}{c}\frac{\partial f}{\partial t} \qquad (5.10.11)$$

$$\mathbf{A}' = \mathbf{A} + \nabla f \qquad (5.10.12)$$

Then

$$\mathbf{E}' = -\frac{1}{c}\frac{\partial \mathbf{A}'}{\partial t} - \nabla \phi'$$

$$= -\frac{1}{c}\frac{\partial}{\partial t}(\mathbf{A} + \nabla f) - \nabla\left(\phi - \frac{1}{c}\frac{\partial f}{\partial t}\right)$$

$$= -\frac{1}{c}\frac{\partial \mathbf{A}}{\partial t} - \frac{1}{c}\frac{\partial}{\partial t}\nabla f - \nabla \phi + \frac{1}{c}\frac{\partial}{\partial t}\nabla f$$

$$= -\frac{1}{c}\frac{\partial \mathbf{A}}{\partial t} - \nabla \phi = \mathbf{E}$$

$$\mathbf{B}' = \nabla \times \mathbf{A}'$$

$$= \nabla \times (\mathbf{A} + \nabla f)$$

$$= \nabla \times \mathbf{A} + \nabla \times \nabla f$$

$$= \nabla \times \mathbf{A} = \mathbf{B}$$

There is a basic indeterminacy in ϕ and **A**. There is a set of ϕ's and **A**'s that describes conveniently the electromagnetic field. The particular ϕ and **A** we consider depend on the choice of a gauge.

Lorenz Gauge. Let us call

$$\chi(\mathbf{x}, t) = \nabla \cdot \mathbf{A} + \frac{1}{c}\frac{\partial \phi}{\partial t} \qquad (5.10.13)$$

If we change **A** into $(\mathbf{A} + \nabla f)$ and ϕ into $[\phi - (1/c)(\partial f/\partial t)]$, we get

$$\chi'(\mathbf{x}, t) = \nabla \cdot (\mathbf{A} + \nabla f) + \frac{1}{c}\frac{\partial}{\partial t}\left(\phi - \frac{1}{c}\frac{\partial f}{\partial t}\right)$$

$$= \chi(\mathbf{x}, t) + \nabla^2 f - \frac{1}{c^2}\frac{\partial^2 f}{\partial t^2} \qquad (5.10.14)$$

$\chi'(\mathbf{x}, t)$ is the new expression that $\chi(\mathbf{x}, t)$ acquires because of the changes of **A** and ϕ:

$$\mathbf{A} \to \mathbf{A} + \nabla f$$

$$\phi \to \phi - \frac{1}{c}\frac{\partial f}{\partial t} \qquad (5.10.15)$$

These changes do not affect the fields $\mathbf{E}$ and $\mathbf{B}$. If we impose the condition $\chi' = 0$, we obtain for f the equation

$$\chi(\mathbf{x}, t) = -\nabla^2 f + \frac{1}{c^2} \frac{\partial^2 f}{\partial t^2} \tag{5.10.16}$$

This equation has a solution for f that renders $\chi' = 0$. The important point here is not what this solution is, but that such a solution exists; as a matter of fact, it is possible that more than one function f is a solution of Eq. (5.10.16). This means that, by taking advantage of the freedom in the choice of $\mathbf{A}$ and ϕ, it is possible to make $\chi = 0$.

The *Lorenz gauge* sets

$$\nabla \cdot \mathbf{A} + \frac{1}{c} \frac{\partial \phi}{\partial t} = 0 \tag{5.10.17}$$

It should be noted that, since more than one function f may be a solution of (5.10.16), the arbitrariness in the choice of $\mathbf{A}$ and ϕ may *not* be completely removed by imposing the Lorentz gauge.

The relations (5.10.9) and (5.10.10) can be written

$$\left(\nabla^2 \phi - \frac{1}{c^2} \frac{\partial^2 \phi}{\partial t^2} \right) + \frac{1}{c} \frac{\partial \chi}{\partial t} = -4\pi\rho \qquad \left(\nabla^2 \mathbf{A} - \frac{1}{c^2} \frac{\partial^2 \mathbf{A}}{\partial t^2} \right) - \nabla \chi$$

Because of (5.10.17) they become

$$= -\frac{4\pi}{c} \mathbf{j}$$

$$\nabla^2 \phi - \frac{1}{c^2} \frac{\partial^2 \phi}{\partial t^2} = -4\pi\rho \quad (5.10.18) \qquad \nabla^2 \mathbf{A} - \frac{1}{c^2} \frac{\partial^2 \mathbf{A}}{\partial t^2} = -\frac{4\pi}{c} \mathbf{j}$$

We call the operator

$$\tag{5.10.19}$$

$$\Box^2 = \nabla^2 - \frac{1}{c^2} \frac{\partial^2}{\partial t^2} \qquad (5.10.20)$$

the *d'Alembertian*. Then

$$\Box^2 \phi = -4\pi\rho \qquad (5.10.21) \qquad \Box^2 \mathbf{A} = -\frac{4\pi}{c} \mathbf{j} \qquad (5.10.22)$$

The Lorentz gauge has the advantage of introducing complete symmetry between the scalar and vector potentials.

Let us review now how we integrate Poisson's equation in the stationary case:

$$\nabla^2 \phi(\mathbf{x}) = -4\pi\rho(\mathbf{x}) \tag{5.10.23}$$

We consider Green's theorem

$$\int d\tau (G\nabla^2\phi - \phi\nabla^2 G) = \int dS \left(G\frac{\partial\phi}{\partial n} - \phi\frac{\partial G}{\partial n} \right) \qquad (5.10.24)$$

and use the Green's function (see Fig. 5.8):

$$G(\mathbf{x}, \mathbf{x}') = \frac{1}{|\mathbf{x} - \mathbf{x}'|} \qquad (5.10.25)$$

for which

$$\nabla^2 G(\mathbf{x}, \mathbf{x}') = \nabla^2 \frac{1}{|\mathbf{x} - \mathbf{x}'|} = -4\pi\delta(\mathbf{x} - \mathbf{x}') \qquad (5.10.26)$$

We note that the integral on the right side of Eq. (5.10.24) is zero: If $\mathbf{x}'$ is the variable of integration for $|\mathbf{x}'| \to \infty$, $\phi = G = 0$. We can write

$$\int d\tau' (G\nabla^2\phi - \phi\nabla^2 G) = 0$$

$$= \int \left[\frac{-4\pi\rho(\mathbf{x}')}{|\mathbf{x} - \mathbf{x}'|} + \phi(\mathbf{x}')4\pi\delta(\mathbf{x} - \mathbf{x}') \right] d\tau'$$

$$= -\int \frac{4\pi\rho(\mathbf{x}')}{|\mathbf{x} - \mathbf{x}'|}d\tau' + 4\pi\phi(\mathbf{x}) \qquad (5.10.27)$$

and we obtain

$$\phi(\mathbf{x}) = \int \frac{\rho(\mathbf{x})'}{|\mathbf{x} - \mathbf{x}'|}d\tau' \qquad (5.10.28)$$

Expression (5.10.28) is the complete solution of Poisson's equation.

In the case of time-dependent phenomena, we have to solve the equation

$$\Box^2\phi(\mathbf{x}, t) = -4\pi\rho(\mathbf{x}, t) \qquad (5.10.29)$$

The appropriate Green's function to use in density with the above equation is given by

$$G(\mathbf{x}, t; \mathbf{x}', t') = \frac{1}{|\mathbf{x} - \mathbf{x}'|}\delta\left(t - t' - \frac{|\mathbf{x} - \mathbf{x}'|}{c} \right)$$

$$= \frac{1}{r}\delta\left(t - t' - \frac{r}{c} \right) \qquad (5.10.30)$$

Figure 5.8 should be consulted for the various position coordinates. The choice of Eq. (5.10.30) as Green's function is based on the following claim.

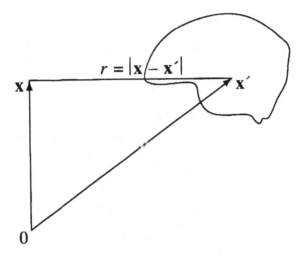

FIGURE 5.8

CLAIM

$$\Box^2 G(\mathbf{x}, t; \mathbf{x}', t') = \Box^2 \left[\frac{1}{r} \delta \left(t - t' - \frac{r}{c} \right) \right]$$

$$= -4\pi \delta(\mathbf{x} - \mathbf{x}') \delta(t - t') \qquad (5.10.31)$$

PROOF Consider the function

$$\delta = \delta \left(t - t' - \frac{r}{c} \right) \qquad (5.10.32)$$

We can write

$$\boldsymbol{\nabla} \left(\frac{1}{r} \delta \right) = \delta \boldsymbol{\nabla} \frac{1}{r} + \frac{1}{r} \boldsymbol{\nabla} \delta$$

$$\nabla^2 \left(\frac{1}{r} \delta \right) = \delta \nabla^2 \frac{1}{r} + \frac{1}{r} \nabla^2 \delta + 2 \boldsymbol{\nabla} \delta \cdot \boldsymbol{\nabla} \frac{1}{r}$$

But

$$\frac{\partial \delta}{\partial t} = \delta'$$

$$\frac{\partial \delta}{\partial r} = -\frac{1}{c} \delta'$$

Then

$$\nabla^2 \delta = \frac{\partial^2 \delta}{\partial r^2} + \frac{2}{r}\frac{\partial \delta}{\partial r} = \frac{1}{c^2}\delta'' - \frac{2}{rc}\delta'$$

$$\nabla \delta = \frac{\partial \delta}{\partial r}\frac{\mathbf{r}}{r} = -\frac{\mathbf{r}}{rc}\delta'$$

and

$$\nabla^2\left(\frac{1}{r}\delta\right) = \delta\nabla^2\frac{1}{r} + \frac{1}{r}\nabla^2\delta + 2\nabla\delta\cdot\nabla\frac{1}{r}$$

$$= \partial\nabla^2\frac{1}{r} + \frac{1}{r}\left(\frac{1}{c^2}\delta'' - \frac{2}{rc}\delta'\right) + 2\left(-\frac{\mathbf{r}}{rc}\delta'\right)\cdot\left(-\frac{\mathbf{r}}{r^3}\right)$$

$$= -\delta\left(t - t' - \frac{r}{c}\right)4\pi\delta(\mathbf{x} - \mathbf{x}') + \frac{1}{rc^2}\delta'' - \frac{2}{r^2c}\delta' + \frac{2}{r^2c}\delta'$$

$$= -4\pi\delta\left(t - t' - \frac{r}{c}\right)\delta(\mathbf{x} - \mathbf{x}') + \frac{1}{rc^2}\delta''\left(t - t' - \frac{r}{c}\right)$$

$$\Box^2\left(\frac{1}{r}\delta\right) = \left(\nabla^2 - \frac{1}{c^2}\frac{\partial^2}{\partial t^2}\right)\left(\frac{1}{r}\delta\right) = \nabla^2\left(\frac{1}{r}\delta\right) - \frac{1}{c^2}\frac{\partial^2}{\partial t^2}\frac{\delta}{r}$$

$$= -4\pi\delta\left(t - t' - \frac{r}{c}\right)\delta(\mathbf{x} - \mathbf{x}') + \frac{1}{rc^2}\delta''\left(t - t' - \frac{r}{c}\right)$$

$$- \frac{1}{rc^2}\delta''\left(t - t' - \frac{r}{c}\right)$$

$$= -4\pi\delta\left(t - t' - \frac{r}{c}\right)\delta(\mathbf{x} - \mathbf{x}')$$

$$= -4\pi\delta(t - t')\delta(\mathbf{x} - \mathbf{x}')$$

which concludes the proof of Eq. (5.10.31).

Now we introduce a new claim.

CLAIM The solution of

$$\Box^2\phi(\mathbf{x}, t) = -4\pi\rho(\mathbf{x}, t) \tag{5.10.33}$$

is given by

$$\phi(\mathbf{x}, t) = \phi_0(\mathbf{x}, t) + \int \frac{\rho[\mathbf{x}', t - (|\mathbf{x} - \mathbf{x}'|/c)]}{|\mathbf{x} - \mathbf{x}'|}d\tau' \tag{5.10.34}$$

where

$$\Box^2 \phi_0(\mathbf{x}, t) = 0$$

PROOF

$$\Box^2 \phi(\mathbf{x}, t) = \Box^2 \phi_0(\mathbf{x}, t) + \Box^2 \int \frac{\rho[\mathbf{x}', t - (|\mathbf{x} - \mathbf{x}'|/c)]}{|\mathbf{x} - \mathbf{x}'|} d\tau'$$

$$= \int \Box_{x,t}^2 \frac{\rho[\mathbf{x}', t - (|\mathbf{x} - \mathbf{x}'|/c)]}{|\mathbf{x} - \mathbf{x}'|} d\tau'$$

$$= \int d\tau' \Box_{x,t}^2 \int dt' \delta \left(t - t' - \frac{|\mathbf{x} - \mathbf{x}'|}{c} \right) \frac{\rho(\mathbf{x}', t')}{|\mathbf{x} - \mathbf{x}'|}$$

$$= \int d\tau' \int dt' \rho(\mathbf{x}', t')$$

$$\times \Box_{x,t}^2 \left[\frac{1}{|\mathbf{x} - \mathbf{x}'|} \delta \left(t - t' - \frac{|\mathbf{x} - \mathbf{x}'|}{c} \right) \right]$$

$$= \int d\tau' \int dt' \rho(\mathbf{x}', t')$$

$$\times \left[-4\pi \delta(\mathbf{x} - \mathbf{x}') \delta(t - t') \right] = -4\pi \rho(\mathbf{x}, t)$$

In the above derivation that proves the claim, we introduce at one point the subscripts $\mathbf{x}$ and t to indicate the variables on which $\Box^2$ operates.

The solutions of Eqs. (5.10.21) and (5.10.22) are then given by

$$\phi(\mathbf{x}, t) = \phi_0(\mathbf{x}, t) + \int \frac{\rho(\mathbf{x}', t_{\text{ret}})}{|\mathbf{x} - \mathbf{x}'|} d\tau'$$

$$\mathbf{A}(\mathbf{x}, t) = \mathbf{A}_0(\mathbf{x}, t) + \int \frac{1}{c} \frac{\mathbf{j}(\mathbf{x}', t_{\text{ret}})}{|\mathbf{x} - \mathbf{x}'|} d\tau'$$

(5.10.35)

where

$$\Box^2 \phi_0(\mathbf{x}, t) = 0$$

$$\Box^2 \mathbf{A}_0(\mathbf{x}, t) = 0$$

(5.10.36)

and

$$t_{\text{ret}} = t - \frac{|\mathbf{x} - \mathbf{x}'|}{c}$$

(5.10.37)

The subscript ret stands for *retarded*.

There is a relation cause and effect between charges (and currents) and ϕ (and $\mathbf{A}$); if there is a change in the currents or charges at $\mathbf{x}'$, it has an effect at point $\mathbf{x}$ after a time r/c where $r = |\mathbf{x} - \mathbf{x}'|$.

Coulomb Gauge. The *Coulomb gauge* can simply be expressed by the condition

$$\boldsymbol{\nabla} \cdot \mathbf{A} = 0 \qquad (5.10.38)$$

Equations (5.10.9) and (5.10.10) become

$$\nabla^2 \phi = -4\pi\rho \qquad (5.10.39)$$

The solution of this equation is simply

$$\phi(\mathbf{x}, t) = \int d\tau' \frac{\rho(\mathbf{x}', t)}{|\mathbf{x} - \mathbf{x}'|} \qquad (5.10.41)$$

This result indicates that the scalar potential is the instantaneous Coulomb potential due to the charge density at time t; this is the reason for the name

Coulomb gauge

$$\nabla^2 \mathbf{A} - \frac{1}{c^2} \frac{\partial^2 \mathbf{A}}{\partial t^2} - \boldsymbol{\nabla} \frac{1}{c} \frac{\partial \phi}{\partial t}$$
$$= -\frac{4\pi}{c} \mathbf{j} \qquad (5.10.40)$$

Then

$$\Box^2 \mathbf{A} = -\frac{4\pi}{c} \mathbf{j} + \frac{1}{c} \boldsymbol{\nabla} \frac{\partial \phi}{\partial t} \qquad (5.10.42)$$

The condition $\boldsymbol{\nabla} \cdot \mathbf{A} = 0$ is valid in the whole space at all times:

$$\Box^2 \boldsymbol{\nabla} \cdot \mathbf{A} = -\frac{4\pi}{c} \boldsymbol{\nabla} \cdot \mathbf{j} + \frac{1}{c} \frac{\partial}{\partial t} \nabla^2 \phi$$
$$-\frac{4\pi}{c} \left(\boldsymbol{\nabla} \cdot \mathbf{j} + \frac{\partial \rho}{\partial t} \right) = 0$$
$$(5.10.43)$$

We can now write

$$\Box^2 \mathbf{A}(\mathbf{x}, t) = -\frac{4\pi}{c} \mathbf{j}(\mathbf{x}, t) + \frac{1}{c} \boldsymbol{\nabla} \int d\tau' \frac{\partial \rho(\mathbf{x}', t)/\partial t}{|\mathbf{x} - \mathbf{x}'|}$$
$$= -\frac{4\pi}{c} \mathbf{j}(\mathbf{x}, t) + \frac{4\pi}{c} \boldsymbol{\nabla} \int d\tau' \frac{\partial \rho(\mathbf{x}', t)/\partial t}{4\pi |\mathbf{x} - \mathbf{x}'|}$$

$$= -\frac{4\pi}{c} \left\{ \mathbf{j}(\mathbf{x}, t) - \nabla \int d\tau' \frac{\partial \rho(\mathbf{x}', t)/\partial t}{4\pi |\mathbf{x} - \mathbf{x}'|} \right\}$$

$$= -\frac{4\pi}{c} \mathbf{j}_{tr}(\mathbf{x}, t) \qquad\qquad (5.10.44)$$

where

$$\mathbf{j}_{tr}(\mathbf{x}, t) = \mathbf{j}(\mathbf{x}, t) - \nabla \int d\tau' \frac{\partial \rho(\mathbf{x}', t)/\partial t}{4\pi |\mathbf{x} - \mathbf{x}'|} \qquad (5.10.45)$$

The subscript tr stands for *transverse*. We note that

$$\nabla \cdot \mathbf{j}_{tr} = 0 \qquad\qquad (5.10.46)$$

The potentials in the Coulomb gauge case are then given by

$$\phi(\mathbf{x}, t) = \int d\tau' \frac{\rho(\mathbf{x}', t)}{|\mathbf{x} - \mathbf{x}'|}$$

$$\mathbf{A}(\mathbf{x}, t) = \mathbf{A}_0(\mathbf{x}, t) + \int \frac{1}{c} \frac{\mathbf{j}_{tr}\{\mathbf{x}', t - [|\mathbf{x} - \mathbf{x}'|/c]\}}{|\mathbf{x} - \mathbf{x}'|} d\tau' \qquad (5.10.47)$$

where

$$\square^2 \mathbf{A}_0(\mathbf{x}, t) = 0 \qquad\qquad (5.10.48)$$

and

$$\mathbf{j}_{tr}(\mathbf{x}, t) = \mathbf{j}(\mathbf{x}, t) - \nabla \int d\tau' \frac{\partial \rho(\mathbf{x}', t)/\partial t}{4\pi |\mathbf{x} - \mathbf{x}'|} \qquad (5.10.49)$$

We note that the vector potential is expressed in terms of the transverse current $\mathbf{j}_{tr}$; for this reason the Coulomb gauge is also called the *transverse gauge*.

The use of the Coulomb gauge is particularly convenient when no sources are present. If $\rho = 0$, we see from Eq. (5.10.41) that also $\phi = 0$, and the electromagnetic fields are simply given by

$$\mathbf{E} = -\frac{1}{c} \frac{\partial \mathbf{A}}{\partial t}$$

$$\mathbf{B} = \nabla \times \mathbf{A} \qquad\qquad (5.10.50)$$

CHAPTER 5 EXERCISES

5.1. A capacitor, consisting of two parallel plates, has the space between the plates filled with a material of dielectric constant K and conductivity σ. The plates are connected to the terminals of a battery. At time $t = 0$ the connection is broken. Calculate the time it takes the voltage across the plates to drop to $1/e$ of its original value.

5.2. A large sphere of radius r_b is made of a material of dielectric constant K and conductivity σ, and is concentric with a small sphere of radius r_a. A charge q_0 is distributed over the surface of the small sphere at time $t = 0$. Calculate the joule heat produced by the passage of the charge from the small to the large sphere and prove that this heat energy is equal to the decrease of the electrostatic energy.

5.3. Consider two charges that move along two mutually perpendicular trajectories, as in Fig. P5.3 and the magnetic forces that the charges exercise on each other. Does action equal reaction in this case? Is the total momentum of this system of charges conserved, considering that no external force is applied to it? Explain.

5.4. A particle of mass m and charge q is suspended on a massless string of length L. At a distance d from the mass point, there is an infinite plane consisting of a perfect conductor. The particle is set into small pendular oscillations, q, m, and the amplitude of the oscillations a are such that the velocity of the particle is much less than the velocity of light.

(a) Compute the frequency of the pendular oscillations if the amplitude a is small compared to L. Neglect damping and gravity.

(b) What are the E and H fields of the Poynting vector $\mathbf{N}$ at a point in the plane of oscillations at distance r? Consider $r \gg a$, λ, d, but λ/d arbitrary.

q_1

q_2

FIGURE P5.3

5.5. Consider the formulation of the conservation laws in light of the notions introduced in this chapter. Consider in particular the conservation of linear momentum. Which is the right expression, the one we learned in classical (Newtonian) mechanics or the one proposed in this chapter that includes the momentum of the electromagnetic field? One could ask: How can the law of momentum be relevant if we choose momentum so as to make the law hold true? Is the law just a matter of definition? Explain.

5.6. A long, straight wire of conductivity σ and radius a carries a current I.

   (a) Find the Poynting vector at the surface of the wire.
   (b) Verify that the energy dissipated per unit time as heat in the wire can be obtained by making use of the Poynting vector derived in Part (a).

5.7. A radio broadcasting station is to be located on a elliptical island whose major axis is in the north-south direction. Using two antennas, both located on the south end of the island and operated with a phase difference between them, describe how the energy northward across the island can be maximized, with little or no energy radiated southward over the water.

5.8. The electric field of a plane electromagnetic wave traveling in vacuum is given by

$$E_x = E_z = 0$$

$$E_y = A \cos \left(\omega t - \frac{2\pi}{\lambda} z \right)$$

A square wire loop of side a is placed in the yz plane in such a way that one of its sides coincides with the y axis and the other with the z axis.

   (a) What is the induced *emf* in the loop?
   (b) For what value of a is the induced rms emf a maximum?

5.9. The classical electromagnetic energy density per unit frequency range of black-body radiation is given by

$$u(\nu) = C\nu^2 T$$

where ν is the frequency and T the absolute temperature. Derive the explicit expression for the constant C.

5.10. The energy density of black-body radiation is given by

$$u = \frac{4\sigma}{c}T^4$$

where $\sigma =$ Stefan constant $= 5.672 \times 10^{-5}\, \text{erg/cm}^2\text{s deg}^4$.

(a) Calculate the free energy per unit volume of the radiation.

(b) Use the expression for the free energy to prove that the pressure exerted on the cavity walls is

$$p = \frac{1}{3}u$$

(c) Consider the radiation field to consist of photons of energy $h\nu$, and use statistical arguments to show that each photon must have a momentum of h/λ in order for the radiation pressure to be given by the expression above.

5.11. Find the pressure-volume relationship for the adiabatic compression of black-body radiation, knowing that the pressure exerted on the cavity walls of a black-body is given by

$$p = \frac{1}{3}u$$

where $u =$ energy density of black-body radiation $= aT^4$

$$\left(a = \frac{4\sigma}{c}, \sigma = \text{Stefan constant} = 5.672 \times 10^{-5}\,\text{erg/cm}^2\text{s deg}^4\right)$$

5.12. The solar radiation near the top of the earth's atmosphere is about 2 cal per min per cm^2.

$$1\,\text{cal} = 4.2\,\text{J}$$
$$\text{radius of the earth} = 6 \times 10^3\,\text{km}$$

(a) How large are the amplitudes of the electric and magnetic field intensities?

(b) If the earth is a perfect absorber, what is the total force on the earth due to the solar radiation?

5.13. A beam of plane polarized light with propagation direction along the z axis is incident normally upon an anisotropic dielectric slab having principal axes x, y and z and with dielectric constants along

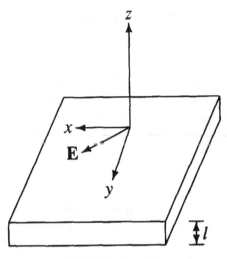

FIGURE P5.13

the three directions given by the real numbers K_x, K_y, and K_z (see Fig. P5.13). The plane of polarization (containing the electric vector) forms an angle θ with the x axis. The thickness of the slab in the z direction is l.

(a) Describe the polarization of the emergent beam.

(b) Find the conditions for the emergent beam to be circularly polarized.

5.14. A flat surface separates a medium that has an index of refraction $n = \sqrt{3}$ from air.

(a) Find the Brewster angle i_B.

(b) Find the ratio E'/E_0, where

$$E_0 = \text{amplitude of the incident wave}$$

$$E_0' = \text{amplitude of the reflected wave}$$

for the two cases of **E** perpendicular to the plane of incidence and **E** in the plane of incidence and for the following angles of incidence:

$$i = i_B - 20, \quad i_B, \quad i_B + 20$$

(c) Comment on the polarization properties and on the phase of the reflected wave.

5.15. A plane sinusoidal electromagnetic wave in vacuum is incident normally on the plane surface of a material with a dielectric constant K.

    (a) What fraction of the incident energy flux is reflected at this surface?

    (b) Suppose that the dielectric is a sheet with a thickness equal to the wavelength λ in the dielectric. Calculate the net fraction of the incident energy flux reflected backward.

5.16. An isotropic light source, placed under water at a distance d from the surface, yields an illuminated circular area of radius 6 m. Calculate the distance d knowing that the dielectric constant of water at optical frequencies is 1.75.

5.17. A rigid fiber, made of transparent dielectric material, is used to guide electromagnetic radiation by taking advantage of the condition of total internal reflection (see Fig. P5.17). Calculate the minimum value that the dielectric constant K of this medium can have in order to trap an incoming wave into the fiber.

5.18. (a) Consider a metal that has dielectric constant K, permeability μ, and conductivity σ. A plane wave polarized in the x direction travels in the medium in the direction of the z axis:

$$\mathbf{E} = E_0\, e^{i(k_z - \omega t)}\mathbf{i}, \quad \mathbf{i} = \text{unit vector in the } x \text{ direction}$$

Find an expression for the $\mathbf{B}$ vector.

    (b) k is a complex quantity:

$$k = |\mathbf{k}|\, e^{i\phi}$$

Find expressions for $|\mathbf{k}|$ and ϕ. What is ϕ when $\sigma \to \infty$?

    (c) Find the ratio of the magnetic energy density to the electric energy density in the metal.

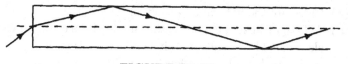

FIGURE P5.17

5.19. A plane electromagnetic wave traveling in the x direction and polarized in the y direction is defined by the two fields

$$E_y = A \sin \omega \left(t - \frac{x}{c} \right)$$

$$H_z = A \sin \omega \left(t - \frac{x}{c} \right)$$

At $x = 0$, the wave falls with normal incidence on the plane surface of a conductor perpendicular to the x axis. The conductor extends to infinity for $x > 0$. Calculate the radiation pressure that the wave exercises on the conductor.

5.20. **(a)** Derive the expressions for the electric and magnetic fields of the principal wave, that is, the lowest mode, in a coaxial line.

(b) Derive an approximate formula for the energy loss at the metallic surface, for large conductivity, in terms of the tangential magnetic field.

(c) From Parts (a) and (b), calculate the attenuation of the principal wave.

5.21. The vector field **E** of an electromagnetic wave produces a current density **j** inside an absorbing body. The accompanying field **H** of the wave produces a force whose density is given by

$$\mathbf{f} = \frac{\mathbf{j} \times \mathbf{H}}{c}$$

Use the Maxwell equations to show that the time average of the radiation pressure of the exercise 5.19 is equal to other volume integral

$$p = \int_0^\infty f_x \, dx$$

5.22. The Q of a cavity, to be used as a container of electromagnetic radiation, is 2π times the ratio of the average energy stored in the cavity to the energy lost per cycle. A *mode* is the configuration of the fields inside the cavity, when the frequency of the radiation corresponds to a cavity resonance. Consider a rectangular cavity of dimensions a, b, and c along the x, y, and z directions, respectively.

(a) Calculate the natural frequency of the cavity.

(b) A TE_{012} transverse mode is that mode of the cavity in which the electric field is parallel to a in the plane of sides a and b, varying as $\sin(\pi y/b)$ and as $\sin(2\pi z/c)$ and independent of x. Compute the Q for the transverse electric mode TE_{012}.

5.23. **(a)** Demonstrate that electromagnetic waves can propagate along a metal pipe of rectangular cross section with perfectly conducting walls.

(b) Calculate the phase velocity and the group velocity of such waves.

(c) Calculate the cutoff frequency below which the propagation of these waves is impossible.

5.24. The wave equations for the potentials ϕ and $\mathbf{A}$ when using the Lorenz gauge and the wave equation for the potential $\mathbf{A}$ when using the Coulomb gauge have the basic structure given by the equation

$$\nabla^2 \psi(\mathbf{x}, t) - \frac{1}{c^2} \frac{\partial^2 \psi(\mathbf{x}, t)}{\partial t^2} = -4\pi f(\mathbf{x}, t)$$

where $f(\mathbf{x}, t)$ is a known source distribution. Show how this equation can be integrated by the use of an advanced Green's function:

$$G(\mathbf{x}, t; \mathbf{x}', t') = \frac{1}{|\mathbf{x} - \mathbf{x}'|} \delta\left(t - t' + \frac{|\mathbf{x} - \mathbf{x}'|}{c}\right)$$

Comment on the result obtained for $\psi(\mathbf{x}, t)$.

6

Theory of Relativity: I

6.1. Principle of Relativity in Mechanics and Electrodynamics

6.1.1. *Galileian Transformation*

Given two coordinate systems S and S', with S' moving with velocity $\mathbf{v}$ with respect to S,

$$\mathbf{x}' = \mathbf{x} - \mathbf{v}t \tag{6.1.1a}$$

$$t' = t \tag{6.1.1b}$$

These equations in which the primed (unprimed) coordinates refer to $S'(S)$ express the *Galileian transformation*.

Let us consider now the arbitrary motion of a particle with coordinate $\mathbf{x}$, $\mathbf{x}'$; by differentiating Eq. (6.1.1a), we obtain

$$\frac{d\mathbf{x}'}{dt'} = \frac{d\mathbf{x}}{dt} - \mathbf{v} \tag{6.1.2}$$

or

$$\mathbf{u}' = \mathbf{u} - \mathbf{v} \qquad (6.1.3)$$

where $\mathbf{u}'(\mathbf{u})$ is the velocity of the particle with respect to the primed (unprimed) coordinate system. Relation (6.1.3) expresses the *addition theorem* of velocities.

Now let us assume that a particle of mass m is acted on by a force $\mathbf{F}$. In the system S, the particle will achieve an acceleration given by

$$m\frac{d^2\mathbf{x}}{dt^2} = \mathbf{F} \qquad (6.1.4)$$

But

$$\frac{d^2\mathbf{x}}{dt^2} = \frac{d^2\mathbf{x}'}{dt'^2} \qquad (6.1.5)$$

and, since the Newtonian mechanics masses are absolute quantities,

$$m = m' \qquad (6.1.6)$$

We have

$$m'\frac{d^2\mathbf{x}'}{dt'^2} = \mathbf{F} \qquad (6.1.7)$$

Therefore, the Newtonian fundamental equations of motion are invariant under a Galileian transformation, and the two systems of coordinates related to each other by such transformation are equivalent. This equivalent means that it is not possible to decide which of the two systems that undergo relative (uniform) motion is stationary and which is moving. This face expresses the *principle of Galileian relativity*.

Let us now consider a system of coordinates S and another system S', which travels with respect to S in the x direction with velocity v. Let a light signal, observed in S, travel with the same speed c in all directions, and in particular in the $+x$ and $-x$ directions; when observed in S', it will travel, according to (6.1.3), with velocity $c - v$ in the $+x$ direction and with velocity $c + v$ in the $-x$ direction. The equivalence of the two systems, valid when considering purely mechanical phenomena, is destroyed when we deal with electromagnetic phenomena.

Since the principle of Galileian relativity applies to the laws of mechanics, but does not apply to electrodynamics, we are forced to choose among the alternatives given in Table 6.1.

Table 6.1.

Hypotheses	Fundamental experiments
1. The laws of mechanics and the Maxwell equations are basically correct. A principle of relativity is valid for mechanics, but not for electrodynamics. There exists a unique privileged frame of reference, the "ether frame," in which the Maxwell equations are valid and in which light propagates with velocity c.	1. Attempt to locate a preferred inertial frame[a] for the laws of electrodynamics.
2. A principle of relativity exists for both mechanics and electrodynamics, but electrodynamics is not correct in the Maxwell formulation.	2. Attempt to observe deviations from the laws of classical electrodynamics.
3. A principle of relativity exists for both mechanics and electrodynamics, but the laws of mechanics in the Newtonian form need modifications.	3. Attempt to observe deviations from classical mechanics.

[a] *Inertial frame* is another expression for *unaccelerated frame*.

6.2. The Search for an Absolute Frame Tied to the Ether

All experimental evidence is in favor of the existence of a principle of relativity for the Maxwell equations and against the existence of the ether.

Trouton and Noble Experiment. Let us consider two charges e_1 and $-e_2$ at rest in a frame S' moving with velocity $\mathbf{v}$ with respect to a frame S.

Measuring things in S, the charge $-e_2$ is feeling not only the electric field due to e_1, but also the magnetic field due to the motion of e_1. The magnetic field due to a current elemet $\mathbf{j}(\mathbf{x})d^3\mathbf{x}$ is given by

$$B(\mathbf{x}_p) = \frac{1}{c} \int d^3\mathbf{x}' j(\mathbf{x})' \times \boldsymbol{\nabla}_{x'} \frac{1}{|\mathbf{x}' - \mathbf{x}_p|}$$

$$= \boldsymbol{\nabla}_{x_p} \times \frac{1}{c} \int d^3\mathbf{x}' \frac{\mathbf{j}(\mathbf{x})'}{|\mathbf{x}' - \mathbf{x}_p|} \qquad (6.2.1)$$

In our case

$$\mathbf{j}(\mathbf{x}') = e_1 \delta(\mathbf{x} - \mathbf{x}')\mathbf{v}$$

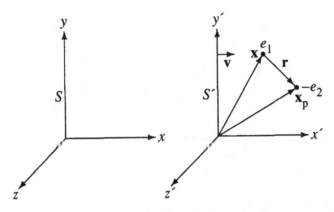

FIGURE 6.1

Then (see Fig. 6.1)

$$\mathbf{B}_1(\mathbf{x}_p) = \frac{e_1 \mathbf{v}}{c} \times \boldsymbol{\nabla}_x \frac{1}{|\mathbf{x} - \mathbf{x}_p|} = \frac{e_1 \mathbf{v} \times \mathbf{r}}{cr^3} \qquad (6.2.2)$$

There is a force acting on the charge $(-e_2)$ given by

$$\mathbf{F} \text{ on } (-e_2) = -e_2 \left(\frac{\mathbf{v}}{c} \times \mathbf{B}_1 \right) = -e_1 e_2 \frac{(\mathbf{v}/c) \times [(\mathbf{v}/c) \times \mathbf{r}]}{r^3} \qquad (6.2.3)$$

This force is proportional to $(v/c)^2$ and lies in the plane of $\mathbf{r}$ and $\mathbf{v}$.

In a similar way, we find that the $\mathbf{B}$ field at the location of e_1 due to the charge $-e_2$ is given by

$$\mathbf{B}_2(\mathbf{x}) = \frac{-e_2(\mathbf{v}/c) \times (-\mathbf{r})}{r^3} = e_2 \frac{(\mathbf{v}/c) \times \mathbf{r}}{r^3} \qquad (6.2.4)$$

and

$$\mathbf{F} \text{ on } e_1 = e_1[(\mathbf{v}/c) \times \mathbf{B}_2] = e_1 e_2 \frac{(\mathbf{v}/c) \times [(\mathbf{v}/c) \times \mathbf{r}]}{r^3}$$

$$= -\mathbf{F} \text{ on } (-e_2) \qquad (6.2.5)$$

Both forces are in the plane of $\mathbf{r}$ and $\mathbf{v}$ and should produce a torque. We note that the Maxwell relations and the preceding formulas, which are derived from them, are valid in the ether frame; therefore, the velocity of the charges that enters the calculations is that with respect to the ether frame.

Trouton and Noble performed a sensitive experiment that used a suspended parallel plate condenser. This experiment, reported in 1903,[1] and repeated at later times, did not give any indication of the existence of a torque. Two different conclusions could be reached.

(1) The earth is at rest in the preferred frame of reference in which the two charges are at absolute rest.

(2) There is no preferred frame of reference.

Considering how unlikely conclusion 1 is, the Trouton and Noble experiment disproves the possibility of locating a preferential inertial frame for the law of electrodynamics, thus leaving only the second conclusion.

Michelson–Morley Experiment. To get at the essence of this experiment, we shall give examples of spaceships and pulses of light, rather than the historical methods used by Michelson and Morley. We shall take their point of view and assume that the ether does exist and in it light travels with velocity c.

A spaceship is at "absolute rest" as in Fig. 6.2a; in it are a source of light pulse and a mirror, at distance l from each other. The time it takes a pulse emitted by the source to arrive at the mirror and return to the source is

$$t = \frac{2l}{c} \tag{6.2.6}$$

Assume now that the situation has changed to that in Fig. 6.2b; that is, the spaceship is moving with absolute velocity v. The source of light and the mirror are lined up in the direction of v. The spaceship travels through the ether; the velocity of the light with respect to the ether is c. The velocity of a pulse of light, in its forward motion toward the mirror, with respect to the spaceship is $c - v$; when the pulse, having been reflected by the mirror, returns to the source, its relative velocity with respect to the spaceship is $c + v$. The time for the light pulse's round trip (source–mirror–source) is

$$t_1 = \frac{l}{c - v} + \frac{l}{c + v} = \frac{2lc}{c^2 - v^2} \tag{6.2.7}$$

Finally, the experiment is repeated with the direction source–mirror made perpendicular to the direction of motion of the spaceship as in Fig. 6.2c. In

---

[1] *Phil. Trans.* A*202*, 165 (1903) and *Proc. Roy. Soc.* (London) *73*, 132 (1903).

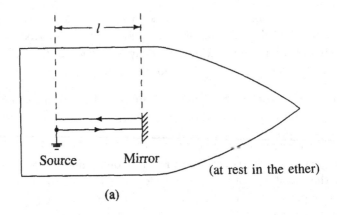

(a)

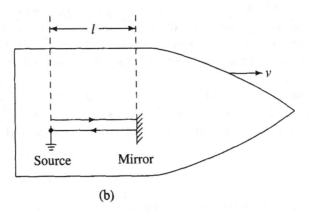

(b)

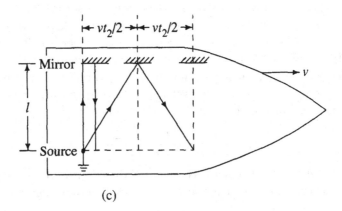

(c)

FIGURE 6.2

this case the time for the light pulse's round trip source–mirror–source is given by

$$t_2 = \frac{2l}{\sqrt{c^2 - c^2}} \qquad (6.2.8)$$

The Michelson–Morley experiment consisted in measuring the difference between times t_1 and t_2. Spaceship earth was used and no difference between t_1 and t_2 was found. Indeed, it was always $t_1 = t_2$, even when the experiment was repeated at various times of the day and at different seasons of the year. This experiment was an attempt to locate a preferential frame of reference, the ether's frame. Such a frame was not found.

6.3. Einstein's Postulates

Both the Trouton and Noble experiment and the Michelson and Morley experiment disproved the existence of a preferred frame of reference for the electromagnetic phenomena. Since no deviations from the laws of classical electrodynamics were to be found, the only logical course of action was the formulation of a principles of relativity valid for both mechanics and electrodynamics.

Such a principle, known as the *principle of special relativity*, was expressed by Einstein in the following two postulates (1905):

(1) The laws of mechanics and the laws of electrodynamics are the same in all inertial frames. The propagation of light with velocity c in vacuum is included in these laws.

(2) No experiment can be designed that would indicate a state of absolute uniform motion or a preferred inertial frame of reference.

6.4. Lorentz Transformation

The principle of relativity, as enunciated by Einstein, implies the absence of any phenomenon that depends on a particular inertial frame. Let us consider two inertial frames, S and S', and let S' move with respect to S with a velocity v, which, for simplicity, we assume to be pointing in the $+z$ direction, as in Fig. 6.3.

We shall be looking for a transformation, which will replace the Galilean transformation and that will be consistent with the principle of Einsteinian relativity. Such a transformation will allow us to go from the coordinates and times in the S frame $(\mathbf{x}, t)$ to the coordinates and times

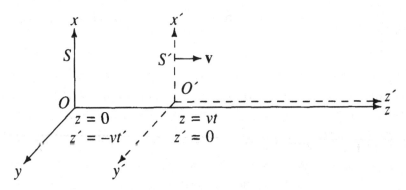

FIGURE 6.3

in the frame S' ($\mathbf{x}'$, t'); the Maxwell equations, when expressed in terms of the coordinates and times in the two frames, will have the same form:

$$\nabla \cdot \mathbf{E} = 4\pi\rho \rightarrow \nabla' \cdot \mathbf{E} = 4\pi\rho'$$

$$\nabla \times E = -\frac{1}{c}\frac{\partial \mathbf{B}}{\partial t} \rightarrow \nabla' \times E' = \frac{1}{c}\frac{\partial \mathbf{B}'}{\partial t'}$$

$$\nabla \cdot \mathbf{B} = 0 \rightarrow \nabla' \cdot \mathbf{B}' = 0$$

$$\nabla \times \mathbf{B} - \frac{1}{c}\frac{\partial \mathbf{E}}{\partial t} = \frac{4\pi}{c}\mathbf{j} \rightarrow \nabla' \times \mathbf{B}' - \frac{1}{c}\frac{\partial \mathbf{E}'}{\partial t'} = \frac{4\pi}{c}\mathbf{j}'$$

A transformation of this type is called *special Lorentzian transformation* and has the following properties:

(1) It must be linear. If it were quadratic, we would have two values of z' for one value of z. More complicated relations are ruled out because they would make the origin of the coordinates physically distinct from all the other points.

(2) If the transformation that connects S to S' has a certain mathematical structure, the transformation that connects S' to S must have the same structure, taking into account the different sign of the velocity.

(3) The velocity of the electromagnetic radiation in vacuum is the same in any inertial frame (and in particular in S and S') and in any direction and equal to the constant value c.

(4) All the transformations must form a *group*. In fact, they must have the

four properties that define a group:

(a) If A is the transformation

$$S \xrightarrow{\mathbf{v'}} S'$$

and B is the transformation

$$S' \xrightarrow{\mathbf{v''}} S''$$

there is a transformation C, which we call the product AB, $C = AB$, given by

$$S \xrightarrow{\mathbf{v'+v''}} S''$$

(b) There is an identity transformation, which we designate with the capital letter E and which corresponds to the transformation of a frame into itself.

(c) Given three transformations $A, B,$ and C,

$$A(BC) = (AB)C$$

(d) Given a transformation X, there will always be a transformation Y (the inverse of X) such that

$$XY = E$$

The inverse of a transformation $S \to S'$ is designated $S' \to S$.

Using requirements 1 and 2, we obtain

$$
\begin{array}{ll}
S \to S' & S' \to S' \\
z = \gamma(v)(z' + vt') & z' = \delta(-v)(z - vt) \\
x = \varepsilon(v)x' & x' = \varepsilon(-v)x \\
y = \varepsilon(v)y' & y' = \varepsilon(-v)y
\end{array}
\tag{6.4.1}
$$

The transformation for z derives from the fact that it must be linear and that, if $z' = 0$, $z = vt$. We see also that we must have

$$\gamma(v) = \gamma(-v) \tag{6.4.2}$$

Let us consider a ruler at rest in S, on the z axis, and with the end point coordinates 0 and L in S. At the time $t' = 0$ for an observer O' in S', the end-point coordinates will be 0 and $L/\gamma(v)$; for such an observer the length of the ruler will be $L/\gamma(v)$. Let us then set the ruler at rest in S' and still on the $z(z')$ axis; in S' the coordinates of the end points of the ruler will be 0 and L, but for an observer O in S at the time $t = 0$ they will be 0 and $L/\gamma(-v)$ and the length of the ruler is $L/\gamma(-v)$. Since the measurements made by the two observes O' and O must be the same, relation (6.4.2) is proved.

We have also

$$x = \frac{1}{\varepsilon(-v)}x' = \varepsilon(v)x' \qquad (6.4.3)$$

or

$$\varepsilon(v)\varepsilon(-v) = 1 \qquad (6.4.4)$$

and, because of symmetry,

$$\varepsilon(v) = \varepsilon(-v) = 1 \qquad (6.4.5)$$

Therefore, transformation (6.4.1) can be written

$$\begin{aligned}
z &= \gamma(v)(z' + vt') & z' &= \gamma(v)(z - vt) \\
x &= x' & x' &= x \qquad (6.4.6) \\
y &= y' & y' &= y
\end{aligned}$$

Assume now that the two frames of reference coincide at time $t = 0$, and that at time $t = 0$ a light pulse is sent to a point P where it triggers an explosion. P has coordinates $(0, 0, z)$ in S and $(0, 0, z')$ in S'; see Fig. 6.4.

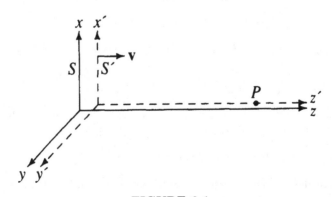

FIGURE 6.4

For an observer in S, the explosion occurs at point z at time z/c, $z = ct$. For an observer in S', the explosion occurs at point z' at time z'/c, $z' = ct'$. These two statements must be consistent with the transformation (6.4.6):

$$ct = \gamma(z' + vt'), \quad ct' = \gamma(z - vt)$$

or, since $z' = ct'$ and $z = ct$,

$$ct = \gamma(ct' + vt'), \quad ct' = \gamma(ct - vt)$$

and

$$\gamma(c + v)t' = ct, \quad \gamma(c - v)t = ct'$$

From the above relations we derive

$$\gamma^2(c^2 - v^2)tt' = c^2 tt'$$

or

$$\gamma^2 = \frac{c^2}{c^2 - v^2} = \frac{1}{1 - (v^2/c^2)}$$

$$\gamma = \frac{1}{\sqrt{1 - (v^2/c^2)}} \tag{6.4.7}$$

We must now find the transformation properties of the time coordinate. We know that

$$z = \gamma(z' + vt')$$
$$z' = \gamma(z - vt)$$

Then

$$z' = \gamma[\gamma(z' + vt') - vt]$$

and

$$t = \gamma\left(t' + \frac{v}{c^2}z'\right) \tag{6.4.8}$$

Analogously,

$$z = \gamma[\gamma(z - vt) + vt']$$

gives

$$t' = \gamma\left(t - \frac{v}{c^2}z\right) \tag{6.4.9}$$

The special Lorentz transformation can now be expressed as follows:

$$x = x' \qquad\qquad x' = x$$
$$y = y' \qquad\qquad y' = y$$
$$z = \gamma(z' + vt') \qquad z' = \gamma(z - vt) \qquad\qquad (6.4.10)$$
$$t = \gamma\left(t' + \frac{v}{c^2}z'\right) \qquad t' = \gamma\left(t - \frac{v}{c^2}z\right)$$
$$S \to S' \qquad\qquad S' \to S$$

Lorentz did not derive these equations; rather, he thought that the scale of lengths had to be changed in order to explain the Michelson–Morley experiment. (We shall treat this subject in the next section.) The above equations were derived by Larmor, before the turn of the century. It was Einstein who grasped their generality.

Note the invariance

$$x'^2 + y'^2 + z'^2 - c^2 t'^2 = x^2 + y^2 + z^2 - c^2 t^2 \qquad\qquad (6.4.11)$$

6.5. Lorentz Contraction, Time Dilation, and Addition of Velocities

Lorentz Contraction. Let us consider a rod at rest in S; its length is (see Fig. 6.5)

$$L = z_B - z_A \qquad\qquad (6.5.1)$$

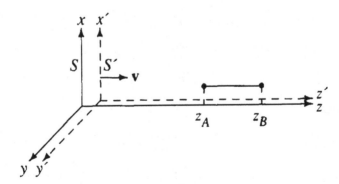

FIGURE 6.5

An observer in S' measures the length of the rod at time t' and finds

$$L' = z'_B - z'_A \tag{6.5.2}$$

On the other hand,

$$z_A = \gamma(z'_A + vt'_A)$$
$$z_B = \gamma(z'_B + vt'_B) \tag{6.5.3}$$

and, being $t'_A = t'_B = t'$,

$$L' = z'_B - z'_A = \frac{1}{\gamma}(z_B - z_A) = L\sqrt{1 - \frac{v^2}{c^2}} \tag{6.5.4}$$

The rod is seen by an observer moving with S' as contracted by the factor $1/\gamma$. Since the contraction factor depends on v^2, a rod moving with S' would be seen contracted by the same factor by an observer at rest in S.

We note that if we replace in Eq. (6.2.7) l by the contracted value $l\sqrt{1 - (v^2/c^2)}$ we obtain for t_1 a value equal to t_2 given by (6.2.8).

Time Dilation. Let us assume that we have a clock with period T at rest in S:

$$T = t_B - t_A \tag{6.5.5}$$

Let us call T' the period in S'. We want to express the fact that the clock is at rest in S:

$$(xyz)_{t_B} = (xyz)_{t_A} \tag{6.5.6}$$

Then

$$T' = t'_B - t'_A = \gamma\left(t_B - \frac{v}{c^2}z_B - t_A + \frac{v}{c^2}z_A\right)$$
$$= \gamma(t_B - t_A) = \gamma T = \frac{1}{\sqrt{1 - (v^2/c^2)}}T \tag{6.5.7}$$

The clock, as seen from the moving system, runs slower. A moving clock is late and lags with respect to a fixed clock by a factor γ:

$$\begin{pmatrix} \text{time in frame of} \\ \text{reference at rest} \\ \text{with clock} \end{pmatrix} = \begin{pmatrix} \text{time in frame of} \\ \text{reference moving} \\ \text{with respect to} \\ \text{clock} \end{pmatrix} \times \sqrt{1 - \frac{v^2}{c^2}} \tag{6.5.8}$$

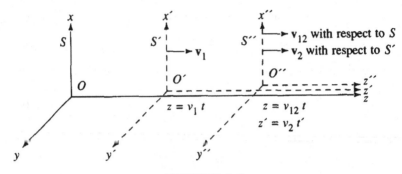

FIGURE 6.6

Addition of Velocities. Let us consider, as in Fig. 6.6, three frames of reference and a point P that has coordinates

$$z, t \text{ in } S$$
$$z', t' \text{ in } S'$$
$$z'', t'' \text{ in } S''$$

We have

$$z' = \gamma_1(z - v_1 t)$$
$$t' = \gamma_1 \left(t - \frac{v_1}{c^2} z \right) \tag{6.5.9}$$

$$z'' = \gamma^2 \left(z' - v_2 t' \right)$$
$$t'' = \gamma^2 \left(t' - \frac{v_2}{c_2} z' \right) \tag{6.5.10}$$

where $\gamma_i = \frac{1}{\sqrt{1-(v_i^2/c^2)}}$. Therefore,

$$z'' = \gamma_2 \left[\gamma_1(z - v_1 t) - v_2 \gamma_1 \left(t - \frac{v_1}{c^2} z \right) \right]$$
$$= \gamma_1 \gamma_2 \left[z \left(1 + \frac{v_1 v_2}{c^2} \right) - (v_1 + v_2) t \right]$$

$$t'' = \gamma_2 \left[\gamma_1 \left(t - \frac{v_1}{c^2} z \right) - \frac{v_2}{c^2} \gamma_1 (z - v_1 t) \right] \tag{6.5.11}$$
$$= \gamma_1 \gamma_2 \left[\left(1 + \frac{v_1 v_2}{c^2} \right) t - \frac{v_1 + v_2}{c^2} z \right]$$

On the other hand, we should have

$$z'' = \gamma_{12}(z - v_{12}t)$$
$$t'' = \gamma_{12}\left(t - \frac{v_{12}}{c^2}z\right) \tag{6.5.12}$$

where $\gamma_{12} = \frac{1}{\sqrt{1-(v_{12}^2/c^2)}}$. Therefore,

$$\gamma_{12} = \gamma_1\gamma_2\left(1 + \frac{v_1v_2}{c^2}\right) \tag{6.5.13}$$

$$\gamma_{12}v_{12} = \gamma_1\gamma_2(v_1 + v_2) \tag{6.5.14}$$

from which we derive

$$v_{12} = \frac{v_1 + v_2}{1 + (v_1v_2/c^2)} \tag{6.5.15}$$

Note that, if we set $v_2 = c$, we obtain $v_{12} = c$, and if we set $v_{12} = c$, we obtain $v_2 = c$, as expected.

6.6. Minkowski Notation

The Lorentz transformation is expressed by the equations

$$
\begin{aligned}
x &= x' & x' &= x \\
y &= y' & y' &= y \\
z &= \gamma(z' + vt') & z' &= \gamma(z - vt) \\
t &= \gamma\left(t' + \frac{v}{c^2}z'\right) & t' &= \gamma\left(t - \frac{v}{c^2}z\right) \\
S &\to S' & S' &\to S
\end{aligned}
\tag{6.6.1}
$$

We change the notation, as follows:

$$
\begin{aligned}
x_0 &= ct \\
x_0' &= ct' \\
x, y, z &= x_1, x_2, x_3 \\
x', y', z' &= x_1', x_2', x_3' \\
\frac{v}{c} &= \beta, \quad vt = \frac{v}{c}ct = \beta x_0 \\
\gamma &= \frac{1}{\sqrt{1 - \beta^2}}
\end{aligned}
\tag{6.6.2}
$$

With this new notation, Eqs. (6.6.1) become

$$x_1 = x_1' \qquad x_1' = x_1$$
$$x_2 = x_2' \qquad x_2' = x_2$$
$$x_3 = \frac{x_3' + \beta x_0'}{\sqrt{1 - \beta^2}} \qquad x_3' = \frac{x_3 - \beta x_0}{\sqrt{1 - \beta_2}} \qquad (6.6.3)$$
$$x_0 = \frac{x_0' + \beta x_3'}{\sqrt{1 - \beta^2}} \qquad x_0' = \frac{x_0 - \beta x_3}{\sqrt{1 \quad \beta^2}}$$

We can neglect the coordinates x_1 and x_2 that do not undergo any modification. Then the possible punctual events in the unprimed frame can be represented in a space–time diagram (x_0, x_3), as in Fig. 6.7.

The motion of a point is represented by a curve whose tangent forms with the time axis the angle

$$\theta = \tan^{-1} \frac{dx_3}{dx_0} = \tan^{-1} \frac{v}{c} \qquad (6.6.4)$$

where $v = $ instantaneous velocity of the point. Since $v \leq c$, the angle θ is always smaller than $45°$. The motion of a ray of light is represented by a straight line inclined by $45°$.

The system of coordinates related to a frame S', moving in the $x_3 = z$ direction with respect to S, is indicated in Fig. 6.8. Note that

$$x_0' = 0, \quad \text{if } x_0 - \beta x_3 = 0, x_0 = \beta x_3, \frac{x_0}{x_3} = \beta$$

$$x_3' = 0, \quad \text{if } x_3 - \beta x_0 = 0, x_3 = \beta x_0, \frac{x_3}{x_0} = \beta$$

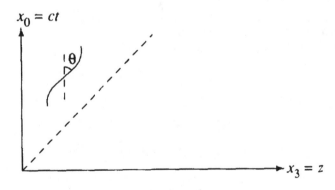

FIGURE 6.7

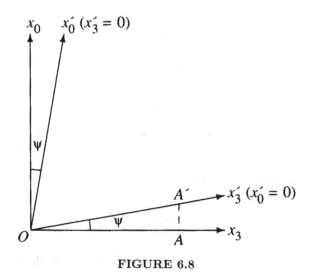

FIGURE 6.8

In Fig. 6.8,

$$\tan \psi = \beta = \frac{v}{c} \tag{6.6.5}$$

We see immediately that the contemporaneity of events is relative. To an observer at rest in S', all the punctual events on the x_3' axis are contemporary; to an observer at rest in S they evolve in time. For the latter the event A' follows event O, by an interval of time equal to $(AA')/c$.

6.7. General Lorentz Transformation

We want now to find a general transformation that allows us to go from a system of coordinates to an equivalent one. The word "general" refers to the following generalizations:

(1) The directions of the velocity **v** with which one frame (S') moves with respect to another frame (S) will be arbitrary.

(2) The axes of S will be allowed to rotate with respect to those of S'.

On the other hand, translations and reflections in both space and time will continue to be ruled out.

Let us consider the two systems of coordinates.

$$\begin{aligned}
x_1 &= x & x_1' &= x' \\
x_2 &= y & x_2' &= y' \\
x_3 &= z & x_3' &= z' \\
x_4 &= ict & x_4' &= ict'
\end{aligned} \qquad (6.7.1)$$

Assume that the origins of the two systems coincide in the four-dimensional space. We proceed.

(1) The transformation must be linear:

$$x_i' = \sum_{k=1}^{4} a_{ik} x_k \qquad (6.7.2)$$

where the a_{ik} coefficients can be considered arbitrary at the outset.

(2) The velocity of light is the same in the two sytems:

$$x^2 + y^2 + z^2 - c^2 t^2 = \text{invariant} \qquad (6.7.3)$$

or

$$\sum_{i=1}^{4} x_i^2 = \sum_{i=1}^{4} x_1'^2 \qquad (6.7.4)$$

This condition will impose some restrictions on the coefficients a_{ik}:

$$\begin{aligned}
\sum_{i=1}^{4} x_i'^2 &= \sum_{i=1}^{4} \left[\left(\sum_{k=1}^{4} a_{ik} x_k \right) \left(\sum_{l=1}^{4} a_{il} x_l \right) \right] \\
&= \sum_{k=1}^{4} \sum_{l=1}^{4} x_k x_l \sum_{i=1}^{4} a_{ik} a_{il}
\end{aligned}$$

Then we must have

$$\sum_{i=1}^{4} a_{ik} a_{il} = \delta_{kl} \qquad (6.7.5)$$

Using Eq. (6.7.2), multiply both members by a_{il} and sum over i:

$$\sum_{i=1}^{4} a_{il} x_i' = \sum_{i=1}^{4} a_{il} \sum_{i=1}^{4} a_{il} x_k = \sum_{k=1}^{4} x_k \sum_{i=1}^{4} a_{il} a_{ik} = x_l \qquad (6.7.6)$$

Then

$$x_l = \sum_{i=1}^{4} a_{il} x_i'$$

and

$$x_i = \sum_{k=1}^{4} a_{ki} x_k' \tag{6.7.7}$$

But

$$\sum_{i=1}^{4} x_i^2 = \sum_{i=1}^{4} \left[\left(\sum_{k=1}^{4} a_{ki} x_k' \right) \left(\sum_{l=1}^{4} a_{li} x_l' \right) \right]$$

$$= \sum_{k=1}^{4} \sum_{l=1}^{4} x_k' x_l' \sum_{i=1}^{4} a_{ki} a_{li}$$

Then we must have

$$\sum_{i=1}^{4} a_{ki} a_{li} = \delta_{kl} \tag{6.7.8}$$

The properties (6.7.5) and (6.7.8) define the coefficients a_{ik}, which form a 4×4 matrix. It is left to the reader to show that the determinant of this matrix can be $+1$ or -1. We choose it to be $+1$, imposing an additional condition whose meaning we shall explain later.

The transformations (6.7.2) form a group. In fact, take the two transformations

$$x_k' = \sum_{l=1}^{4} a_{kl} x_l \tag{6.7.9}$$

$$x_i'' = \sum_{k=1}^{4} b_{ik} x_k' = \sum_{l=1}^{4} c_{il} x_l \tag{6.7.10}$$

Then

$$x_i'' = \sum_{k=1}^{4} b_{ik} x_k' = \sum_{k=1}^{4} b_{ik} \sum_{l=1}^{4} a_{kl} x_l$$

$$= \sum_{l=1}^{4} x_l \left(\sum_{k=1}^{4} b_{ik} a_{kl} \right) = \sum_{l} c_{il} x_l \tag{6.7.11}$$

where

$$c_{il} = \sum_{k=1}^{4} b_{ik} a_{kl} \tag{6.7.12}$$

Now we show that the coefficients c_{il} have the same properties as the coefficients a_{ik} and b_{ik}:

$$\sum_{i=1}^{4} c_{il} c_{in} = \sum_{i=1}^{4} \left[\left(\sum_{k=1}^{4} b_{ik} a_{kl} \right) \left(\sum_{m=1}^{4} b_{im} a_{mn} \right) \right]$$

$$= \sum_{k=1}^{4} \sum_{m=1}^{4} a_{kl} a_{mn} \sum_{i=1}^{4} b_{ik} b_{im} = \sum_{k=1}^{4} \sum_{m=1}^{4} a_{kl} a_{mn} \delta_{km}$$

$$= \sum_{k=1}^{4} a_{kl} a_{kn} = \delta_{ln} \tag{6.7.13}$$

$$\sum_{l=1}^{4} c_{il} c_{nl} = \sum_{l=1}^{4} \left[\left(\sum_{k=1}^{4} b_{ik} a_{kl} \right) \left(\sum_{m=1}^{4} b_{nm} a_{ml} \right) \right]$$

$$= \sum_{k=1}^{4} \sum_{m=1}^{4} b_{ik} b_{nm} \sum_{l=1}^{4} a_{kl} a_{ml} = \sum_{k=1}^{4} \sum_{m=1}^{4} b_{ik} b_{nm} \delta_{km}$$

$$= \sum_{k=1}^{4} b_{ik} b_{nk} = \delta_{in} \tag{6.7.14}$$

We note at this point that the coefficients a_{ik} are all real, with the exception of a_{14}, a_{24}, a_{34}, a_{41}, a_{42} and a_{43}, which are pure imaginary; this feature is preserved when going from one system to another.

Let us consider the following "pure" transformations.

Ordinary Rotation. Time does not change, and the measure of time does not change. The transformations are of the type

$$\begin{pmatrix} b_{11} & b_{12} & b_{13} & 0 \\ b_{21} & b_{22} & b_{23} & 0 \\ b_{31} & b_{32} & b_{33} & 0 \\ 0 & 0 & 0 & 1 \end{pmatrix} \tag{6.7.15}$$

The 3×3 matrix of the b_{ik} $(i, k = 1, 2, 3)$ coefficients represents a coordinate transformation in real space and is a real orthogonal matrix.

Pure Lorentz Transformation. Consider two inertial frames S and S', with S' moving respect to S with a constant velocity $\mathbf{v}$. Let the subscripts $\parallel$ and $\perp$ indicate components of vectors parallel and perpendicular to $\mathbf{v}$, respectively. We have (see Fig. 6.9)

$$\mathbf{r}'_{\parallel} = \frac{\mathbf{r}_{\parallel} - \mathbf{v}t}{\sqrt{1 - \beta^2}}$$

$$\mathbf{r}'_{\perp} = \mathbf{r}_{\perp} \tag{6.7.16}$$

$$t' = \frac{t - [(\mathbf{v} \cdot \mathbf{r})/c^2]}{\sqrt{1 - \beta^2}}$$

All rotations are excluded when dealing with pure Lorentz transformations; in particular, spatial rotations about the direction of $\mathbf{v}$ are excluded, are $\mathbf{r}'$ can be put in the form

$$\mathbf{r}' = a\mathbf{r} + b\mathbf{v}$$

since it lies in the $(\mathbf{r}, \mathbf{v})$ plane. It follows from Eq. (6.7.16)

$$\mathbf{r}'_{\perp} = a\mathbf{r}_{\perp}, \quad a = 1 \tag{6.7.17}$$

and

$$\mathbf{r}'_{\parallel} = \mathbf{r}_{\parallel} + b\mathbf{v} = \frac{\mathbf{r}_{\parallel} - \mathbf{v}t}{\sqrt{1 - \beta^2}} \tag{6.7.18}$$

CLAIM

$$b = \frac{\mathbf{r} \cdot \mathbf{v}}{v^2} \left(\frac{1}{\sqrt{1 - \beta^2}} - 1 \right) - \frac{t}{\sqrt{1 - \beta^2}} \tag{6.7.19}$$

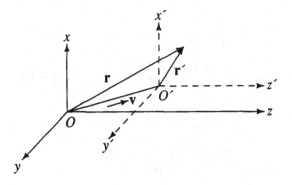

FIGURE 6.9

PROOF

$$\mathbf{r}'_\| = \mathbf{r}_\| + \left[\frac{\mathbf{r}\cdot\mathbf{v}}{v^2}\left(\frac{1}{\sqrt{1-\beta^2}}-1\right)-\frac{t}{\sqrt{1-\beta^2}}\right]\mathbf{v}$$

$$= \mathbf{r}_\| + \frac{\mathbf{r}\cdot\mathbf{v}}{v^2}\frac{\mathbf{v}}{\sqrt{1-\beta^2}} - \frac{\mathbf{r}\cdot\mathbf{v}}{v^2}\mathbf{v} - \frac{t\mathbf{v}}{\sqrt{1-\beta^2}}$$

$$= \mathbf{r}_\| + \frac{\mathbf{r}_\|}{\sqrt{1-\beta^2}} - \mathbf{r}_\| - \frac{t\mathbf{v}}{\sqrt{1-\beta^2}} = \frac{\mathbf{r}_\|}{\sqrt{1-\beta^2}}\frac{\mathbf{v}t}{\sqrt{1-\beta^2}}$$

Q.E.D

We can then write

$$\mathbf{r}' = \mathbf{r} + \frac{\mathbf{v}(\mathbf{r}\cdot\mathbf{v})}{v^2}\left(\frac{1}{\sqrt{1-\beta^2}}-1\right)-\frac{\mathbf{v}t}{\sqrt{1-\beta^2}}$$

$$t' = \frac{t-[(\mathbf{r}\cdot\mathbf{v})/c^2]}{\sqrt{1-\beta^2}}$$

(6.7.20)

and

$$x' = x + \frac{v_x(v_x x + v_y y + v_z z)}{v^2}\left(\frac{1}{\sqrt{1-\beta^2}}-1\right)$$

$$- \frac{v_x t}{\sqrt{1-\beta^2}}$$

$$y' = y + \frac{v_y(v_x x + v_y y + v_z z)}{v^2}\left(\frac{1}{\sqrt{1-\beta^2}}-1\right)$$

$$- \frac{v_y t}{\sqrt{1-\beta^2}}$$

(6.7.21)

$$z' = z + \frac{v_z(v_x x + v_y y + v_z z)}{v^2}\left(\frac{1}{\sqrt{1-\beta^2}}-1\right)$$

$$- \frac{v_z t}{\sqrt{1-\beta^2}}$$

$$t' = \frac{t-[(v_x x + v_y y + v_z z)/c^2]}{\sqrt{1-\beta^2}}$$

Then the matrix of the coefficients is given by

$$
\begin{vmatrix}
1 + \dfrac{v_x^2}{v^2}\chi & \dfrac{v_x v_y}{v^2}\chi & \dfrac{v_x v_z}{v^2}\chi & \dfrac{iv_x}{c\sqrt{1-\beta^2}} \\[2ex]
\dfrac{v_x v_y}{v^2}\chi & 1 + \dfrac{v_y^2}{v^2}\chi & \dfrac{v_y v_z}{v^2}\chi & \dfrac{iv_y}{c\sqrt{1-\beta^2}} \\[2ex]
\dfrac{v_x v_z}{v^2}\chi & \dfrac{v_y v_z}{v^2}\chi & 1 + \dfrac{v_z^2}{v^2}\chi & \dfrac{iv_z}{c\sqrt{1-\beta^2}} \\[2ex]
-\dfrac{iv_x}{c\sqrt{1-\beta^2}} & -\dfrac{iv_y}{c\sqrt{1-\beta^2}} & -\dfrac{iv_z}{c\sqrt{1-\beta^2}} & \dfrac{1}{\sqrt{1-\beta^2}}
\end{vmatrix}
\qquad (6.7.22)
$$

where

$$
\chi = \frac{1}{\sqrt{1-\beta^2}} - 1 \qquad (6.7.23)
$$

If $v_x = v_y = 0$,

$$
\begin{vmatrix}
1 & 0 & 0 & 0 \\[1ex]
0 & 1 & 0 & 0 \\[1ex]
0 & 0 & \dfrac{1}{\sqrt{1-\beta^2}} & \dfrac{i\beta}{\sqrt{1-\beta^2}} \\[2ex]
0 & 0 & -\dfrac{i\beta}{\sqrt{1-\beta^2}} & \dfrac{1}{\sqrt{1-\beta^2}}
\end{vmatrix}
\qquad (6.7.24)
$$

It is always possible to decompose in a unique way the most general transformation

$$
x_i' = \sum_{k=1}^{4} a_{ik} x_k \qquad (6.7.25)
$$

into the product of a pure Lorentz transformation and a pure spatial transformation. The proof of this statement follows.

The origin O' of the prime frame S' has coordinates in S':

$$
x_1' = x_2' = x_3' = 0 \qquad (6.7.26)
$$

The origin O' of the frame S' has the following coordinates in S:

$$
x_i = a_{4i} x_4', \quad i = 1, 2, 3, 4 \qquad (6.7.27)
$$

or

$$x_1 = a_{41}x_4' = \frac{a_{41}}{a_{44}}x_4 = \frac{a_{41}}{a_{44}}ict$$

$$x_2 = a_{42}x_4' = \frac{a_{42}}{a_{44}}x_4 = \frac{a_{42}}{a_{44}}ict$$

$$x_3 = a_{43}x_4' = \frac{a_{43}}{a_{44}}x_4 = \frac{a_{43}}{a_{44}}ict$$

$$x_4 = a_{44}x_4'$$

(6.7.28)

The relative velocity of S' with respect to S has the following components:

$$v_1 = ic\frac{a_{41}}{a_{44}}$$

$$v_2 = ic\frac{a_{42}}{a_{44}}$$

$$v_3 = ic\frac{a_{43}}{a_{44}}$$

(6.7.29)

Consider now a frame S'' that moves with respect to S with a constant velocity of components v_1, v_2, and v_3 given above. We can pass from S to S'' by a means of a pure Lorentz transformation defined by v_1, v_2, and v_3. The frame S'' and the frame S' have the origin of the coordinates in common. S'' differs from S' only by an ordinary spatial rotation about their common origin.

Let b_{ik} be the coefficients of this rotation and c_{ik} the coefficients of the pure Lorentz transformation. We can write

$$a_{il} = \sum_{s=1}^{4} b_{is}c_{sl}$$

(6.7.30)

Multiplying by c_{kl} and summing over l, we obtain

$$\sum_{l=1}^{4} c_{kl}a_{il} = \sum_{l=1}^{4} c_{kl}\sum_{s=1}^{4} b_{is}c_{sl}$$

$$= \sum_{s=1}^{4} b_{is}\sum_{l=1}^{4} c_{kl}c_{sl} = \sum_{s=1}^{4} b_{is}\delta_{ks} = b_{ik}$$

(6.7.31)

That is, we obtain the coefficients of the pure rotation.

We can now make the following observations:

(1) The determinant of a pure Lorentz transformation matrix is always $+1$, as is easy to verify from Eq. (6.7.24) or, with more difficulty, from Eq. (6.7.22).

(2) Given a general Lorentz transformation, the sign of the determinant is then determined by the spatial part. By limiting ourselves to determinants equal to $+1$, we are considering only *proper* rotations. Reflections or inversions through the origin of coordinates are excluded.

(3) It is not possible to obtain the transformation represented by the matrix Eq. (6.7.22) as the succession of three *special* transformations of type (6.7.24) in the x, y, and z directions.

(4) The product of two special Lorentz transformations is not in general a pure Lorentz transformation, but includes a proper rotation.

6.8. Scalars, Vectors, and Tensors in Four Dimensions

A *scalar* quantity is, in general, invariant under a Lorentz transformation. A product of two four-vectors is a scalar:

$$\sum_{\mu=1}^{4} q_\mu^{(1)} q_\mu^{(2)} = \text{invariant} \tag{6.8.1}$$

A *four-vector* is defined by four components, three real and one imaginary. A position four-vector is given by

$$x_\mu \equiv (x_1, x_2, x_3, x_4 = ict) \tag{6.8.2}$$

The components in two inertial frames S and S' are related as follows:

$$x_\nu = \sum_\mu a_{\mu\nu} x_\mu', \quad x_\mu' = \sum_\nu a_{\mu\nu} x_\nu \tag{6.8.3}$$

Consider two points

$$P(x_1, x_2, x_3, x_4) \quad \text{and} \quad Q(y_1, y_2, y_3, y_4)$$

The length of the four-vector PQ in the S frame is

$$l = \sqrt{(y_1 - x_1)^2 + (y_2 - x_2)^2 + (y_3 - x_3)^2 + (y_4 - x_4)^2} \tag{6.8.4}$$

If l' is the length of the four-vector in S',

$$l'^2 = \sum_{i=1}^{4}(y_i' - x_i')^2$$

$$= \sum_{i=1}^{4}\left[\sum_{k=1}^{4}a_{ik}(y_k - x_k)\right]\left[\sum_{n=1}^{4}a_{in}(y_n - x_n)\right]$$

$$= \sum_{k=1}^{4}\sum_{n=1}^{4}(y_k - x_k)(y_n - x_n)\sum_{i=1}^{4}a_{ik}a_{in} = \sum_{k=1}^{4}(y_k - x_k)^2 \quad (6.8.5)$$

This shows that the length or module of a four-vector is invariant. In particular, if $l^2 > 0$ the four-vector is called *spatial*, if $l^2 < 0$ the four-vector is called *temporal*, and if $l^2 = 0$, the four-vector is called *lightlike*.

In three dimensions it is always possible to bring *any* coordinate axis in the direction of a certain vector by means of an appropriate rotation of the coordinate system. In four dimensions this is not possible because of the special role of the fourth dimension, associated to an *imaginary* axis. If, for a certain vector, $l^2 > 0$, this relation will continue to be valid regardless of the orientation of the axes. In particular, if a coordinate axis is made to fall in the direction of the vector, the vector has only one component different from zero that is real; this means that we can line up only a spatial axis in the direction of a spatial vector. If $l^2 < 0$, we can only line up the time axis in the direction of the vector.

A *four-dimensional tensor* is given in the following form:

$$\begin{vmatrix} T_{11} & T_{12} & T_{13} & T_{14} \\ T_{21} & T_{22} & T_{23} & T_{24} \\ T_{31} & T_{32} & T_{33} & T_{34} \\ T_{41} & T_{42} & T_{43} & T_{44} \end{vmatrix} \quad (6.8.6)$$

It has 16 components, which transform as follows:

$$T_{ik}' = \sum_{l=1}^{4}\sum_{m=1}^{4}a_{il}a_{km}T_{lm} \quad (6.8.7)$$

where $i, k = 1, 2, 3, 4$. If $T_{ik} = T_{ki}$, the tensor is *symmetric*; if $T_{ik} = -T_{ki}$, the tensor is *antisymmetric*. Every tensor can be decomposed into the sum of a symmetric tensor and an antisymmetric tensor:

$$T_{ik} = \frac{1}{2}(T_{ik} + T_{ki}) + \frac{1}{2}(T_{ik} - T_{ki}) \tag{6.8.8}$$

where

$$\frac{1}{2}(T_{ik} + T_{ki}) = \text{symmetric tensor}$$

$$\frac{1}{2}(T_{ik} - T_{ki}) = \text{antisymmetric tensor}$$

An antisymmetric tensor has only six independent components:

$$\begin{pmatrix} 0 & T_{12} & T_{13} & T_{14} \\ -T_{12} & 0 & T_{23} & T_{24} \\ -T_{13} & -T_{23} & 0 & T_{34} \\ -T_{14} & -T_{24} & -T_{34} & 0 \end{pmatrix} \tag{6.8.9}$$

We shall now examine the *fields*, entities that are functions of position and time, in four dimensions. A *scalar field S* is defined in such a way that, while its functional form may change under a Lorentz transformation, its value at a certain point in four-dimensional space remains unchanged:

$$S(x_\mu) = S'(X'_\mu) \tag{6.8.10}$$

A *vector field* is characterized by four components in the space–time system (see Fig. 6.10). The components of the vector are different in the different system of coordinates; these components are examples of "scalars that vary."

We define the *gradient operator* in four dimensions:

$$\frac{\partial}{\partial x_\mu} \equiv \left(\frac{\partial}{\partial x_1}, \frac{\partial}{\partial x_2}, \frac{\partial}{\partial x_3}, \frac{\partial}{\partial x_4} = \frac{1}{ic}\frac{\partial}{\partial t} \right) \tag{6.8.11}$$

If we operate with this operator on a scalar field, we obtain

$$\frac{\partial S'(x')}{\partial x'_\mu} = \frac{\partial S(x)}{\partial x'_\mu} = \sum_\nu \frac{\partial S(x)}{\partial x_\nu}\frac{\partial x_\nu}{\partial x'_\mu} = \sum_\nu a_{\mu\nu}\frac{\partial S(x)}{\partial x_\nu} \tag{6.8.12}$$

The "product" of $\partial/\partial x'_\mu$ and $S'(x')$ is a vector field; then $\partial/\partial x_\mu$ is a vector. The *divergence* of a vector field in four dimensions is a scalar field:

$$\sum_{\mu=1}^{4} \frac{\partial V_\mu}{\partial x_\mu} = \text{scalar field} \tag{6.8.13}$$

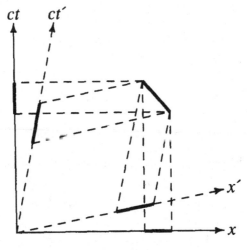

FIGURE 6.10

A *tensor field* $T_{\mu\nu}(x)$ is such that

$$T'_{\mu\nu}(x') \sum_{\sigma=1}^{4} \sum_{\tau=1}^{4} a_{\mu\sigma} a_{\nu\tau} T_{\sigma\tau}(x) \tag{6.8.14}$$

The cross product of two vector fields is an antisymmetric tensor field. For one component,

$$\begin{aligned}
A'_{\mu} B'_{\nu} - A'_{\nu} B'_{\mu} &= \sum_{\sigma=1}^{4} a_{\mu\sigma} A_{\sigma} \sum_{\tau=1}^{4} a_{\nu\tau} B_{\tau} \\
&\quad - \sum_{\tau=1}^{4} a_{\nu\tau} A_{\tau} \sum_{\sigma=1}^{4} a_{\mu\sigma} B_{\sigma} \\
&= \sum_{\sigma=1}^{4} \sum_{\tau=1}^{4} a_{\mu\sigma} a_{\nu\tau} (A_{\sigma} B_{\tau} - A_{\tau} B_{\sigma})
\end{aligned} \tag{6.8.15}$$

The curl of a vector field is also an antisymmetric tensor field; a typical component is

$$\frac{\partial V_{\mu}}{\partial x_{\nu}} - \frac{\partial V_{\nu}}{\partial x_{\mu}} \tag{6.8.16}$$

Given a tensor field $T_{\mu\nu}(x)$, the four components $\sum_{\nu=1}^{4}(\partial T_{\mu\nu}/\partial x_{\nu})$ form a four-vector field:

$$\sum_{\nu=1}^{4} \frac{\partial T_{\mu\nu}}{\partial x_{\nu}} = V_{\mu} \tag{6.8.17}$$

In fact,

$$\frac{\partial T'_{\tau\lambda}}{\partial x'_{\lambda}} = \sum_{\nu=1}^{4} a_{\lambda\nu} \frac{\partial T'_{\tau\lambda}}{\partial x_{\nu}} = \sum_{\nu=1}^{4} a_{\lambda\nu} \sum_{\mu=1}^{4} \sum_{\sigma=1}^{4} a_{\tau\mu} a_{\lambda\sigma} \frac{\partial T_{\mu\sigma}}{\partial x_{\nu}}$$

and

$$\sum_{\lambda=1}^{4} \frac{\partial T'_{\tau\lambda}}{\partial x'_{\lambda}} = \sum_{\lambda=1}^{4} \sum_{\nu=1}^{4} a_{\lambda\nu} \sum_{\mu=1}^{4} \sum_{\sigma=1}^{4} a_{\tau\mu} a_{\lambda\sigma} \frac{\partial T_{\mu\sigma}}{\partial x_{\nu}}$$

$$= \sum_{\nu=1}^{4} \sum_{\mu=1}^{4} a_{\tau\mu} \sum_{\sigma=1}^{4} \sum_{\lambda=1}^{4} a_{\lambda\nu} a_{\lambda\sigma} \frac{\partial T_{\mu\sigma}}{\partial x_{\nu}}$$

$$= \sum_{\nu=1}^{4} \sum_{\mu=1}^{4} a_{\tau\mu} \sum_{\sigma=1}^{4} \delta_{\nu\sigma} \frac{\partial T_{\mu\sigma}}{\partial x_{\nu}} = \sum_{\nu=1}^{4} \sum_{\mu=1}^{4} a_{\tau\mu} \frac{\partial T_{\mu\nu}}{\partial x_{\nu}}$$

$$= \sum_{\mu=1}^{4} a_{\tau\mu} \left(\sum_{\nu=1}^{4} \frac{\partial T_{\mu\nu}}{\partial x_{\nu}} \right)$$

In a similar manner, it is possible to show that the four components $\sum_{\tau=1}^{4}(\partial T_{\tau\lambda}/\partial x_{\tau})$ form a four-vector field.

D'Alembertian Operator

$$\Box^2 = \sum_{\mu=1}^{4} \frac{\partial^2}{\partial x_{\mu}} \tag{6.8.18}$$

is a scalar operator. It does not change the nature of the field it operates on:

$$\Box^2 S_1(x) = S_2(x) = \text{scalar field} \tag{6.8.19}$$

$$\Box^2 V_{\mu}^{(1)}(x) = V_{\mu}^{(2)}(x) = \text{vector field} \tag{6.8.20}$$

The first postulate of Einstein's principle of special relativity tells us that all the laws of physics must have the same form in different Lorentz (inertial) frames. The equations that describe these laws must be written in such ways that their forms are independent of the choice of the inertial frame; when this is the case the equations are said to be expressed in *Lorentz-covariant form.*

Covariant relations have the same transformation properties under Lorentz transformations, as exemplified below. Let us consider the following four relations:

(a) $S^{(1)}(x) = S^{(2)}(x)$

(b) $V_\mu^{(1)}(x) = V_\mu^{(2)}(x)$

(c) $T_{\mu\nu}^{(1)}(x) = T_{\mu\nu}^{(2)}(x)$

(d) $S(x) = V_4(x)$

expressing the fact that in a certain system of coordinates

(a) Two scalars are equal.

(b) Two vectors are equal.

(c) Two tensors are equal.

(d) A scalar is equal to the fourth component of a vector.

Consider a new system of coordinates:

(a) Since $S(x) = S'(x')$,

$$S^{(1)'}(x') = S^{(2)'}(x')$$

(b) Since

$$V_\mu'(x') = \sum_{\nu=1}^{4} a_{\mu\nu} V_\nu(x)$$

it follows that

$$V_\mu^{(1)'}(x') - V_\mu^{(2)'}(x') = \sum_{\nu=1}^{4} a_{\mu\nu}[V_\nu^{(1)}(x) - V_\nu^{(2)}(x)] = 0$$

The difference in brackets is zero for any ν; then

$$V_\mu^{(1)'}(x') = V_\mu^{(2)'}(x')$$

(c) Since

$$T_{\mu\nu}'(x') = \sum_{\sigma=1}^{4}\sum_{\tau=1}^{4} a_{\mu\sigma} a_{\nu\tau} T_{\sigma\tau}(x)$$

we can write

$$T^{(1)\prime}_{\mu\nu}(x') - T^{(2)\prime}_{\mu\nu}(x') = \sum_{\sigma=1}^{4}\sum_{\tau=1}^{4} a_{\mu\sigma}a_{\nu\tau}[T^{(1)}_{\sigma\tau}(x) - T^{(2)}_{\sigma\tau}(x)]$$

Then

$$T^{(1)\prime}_{\mu\nu}(x') = T^{(2)\prime}_{\mu\nu}(x')$$

(d) We can write

$$S'(x') = \sum_{\nu=1}^{4} a_{\nu 4}V'_{\nu}(x') = a_{44}V'_{4}(x') + \sum_{\nu=1}^{3} a_{\nu 4}V'_{\nu}(x')$$

Relation (d) is not covariant; that is, it is not valid in another coordinate system.

6.9. Four-Velocity, Four-Acceleration, and Proper Time

We can study the motion of a point by giving its four coordinates as a function of a parameter called *proper time* τ, which is the time given by a clock that is tied to the point:

$$x_i = x_i(\tau) \tag{6.9.1}$$

The derivatives of the coordinates x_i with respect to proper time are

$$u_i = \frac{dx_i(\tau)}{d\tau} \tag{6.9.2}$$

and represent the components of the *four-velocity* vector. The fourth component of the four-velocity

$$u_4 = \frac{dx_4}{d\tau} = ic\frac{dt}{d\tau} \tag{6.9.3}$$

represents, apart from (ic), the variation of ordinary time for an observer at rest with the moving point. For such an observer

$$\sum_{\mu=1}^{4} dx'^2_{\mu} = \sum_{i=1}^{3} dx'^2_i - c^2 dt'^2 = -c^2(d\tau)^2 \tag{6.9.4}$$

That is, the quantity $d\tau$ is invariant.

We can write

$$d\tau = dt\sqrt{1 - \frac{v^2}{c^2}} \tag{6.9.5}$$

where $\mathbf{v}$ = instantaneous velocity of the point. Therefore,

$$v_1 = \frac{dx_1}{dt} = \frac{dx_1}{d\tau}\frac{d\tau}{dt} = u_1\sqrt{1 - \frac{v^2}{c^2}}$$

$$v_2 = \frac{dx_2}{dt} = \frac{dx_2}{d\tau}\frac{d\tau}{dt} = u_2\sqrt{1 - \frac{v^2}{c^2}}$$

$$v_3 = \frac{dx_3}{dt} = \frac{dx_3}{d\tau}\frac{d\tau}{dt} = u_3\sqrt{1 - \frac{v^2}{c^2}} \tag{6.9.6}$$

$$v_4 = \frac{dx_4}{dt} = \frac{dx_4}{d\tau}\frac{d\tau}{dt} = u_4\sqrt{1 - \frac{v^2}{c^2}}$$

and, putting $\beta = v/c$,

$$u_1 = \frac{v_1}{\sqrt{1 - \beta^2}}$$

$$u_2 = \frac{v_2}{\sqrt{1 - \beta^2}}$$

$$u_3 = \frac{v_3}{\sqrt{1 - \beta^2}} \tag{6.9.7}$$

$$u_4 = \frac{ic}{\sqrt{1 - \beta^2}}$$

The four-velocity is a temporal vector. This can be seen easily if we consider a system at rest with the moving point; in this system

$$u_1 = u_2 = u_3 = 0$$
$$u_4 = 4ic$$

and

$$\sum_{i=1}^{4} u_i^2 = -c^2 < 0 \tag{6.9.8}$$

We define *four-acceleration* as a four-vector whose components are given by

$$b_i = \frac{du_i}{d\tau} \tag{6.9.9}$$

We can write

$$b_1 = \left(\frac{d}{dt} \frac{v_1}{\sqrt{1 - \beta^2}} \right) \frac{dt}{d\tau} = \frac{\dot{v}_1}{1 - \beta^2} + \frac{v_1 (\mathbf{v} \cdot \dot{\mathbf{v}})}{c^2 (1 - \beta^2)^2}$$

$$b_2 = \left(\frac{d}{dt} \frac{v_2}{\sqrt{1 - \beta^2}} \right) \frac{dt}{d\tau} = \frac{\dot{v}_2}{1 - \beta^2} + \frac{v_2 (\mathbf{v} \cdot \dot{\mathbf{v}})}{c^2 (1 - \beta^2)^2}$$

$$b_3 = \left(\frac{d}{dt} \frac{v_3}{\sqrt{1 - \beta^2}} \right) \frac{dt}{d\tau} = \frac{\dot{v}_3}{1 - \beta^2} + \frac{v_3 (\mathbf{v} \cdot \dot{\mathbf{v}})}{c^2 (1 - \beta^2)^2}$$

$$b_4 = \left(\frac{d}{dt} \frac{ic}{\sqrt{1 - \beta^2}} \right) \frac{dt}{d\tau} = \frac{i}{c} \frac{(\mathbf{v} \cdot \dot{\mathbf{v}})}{(1 - \beta^2)^2}$$

$$(6.9.10)$$

In the system at rest with the moving point, $\mathbf{v} = 0$ and

$$b_1 = \dot{v}_1$$
$$b_2 = \dot{v}_2$$
$$b_3 = \dot{v}_3$$
$$b_4 = 0$$

$$(6.9.11)$$

The temporal component is zero and the three spatial components are the components of ordinary acceleration: the four-acceleration is a spatial vector.

6.10. Lorentz-Covariant Form of the Potential Equations

The potential equations in the Lorenz gauge are given by

$$\Box^2 \phi = -4\pi \rho$$
$$\Box^2 \mathbf{A} = -\frac{4\pi}{c} \mathbf{j}$$

$$(6.10.1)$$

The Lorentz gauge is given by

$$\mathbf{\nabla} \cdot \mathbf{A} + \frac{1}{c} \frac{\partial \phi}{\partial t} = 0 \qquad (6.10.2)$$

We now define a four-vector:

$$A_\mu \equiv (\mathbf{A}, A_4 = i\phi) \qquad (6.10.3)$$

The Lorentz gauge can now be expressed as follows:

$$\sum_{\mu=1}^{4} \frac{\partial A_\mu}{\partial x_\mu} = 0 \tag{6.10.4}$$

In fact,

$$\sum_{\mu=1}^{4} \frac{\partial A_\mu}{\partial x_\mu} = \boldsymbol{\nabla} \cdot \mathbf{A} + \frac{1}{ic} \frac{\partial i\phi}{\partial t} = \boldsymbol{\nabla} \cdot \mathbf{A} + \frac{1}{c} \frac{\partial \phi}{\partial t} = 0$$

The Lorentz gauge is a feature that will continue to be valid after the system has undergone a Lorentz transformation:

$$\sum_{\mu=1}^{4} \frac{\partial A'_\mu}{\partial x'_\mu} = \boldsymbol{\nabla}' \cdot \mathbf{A}' + \frac{1}{c} \frac{\partial \phi'}{\partial t'} = 0 \tag{6.10.5}$$

where

$$A'_\mu \equiv (\mathbf{A}', i\phi') \tag{6.10.6}$$

The D'Alembertian operator can be expressed as follows:

$$\Box^2 = \nabla^2 - \frac{1}{c^2} \frac{\partial^2}{\partial t^2} = \frac{\partial^2}{\partial x_1^2} + \frac{\partial^2}{\partial x_2^2} + \frac{\partial^2}{\partial x_3^2} + \frac{\partial^2}{\partial x_4^2} \tag{6.10.7}$$

We define at this point the *four-current* vector

$$j_\mu = (\mathbf{j}, ic\rho) \tag{6.10.8}$$

The potential Eqs. (6.10.1) can then be expressed as follows:

$$\Box^2 A_\mu = -\frac{4\pi}{c} j_\mu \tag{6.10.9}$$

The potential four-vector is defined by (6.10.3). The electric field is given by

$$\mathbf{E} = -\boldsymbol{\nabla}\phi - \frac{1}{c} \frac{\partial \mathbf{A}}{\partial t} \tag{6.10.10}$$

Therefore,

$$E_i = -\frac{\partial \phi}{\partial x_i} - \frac{1}{c} \frac{\partial A_i}{\partial t} = i\frac{\partial A_4}{\partial x_i} - i\frac{\partial A_i}{\partial x_4}, \quad i = 1, 2, 3$$

and

$$-iE_i = \frac{\partial A_4}{\partial x_i} - \frac{\partial A_i}{\partial x_4}, \quad i = 1, 2, 3 \tag{6.10.11}$$

The magnetic induction **B** is given by

$$\mathbf{B} = \nabla \times \mathbf{A} \tag{6.10.12}$$

Therefore,

$$B_i = \frac{\partial A_l}{\partial x_k} - \frac{\partial A_k}{\partial x_l} \tag{6.10.13}$$

where (ikl) is a cyclic permutation of $(1, 2, 3)$.

Let us define a new (antisymmetric) tensor

$$F_{\mu\nu}(x) = \frac{\partial A_\nu}{\partial x_\mu} - \frac{\partial A_\mu}{\partial A_\nu} \tag{6.10.14}$$

where $\mu, \nu = 1, 2, 3, 4$. We shall now find the six independent components of this tensor:

$$F_{12} = \frac{\partial A_2}{\partial x_1} - \frac{\partial A_1}{\partial x_2} = B_3$$

$$F_{23} = \frac{\partial A_3}{\partial x_2} - \frac{\partial A_2}{\partial x_3} = B_1$$

$$F_{31} = \frac{\partial A_1}{\partial x_3} - \frac{\partial A_3}{\partial x_1} = B_2$$

$$F_{14} = \frac{\partial A_4}{\partial x_1} - \frac{\partial A_1}{\partial x_4} = \frac{\partial (i\phi)}{\partial x_1} - \frac{\partial A_1}{ic\partial t}$$

$$= -i \left(-\frac{\partial \phi}{\partial x_1} - \frac{1}{c} \frac{\partial A_1}{\partial t} \right) = -iE_1$$

$$F_{24} = \frac{\partial A_4}{\partial x_2} - \frac{\partial A_2}{\partial x_4} = -iE_2$$

$$F_{34} = \frac{\partial A_4}{\partial x_3} - \frac{\partial A_3}{\partial x_4} = -iE_3$$

We can summarize the results of this section as follows:

$$\Box^2 A_\mu = -\frac{4\pi}{c} j_\mu$$

$$\sum_\mu \frac{\partial A_\mu}{\partial x_\mu} = 0 \qquad (6.10.15)$$

$$F_{\mu\nu} = \frac{\partial A_\nu}{\partial x_\mu} - \frac{\partial A_\mu}{\partial x_\nu}$$

where

$$A_\mu \equiv (\mathbf{A}, i\phi)$$
$$j_\mu \equiv (\mathbf{j}, ic\rho)$$
$$F_{12} = B_3, \quad F_{14} = -iE_1 \qquad (6.10.16)$$
$$F_{23} = B_1, \quad F_{24} = -iE_2$$
$$F_{31} = B_2, \quad F_{34} = -iE_3$$

6.11. Plane Waves

In Eqs. (6.10.15), we have solutions that are waves if $j_\mu = 0$. These solutions are of the type

$$A_\mu(\mathbf{x}, t) = A_\mu^{(0)} e^{i(\mathbf{k} \cdot \mathbf{x} - \omega t)} \qquad (6.11.1)$$

A_μ is a four-vector:

$$A_\mu(\mathbf{x}, t) = \sum_\nu a_{\nu\mu} A_\nu'(\mathbf{x}', t') \qquad (6.11.2)$$

Since $\omega = kc$ with $k = |\mathbf{k}|$, we can express the phase part of A as follows:

$$\exp[i(\mathbf{k} \cdot \mathbf{x} - \omega t)]$$
$$= \exp(i\mathbf{k} \cdot \mathbf{x} - i\omega t) = \exp(i\mathbf{k} \cdot \mathbf{x} - ikct)$$
$$= \exp(i\mathbf{k} \cdot \mathbf{x} - kx_4) = \exp(i\mathbf{k} \cdot \mathbf{x} + ik_4 x_4)$$
$$= \exp\left(i \sum_{\lambda=1}^{4} k_\lambda x_\lambda \right) \qquad (6.11.3)$$

where we have set

$$k = |\mathbf{k}| = -ik_4 \qquad (6.11.4)$$

We can rewrite the expression (6.11.1) as follows:

$$A_\mu(\mathbf{x}, t) = A_\mu^{(0)} \exp(i\mathbf{k} \cdot \mathbf{x} - i\omega t)$$

$$= A_\mu^{(0)} \exp\left(i \sum_{\lambda=1}^{4} k_\lambda x_\lambda\right)$$

$$= \left(\sum_\nu a_{\nu\mu} A_\nu^{(0)'}\right) \exp\left(i \sum_{\nu=1}^{4} k_\nu' x_\nu'\right) \tag{6.11.5}$$

The two members of this relation must be equal; we have, separately,

$$A_\mu^0 = \sum_{\nu=1}^{4} a_{\nu\mu} A_\nu^{(0)'}$$

$$\exp\left(i \sum_{\lambda=1}^{4} k_\lambda x_\lambda\right) = \exp\left(i \sum_{\nu=1}^{4} k_\nu' x_\nu'\right) \tag{6.11.6}$$

We can then say that the phase factor $\exp(i \sum_{\lambda=1}^{4} k_\lambda x_\lambda)$ is a Lorentz invariant and

$$k_\lambda \equiv \left(\mathbf{k}, k_4 = \frac{i\omega}{c}\right) \tag{6.11.7}$$

is a four-vector. We can also write

$$\sum_{\lambda=1}^{4} k_\lambda x_\lambda = \sum_{\lambda=1}^{4} k_\lambda \sum_{\nu=1}^{4} a_{\nu\lambda} x_\nu'$$

$$= \sum_{\nu=1}^{4} x_\nu' \left(\sum_{\lambda=1}^{4} a_{\nu\lambda} k_\lambda\right) = \sum_{\nu=1}^{4} k_\nu' x_\nu' \tag{6.11.8}$$

because

$$k_\nu' = \sum_{\lambda=1}^{4} a_{\nu\lambda} k_\lambda \tag{6.11.9}$$

We have the following dispersion relations:

In vacuum:

$$\omega = c|k| \tag{6.11.10}$$

In an isotropic medium:

$$\omega = \frac{c}{n}|\mathbf{k}| \tag{6.11.11}$$

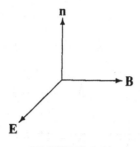

FIGURE 6.11

where n = index of refraction

In an anisotropic medium:

$$\omega = \frac{c}{n(\text{angle of } \mathbf{k})}|\mathbf{k}| \tag{6.11.12}$$

Whatever function ω is of $\mathbf{k}$,

$$\omega = \omega(\mathbf{k}) \tag{6.11.13}$$

We can always write

$$k_\mu \equiv \left(\mathbf{k}, k_4 = \frac{i\omega}{c}\right) \tag{6.11.14}$$

Let us consider now a plane wave in vacuum whose $\mathbf{E}$ and $\mathbf{B}$ fields are oriented as in Fig. 6.11. Let $\mathbf{n}$ be the unit vector in the $\mathbf{k}$ direction and $\omega = 2\pi\nu$; then

$$\mathbf{k} = \frac{\omega}{c}\mathbf{n} = \frac{2\pi\nu}{c}\mathbf{n} \tag{6.11.15}$$

Let

$$\mathbf{E} = \mathbf{E}_0 e^{i\phi}$$
$$\mathbf{B} = \mathbf{B}_0 e^{i\phi} \tag{6.11.16}$$

where

$$\phi = \mathbf{k} \cdot \mathbf{x} - \omega t = \frac{2\pi\nu}{c}(\mathbf{n} \cdot \mathbf{x}) - 2\pi\nu t = 2\pi\nu\left(\frac{\mathbf{n} \cdot \mathbf{x}}{c} - t\right) \tag{6.11.17}$$

In the primed systems,

$$\mathbf{E}' = \mathbf{E}_0' e^{i\phi'}$$
$$\mathbf{B}' = \mathbf{B}_0' e^{i\phi'} \tag{6.11.18}$$

where

$$\phi' = \mathbf{k}' \cdot \mathbf{x}' - \omega' t' = 2\pi \nu' \left(\frac{\mathbf{n}' \cdot \mathbf{x}'}{c} - t' \right) \qquad (6.11.19)$$

We expect

$$\phi = \phi'$$

or

$$\nu \left(t - \frac{n_x x + n_y y + n_z z}{c} \right) = \nu' \left(t' - \frac{n_x' x' + n_y' y' + n_z' z'}{c} \right) \qquad (6.11.20)$$

Let us consider the Lorentz transformation

$$
\begin{aligned}
y &= y' & y' &= y \\
z &= z' & z' &= z \\
x &= \gamma(x' + vt') & x' &= \gamma(x - vt) \\
t &= \gamma\left(t' + \frac{v}{c^2}x'\right) & t' &= \gamma\left(t - \frac{v}{c^2}x\right)
\end{aligned}
\qquad (6.11.21)
$$

Then (6.11.20) becomes

$$\nu \left\{ t - \frac{n_x x + n_y y + n_z z}{c} \right\} = \nu' \left\{ \gamma\left(t - \frac{v}{c^2}x\right) \right.$$

$$\left. - \frac{1}{c}[n_x' \gamma(x - vt) + n_y' y + n_z' z] \right\} \qquad (6.11.22)$$

or

$$\nu = \nu' \left[\gamma + \gamma \frac{v}{c} n_x' \right] = \nu' \frac{1 + \beta n_x'}{\sqrt{1 - \beta^2}}$$

$$\nu n_x = \nu' \left[\gamma \frac{v}{c} + \gamma n_x' \right] = \nu' \frac{\beta + n_x'}{\sqrt{1 - \beta^2}} \qquad (6.11.23)$$

$$\nu n_y = \nu' n_y'$$

$$\nu n_z = \nu' n_z'$$

From the first relation

$$\frac{\nu'}{\nu} = \frac{\sqrt{1 - \beta^2}}{1 + \beta n_x'} \qquad (6.11.24)$$

Therefore,

$$n_x = \frac{\nu'}{\nu}\frac{\beta + n'_x}{\sqrt{1-\beta^2}} = \frac{\sqrt{1-\beta^2}}{1+\beta n'_x}\frac{\beta + n'_x}{\sqrt{1-\beta^2}} = \frac{\beta + n'_x}{1+\beta n'_x}$$

$$n_y = \frac{\nu'}{\nu}n'_y = \frac{\sqrt{1-\beta^2}}{1+\beta n'_x}n'_y \qquad\qquad (6.11.25)$$

$$n_z = \frac{\nu'}{\nu}n'_z = \frac{\sqrt{1-\beta^2}}{1+\beta n'_x}n'_z$$

We can now summarize these results as follows

$$\nu = \nu'\frac{1+\beta n'_x}{\sqrt{1-\beta^2}} \qquad \nu' = \nu\frac{1-\beta n_x}{\sqrt{1-\beta^2}}$$

$$n_x = \frac{\beta + n'_x}{1+\beta n'_x} \qquad n'_x = \frac{-\beta + n_x}{1-\beta n_x}$$

$$n_y = \frac{\sqrt{1-\beta^2}}{1+\beta n'_x}n'_y \qquad n'_y = \frac{\sqrt{1-\beta^2}}{1-\beta n_x}n_y \qquad (6.11.26)$$

$$n_z = \frac{\sqrt{1-\beta^2}}{1+\beta n'_x}n'_z \qquad n'_z = \frac{\sqrt{1-\beta^2}}{1-\beta n_x}n_z$$

This means that if a plane wave is such that an observer in the frame $S(S')$ measures the frequency $\nu(\nu')$ and the direction cosines $n_x(n'_x)$, $n_y(n'_y)$, and $n_z(n'_z)$ then an observer in the frame $(S'S)$ measures the frequency $\nu'(\nu)$ and the direction cosines $n'_x(n_x)$, $n'_y(n_y)$, and $n'_z(n_z)$.

EXAMPLE: Reflection from Moving Mirror

A beam of light with frequency ν and direction determined by

$$n_x = \cos\theta_0$$

$$n_y = \sin\theta_0$$

$$n_z = 0$$

is reflected by a mirror that moves with velocity v in the x direction with respect to the frame S (see Fig. 6.12). We want to calculate the frequency

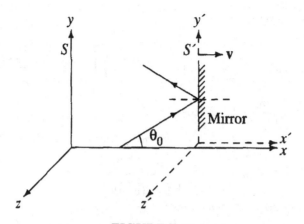

FIGURE 6.12

and the direction of the reflected beam. We divide the problem into three parts:

(1) Transfer to the moving system:

$$\nu' = \nu_0 \frac{1 - \beta \cos \theta_0}{\sqrt{1 - \beta^2}}$$

$$n'_x = \frac{-\beta + \cos \theta_0}{1 - \beta \cos \theta_0}$$

$$n'_y = \frac{\sqrt{1 - \beta^2}}{1 - \beta \cos \theta_0} \sin \theta_0$$

$$(6.11.27)$$

$$n'_z = 0$$

(2) Apply the law of reflection in the system moving with the mirror. This amounts to changing the sign of n'_x. The reflected wave in this system will have

$$\nu' = \nu_0 \frac{1 - \beta \cos \theta_0}{\sqrt{1 - \beta^2}}$$

$$n'_x = -\frac{\cos \theta_0 - \beta}{1 - \beta \cos \theta_0}$$

$$n'_y = \frac{\sqrt{1 - \beta^2}}{1 - \beta \cos \theta_0} \sin \theta_0$$

$$(6.11.28)$$

$$n'_z = 0$$

(3) Return to system S:

$$\nu = \nu' \frac{1 + \beta n'_x}{\sqrt{1 - \beta^2}} = \nu_0 \left(\frac{1 - \beta \cos \theta_0}{\sqrt{1 - \beta^2}} \right) \left(\frac{1 - \beta \dfrac{\cos \theta_0 - \beta}{1 - \beta \cos \theta_0}}{\sqrt{1 - \beta^2}} \right)$$

$$= \frac{\nu_0}{1 - \beta^2} (1 - 2\beta \cos \theta_0 + \beta^2)$$

$$n_x = \frac{\beta + n'_x}{1 + \beta n'_x} = \frac{\beta - \dfrac{\cos \theta_0 - \beta}{1 - \beta \cos \theta_0}}{1 - \beta \left(\dfrac{\cos \theta_0 - \beta}{1 - \beta \cos \theta_0} \right)}$$

$$= \frac{2\beta - \beta^2 \cos \theta_0 - \cos \theta_0}{1 - 2\beta \cos \theta_0 + \beta^2} \qquad (6.11.29)$$

$$n_y = \frac{\sqrt{1 - \beta^2}}{1 + \beta n'_x} n'_y$$

$$= \frac{\sqrt{1 - \beta^2}}{1 - \beta \left(\dfrac{\cos \theta_0 - \beta}{1 - \beta \cos \theta_0} \right)} \frac{\sqrt{1 - \beta^2}}{1 - \beta \cos \theta_0} \sin \theta_0$$

$$= \frac{(1 - \beta^2) \sin \theta_0}{1 - 2\beta \cos \theta_0 + \beta^2}$$

$$n_z = 0$$

EXAMPLE: Doppler Effect

Consider a source at rest in S', emitting radiation of frequency ν' in S'. As before, S' is moving with respect to S with velocity v in the x direction. From Eq. (6.11.26),

$$\nu' = \nu \frac{1 - \beta n_x}{\sqrt{1 - \beta^2}} \qquad (6.11.30)$$

and

$$\nu = \nu' \frac{\sqrt{1 - \beta^2}}{1 - \beta n_x} \qquad (6.11.31)$$

In the case that S' moves in a general direction, βn_x is replaced by

$$\boldsymbol{\beta} \cdot \mathbf{n} = \frac{\mathbf{v} \cdot \mathbf{n}}{c} \qquad (6.11.32)$$

where **n** in the direction of observation in the frame S. Then, in general,

$$\nu = \nu' \frac{\sqrt{1 - \beta^2}}{1 - \dfrac{\mathbf{v} \cdot \mathbf{n}}{c}} \tag{6.11.33}$$

For $v/c \ll 1$, we obtain

$$\nu = \nu' \left(1 + \frac{\mathbf{v} \cdot \mathbf{n}}{c}\right) \tag{6.11.34}$$

If $\mathbf{n} \cdot \mathbf{v} > 0$, the source moves toward the observer; if $\mathbf{n} \cdot \mathbf{v} < 0$, the source moves away from the observer. Formula (6.11.33) gives the usual, nonrelativistic *longitudinal Doppler shift* of the frequency.

A new result is obtained if the source moves in a direction perpendicular to the direction of observation: $\mathbf{v} \cdot \mathbf{n} = 0$. In this case, (6.11.33) gives

$$\nu = \nu' \sqrt{1 - \beta^2} \tag{6.11.35}$$

This *transversal Doppler shift* of the frequency escaped experimental observation for a long time, because it is practically proportional to β^2, but has actually been observed in precise measurements.[2]

EXAMPLE: Fixed Star Aberration

The situation is described in Fig. 6.13. A "fixed" star has in the frame S' the direction determined by

$$n'_x = 0$$
$$n'_y = -1 \tag{6.11.36}$$
$$n'_z = 0$$

Then, for an observer in the frame S,

$$n_x = \frac{\beta + n'_x}{1 + \beta n'_x} = \beta = \frac{v}{c}$$

$$n_y = \frac{n'_y \sqrt{1 - \beta^2}}{1 + \beta n'_x} = -\sqrt{1 - \beta^2} \tag{6.11.37}$$

$$n_z = \frac{n'_z \sqrt{1 - \beta^2}}{1 + \beta n'_x} = 0$$

---

[2] H. J. Hay, J. P. Schiffer, T. E. Cranshaw, and P. A. Egelstaff, *Phys. Rev. Lett.* *4*, 165 (1960).

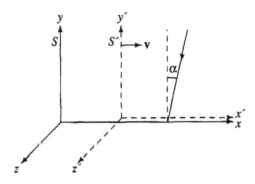

FIGURE 6.13

The angle α is given by

$$\tan \alpha = \frac{n_x}{n_y} = \frac{-\beta}{\sqrt{1 - \beta^2}} = \frac{-v/c}{\sqrt{1 - \left(\frac{v}{c}\right)^2}} \qquad (6.11.38)$$

or

$$\sin \alpha = -\frac{v}{c} \qquad (6.11.39)$$

6.12. The Twin Paradox[3]

John and Daniel are twin brother. John, a physics teacher, does not like to travel; Daniel, an astronaut, likes it. Daniel boards a spaceship, travels to a star at speed v, and, after a brief visit, returns to earth at the same speed.

Special relativity implies that, if Daniel is moving with respect to John, time intervals in his (Daniel's) inertial frame are dilated with respect to time intervals in John's inertial frame by an amount $\gamma = 1/\sqrt{1 - \beta^2}$, with $\beta = v/c$. Therefore, at the end of his trip, Daniel has aged less than John.

However, except during the short intervals of time during which Daniel is accelerating or decelerating, Daniel is in an inertial frame and, therefore, he can say that John is moving, not himself. Therefore, John's time interval should be dilated, says Daniel.

---

[3]R. J. Muller, *Am. J. of Phys.* *40*, 966 (1972).

To eliminate the apparent contradiction, we shall assume that the two brothers perform the following two calculations. Each brother, obviously, makes his calculation in the particular situation in which he finds himself.

John's Calculation

D_j = distance earth–star as measured by me (John)

D_d = distance earth–star as measured by Daniel $= D_j/\gamma$

Daniel travels a total distance $2D_j$ at velocity v ($-v$ during the return). Daniel travels for a time $2D_j/v$. The time indicated by Daniel's clock is

$$t_d = \gamma\left(t_j - \frac{v}{c^2}x_j\right) \qquad (6.12.1)$$

where

t_j = time measured in my (John's) clock

x_j = distance between me and my brother

During Daniel's trip to the star,

$$x_j = vt_j \qquad (6.12.2)$$

Then

$$t_d = \gamma\left(t_j - \frac{v}{c^2}vt_j\right) = \gamma t_j\left(1 - \frac{v^2}{c^2}\right) = \frac{t_j}{\gamma} \qquad (6.12.3)$$

and

$$\Delta t_d = \frac{\Delta t_j}{\gamma} \qquad (6.12.4)$$

Daniel's clock is running slower than John's clock by a factor

$$\frac{1}{\gamma} = \sqrt{1 - \beta^2}$$

At the end of Daniel's trip, I (John) have aged by

$$A_j = \frac{2D_j}{v} = \gamma A_d \qquad (6.12.5)$$

During his trip, Daniel has aged

$$A_d = \frac{2D_j}{\gamma v} \tag{6.12.6}$$

The difference in our ages is then

$$A_j - A_d = \frac{2D_j}{v}\left(1 - \frac{1}{\gamma}\right) = \frac{2D_j}{v}(1 - \sqrt{1 - \beta^2}) \tag{6.12.7}$$

Daniel's Calculation

D_j = distance earth–star measured by John

D_d = distance earth–star measured by me (Daniel) $= \dfrac{D_j}{\gamma}$

During my trip, the time indicated by John's clock is

$$t_j = \gamma\left(t_d - \frac{(-v)(-x_d)}{c^2}\right) = \gamma\left(t_d - \frac{vx_d}{c^2}\right) \tag{6.12.8}$$

where

t_d = time measured in my (Daniel's) clock

d_d = distance between me and my brother

During my trip to the star,

$$x_d = vt_d \tag{6.12.9}$$

Then

$$t_j = \gamma t_d\left(1 - \frac{v^2}{c^2}\right) = \frac{t_d}{\gamma} \tag{6.12.10}$$

and

$$\Delta t_j = \frac{\Delta t_d}{\gamma} \tag{6.12.11}$$

My trip takes a total time $2D_d/v$. At the end of the trip, I (Daniel) have aged

$$A_d = \frac{2D_d}{v} \tag{6.12.12}$$

John has aged

$$A_j = \frac{2D_d}{v\gamma} = \frac{1}{\gamma}A_d \tag{6.12.13}$$

The difference in our ages is

$$A_d - A_j = \frac{2D_d}{v}\left(1 - \frac{1}{\gamma}\right) = \frac{2D_j}{\gamma v}\left(1 - \frac{1}{\gamma}\right)$$
$$= \frac{2D_j}{\gamma v}(1 - \sqrt{1 - \beta^2}) \tag{6.12.14}$$

The two calculations disagree!

Solution of the Paradox. Since John is always in the *same* inertial frame, we have to assume that his calculation is correct. The following argument is a follow-up to Daniel's calculation

When Daniel reaches the star and starts moving back, he changes inertial frames. Just before the turnaround,

$$t_j^f = \gamma\left(t_d - \frac{vD_d}{c^2}\right) \tag{6.12.15}$$

Immediately after the turnaround,

$$t_j^b = \gamma\left(t_d + \frac{vD_d}{c^2}\right) \tag{6.12.16}$$

The difference between these two times is

$$t_j^b - t_j^f = \gamma t_d + \gamma\frac{vD_d}{c^2} - \gamma t_d + \gamma\frac{vD_d}{c^2} = \frac{2\gamma vD_d}{c^2} = \frac{2vD_j}{c^2} \tag{6.12.17}$$

because $D_j = \gamma D_d$. The effect that a change in the inertial frame has on the time of a distant event is proportional to the distance to that event.

We can then ignore the effects of acceleration at the beginning and at the end of Daniel's trip, when the distance between Daniel and John is zero.

When Daniel returns to earth, the difference in Daniel's and John's ages is $(A_d - A_j)$ calculated in (6.12.14) minus the amount due to the change of inertial frames:

$$A_d - A_j - \frac{2vD_j}{c^2} = \left(1 - \frac{1}{\gamma}\right)\frac{2D_j}{\gamma v} - \frac{2vD_j}{c^2}$$

$$= \frac{2D_j}{\gamma v} - \frac{2D_j}{\gamma^2 v} - \frac{2vD_j}{c^2}$$

$$= \frac{2D_j}{\gamma v} - \frac{2D_j}{v}\left(\frac{1}{\gamma^2} + \frac{v^2}{c^2}\right)$$

$$= \frac{2D_j}{v}\left(\frac{1}{\gamma} - 1\right) \qquad (6.12.18)$$

or

$$A_j - A_d = \frac{2D_j}{v}\left(1 - \frac{1}{\gamma}\right) \qquad (6.12.19)$$

which coincides with (6.12.7)!

EXAMPLE

Refer to Fig. 6.14

$$D_j = 7.5 \text{ light years}$$

$$v = \frac{3}{4}c$$

$$\frac{2D_j}{v} = \frac{15 \times c}{(3/4)c} = 20 \text{ years}$$

$$A_j = \frac{2D_j}{v} = 20 \text{ years}$$

$$A_j - A_d = \frac{2D_j}{v}\left(1 - \frac{1}{\gamma}\right) = 20(1 - 0.66) = 6.8 \text{ years}$$

$$A_d = 20 - 6.8 = 13.2 \text{ years}$$

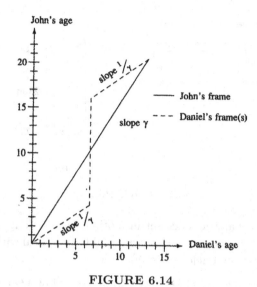

FIGURE 6.14

CHAPTER 6 EXERCISES[4]

6.1. "Ainsi, tounjours poussées vers de nouveaux rivages,
Dans la nuit éternelle emportés sans retour,
Ne pourrons-nous jamais sur l'ocean des ages

Jeter l'ancre un seul jour?"

"As we are pushed toward new shores,
Swept along to the eternal night without return,
Can we ever on the ocean of the ages

Let the anchor down for one day?"

From "Le Lac," by A. de Lamartine.

Comment in terms relevant to the theory of special relativity.

6.2. A free-falling enclosure can provide in principle an inertial frame to bodies inside it because of the absence of relative acceleration between these bodies and the enclosure. In such an environment, the laws of motion appear in their simplest form; a particle at rest remains at

---

[4]Some of the exercises of this Chapter are adapted versions of problems that appear in *Spacetime Physics* (1st and 2nd editions), by E. F. Taylor and J. A. Wheeler, W. H. Freeman Co., San Francisco.

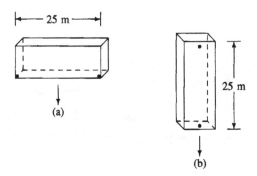

FIGURE P6.2

rest, and a particle, once set in motion, moves along a straight line. In reality, in such a situation we would have to deal with the fact that the gravitational field is not uniform.

(a) A room 25 m long (Fig. P6.2) is free falling near the earth. The free fall starts at time $t = 0$ from a height of 250 m and takes place in a direction perpendicular to the long sides of the room. Two tiny test balls are on the left and right edges of the room at $t = 0$ (see Fig. P6.2a). How close will they have moved at the time the fall ends?

(b) Two test particles are in a room 25 meters long, close to, but not touching, the walls (see Fig. P6.2b). The room is 250 m above the earth and, at time $t = 0$, begins to free fall along a direction parallel to its long sides. By what amount will the particles have moved away from each other at the time the fall ends?

6.3. A frame of reference is recognized as being inertial if the motions of noncolliding test particles are described as taking place in the absence of any force. Consider two frames of reference, a frame S that has been found to be inertial and a frame S'.

(a) If a test particle moves along straight lines in both S and S', can we say that S' is also an inertial frame?

(b) If the above evidence is not enough, what are the simplest possible tests that you would do to find out if S' is inertial?

6.4. A rocket clock emits a pair of light flashes that are detected by two observers, one in the laboratory and the other in the rocket. The data recorded are reported in the first line of the following table. A different rocket carries a clock that emits another pair of flashes. The second line of the table reports the data that relate to this second pair of flashes, and so on. Complete the table (we have set $c = 1$.)

Speed of rocket with respect to laboratory (m/m)	Rocket observer, Δt (m)	Lab observer, Δt (m)	Lab observer, Δs (m)
0.724		29	21
		52	48
	20		99
	24	30	
0.280			7
	20	25	
0.8		40	
0.6	12		

6.5. Let S' be an inertial frame that moves along the x' direction with respect to a (laboratory) frame S with a constant velocity βc. The x axis of S is in the same direction as the x' axis of S'. A ruler, at rest in S', has the length L' in S' and makes an angle θ' with the x' axis.

(a) What is the angle θ that the ruler forms with the x axis in S?
(b) What is the length of the ruler in S?

6.6. A charged particle Q moves along a certain linear path in the laboratory frame S. Two other charged particles q and q' with $q = q'$ are present, one in the line of motion and the other off the line of motion (see Fig. P6.6). At the time of closest approach to q', both particles are at the same distance from Q in S. Explain why, at that time, the particle q experiences a force smaller than it would if Q were at rest, and q' a force greater if Q were at rest.

6.7. A rocket moves in the x direction of the laboratory frame S with velocity βc.

(a) A light flash is sent by an astronaut in the rocket in a direction perpendicular with respect to the direction of motion. What is the angle that the path of the light flash makes with the x axis in S?

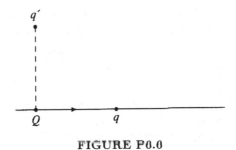

FIGURE P0.0

(b) What is the angle θ that the path of the light flash makes with the x axis in S if the astronaut sends the flash in a direction making an angle θ' with the line of motion?

6.8. A particle moves with velocity $\beta' c$ in the (x', y') plane of a frame S' along a line making an angle θ' with the x' axis. The frame S' moves with velocity $\beta_r c$ with respect to the laboratory frame S along the x direction (which is parallel to the x' direction). What is the angle θ that the velocity of this particle makes with the x axis of S?

6.9. Prove that anything that moves past an observer in an inertial frame at the speed $c =$ speed of light moves past any other inertial observer with the same speed c.

6.10. A charge distribution $\rho(\mathbf{x})$ and a permanent magnet are at rest in a frame S and produce in S an electric field $\mathbf{E}$ and a magnetic induction field $\mathbf{B}$, respectively. Derive an expression for the magnetic induction field $\mathbf{B}'$ measured in the frame S' that moves with respect to the frame S with nonrelativistic velocity $\mathbf{v}$.

6.11. An unstable particle of rest mass m_0 has a mean lifetime to when it is at rest. It moves in vacuum with velocity v, which is not small in comparison with the velocity of light c. What is the probability for this particle to travel a distance L without decaying?

6.12. (a) Two events A and B have a timelike separation. Show that it is possible to find an inertial frame in which A and B occur at the same place.

(b) Two events C and D have a spacelike separation. Show that it is possible to find an inertial frame in which C and D occur at the same time.

6.13. Fizeau found that the measured velocity of light in a liquid that has an index of refraction n and that moves with a velocity $v(v \ll c)$ is given by

$$w = \frac{c}{n} + \left(1 - \frac{1}{n^2}\right) v$$

Show that this result can be obtained from the special theory of relativity.

6.14. Let S be the laboratory inertial frame and S' an inertial frame that moves with respect to S with velocity βc along the $x(x')$ direction. Find the velocity of a clock C in the S frame that is equal and opposite to that of the same clock C in the S' frame such that the time readings of C by two observers, one in S and the other in S', agree.

6.15. Let S be the laboratory inertial frame and S' an inertial frame that moves with respect to S with velocity $\beta_r c$ along the $x(x')$ direction. A particle moves with constant speed $\beta'c$ along the y' direction in the frame S'. Determine the velocity of the particle in S.

6.16. Let S be the laboratory inertial frame and S' an inertial frame that moves with respect to S with velocity βc along the $x(x')$ direction.

(a) Show that, if two events A and B occur at the same place and simultaneously in S, they also occur simultaneously in all other inertial frames. On the other hand, if A and B occur simultaneously, but not at the same place, in S, they do not occur simultaneously in any other inertial frame S'.

(b) Consider two events A and B, and call Δx, Δy, and Δz the components of their distance in S in the x, y, and z directions, respectively. Show that, if they occur simultaneously and $\Delta x = 0$ in S, they will also occur simultaneously in any other inertial frame of type S', even if $\Delta y \neq 0$ and/or $\Delta z \neq 0$.

6.17. Let S be the laboratory inertial frame and S' an inertial frame that moves with respect to S with velocity βc along the $x(x')$ direction, and let the origins of the two frames coincide at the time $t = t' - 0$. Assume that each frame carries with it a string of synchronized clocks equally spaced within the frame along the $x(x')$ axis.

(a) Prove the following. At time $t = 0$ in S, the clocks along the $+x$ axis in S' appear to be set behind to an observer in the S frame, with the clocks that are farther from the origin farther behind. On the other hand, the clocks on the $-x$ axis in S' appear to be set ahead in the S frame, with the clocks that are farther from the origin farther ahead.

(b) Also prove the following. At time $t' = 0$ in S', the clocks along the $+x$ axis in S appear to be set ahead to an observer in the S' frame, with the clocks that are farther from the origin farther ahead. On the other hand, the clocks on the $-x$ axis in S appear to be set behind in the S' frame, with the clocks that are farther from the origin farther behind.

(c) Note the apparent asymmetry between the two frames that seems to violate the principle of relativity and allows us to tell the two frames apart. Present some argument that may solve this paradox.

6.18. Let S be the laboratory inertial frame and S' an inertial frame that moves with respect to S with velocity $\beta_r c$ along the $x(x')$ direction, and let the origins of the two frames coincide at the time $t = t' = 0$. Assume that a ruler of length L moves in the y direction in S with speed βc, with its middle point passing the origin at time $t = t' = 0$. To an observer in S, the ruler is parallel to the x axis at time $t = 0$ and remains so during its motion.

(a) Explain why the ruler appears tilted upward in the positive x' direction to an observer in S'.

(b) Calculate the angle that, according to an observer in S', the ruler forms with the x' axis.

6.19. A thin plate, parallel to the (x, y) plane of the laboratory S frame, moves upward in the z direction, passing this plane at time $t = 0$. The plate has a circular hole 1 m in diameter whose symmetry axis is the z axis. A ruler 1 m long is parallel to the x axis and travels with speed $\beta_r c$ in the $+x$ direction of S. Its center arrives at the origin at the time $t = 0$. Since the ruler is Lorentz contracted in S, it will pass through the hole of the plate that moves upward. Things appear different to an observer in the S' frame of the moving ruler. In S' the ruler will be 1 m in length, but the hole will be contracted in the x direction. How is it possible, under these conditions, for the ruler to go through the hole?

6.20. No particle can travel with a speed greater than c, the speed of light in vacuum. However, it is possible for a particle to travel in a medium with a speed βc greater than the speed $\beta' c$ of light in the medium. Under these conditions a coherent emission called *Cerenkov radiation* is emitted by the particle in a light cone (see Fig. P6.20) whose half-angle is given by

$$\cos \theta = \frac{\beta'}{\beta}$$

(a) Calculate the minimum speed that a particle can have in order to produce Cerenkov radiation in a medium in which $\beta' = \frac{2}{3}$.

(b) Calculate the maximum angle at which the Cerenkov radiation can be produced in such a medium.

6.21. Show that, in the potential equations in the Lorentz gauge, the continuity equation for charge and current density follows from the Lorentz condition.

6.22. An electromagnetic wave is reflected from a perfect mirror at an angle θ (Fig. P6.22).

(a) Calculate the radiation pressure on the mirror. Express it in terms of the energy density of the wave.

(b) Assume a cavity with perfectly reflecting walls filled with a very large number of wave packets of random orientations. Calculate

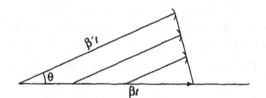

FIGURE P6.20

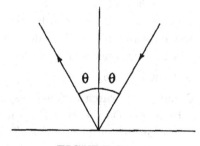

FIGURE P6.22

the average radiation pressure in terms of the average energy density inside the cavity.

6.23. A beam of electromagnetic radiation of frequency ν_1 and intensity I_1 moves in the $+x$ direction and is reflected by a plane mirror that is perpendicular to the x axis and moves in the $+x$ direction with velocity $v = \beta c$. Calculate the frequency ν_2 and the intensity I_2 of the reflected beam.

6.24. A point light source, which is stationary in the S' system, moves at a constant relativistic velocity v, parallel to the x axis in the S system. The two systems coincide at time $t = 0$. A photon whose total energy in S' is $h\nu'$ is emitted at an angle θ' with respect to the x' axis of the S' system. Using the transformation law for the four-momentum vector and recalling that photons have zero rest mass, derive the following:

(a) The formula for the angle θ with respect to the x axis at which the photon is received in the S system.

(b) The formula for the relativistic Doppler shift of the frequency.

6.25. Astronaut Dan boards a spaceship that can reach the speed $0.5c$. The destination is Drepanon, a "fixed" star 11.3 light-years away from Earth. The purpose of the trip is to find out if, as it is suspected, Drepanon has planets. Dan goes full speed, stops for 5 years, as recorded by his clock, at the planet closest to Drepanon, and returns to Earth with the same speed, $0.5c$. According to John, Dan's brother who stayed at home:

(a) At what time does the rocket arrive at Drepanon?

(b) At what time does the rocket leave Drepanon?

(c) At what time does the rocket arrive back to earth?
According to Dan,

(d) At what time does he arrive at Drepanon?

(e) At what time does he leave Drepanon?

(f) At what time does he arrive back to Earth?

(g) As Dan moves outward toward Drepanon, what is the distance Earth–Drepanon, according to an observer in his frame?

6.26. With reference to Exercise 6.25:

(a) One of Dan's outgoing "watch stations", say R, passes the Earth at the same time that Dan reaches Drepanon. What time does the R clock read at this event of passing? What time does the clock on Earth read when this event occurs?

(b) One of Dan's incoming "watch stations", say Z, passes the Earth at the same time in Dan's frame that Dan leaves Drepanon to come back home. What time does the Z clock read at this event of passing? What time does the clock on Earth read when this even occurs?

(c) Set up an (x, y) system of axes with the x axis indicating the "rocket age" and the y axis indicating the "Earth age". Draw the line corresponding to the observations of these two ages by John and the line corresponding to the observations by Dan.

6.27. Plot the events and the world lines for the trip described in Exercise 6.25.

(a) On a space–time diagram, draw the world line for the round trip taken by Dan, as observed in the Earth's frame. On the same diagram draw the world line for John, Dan's brother, who stayed at home. Label the diagram "Earth–Frame".

(b) On another space–time diagram, draw the same world lines as observed in Dan's outgoing frame. This is the frame that travels with Dan to Drepanon and then keeps on going at the same speed as Dan returns to Earth. Label the diagram "Outgoing Spaceship Frame".

Label each event "leave Earth", "arrive Drepanon", and so on. On each segment of each world line, put a number equal to the proper time along that segment.

6.28. The twin paradox can be solved by using the Doppler effect as follows: John remains on earth. Daniel travels with a large speed v to a star and returns to earth with the same speed. Both John and Daniel observe a *very distant* variable star whose light gets alternatively dimmer and then brighter with a frequency ν in the earth (John's) frame and ν' in Daniel's frame. This variable star is *very much* farther away than the length of Daniel's path and is in a direction perpendicular to this path in the earth frame. Both observers will count the same number of pulsations of the variable star during Daniel's round trip. Use this fact and the expression for the quadratic Doppler shift to find a relation between the time A_j by which John has aged and the time A_d by which Daniel has aged during Daniel's round trip.

6.29. Astronaut Dan boards a spaceship that immediately reaches the speed $0.6c$. The destination is Selinunte, a "fixed" star 12 light-years away from Earth. His brother John stays home, waits for 4 years, and then boards another spaceship and reaches his brother at the very time Daniel gets to Selinunte. By how much have the two brothers aged as they arrive at their destination?

6.30. Consider a gas of atoms at temperature T inside an enclosure. The atoms emit light that passes in the z direction through a window of the enclosure and can then be observed as a spectral line in a spectrometer. A stationary atom would emit light at frequency ν_0; but, because of the Doppler effect, the frequency of the light observed from an atom having a z component of the velocity v_z will be different. As a consequence, the light arriving on the spectrometer is characterized by some intensity distribution $I(\nu)\,d\nu$. Calculate:

(a) The mean frequency $\bar{\nu}$ of the light observed in the spectrometer

(b) The root mean square frequency shift

$$(\Delta\nu)_{\text{rms}} = [\overline{(\nu - \bar{\nu}^2)}]^{1/2}$$

of the light observed in the spectrometer

(c) The relative intensity distribution $I(\nu)\,d\nu$ of the light observed in the spectrometer.

7

Theory of Relativity: II

7.1. Lorentz Transformation and E and B Fields

We have previously expressed the Maxwell equations as follows:

$$\Box^2 A_\mu = \frac{4\pi}{c} j_\mu \tag{7.1.1}$$

$$\sum_\mu \frac{\partial A_\mu}{\partial x_\mu} = 0 \tag{7.1.2}$$

$$F_{\mu\nu} = \frac{\partial A_\nu}{\partial x_\mu} - \frac{\partial A_\mu}{\partial x_\nu} \tag{7.1.3}$$

where

$$A_\mu \equiv (\mathbf{A}, i\phi) \tag{7.1.4}$$

$$j_\mu \equiv (\mathbf{j}, ic\rho) \tag{7.1.5}$$

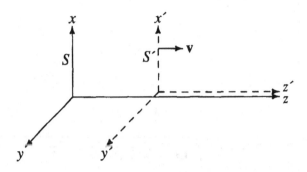

FIGURE 7.1

and

$$F_{12} = B_3, \quad F_{23} = B_1, \quad F_{31} = B_2 \tag{7.1.6}$$

$$F_{14} = -iE_1, \quad F_{24} = -iE_2, \quad F_{34} = -iE_3 \tag{7.1.7}$$

Consider now the Lorentz transformation (see Fig. 7.1) represented by

$$x_i = \sum_{k=1}^{4} a_{ki} x_k', \quad x_i' = \sum_{k=1}^{4} a_{ik} x_k, \tag{7.1.8}$$

That is,

$$
\begin{array}{ll}
x_1 = x_1' & x_1' = x_1 \\
x_{21} = x_2' & x_2' = x_2 \\
x_3 = \dfrac{x_3' - i\beta x_4'}{\sqrt{1 - \beta^2}} & x_3' = \dfrac{x_3 + i\beta x_4}{\sqrt{1 - \beta^2}} \\
x_4 = \dfrac{i\beta x_3' + x_4'}{\sqrt{1 - \beta^2}} & x_4' = \dfrac{-i\beta x_3 + x_4}{\sqrt{1 - \beta^2}}
\end{array}
\tag{7.1.9}
$$

The matrix of the transformation has the coefficients

$$
\begin{array}{llll}
a_{11} = 1, & a_{12} = 0, & a_{13} = 0, & a_{14} = 0 \\
a_{21} = 0, & a_{22} = 1, & a_{23} = 0, & a_{24} = 0 \\
a_{31} = 0, & a_{32} = 0, & a_{33} = \dfrac{1}{\sqrt{1 - \beta^2}}, & a_{34} = \dfrac{i\beta}{\sqrt{1 - \beta^2}} \\
a_{41} = 0, & a_{42} = 0, & a_{43} = \dfrac{-i\beta}{\sqrt{1 - \beta^2}}, & a_{44} = \dfrac{1}{\sqrt{1 - \beta^2}}
\end{array}
\tag{7.1.10}
$$

The four components of a vector and the sixteen components of a tensor transform as follows:

$$V'_\mu = \sum_\sigma a_{\mu\sigma} V_\sigma \tag{7.1.11}$$

$$F'_{\mu\nu} = \sum_{\sigma\tau} a_{\mu\sigma} a_{\nu\tau} F_{\sigma\tau} \tag{7.1.12}$$

Therefore, considering the four-vector j_μ,

$$j'_x = j_x$$

$$j'_y = j_y$$

$$j'_z = \frac{1}{\sqrt{1-\beta^2}} j_z + \frac{i\beta}{\sqrt{1-\beta^2}} ic\rho = \frac{j_z - v\rho}{\sqrt{1-\beta^2}} \tag{7.1.13}$$

$$ic\rho' = \frac{-i\beta j_z}{\sqrt{1-\beta^2}} + \frac{ic\rho}{\sqrt{1-\beta^2}} = \frac{ic\rho - i\beta j_z}{\sqrt{1-\beta^2}}$$

The last equation can be rewritten as

$$\rho' = \frac{\rho - (v/c^2) j_z}{\sqrt{1-\beta^2}} \tag{7.1.14}$$

Note that ρ' may be different from zero, even if $\rho = 0$, provided $j_z \neq 0$. This means that, if a body traversed by a current seems without charge to an observer who moves with it, it will have a charge for an observer who is in a different inertial frame.

Let us now work out the transformation relations for the six components of the antisymmetric tensor representing the **E** and **B** fields:

$$\begin{pmatrix} F_{11} & F_{12} & F_{13} & F_{14} \\ F_{21} & F_{22} & F_{23} & F_{24} \\ F_{31} & F_{32} & F_{33} & F_{34} \\ F_{41} & F_{42} & F_{43} & F_{44} \end{pmatrix} = \begin{pmatrix} 0 & B_3 & -B_2 & -iE_1 \\ -B_3 & 0 & B_1 & -iE_2 \\ B_2 & -B_1 & 0 & -iE_3 \\ iE_1 & iE_2 & iE_3 & 0 \end{pmatrix} \tag{7.1.15}$$

We have in detail

$$F'_{14} = -iE'_1, \quad \mu = 1, \quad \nu = 4$$

$$= a_{11} a_{41} F_{11} + a_{11} a_{42} F_{12} + a_{11} a_{43} F_{13} + a_{11} a_{44} F_{14}$$

$$+ a_{12}a_{41}F_{21} + a_{12}a_{42}F_{22} + a_{12}a_{43}F_{23} + a_{12}a_{44}F_{24}$$

$$+ a_{13}a_{41}F_{31} + a_{13}a_{42}F_{32} + a_{13}a_{43}F_{33} + a_{13}a_{44}F_{34}$$

$$+ a_{14}a_{41}F_{41} + a_{14}a_{42}F_{42} + a_{14}a_{43}F_{43} + a_{14}a_{44}F_{44}$$

$$= a_{43}F_{13} + a_{44}F_{14}$$

$$= -\frac{i\beta}{\sqrt{1-\beta^2}}(-B_2) + \frac{1}{\sqrt{1-\beta^2}}(-iE_1) \tag{7.1.16}$$

$$F'_{24} = -iE'_2, \quad \mu = 2, \quad \nu = 4$$

$$= a_{21}a_{41}F_{11} + a_{21}a_{42}F_{12} + a_{21}a_{43}F_{13} + a_{21}a_{44}F_{14}$$

$$+ a_{22}a_{41}F_{21} + a_{22}a_{42}F_{22} + a_{22}a_{43}F_{23} + a_{22}a_{44}F_{24}$$

$$+ a_{23}a_{41}F_{31} + a_{23}a_{42}F_{32} + a_{23}a_{43}F_{33} + a_{23}a_{44}F_{34}$$

$$+ a_{24}a_{41}F_{41} + a_{24}a_{42}F_{42} + a_{24}a_{43}F_{43} + a_{24}a_{44}F_{44}$$

$$= a_{43}F_{23} + a_{44}F_{24}$$

$$= -\frac{i\beta}{\sqrt{1-\beta^2}}B_1 + \frac{1}{\sqrt{1-\beta^2}}(-iE_2) \tag{7.1.17}$$

$$F'_{34} = -iE'_3, \quad \mu = 3, \quad \nu = 4$$

$$= a_{31}a_{41}F_{11} + a_{31}a_{42}F_{12} + a_{31}a_{43}F_{13} + a_{31}a_{44}F_{14}$$

$$+ a_{32}a_{41}F_{21} + a_{32}a_{42}F_{22} + a_{32}a_{43}F_{23} + a_{32}a_{44}F_{24}$$

$$+ a_{33}a_{41}F_{31} + a_{33}a_{42}F_{32} + a_{33}a_{43}F_{33} + a_{33}a_{44}F_{34}$$

$$+ a_{34}a_{41}F_{41} + a_{34}a_{42}F_{42} + a_{34}a_{43}F_{43} + a_{34}a_{44}F_{44}$$

$$= a_{33}a_{43}F_{33} + a_{33}a_{44}F_{34} + a_{34}a_{43}F_{43} + a_{34}a_{44}F_{34}$$

$$= \frac{1}{1-\beta^2}(-iE_3) + \frac{\beta^2}{1-\beta^2}(iE_3) = -iE_3 \tag{7.1.18}$$

In summary,

$$-E'_1 = \frac{\beta B_2 - E_1}{\sqrt{1-\beta^2}} \quad \text{or} \quad E'_x = \frac{E_x - \beta B_y}{\sqrt{1-\beta^2}}$$

$$-E'_2 = -\frac{\beta B_1 - E_2}{\sqrt{1-\beta^2}} \quad \text{or} \quad E'_y = \frac{E_y + \beta B_x}{\sqrt{1-\beta^2}} \tag{7.1.19}$$

$$-E'_3 = -E_3 \quad \text{or} \quad E'_z = E_z$$

Also,

$$F'_{23} = B'_1, \quad \mu = 2, \quad \nu = 3$$

$$= a_{21}a_{31}F_{11} + a_{21}a_{32}F_{12} + a_{21}a_{33}F_{13} + a_{21}a_{34}F_{14}$$

$$+ a_{22}a_{31}F_{21} + a_{22}a_{32}F_{22} + a_{22}a_{33}F_{23} + a_{22}a_{34}F_{24}$$

$$+ a_{23}a_{31}F_{31} + a_{23}a_{32}F_{32} + a_{23}a_{33}F_{33} + a_{23}a_{34}F_{34}$$

$$+ a_{24}a_{31}F_{41} + a_{24}a_{32}F_{42} + a_{24}a_{33}F_{43} + a_{24}a_{34}F_{44}$$

$$= a_{33}F_{23} + a_{34}F_{24} = \frac{1}{\sqrt{1-\beta^2}}B_1$$

$$+ \frac{i\beta}{\sqrt{1-\beta^2}}(-iE_2) = \frac{B_1 + \beta E_2}{\sqrt{1-\beta^2}} \qquad (7.1.20)$$

$$F'_{31} = B'_2, \quad \mu = 3, \quad \nu = 1$$

$$= a_{31}a_{11}F_{11} + a_{31}a_{12}F_{12} + a_{31}a_{13}F_{13} + a_{31}a_{14}F_{14}$$

$$+ a_{32}a_{11}F_{21} + a_{32}a_{12}F_{22} + a_{32}a_{13}F_{23} + a_{32}a_{14}F_{24}$$

$$+ a_{33}a_{11}F_{31} + a_{33}a_{12}F_{32} + a_{33}a_{13}F_{33} + a_{33}a_{14}F_{34}$$

$$+ a_{34}a_{11}F_{41} + a_{34}a_{12}F_{42} + a_{34}a_{13}F_{43} + a_{34}a_{14}F_{44}$$

$$= a_{33}F_{31} + a_{34}F_{41} = \frac{1}{\sqrt{1-\beta^2}}B_2$$

$$+ \frac{i\beta}{\sqrt{1-\beta^2}}(iE_1) = \frac{B_2 - \beta E_1}{\sqrt{1-\beta^2}} \qquad (7.1.21)$$

$$F'_{12} = B'_3, \quad \mu = 1, \quad \nu = 3$$

$$= a_{11}a_{21}F_{11} + a_{11}a_{22}F_{12} + a_{11}a_{23}F_{13} + a_{11}a_{24}F_{14}$$

$$+ a_{12}a_{21}F_{21} + a_{21}a_{22}F_{22} + a_{12}a_{23}F_{23} + a_{12}a_{24}F_{24}$$

$$+ a_{13}a_{21}F_{31} + a_{13}a_{22}F_{32} + a_{13}a_{23}F_{33} + a_{13}a_{24}F_{34}$$

$$+ a_{14}a_{21}F_{41} + a_{14}a_{22}F_{42} + a_{14}a_{23}F_{43} + a_{14}a_{24}F_{44}$$

$$= a_{22}F_{12} = B_3 \qquad (7.1.22)$$

In summary,

$$B'_1 = \frac{B_1 + \beta E_2}{\sqrt{1-\beta^2}} \quad \text{or} \quad B'_x = \frac{B_x + \beta E_y}{\sqrt{1-\beta^2}}$$

$$B_2' = \frac{B_2 - \beta E_1}{\sqrt{1 - \beta^2}} \quad \text{or} \quad B_y' = \frac{B_y - \beta E_x}{\sqrt{1 - \beta^2}} \tag{7.1.23}$$

$$B_3' = B_3 \qquad \text{or} \quad B_z' = B_z$$

We have then

$$E_z' = E_z, \qquad\qquad\qquad B_z' = B_z$$

$$E_x' = \frac{E_x - (v/c)B_y}{\sqrt{1 - \beta^2}}, \quad B_x' - \frac{B_x + (v/c)E_y}{\sqrt{1 - \beta^2}} \tag{7.1.24}$$

$$E_y' = \frac{E_y + (v/c)B_x}{\sqrt{1 - \beta^2}}, \quad B_y' = \frac{B_y - (v/c)E_x}{\sqrt{1 - \beta^2}}$$

But

$$\mathbf{v} \times \mathbf{B} = \begin{pmatrix} \mathbf{i} & \mathbf{j} & \mathbf{k} \\ 0 & 0 & v \\ B_x & B_y & B_z \end{pmatrix} = -\mathbf{i}vB_y + \mathbf{j}vB_x \tag{7.1.25}$$

and

$$\mathbf{v} \times \mathbf{E} = -\mathbf{i}vE_y + \mathbf{j}vE_x \tag{7.1.26}$$

Let the subscripts $\parallel$ and $\perp$ indicate the components of the fields parallel and perpendicular to $\mathbf{v}$, respectively; we obtain

$$\mathbf{E}_\parallel' = \mathbf{E}_\parallel, \qquad\qquad\qquad \mathbf{B}_\parallel' = \mathbf{B}_\parallel$$

$$\mathbf{E}_\perp' = \frac{\mathbf{E}_\perp + [(\mathbf{v}/c) \times \mathbf{B}]}{\sqrt{1 - (v^2/c^2)}}, \quad \mathbf{B}_\perp' = \frac{\mathbf{B}_\perp + [(\mathbf{v}/c) \times \mathbf{E}]}{\sqrt{1 - (v^2/c^2)}} \tag{7.1.27}$$

For $v \ll c$ (nonrelativistic velocities),

$$\mathbf{E}_\parallel' = \mathbf{E}_\parallel, \qquad\qquad\qquad \mathbf{B}_\parallel' = \mathbf{B}_\parallel$$

$$\mathbf{E}_\perp' = \mathbf{E}_\perp + \frac{\mathbf{v}}{c} \times \mathbf{B}, \quad \mathbf{B}_\perp' = \mathbf{B}_\perp - \frac{\mathbf{v}}{c} \times \mathbf{E} \tag{7.1.28}$$

or

$$\mathbf{E}' = \mathbf{E} + \frac{\mathbf{v}}{c} \times \mathbf{B}, \quad \mathbf{B}' = \mathbf{B} - \frac{\mathbf{v}}{c} \times \mathbf{E} \tag{7.1.29}$$

We can now arrive at the preceding two relations in a different way. Consider two systems of axes, S, the laboratory system, and S', a system that moves with respect to S with constant velocity $\mathbf{v}$. Let q be a charge at rest in S', and let $\mathbf{E}$, $\mathbf{B}$ and $\mathbf{E}'$, $\mathbf{B}'$ be the values of the fields in systems S and S', respectively. The forces measured in the two systems are

$$\mathbf{F} = q\left(\mathbf{E} + \frac{\mathbf{v}}{c} \times \mathbf{B}\right) \tag{7.1.30}$$

$$\mathbf{F}' = q\mathbf{E}' \tag{7.1.31}$$

Since $\mathbf{v} = $ constant, we take

$$\mathbf{F} = \mathbf{F}' \quad \text{and} \quad \mathbf{E}' = \mathbf{E} + \frac{\mathbf{v}}{c} \times \mathbf{B} \tag{7.1.32}$$

Consider now a charge distribution $\rho(\mathbf{x})$ producing an electric field $\mathbf{E}(\mathbf{x})$ and a magnet producing a field $\mathbf{B}(\mathbf{x})$, and rest in S. In S,

$$\mathbf{E}(\mathbf{x}) = -\nabla \int d\tau' \frac{\rho(\mathbf{x}')}{|\mathbf{x} - \mathbf{x}'|} \tag{7.1.33}$$

In the system S', we have

$$\mathbf{j}(\mathbf{x}) = -\rho(\mathbf{x})\mathbf{v}$$

and

$$\begin{aligned}
\mathbf{B}' &= \mathbf{B} + \nabla \times \frac{1}{c} \int d\tau' \frac{\mathbf{j}(\mathbf{x}')}{|\mathbf{x} - \mathbf{x}'|} \\
&= \mathbf{B} + \nabla \times \left(-\frac{\mathbf{v}}{c}\right) \int d\tau' \frac{\rho(\mathbf{x}')}{|\mathbf{x} - \mathbf{x}'|} \\
&= \mathbf{B} + \frac{\mathbf{v}}{c} \times \nabla \int d\tau' \frac{\rho(\mathbf{x}')}{|\mathbf{x} - \mathbf{x}'|} \\
&= \mathbf{B} - \frac{\mathbf{v}}{c} \times \mathbf{E}
\end{aligned} \tag{7.1.34}$$

We have rederived formulas (7.1.29). These relations are, however, not exact. If we make the inverse transformation, we get into trouble. These transformations do not form a group. Two assumptions were incorrect in our approximate treatment:

(1) We assumed the force to be equal in the two frames of reference.

(2) We put $\mathbf{j}' = -\rho\mathbf{v}$ instead of $-\rho\mathbf{v}/\sqrt{1 - \beta^2}$.

Let us now express the Maxwell equations in covariant form by means of the new formalism. We begin with the inhomogeneous equations

$$\nabla \cdot \mathbf{E} = 4\pi\rho$$

$$\nabla \times \mathbf{B} - \frac{1}{c}\frac{\partial \mathbf{E}}{\partial t} = \frac{4\pi}{c}\mathbf{j} \qquad (7.1.35)$$

CLAIM The inhomogeneous Maxwell equations can be expressed as follows:

$$\sum_{\nu} \frac{\partial F_{\mu\nu}}{\partial x_\nu} = \frac{4\pi}{c} j_\mu \qquad (7.1.36)$$

PROOF For $\mu = 1$,

$$\frac{\partial F_{11}}{\partial x_1} + \frac{\partial F_{12}}{\partial x_2} + \frac{\partial F_{13}}{\partial x_3} + \frac{\partial F_{14}}{\partial x_4} = \frac{4\pi}{c} j_1$$

gives

$$\frac{\partial B_3}{\partial x_2} + \frac{\partial B_2}{\partial x_3} - \frac{1}{c}\frac{\partial E_1}{\partial t} = \frac{4\pi}{c} j_1$$

For $\mu = 2$,

$$\frac{\partial F_{21}}{\partial x_1} + \frac{\partial F_{22}}{\partial x_2} + \frac{\partial F_{23}}{\partial x_3} + \frac{\partial F_{24}}{\partial x_4} = \frac{4\pi}{c} j_2$$

gives

$$-\frac{\partial B_3}{\partial x_1} + \frac{\partial B_1}{\partial x_3} - \frac{1}{c}\frac{\partial E_2}{\partial t} = \frac{4\pi}{c} j_2$$

For $\mu = 3$,

$$\frac{\partial F_{31}}{\partial x_1} + \frac{\partial F_{32}}{\partial x_2} + \frac{\partial F_{33}}{\partial x_3} + \frac{\partial F_{34}}{\partial x_4} = \frac{4\pi}{c} j_3$$

gives

$$\frac{\partial B_2}{\partial x_1} - \frac{\partial B_1}{\partial x_2} - \frac{1}{c}\frac{\partial E_3}{\partial t} = \frac{4\pi}{c} j_3$$

For $\mu = 4$,

$$\frac{\partial F_{41}}{\partial x_1} + \frac{\partial F_{42}}{\partial x_2} + \frac{\partial F_{43}}{\partial x_3} + \frac{\partial F_{44}}{\partial x_4} = \frac{4\pi}{c} j_4$$

gives

$$\frac{\partial E_1}{\partial x_1} + \frac{\partial E_2}{\partial x_2} + \frac{\partial E_3}{\partial x_3} = 4\pi\rho \qquad \text{Q.E.D}$$

Let us now consider the homogeneous Maxwell equations

$$\mathbf{\nabla} \cdot \mathbf{B} = 0$$
$$\mathbf{\nabla} \times \mathbf{E} + \frac{1}{c}\frac{\partial \mathbf{B}}{\partial t} = 0 \tag{7.1.37}$$

Let us define a new symbol $\varepsilon_{\alpha\beta\gamma\delta}$ whose value is

$$\begin{cases} +1 & \text{if } \alpha\beta\gamma\delta \text{ is an even permutation of 1234} \\ -1 & \text{if } \alpha\beta\gamma\delta \text{ is an odd permutation of 1234} \\ 0 & \text{if any two indexes are equal} \end{cases} \tag{7.1.38}$$

For example,

$$\varepsilon_{2143} = 1$$
$$\varepsilon_{1243} = -1$$
$$\varepsilon_{2243} = 0$$

Let us define further a tensor

$$\hat{F}_{\mu\nu} = \frac{1}{2}\sum_{\sigma}\sum_{\tau}\varepsilon_{\mu\nu\sigma\tau}F_{\sigma\tau} \tag{7.1.39}$$

The various elements of the tensor $\hat{F}$ are given by

$$\hat{F}_{12} = \frac{1}{2}\sum_{\sigma\tau}\varepsilon_{12\sigma\tau}F_{\sigma\tau} = \frac{1}{2}\varepsilon_{1234}F_{34} + \frac{1}{2}\varepsilon_{1243}F_{43} = -iE_3$$

$$\hat{F}_{21} = \frac{1}{2}\sum_{\sigma\tau}\varepsilon_{21\sigma\tau}F_{\sigma\tau} = \frac{1}{2}\varepsilon_{2134}F_{34} + \frac{1}{2}\varepsilon_{2143}F_{43} = iE_3$$

$$\hat{F}_{13} = \frac{1}{2}\sum_{\sigma\tau}\varepsilon_{13\sigma\tau}F_{\sigma\tau} = \frac{1}{2}\varepsilon_{1324}F_{24} + \frac{1}{2}\varepsilon_{1342}F_{42} = iE_2$$

$$\hat{F}_{31} = \frac{1}{2}\sum_{\sigma\tau}\varepsilon_{31\sigma\tau}F_{\sigma\tau} = \frac{1}{2}\varepsilon_{3124}F_{24} + \frac{1}{2}\varepsilon_{3142}F_{42} = -iE_2$$

$$\hat{F}_{14} = \frac{1}{2}\sum_{\sigma\tau}\varepsilon_{14\sigma\tau}F_{\sigma\tau} = \frac{1}{2}\varepsilon_{1432}F_{32} + \frac{1}{2}\varepsilon_{1423}F_{23} = B_1$$

$$\hat{F}_{41} = \frac{1}{2}\sum_{\sigma\tau}\varepsilon_{41\sigma\tau}F_{\sigma\tau} = \frac{1}{2}\varepsilon_{4123}F_{23} + \frac{1}{2}\varepsilon_{4123}F_{32} = -B_1$$

$$\hat{F}_{34} = \frac{1}{2} \sum_{\sigma\tau} \varepsilon_{34\sigma\tau} F_{\sigma\tau} = \frac{1}{2} \varepsilon_{3412} F_{12} + \frac{1}{2} \varepsilon_{3421} F_{21} = B_3$$

$$\hat{F}_{43} = \frac{1}{2} \sum_{\sigma\tau} \varepsilon_{43\sigma\tau} F_{\sigma\tau} = \frac{1}{2} \varepsilon_{4312} F_{12} + \frac{1}{2} \varepsilon_{4321} F_{21} = -B_3$$

$$\hat{F}_{24} = \frac{1}{2} \sum_{\sigma\tau} \varepsilon_{24\sigma\tau} F_{\sigma\tau} = \frac{1}{2} \varepsilon_{2413} F_{13} + \frac{1}{2} \varepsilon_{2431} F_{31} = B_2$$

$$\hat{F}_{42} = \frac{1}{2} \sum_{\sigma\tau} \varepsilon_{42\sigma\tau} F_{\sigma\tau} = \frac{1}{2} \varepsilon_{4213} F_{13} + \frac{1}{2} \varepsilon_{4231} F_{31} = -B_2$$

$$\hat{F}_{23} = \frac{1}{2} \sum_{\sigma\tau} \varepsilon_{23\sigma\tau} F_{\sigma\tau} = \frac{1}{2} \varepsilon_{2314} F_{14} + \frac{1}{2} \varepsilon_{2341} F_{41} = -iE_1$$

$$\hat{F}_{32} = \frac{1}{2} \sum_{\sigma\tau} \varepsilon_{32\sigma\tau} F_{\sigma\tau} = \frac{1}{2} \varepsilon_{3214} F_{14} + \frac{1}{2} \varepsilon_{3241} F_{41} = iE_1$$

We can represent the two tensors F and $\hat{F}$ in the following condensed form:

$$
\begin{array}{llll}
F_{11} = 0 & F_{12} = B_3 & F_{13} = -B_2 & F_{14} = -iE_1 \\
\hat{F}_{11} = 0 & \hat{F}_{12} = -iE_3 & \hat{F}_{13} = iE_2 & \hat{F}_{14} = B_1 \\
F_{21} = -B_3 & F_{22} = 0 & F_{23} = B_1 & F_{24} = -iE_2 \\
\hat{F}_{21} = iE_3 & \hat{F}_{22} = 0 & \hat{F}_{23} = -iE_1 & F_{24} = B_2 \\
F_{31} = B_2 & F_{32} = -B_1 & F_{33} = 0 & F_{34} = -iE_3 \\
\hat{F}_{31} = -iE_2 & \hat{F}_{32} = iE_1 & \hat{F}_{33} = 0 & \hat{F}_{34} = B_3 \\
F_{41} = iE_1 & F_{42} = iE_2 & F_{43} = iE_3 & F_{44} = 0 \\
\hat{F}_{41} = -B_1 & \hat{F}_{42} = -B_2 & \hat{F}_{43} = -B_3 & \hat{F}_{44} = 0
\end{array}
\qquad (7.1.40)
$$

CLAIM The homogeneous Maxwell equations can be expressed as follows:

$$\sum_{\nu} \frac{\partial \hat{F}_{\mu\nu}}{\partial x_\nu} = 0 \qquad (7.1.41)$$

PROOF For $\mu = 1$,

$$\frac{\partial \hat{F}_{12}}{\partial x_2} + \frac{\partial \hat{F}_{13}}{\partial x_3} + \frac{\partial \hat{F}_{14}}{\partial x_4} = 0$$

gives

$$\frac{\partial E_3}{\partial x_2} - \frac{\partial E_2}{\partial x_3} + \frac{1}{c}\frac{\partial B_1}{\partial t} = 0$$

For $\mu = 2$,

$$\frac{\partial \hat{F}_{21}}{\partial x_1} + \frac{\partial \hat{F}_{23}}{\partial x_3} + \frac{\partial \hat{F}_{24}}{\partial x_4} = 0$$

gives

$$\frac{\partial E_3}{\partial x_1} + \frac{\partial E_1}{\partial x_3} + \frac{1}{c}\frac{\partial B_2}{\partial t} = 0$$

For $\mu = 3$,

$$\frac{\partial \hat{F}_{31}}{\partial x_1} + \frac{\partial \hat{F}_{32}}{\partial x_2} + \frac{\partial \hat{F}_{34}}{\partial x_4} = 0$$

gives

$$\frac{\partial E_2}{\partial x_1} - \frac{\partial E_1}{\partial x_2} + \frac{1}{c}\frac{\partial B_3}{\partial t} = 0$$

For $\mu = 4$,

$$\frac{\partial \hat{F}_{41}}{\partial x_1} + \frac{\partial \hat{F}_{42}}{\partial x_2} + \frac{\partial \hat{F}_{43}}{\partial x_3} = 0$$

gives

$$\frac{\partial B_1}{\partial x_1} + \frac{\partial B_2}{\partial x_2} + \frac{\partial B_3}{\partial x_3} = 0$$

Q.E.D

In particular, the relation

$$\frac{\partial \hat{F}_{12}}{\partial x_2} + \frac{\partial \hat{F}_{13}}{\partial x_3} + \frac{\partial \hat{F}_{14}}{\partial x_4} = 0$$

can be written

$$\frac{\partial F_{34}}{\partial x_2} + \frac{\partial F_{42}}{\partial x_3} + \frac{\partial F_{34}}{\partial x_4} = 0$$

Then the relations

$$\sum_{\nu} \frac{\partial \hat{F}_{\mu\nu}}{\partial x_{\nu}} = 0$$

can be rewritten in terms of the F elements:

$$\frac{\partial F_{\mu\nu}}{\partial x_{\sigma}} + \frac{\partial F_{\nu\sigma}}{\partial x_{\mu}} + \frac{\partial F_{\sigma\mu}}{\partial x_{\nu}} = 0 \qquad (7.1.42)$$

We can then write the four Maxwell equations as follows:

$$\left.\begin{array}{l} \nabla \cdot \mathbf{E} = 4\pi\rho \\[2mm] \nabla \times \mathbf{B} - \dfrac{1}{c}\dfrac{\partial \mathbf{E}}{\partial t} = \dfrac{4\pi}{c}\mathbf{j} \end{array}\right\} \rightarrow \sum_\nu \dfrac{\partial F_{\mu\nu}}{\partial x_\nu} = \dfrac{4\pi}{c}j_\mu \qquad (7.1.43)$$

$$\left.\begin{array}{l} \nabla \cdot \mathbf{B} = 0 \\[2mm] \nabla \times \mathbf{E} + \dfrac{1}{c}\dfrac{\partial \mathbf{B}}{\partial t} = 0 \end{array}\right\} \begin{array}{c} \rightarrow \quad \sum_\nu \dfrac{\partial \hat{F}_{\mu\nu}}{\partial x_\nu} = 0 \\[4mm] \text{or} \quad \dfrac{\partial F_{\mu\nu}}{\partial x_\sigma} + \dfrac{\partial F_{\nu\sigma}}{\partial x_\mu} + \dfrac{\partial F_{\sigma\mu}}{\partial x_\nu} = 0 \end{array} \qquad (7.1.44)$$

7.2. Charged Mass Point in Electromagnetic Field Minkowski Force

The postulates of relativity are general and are also valid in mechanics. We shall now formulate the relations that express the laws of mechanics in covariant form.

Newton's law for a charged particle of mass m is given by

$$\mathbf{F} = m\mathbf{a} = m\frac{d^2\mathbf{x}}{dt^2} = q\left(\mathbf{E} + \frac{\mathbf{v}}{c} \times \mathbf{B}\right) \qquad (7.2.1)$$

This law is approximate and incompatible with relativity. It is approximately valid for $v/c \ll 1$. Whatever the exact equation is, when $v/c \ll 1$, it must reduce to Newton's equation.

In the previous chapter we defined a four-vector called *four-velocity*:

$$u_\mu \equiv \left(\frac{\mathbf{v}}{\sqrt{1-\beta^2}}, \frac{ic}{\sqrt{1-\beta^2}}\right) \qquad (7.2.2)$$

whose four components are related as follows:

$$\sum_{i=1}^{4} u_i^2 = -c^2 \qquad (7.2.3)$$

Let us now consider a moving point whose instantaneous velocity in a system of coordinate S is $\mathbf{v}(t)$. In the four-dimensional space defined by the coordinates x_1 to x_4, the length of an infinitesimal element is given by

$$ds^2 = dx^2 + dy^2 + dz^2 - c^2 dt^2 \qquad (7.2.4)$$

In a coordinate system S' in which the moving point is at rest, the length of an infinitesimal element is given by

$$ds'^2 = -c^2 d\tau^2 \qquad (7.2.5)$$

τ is the time measured by a clock tied to the moving body and is called *proper time*. The equality $ds^2 = ds'^2$ leads to

$$dx^2 + dy^2 + dz^2 - c^2 dt^2 = -c^2 d\tau^2 \qquad (7.2.6)$$

and to

$$d\tau = dt \sqrt{1 - \frac{v_2}{c^2}} \qquad (7.2.7)$$

Relation (7.2.7) shows that the proper time of a moving point is Lorentz invariant.

We can study the movement of the point by giving its four coordinates as a function of proper time:

$$x_i = x_i(\tau) \qquad (7.2.8)$$

The four-velocity components are the derivatives of these coordinates with respect to proper time:

$$u_i = \frac{dx_i}{d\tau} = \frac{dx_i}{dt} \frac{1}{\sqrt{1 - \beta^2}} = \frac{v_i}{\sqrt{1 - \beta^2}}, \quad i = 1, 2, 3 \qquad (7.2.9)$$

$$u_4 = \frac{dx_4}{d\tau} = \frac{ic\, dt}{d\tau} = \frac{ic}{\sqrt{1 - \beta^2}} \qquad (7.2.10)$$

Let us now define a four-vector called the *Minkowski force*:

$$F_\mu = m \frac{du_\mu}{d\tau} \qquad (7.2.11)$$

and let us propose the following relation:

$$F_\mu = m \frac{du_\mu}{d\tau} = \frac{e}{c} \sum_v F_{\mu\nu} u_\nu \qquad (7.2.12)$$

where m = rest mass = constant. Let us write down the four components of this equation.

For $\mu = 1$,

$$m \frac{du_1}{d\tau} = \frac{e}{c} [F_{12} u_2 + F_{13} u_3 + F_{14} u_4]$$

or

$$\frac{m}{\sqrt{1-\beta^2}}\frac{d}{dt}\frac{v_1}{\sqrt{1-\beta^2}} = \frac{e}{c}\left(\frac{1}{\sqrt{1-\beta^2}}\right)[B_3 v_2 - B_2 v_3 - iE_1 ic]$$

or

$$\frac{d}{dt}\frac{mv_1}{\sqrt{1-\beta^2}} = \frac{e}{c}[(\mathbf{v}\times\mathbf{B})_1 + E_1 c] = e\left[E_1 + \left(\frac{\mathbf{v}}{c}\times\mathbf{B}\right)_1\right]$$

For $\mu = 2$,

$$m\frac{du_2}{d\tau} = \frac{e}{c}[F_{21}u_1 + F_{23}u_3 + F_{24}u_4]$$

or

$$\frac{m}{\sqrt{1-\beta^2}}\frac{d}{dt}\frac{v_2}{\sqrt{1-\beta^2}} = \frac{e}{c}\left(\frac{1}{\sqrt{1-\beta^2}}\right)[B_1 v_3 - B_3 v_1 - iE_2 ic]$$

or

$$\frac{d}{dt}\frac{mv_2}{\sqrt{1-\beta^2}} = \frac{e}{c}[(\mathbf{v}\times\mathbf{B})_2 + E_2 c] = e\left[E_2 + \left(\frac{\mathbf{v}}{c}\times\mathbf{B}\right)_2\right]$$

For $\mu = 3$,

$$m\frac{du_3}{d\tau} = \frac{e}{c}[F_{31}u_1 + F_{32}u_2 + F_{34}u_4]$$

or

$$\frac{m}{\sqrt{1-\beta^2}}\frac{d}{dt}\frac{v^3}{\sqrt{1-\beta^2}} = \frac{e}{c}\left(\frac{1}{\sqrt{1-\beta^2}}\right)[B_2 v_1 - B_1 v_2 - iE_3 ic]$$

or

$$\frac{d}{dt}\frac{mv_3}{\sqrt{1-\beta^2}} = \frac{e}{c}[(\mathbf{v}\times\mathbf{B})_3 + E_3 c] = e\left[E_3 + \left(\frac{\mathbf{v}}{c}\times\mathbf{B}\right)_3\right]$$

For $\mu = 4$,

$$m\frac{du_4}{d\tau} = \frac{e}{c}[F_{41}u_1 + F_{42}u_2 + F_{43}u_3]$$

or

$$\frac{m}{\sqrt{1-\beta^2}}\frac{d}{dt}\frac{ic}{\sqrt{1-\beta^2}} = \frac{e}{c}\left(\frac{1}{\sqrt{1-\beta^2}}\right)[iE_1 v_1 + iE_2 v_2 - iE_3 v_3]$$

or

$$\frac{d}{dt}\frac{mc^2}{\sqrt{1-\beta^2}} = e(\mathbf{E}\cdot\mathbf{v})$$

We can then write the following relations:

$$\frac{d}{dt}\frac{m\mathbf{v}}{\sqrt{1-\beta^2}} = e\left[\mathbf{E} + \frac{\mathbf{v}}{c} \times \mathbf{B}\right] \qquad (7.2.13)$$

$$\frac{d}{dt}\frac{mc^2}{\sqrt{1-\beta^2}} = e(\mathbf{E}\cdot\mathbf{v}) \qquad (7.2.14)$$

We can rewrite Eq. (7.2.14) in a different way:

$$\frac{d}{dt}\frac{mc^2}{\sqrt{1-\beta^2}} = mc^2\frac{d}{dt}\frac{1}{\sqrt{1-(v^2/c^2)}} = \frac{m}{1-(v^2/c^2)^{3/2}}\mathbf{v}\cdot\frac{d\mathbf{v}}{dt}$$

Therefore Eq. (7.2.14) becomes

$$\frac{m\mathbf{v}\cdot(d\mathbf{v}/dt)}{(1-\beta^2)^{3/2}} = e(\mathbf{E}\cdot\mathbf{v}) \qquad (7.2.15)$$

Let us now take relation (7.2.13) and multiply it by $\mathbf{v}$:

$$\mathbf{v}\cdot\frac{d}{dt}\frac{m\mathbf{v}}{\sqrt{1-\beta^2}} = e(\mathbf{E}\cdot\mathbf{v})$$

or

$$\mathbf{v}\cdot\frac{d\mathbf{v}}{dt}\frac{m}{\sqrt{1-\beta^2}} + \frac{m\mathbf{v}\cdot(d\mathbf{v}/dt)\beta^2}{(1-\beta^2)^{3/2}} = \frac{m\mathbf{v}\cdot(d\mathbf{v}/dt)}{(1-\beta^2)^{3/2}} \qquad (7.2.16)$$

Therefore, (7.2.14) does not contain information that is not in (7.2.13).
We set

$$\mathbf{p} = \frac{m\mathbf{v}}{\sqrt{1-\beta^2}} = \text{momentum} \qquad (7.2.17)$$

$$\mathbf{F} = e\left(\mathbf{E} + \frac{\mathbf{v}}{c} \times \mathbf{B}\right) = \text{force} \qquad (7.2.18)$$

and define a new four-vector:

$$p_\mu \equiv \left(\frac{m\mathbf{v}}{\sqrt{1-\beta^2}}, \frac{imc}{\sqrt{1-\beta^2}}\right)$$

$$= energy-momentum\ four\text{-}vector \qquad (7.2.19)$$

We can now justify our terminology, $e(\mathbf{E}\cdot\mathbf{v})$ is the work done by $\mathbf{E}$ in the unit time. Because of (7.2.14), we identify the quantity $mc^2/\sqrt{1-\beta^2}$ with the energy:

$$mc^2(1 - \beta^2)^{-1/2} = mc^2 + \frac{1}{2}mv^2 + \cdots \tag{7.2.20}$$

If the masses were indestructible, the term mc^2 would not be important. We can now write

$$p_\mu = \left(\frac{m\mathbf{v}}{\sqrt{1 - \beta^2}}, \frac{iE}{c} \right) \equiv \left(\mathbf{p}, \frac{iE}{c} \right) \tag{7.2.21}$$

where $E =$ energy. We also have

$$E = \frac{mc^2}{\sqrt{1 - \beta^2}} \tag{7.2.22}$$

$$\mathbf{p} = \frac{m\mathbf{v}}{\sqrt{1 - \beta^2}} = \frac{E\mathbf{v}}{c^2} \tag{7.2.23}$$

$$\mathbf{v} = \frac{\mathbf{p}c^2}{E} \tag{7.2.24}$$

and

$$\sum_\mu p_\mu^2 = p^2 - \frac{E^2}{c^2} = \frac{m^2 v^2}{1 - \beta^2} - \frac{m^2 c^2}{1 - \beta^2} = -m^2 c^2 < 0 \tag{7.2.25}$$

p_μ is a timelike four-vector.

From Eq. (7.2.23), we derive

$$E^2 = (pc)^2 + (mc^2)^2 \tag{7.2.26}$$

We also have

$$\frac{v}{c} = \frac{pc}{E} \tag{7.2.27}$$

and, since $v < c, pc < E$. The fact that the velocity of a particle is always smaller than the velocity c of the electromagnetic radiation is a crucial point in relativity. A particle can never be accelerated to a velocity greater than c. Even if we use a very large amount of work, we can only bring v very close to c; every finite amount of work will always result in a velocity v smaller than c.

We have plotted E/mc^2 versus β in Fig. 7.2.

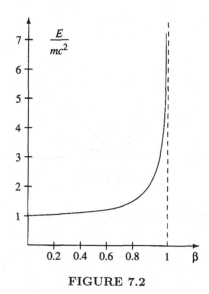

FIGURE 7.2

7.3. Gauss's Theorem in Four Dimensions

In three dimensions, Gauss's theorem is expressed as follows:

$$\int_V d\tau \frac{\partial f}{\partial x_k} = \int_S dS \, n_k f \qquad (7.3.1)$$

where the function f is a scalar or a component of a vector. This theorem expresses a relation between a volume integral and a surface integral. n_k represents the kth component of the unit vector $\mathbf{n}$ perpendicular to the surface element dS. We want now to extend this theorem to four dimensions.

Let us draw the space-time axes as in Fig. 7.3. The cone represented by $|x| = ct$ is called the *light cone*. We consider two types of vectors:

(1) Vectors like a, making an angle smaller than 45° with respect to the x axis, called *spacelike*,

(2) Vectors like b, making an angle larger than 45° with respect to the x axis, called *timelike*

We can perform a Lorentz transformation such that the temporal axis is parallel to a timelike vector; the vector would then have components $(0, 0, 0, \xi)$. We can perform a Lorentz transformation such that one spatial axis is parallel to a spacelike vector; the vector would then have components $(\xi, 0, 0, 0)$ or $(0, \xi, 0, 0)$, or $(0, 0, \xi, 0)$. The properties of timelikeness or spacelikeness are invariant under a Lorentz transformation.

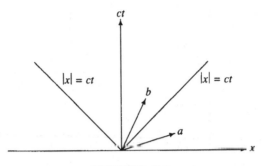

FIGURE 7.3

The four-velocity vector is a timelike vector:

$$u_\mu \equiv \frac{dx_\mu}{d\tau} \equiv \left(\frac{v_1}{\sqrt{1 - \beta^2}}, \frac{v_2}{\sqrt{1 - \beta^2}}, \frac{v_3}{\sqrt{1 - \beta^2}}, \frac{v_4}{\sqrt{1 - \beta^2}} \right) \qquad (7.3.2)$$

$$\sum_{\mu=1}^{4} u_\mu = -c^2 < 0 \qquad (7.3.3)$$

The four-acceleration vector

$$a_\mu = \frac{du_\mu}{d\tau} \qquad (7.3.4)$$

is orthogonal to u_μ:

$$\sum_\mu u_\mu \frac{du_\mu}{d\tau} = \frac{1}{2} \frac{d}{d\tau} \left(\sum_\mu u_\mu^2 \right) = 0 \qquad (7.3.5)$$

since $\sum_\mu u_\mu^2 = -c^2$ is an invariant.

A vector orthogonal to a timelike vector is a spacelike vector; therefore, the four-acceleration vector is a spacelike vector.

We define a *spacelike surface* as a surface (see Fig. 7.4) such that a vector normal to it is always a timelike vector. This property of a surface in four-dimensional space is invariant under a Lorentz transformation. In the case of the spacelike surface in Fig. 7.4, we have

$$s_\mu = \text{space vector} : \sum_\mu s_\mu^2 > 0$$

$$n_\mu = \text{timelike vector} : \sum_\mu n_\mu^2 > 0$$

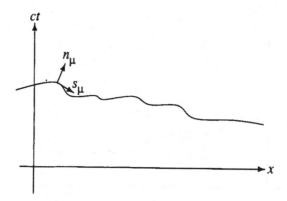

FIGURE 7.4

We define the vector n_μ as a *timelike unit vector* by setting

$$\sum_\mu n_\mu^2 = -1 \qquad (7.3.6)$$

It is possible to perform a Lorentz transformation such that, as a consequence, n_μ would have the following coordinates:

$$n_\mu \equiv (0, 0, 0, i) \qquad (7.3.7)$$

The Gauss theorem in four dimensions expresses the fact that a volume integral can be always transformed into a surface integral. Consider a volume bound by two spacelike surfaces S_1 and S_2 as in Fig. 7.5; the volume closes at infinity in spacelike directions. If the contributions on the surface along spacelike directions vanish in the limit $\to \infty$, Gauss's theorem takes the form

$$\int_V d^4x \, \frac{\partial f}{\partial x_\mu} = \int dS_2 n_\mu^{(2)} f - \int dS_1 n_\mu^{(1)} f \qquad (7.3.8)$$

Let us consider an application of this theorem. Let f be one of the four components of the four-current

$$j_\mu \equiv (j, ic\rho) \qquad (7.3.9)$$

The divergence in four dimensions of this four-vector is zero:

$$\sum_\mu \frac{\partial j_\mu}{\partial x_\mu} = \boldsymbol{\nabla} \cdot \mathbf{j} + \frac{\partial \rho}{\partial t} = 0 \qquad (7.3.10)$$

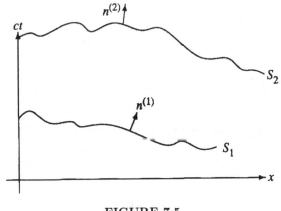

FIGURE 7.5

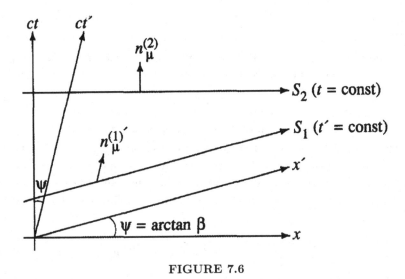

FIGURE 7.6

in accordance with the equation of continuity. We choose the two surfaces S_1 and S_2 as in Fig. 7.6:

$$dS_2 = dx\,dy\,dz, \qquad n_\mu^{(2)} \equiv (0,0,0,i)$$
$$dS_1 = dx'\,dy'\,dz', \qquad n_\mu^{(1)'} \equiv (0,0,0,i)$$

$$(7.3.11)$$

We apply now Gauss's theorem to each component of the four-current vector:

$$\int d^4x \frac{\partial j_\mu}{\partial x_\mu} = \int dS_2 n_\mu^{(2)} j_\mu - \int dS_1 n_\mu^{(1)} j_\mu \qquad (7.3.12)$$

Summing up,

$$\int d^4x \sum_\mu \frac{\partial j_\mu}{\partial x_\mu} = 0$$

$$= \int_{t=\text{const}} dx\, dy\, dz \sum_\mu n_\mu^{(2)} j_\mu$$

$$- \int_{t'=\text{const}} dx'\, dy'\, dz' \sum_\mu n_\mu^{(1)'} j_\mu'$$

$$= \int_{t=\text{const}} dx\, dy\, dz\, i(ic\rho)$$

$$- \int_{r'=\text{const}} dx'\, dy'\, dz' i(ic\rho') \qquad (7.3.13)$$

or

$$Q = \int_{t=\text{const}} \rho(\mathbf{x}, t) d^3\mathbf{x} = Q' = \int_{t'=\text{const}} \rho'(\mathbf{x}', t') d^3\mathbf{x}' \qquad (7.3.14)$$

This result has two important consequences:

(1) The total charge Q is conserved in time:

$$Q(t) = Q(t') \qquad (7.3.15)$$

This can be seen by taking the surface S_2 parallel to the surface S_1 in the diagram of Fig. 7.6.

(2) The total charge Q is also a Lorentz invariant. Two observers, in uniform motion with respect to each other, observe the same total charge. An observer adds up all the charges in a surface $t = \text{const}$, which he calls space; the other observer adds up all the charges in a surface $t' = \text{const}$, which he calls space.

Some relevant points can now be made:

(1) In general, if we have any four-vector, such that

$$\sum_{\mu} \frac{\partial f_{\mu}}{\partial x_{\mu}} = 0 \qquad (7.3.16)$$

then $\int d^3x\, f_4(\mathbf{x}, t)$ is invariant.

(2) This result can be generalized. If $T_{\mu\nu}$ is a second-rank tensor and if

$$\sum_{\mu} \frac{\partial T_{\mu\nu}}{\partial x_{\mu}} = 0 \qquad (7.3.17)$$

then $\int d^3\mathbf{x}\, T_{4\nu}$ is a four-vector.

In fact, let p^{μ} be an arbitrary four-vector, constant in space and time. Let us consider the four-vector

$$f_{\mu} = \sum_{\nu=1}^{4} p_{\nu} T_{\mu\nu} \qquad (7.3.18)$$

in which $T_{\mu\nu}$ satisfies relation (7.3.17). Then

$$\sum_{\mu} \frac{\partial f_{\mu}}{\partial x_{\mu}} = \sum_{\nu} p_{\nu} \sum_{\mu} \frac{\partial T_{\mu\nu}}{\partial x_{\mu}} = 0$$

and, therefore,

$$\int d^3\mathbf{x}\, f_4 = \int d^3\mathbf{x} \sum_{\nu} p_{\nu} T_{4\nu}$$

$$= \sum_{\nu} p_{\nu} \int T_{4\nu} d^3\mathbf{x} = \sum_{\nu} p'_{\nu} \int T'_{4\nu} d^3\mathbf{x}'$$

is an invariant. But p_{μ} is an arbitrary four-vector; therefore $\int T_{4\nu} d^3\mathbf{x}$ is also a four-vector.

7.4. Electromagnetic Energy–Momentum Tensor

In this section we shall treat the interaction between an electromagnetic field and some charges and currents. Charges and currents are described in terms of

$$\begin{aligned} \rho(\mathbf{x}, t) &= \text{charge density} \\ \mathbf{j}(\mathbf{x}, t) &= \text{current density} \end{aligned} \qquad (7.4.1)$$

For a point charge at position $\mathbf{x}_p(t)$, moving with velocity $\mathbf{v}$, we get

$$\rho(\mathbf{x}, t) = e\delta^{(3)}(\mathbf{x} - \mathbf{x}_p(t))$$
$$\mathbf{j}(\mathbf{x}, t) = \mathbf{v}\rho(\mathbf{x}, t) = e\mathbf{v}\delta^{(3)}(\mathbf{x} - \mathbf{x}_p(t)) \tag{7.4.2}$$

We need to introduce the concept of *force density*. We have already introduced the Minkowski force:

$$F_\mu = \frac{dp_\mu}{d\tau} = \frac{e}{c} \sum_\nu F_{\mu\nu} u_\nu \tag{7.4.3}$$

where

$$p_\mu \equiv \left(\mathbf{p}, \frac{iE_p}{c} \right), \quad E_p = \text{energy}$$

$$u_\nu \equiv \left(\frac{\mathbf{v}}{\sqrt{1 - \beta^2}}, \frac{ic}{\sqrt{1 - \beta^2}} \right)$$

$F_{\mu\nu} \equiv$ tensor representing the $\mathbf{E}$ and $\mathbf{B}$ fields

From this equation we have derived

$$\mathbf{F} = \frac{d\mathbf{p}}{dt} = e\left(\mathbf{E} + \frac{\mathbf{v}}{c} \times \mathbf{B} \right) \tag{7.4.4}$$

$$\frac{dE_p}{dt} = e(\mathbf{E} \cdot \mathbf{v}) \tag{7.4.5}$$

where E_p designates the energy, the subscript p having been introduced to distinguish the energy from the field E.

We define the following four-vector:

$$f_\mu = \frac{1}{c} \sum_\nu F_{\mu\nu} j_\nu \tag{7.4.6}$$

and evaluate it for the case in which the four-current is given by

$$j_\mu \equiv (\mathbf{j}, ic\rho) \equiv [e\mathbf{v}\delta^{(3)}(\mathbf{x} - \mathbf{x}_p(t)), ice\delta^{(3)}(\mathbf{x} - \mathbf{x}_p(t))] \tag{7.4.7}$$

Therefore,

$$\int d^3\mathbf{x}\, f_1 = \frac{1}{c} \int d^3\mathbf{x}(F_{12}j_2 + F_{13}j_3 + F_{14}j_4)$$

$$= \frac{1}{c} \int d^3\mathbf{x}(B_3j_2 - B_2j_3 + E_1c\rho)$$

$$= \frac{1}{c} \int d^3\mathbf{x}[(\mathbf{j} \times \mathbf{B})_1 + E_1 c\rho]$$

$$= \frac{e}{c} \int d^3\mathbf{x}[(\mathbf{v} \times \mathbf{B})_1 + E_1 c]\delta^{(3)}(\mathbf{x} - \mathbf{x}_p(t))$$

$$= \frac{e}{c}[(\mathbf{v} \times \mathbf{B})_1 + E_1 c]_{\mathbf{x}=\mathbf{x}_p(t)}$$

Analogously,

$$\int d^3\mathbf{x}\, f_2 = \frac{e}{c}[(\mathbf{v} \times \mathbf{B})_2 + E_2 c]_{\mathbf{x}=\mathbf{x}_p(t)}$$

$$\int d^3\mathbf{x}\, f_3 = \frac{e}{c}[(\mathbf{v} \times \mathbf{B})_3 + E_3 c]_{\mathbf{x}=\mathbf{x}_p(t)}$$

and

$$\int d^3\mathbf{x}\, f_4 = \frac{1}{c} \int d^3\mathbf{x}(F_{41} j_1 + F_{42} j_2 + F_{43} j_3)$$

$$= \frac{ie}{c} \int d^3\mathbf{x}[E_1 v_1 \delta^{(3)}(\mathbf{x} - \mathbf{x}_p(t))$$

$$+ E_2 v_2 \delta^{(3)}(\mathbf{x} - \mathbf{x}_p(t)) + E_3 v_3 \delta^{(3)}(\mathbf{x} - \mathbf{x}_p(t))]$$

$$= \frac{ie}{c}(\mathbf{E} \cdot \mathbf{v})_{\mathbf{x}=\mathbf{x}_p(t)}$$

Therefore,

$$\int d^3\mathbf{x}\, \mathbf{f} = e\left(\mathbf{E} + \frac{\mathbf{v}}{c} \times \mathbf{B}\right) = \mathbf{F} = \frac{d\mathbf{p}}{dt}$$

$$\int d^3\mathbf{x}\, f_4 = \frac{ie}{c}(\mathbf{E} \cdot \mathbf{v}) = \frac{i}{c}\frac{dE_p}{dt}$$

(7.4.8)

These equations relate the integral over space of a four-vector (f_μ) to the derivative with respect to time of another four-vector (p_μ). The four-vector f_μ, defined by Eq. (7.4.6), gives, in its space-part $\mathbf{f}$, the rate of change of the mechanical momentum per unit volume and, in its time-part $f_4, i/c$ times the rate of change of the mechanical energy per unit volume. Therefore, the four-vector f_μ can be thought of as a *force-density four-vector* and definition (7.4.6) as the covariant form of the Lorentz force equation.

We want now to evaluate the four-vector f_μ in the case of a general four-current. We start from the covariant expressions for the Maxwell equations:

$$\sum_\sigma \frac{\partial F_{\nu\sigma}}{\partial x_\sigma} = \frac{4\pi}{c} j_\nu$$

$$\sum_\nu \frac{\partial \hat{F}_{\mu\nu}}{\partial x_\nu} = 0$$

(7.4.9)

From the first of (7.4.9),

$$j_\nu = \frac{c}{4\pi} \sum_\sigma \frac{\partial F_{\nu\sigma}}{\partial x_\sigma}$$

(7.4.10)

The second of (7.4.9) can be expressed as follows:

$$\frac{\partial \hat{F}_{12}}{\partial x_2} + \frac{\partial \hat{F}_{13}}{\partial x_3} + \frac{\partial \hat{F}_{14}}{\partial x_4} = 0 \rightarrow \frac{\partial F_{34}}{\partial x_2} + \frac{\partial F_{42}}{\partial x_3} + \frac{\partial F_{23}}{\partial x_4} = 0$$

$$\frac{\partial \hat{F}_{21}}{\partial x_1} + \frac{\partial \hat{F}_{23}}{\partial x_3} + \frac{\partial \hat{F}_{24}}{\partial x_4} = 0 \rightarrow \frac{\partial F_{43}}{\partial x_1} + \frac{\partial F_{14}}{\partial x_3} + \frac{\partial F_{31}}{\partial x_4} = 0$$

$$\frac{\partial \hat{F}_{31}}{\partial x_1} + \frac{\partial \hat{F}_{32}}{\partial x_2} + \frac{\partial \hat{F}_{34}}{\partial x_4} = 0 \rightarrow \frac{\partial F_{24}}{\partial x_1} + \frac{\partial F_{41}}{\partial x_2} + \frac{\partial F_{12}}{\partial x_4} = 0$$

$$\frac{\partial \hat{F}_{41}}{\partial x_1} + \frac{\partial \hat{F}_{42}}{\partial x_2} + \frac{\partial \hat{F}_{43}}{\partial x_3} = 0 \rightarrow \frac{\partial F_{32}}{\partial x_1} + \frac{\partial F_{13}}{\partial x_2} + \frac{\partial F_{21}}{\partial x_3} = 0$$

or

$$\frac{\partial F_{\mu\nu}}{\partial x_\sigma} + \frac{\partial F_{\nu\sigma}}{\partial x_\mu} + \frac{\partial F_{\sigma\mu}}{\partial x_\nu} = 0$$

or

$$\frac{\partial F_{\mu\nu}}{\partial x_\sigma} + \frac{\partial F_{\sigma\mu}}{\partial x_\nu} = -\frac{\partial F_{\nu\sigma}}{\partial x_\mu}$$

(7.4.11)

We have then

$$f_\mu = \frac{1}{c} \sum_\nu F_{\mu\nu} j_\nu$$

$$= \frac{1}{4\pi} \sum_\nu \sum_\sigma \left(F_{\mu\nu} \frac{\partial F_{\nu\sigma}}{\partial x_\sigma} \right)$$

$$= \frac{1}{4\pi} \sum_\nu \sum_\sigma \left[\frac{\partial}{\partial x_\sigma} (F_{\mu\nu} F_{\nu\sigma}) - \frac{\partial F_{\mu\nu}}{\partial x_\sigma} F_{\nu\sigma} \right]$$

$$= \frac{1}{4\pi} \left[\sum_\sigma \frac{\partial}{\partial x_\sigma} \left(\sum_\nu F_{\mu\nu} F_{\nu\sigma} \right) \right.$$

$$\left. - \frac{1}{2} \sum_\sigma \sum_\nu \left(\frac{\partial F_{\mu\nu}}{\partial x_\sigma} F_{\nu\sigma} + \frac{\partial F_{\mu\nu}}{\partial x_\sigma} F_{\nu\sigma} \right) \right]$$

$$= \frac{1}{4\pi} \left\{ \sum_\sigma \frac{\partial}{\partial x_\sigma} \left(\sum_\nu F_{\mu\nu} F_{\nu\sigma} \right) \right.$$

$$\left. - \left[\frac{1}{2} \sum_\sigma \sum_\nu \frac{\partial F_{\mu\nu}}{\partial x_\sigma} F_{\nu\sigma} + \frac{1}{2} \sum_\sigma \sum_\nu \frac{\partial F_{\mu\sigma}}{\partial x_\nu} F_{\sigma\nu} \right] \right\}$$

$$= \frac{1}{4\pi} \left\{ \sum_\sigma \frac{\partial}{\partial x_\sigma} \left(\sum_\nu F_{\mu\nu} F_{\nu\sigma} \right) \right.$$

$$\left. - \left[\frac{1}{2} \sum_\sigma \sum_\nu \frac{\partial F_{\mu\nu}}{\partial x_\sigma} F_{\nu\sigma} + \frac{1}{2} \sum_\sigma \sum_\nu \frac{\partial F_{\sigma\mu}}{\partial x_\nu} F_{\nu\sigma} \right] \right\}$$

$$= \frac{1}{4\pi} \left\{ \sum_\sigma \frac{\partial}{\partial x_\sigma} \left(\sum_\nu F_{\mu\nu} F_{\nu\sigma} \right) \right.$$

$$\left. - \left[\frac{1}{2} \sum_\sigma \sum_\nu F_{\nu\sigma} \left(\frac{\partial F_{\mu\nu}}{\partial x_\sigma} + \frac{\partial F_{\sigma\mu}}{\partial x_\nu} \right) \right] \right\}$$

$$= \frac{1}{4\pi} \left\{ \sum_\sigma \frac{\partial}{\partial x_\sigma} \left(\sum_\nu F_{\mu\nu} F_{\nu\sigma} \right) + \frac{1}{2} \sum_\sigma \sum_\nu F_{\nu\sigma} \left(\frac{\partial F_{\nu\sigma}}{\partial x_\mu} \right) \right\}$$

$$= \frac{1}{4\pi} \left\{ \sum_\sigma \frac{\partial}{\partial x_\sigma} \left(\sum_\nu F_{\mu\nu} F_{\nu\sigma} \right) + \frac{1}{4} \frac{\partial}{\partial x_\mu} \left[\sum_\nu \sum_\sigma (F_{\nu\sigma})^2 \right] \right\}$$

$$= \frac{1}{4\pi} \sum_\sigma \frac{\partial}{\partial x_\sigma} \left[\sum_\nu F_{\mu\nu} F_{\nu\sigma} + \frac{1}{4} \delta_{\mu\sigma} \sum_\lambda \sum_\nu (F_{\lambda\nu})^2 \right]$$

$$= \sum_\sigma \frac{\partial T_{\mu\sigma}}{\partial x_\sigma}$$

or

$$f_\mu = \sum_\sigma \frac{\partial T_{\mu\sigma}}{\partial x_\sigma} \tag{7.4.12}$$

where

$$T_{\mu\sigma} = \frac{1}{4\pi} \left[\sum_{\nu} F_{\mu\nu} F_{\nu\sigma} + \frac{1}{4} \delta_{\mu\sigma} \sum_{\lambda\nu} (F_{\lambda\nu})^2 \right] \qquad (7.4.13)$$

Now we evaluate the elements of the tensor $T_{\mu\sigma}$. We start by introducing the following convention:

(1) Whenever the indexes i and k are used as subscripts, then $i, k = 1, 2, 3$.
(2) Whenever the indexes $\lambda, \mu, \sigma,$ and ν (Greek letters) are used, $\lambda, \mu, \sigma, \nu = 1, 2, 3, 4$.

Now

$$\sum_{\lambda=1}^{4} \sum_{\nu=1}^{4} (F_{\lambda\nu})^2 = F_{11}^2 + F_{12}^2 + F_{13}^2 + F_{14}^2 + F_{21}^2 + F_{22}^2 + F_{23}^2 + F_{24}^2$$

$$+ F_{31}^2 + F_{32}^2 + F_{33}^2 + F_{34}^2 + F_{41}^2 + F_{42}^2 + F_{43}^2 + F_{44}^2$$

$$= 0 + B_3^2 + B_2^2 - E_1^2 + B_3^2 + 0 + B_1^2 - E_2^2$$

$$+ B_2^2 + B_1^2 + 0 - E_3^2 - E_1^2 - E_2^2 - E_3^2 + 0$$

$$= 2(B^2 - E^2)$$

$$T_{11} = \frac{1}{4\pi} \left[F_{12}F_{21} + F_{13}F_{31} + F_{14}F_{41} + \frac{1}{2}(B^2 - E^2) \right]$$

$$= \frac{1}{4\pi} \left[-B_3^2 - B_2^2 + E_1^2 + \frac{1}{2}(B_1^2 + B_2^2 + B_3^2 - E_1^2 - E_2^2 - E_3^2) \right]$$

$$= \frac{1}{4\pi} \left[E_1^2 + B_1^2 - \frac{1}{2}(E^2 + B^2) \right]$$

$$T_{22} = \frac{1}{4\pi} \left[F_{21}F_{12} + F_{23}F_{32} + F_{24}F_{42} + \frac{1}{2}(B^2 - E^2) \right]$$

$$= \frac{1}{4\pi} \left[-B_3^2 - B_1^2 + E_2^2 + \frac{1}{2}(B^2 - E^2) \right]$$

$$= \frac{1}{4\pi} \left[E_2^2 + B_2^2 - \frac{1}{2}(E^2 + B^2) \right]$$

$$T_{33} = \frac{1}{4\pi} \left[F_{31}F_{13} + F_{32}F_{23} + F_{34}F_{43} + \frac{1}{2}(B^2 - E^2) \right]$$

$$= \frac{1}{4\pi}\left[-B_2^2 - B_1^2 + E_3^2 + \frac{1}{2}(B^2 - E^2)\right]$$

$$= \frac{1}{4\pi}\left[E_3^2 + B_3^2 - \frac{1}{2}(E^2 + B^2)\right]$$

$$T_{12} = \frac{1}{4\pi}[F_{13}F_{32} + F_{14}F_{42}] = \frac{1}{4\pi}(B_1 B_2 - E_1 E_2) = T_{21}$$

$$T_{13} = \frac{1}{4\pi}[F_{12}F_{23} + F_{14}F_{43}] = \frac{1}{4\pi}(B_1 B_3 + E_1 E_3) = T_{31}$$

$$T_{23} = \frac{1}{4\pi}[F_{21}F_{13} + F_{24}F_{43}] = \frac{1}{4\pi}(B_2 B_3 + E_2 E_3) = T_{32}$$

Therefore,

$$T_{ik} = \frac{1}{4\pi}\left[E_i E_k + B_i B_k - \frac{1}{2}\delta_{ik}(E^2 + B^2)\right] \qquad (7.4.14)$$

Also,

$$T_{14} = \frac{1}{4\pi}(F_{12}F_{24} + F_{13}F_{34}) = -\frac{i}{4\pi}(B_3 E_2 - B_2 E_3)$$

$$= -\frac{i}{4\pi}(\mathbf{E} \times \mathbf{B})_1 = T_{41}$$

$$T_{24} = \frac{1}{4\pi}(F_{21}F_{14} + F_{23}F_{34}) = -\frac{i}{4\pi}(B_1 E_3 - B_3 E_1)$$

$$= -\frac{i}{4\pi}(\mathbf{E} \times \mathbf{B})_2 = T_{42}$$

$$T_{34} = \frac{1}{4\pi}(F_{31}F_{14} + F_{32}F_{24}) = -\frac{i}{4\pi}(B_2 E_1 - B_1 E_2)$$

$$= -\frac{i}{4\pi}(\mathbf{E} \times \mathbf{B})_3 = T_{43}$$

$$T_{44} = \frac{1}{4\pi}\left[F_{41}F_{14} + F_{42}F_{24} + F_{43}F_{34} + \frac{1}{2}(B^2 - E^2)\right]$$

$$= \frac{1}{8\pi}[(E)^2 + (B)^2] = W = \begin{matrix} \text{energy density of the} \\ \text{electromagnetic field} \end{matrix} \qquad (7.4.15)$$

Therefore,

$$T_{i4} = -\frac{i}{4\pi}(\mathbf{E} \times \mathbf{B})_i = -\frac{i}{c}N_i = -icG_i, \quad i = 1,2,3 \qquad (7.4.16)$$

where

$$\mathbf{N} = \frac{c}{4\pi}(\mathbf{E} \times \mathbf{B}) = \text{Poynthing vector} \qquad (7.4.17)$$

$$\mathbf{G} = \frac{1}{4\pi c}(\mathbf{E} \times \mathbf{B}) = \text{momentum density} \qquad (7.4.18)$$

Therefore, the tensor T is given by

$$T = \begin{pmatrix} T_{11} & T_{12} & T_{13} & -icG_1 = -\dfrac{i}{c}N_1 \\[2mm] T_{21} & T_{22} & T_{23} & -icG_2 = -\dfrac{i}{c}N_2 \\[2mm] T_{31} & T_{32} & T_{33} & -icG_3 = -\dfrac{i}{c}N_3 \\[2mm] -icG_1 = -\dfrac{i}{c}N_1 & -icG_2 = -\dfrac{i}{c}N_2 & -icG_3 = -\dfrac{i}{c}N_3 & W \end{pmatrix}$$

$$(7.4.19)$$

We can now rewrite relation (7.4.12) as follows:

$$f_i = \sum_{k=1}^{3} \frac{\partial T_{ik}}{\partial x_k} + \frac{\partial T_{i4}}{\partial x_4} = \sum_{k=1}^{3} \frac{\partial T_{ik}}{\partial x_k} - \frac{dG_i}{dt}, \quad i = 1,2,3$$

$$f_4 = \sum_{k=1}^{3} \frac{\partial T_{4k}}{\partial x_k} + \frac{\partial T_{44}}{\partial x_4} = -\frac{i}{c}\sum_{k=1}^{3} \frac{\partial N_k}{\partial x_k} - \frac{i}{c}\frac{\partial W}{\partial t}$$

$$(7.4.20)$$

Integrating over space coordinates gives

$$\int d^3\mathbf{x}\,\mathbf{f} = \frac{d\mathbf{p}}{dt} = \int d^3\mathbf{x}\,\boldsymbol{\nabla}T - \frac{d}{dt}\int d^3\mathbf{x}\,\mathbf{G}$$

$$= \int d^3\mathbf{x}\,\boldsymbol{\nabla}T - \frac{d}{dt}\mathcal{P}^{\text{elm}} \qquad (7.4.21)$$

where $\boldsymbol{\nabla}T$ is a vector defined as

$$\boldsymbol{\nabla}T \equiv \left(\frac{\partial T_{11}}{\partial x_1} + \frac{\partial T_{12}}{\partial x_2} + \frac{\partial T_{13}}{\partial x_3}, \frac{\partial T_{21}}{\partial x_1} + \frac{\partial T_{22}}{\partial x_2} \right.$$

$$\left. + \frac{\partial T_{23}}{\partial x_3}, \frac{\partial T_{31}}{\partial x_1} + \frac{\partial T_{32}}{\partial x_2} + \frac{\partial T_{33}}{\partial x_3} \right) \qquad (7.4.22)$$

and

$$\mathcal{P}^{\mathrm{elm}} = \int d^3\mathbf{x}\,\mathbf{G} = \text{electromagnetic total momentum} \qquad (7.4.23)$$

Also

$$\int d^2\mathbf{x}\, f_4 = \frac{i}{c}\frac{dE_p}{dt} = -\frac{i}{c}\int d^3\mathbf{x}(\boldsymbol{\nabla}\cdot\mathbf{N}) - \frac{i}{c}\frac{d}{dt}\int d^3\mathbf{x}\,W$$

$$= -\frac{i}{c}\int d^3\mathbf{x}(\boldsymbol{\nabla}\cdot\mathbf{N}) - \frac{i}{c}\frac{d}{dt}\mathcal{E}^{\mathrm{elm}} \qquad (7.4.24)$$

where

$$\mathcal{E}^{\mathrm{elm}} = \int d^3\mathbf{x}\,W = \text{total electromagnetic energy} \qquad (7.4.25)$$

Then we get

$$\frac{d(\mathbf{p}+\mathcal{P}^{\mathrm{elm}})}{dt} = \int d^3x\,\boldsymbol{\nabla}T = \int dS\,\mathbf{n}T \qquad (7.4.26)$$

$$\frac{d(E_p+\mathcal{E}^{\mathrm{elm}})}{dt} = -\int d^3\mathbf{x}(\boldsymbol{\nabla}\cdot\mathbf{N}) = -\int dS(\mathbf{n}\cdot\mathbf{N}) \qquad (7.4.27)$$

We can now make the following observations:

(1) $\mathbf{n}$ is the outward normal to the closed surface S. In relation (7.4.26), $(\mathbf{n}T)_k = \sum_i n_i T_{ki}$ represents the kth component of the flow per unit area of momentum across the surface S *into* the volume bound by S. $\mathbf{n}T$ is then the force per unit area transmitted across S and acting on whatever is inside S.

(2) In (7.4.27), $\mathbf{n}\cdot\mathbf{N}$ represents the energy leaving the volume considered per unit time per unit area. The total energy inside the volume is

$$E_{\mathrm{tot}} = \mathcal{E}^{\mathrm{elm}} + E_p = \text{energy of the field} + \text{mechanical energy}$$

If $\int \mathbf{n}\cdot\mathbf{N}\,dS = 0$, then $E_{\mathrm{tot}} = $ constant. In these conditions, if energy is gained by the particles by a certain amount, energy will be lost by the fields by the same amount.

(3) If we consider the entire space, the fields will go to zero at the surface that is at infinity, and in both equations we shall have zero at the right members.

7.5. Green's Functions for the Potential Equations

The potential equations in the Lorentz gauge, presented in Sec. 5.10, are written as follows:

$$\left(\nabla^2 - \frac{1}{c^2}\frac{\partial}{\partial t^2}\right)\mathbf{A} = -\frac{4\pi}{c}\mathbf{j}$$

$$\left(\nabla^2 - \frac{1}{c^2}\frac{\partial}{\partial t^2}\right)\phi = -4\pi\rho \tag{7.5.1}$$

These equations are a generalization of Poisson's equations. Each general solution is given by the general solution of the homogeneous equation plus a particular solution of the inhomogeneous equation.

In covariant form, Eqs. (7.5.1) are written as

$$\Box^2 A_\mu = -\frac{4\pi}{c}j_\mu \tag{7.5.2}$$

where

$$A_\mu \equiv (\mathbf{A}, i\phi) \tag{7.5.3}$$

$$j_\mu \equiv (\mathbf{j}, ic\rho) \tag{7.5.4}$$

and

$$\sum_\mu \frac{\partial A_\mu}{\partial x_\mu} = 0, \quad \sum_\mu \frac{\partial j_\mu}{\partial x_\mu} = 0 \tag{7.5.5}$$

We shall use the following notation:

$$x_0 = ct$$
$$d^4x = d^3\mathbf{x}\, dx_0 \tag{7.5.6}$$
$$\delta^4(x) = \delta^{(3)}(\mathbf{x})\delta(x_0)$$

We define the Green's function $G(\mathbf{x}, x_0; \mathbf{x}', x_0')$ in the following way:

$$\Box_x^2 G(\mathbf{x}, x_0; \mathbf{x}', x_0') = -4\pi\delta^{(4)}(x - x')$$
$$= -4\pi\delta^{(3)}(\mathbf{x} - \mathbf{x}')\delta(x_0 - x_0') \tag{7.5.7}$$

Assume that we have formed a G function that obeys Eq. (7.5.7):

$$\int d^4x' G(x, x')f(x') = \int d^3\mathbf{x}'\, dx_0'\, G(\mathbf{x}, x_0; \mathbf{x}', x_0')f(\mathbf{x}', x_0') \tag{7.5.8}$$

Let us operate as follows:

$$\Box_x^2 \int d^4x' G(x, x') f(x')$$

$$= \Box_x^2 \int d^3\mathbf{x}' \, dx_0' G(\mathbf{x}, x_0; \mathbf{x}', x_0') f(\mathbf{x}', x_0')$$

$$= -4\pi \int d^3\mathbf{x}' \, dx_0' \delta^{(3)}(\mathbf{x} - \mathbf{x}') \delta(x_0 - x_0') f(\mathbf{x}', x_0')$$

$$= -4\pi f(\mathbf{x}, x_0) = -4\pi f(x) \qquad (7.5.9)$$

Now

$$\Box^2 A_\mu(x) = -\frac{4\pi}{c} j_\mu(x) \qquad (7.5.10)$$

Therefore,

$$A_\mu(x) = \frac{1}{c} \int d^4x' G(x, x') j_\mu(x') + A_\mu^0(x) \qquad (7.5.11)$$

where

$$\Box_x^2 A_\mu^0(x) = 0$$

The first term in the right member is a particular solution of the inhomogeneous equation, and the second term is the general solution of the homogeneous equation.

It is not clear at this point if we have solved the problem. There is still the Lorentz condition to take into account,

$$\sum_\mu \frac{\partial A_\mu}{\partial x_\mu} = 0 \qquad (7.5.12)$$

and the continuity equation,

$$\sum_\mu \frac{\partial j_\mu}{\partial x_\mu} = 0 \qquad (7.5.13)$$

Let us assume that G is a function of $(x - x')$ only. Then

$$\sum_\mu \frac{\partial A_\mu}{\partial x_\mu} = \frac{1}{c} \int d^4x' \sum_\mu \left[\frac{\partial}{\partial x_\mu} G(x - x') j_\mu(x') \right] + \sum_\mu \frac{\partial A_\mu^{(0)}}{\partial x_\mu}$$

$$= -\frac{1}{c} \int d^4x' \sum_\mu \left\{ \left[\frac{\partial}{\partial x'_\mu} G(x - x') \right] j_\mu(x') \right\} + \sum_\mu \frac{\partial A_\mu^0}{\partial x_\mu}$$

$$= \frac{1}{c} \int d^4x' G(x - x') \sum_\mu \frac{\partial j_\mu(x')}{\partial x'_\mu} + \sum_\mu \frac{\partial A_\mu^0}{\partial x_\mu} \qquad (7.5.14)$$

having integrated by parts and having assumed that at the boundaries G and j_μ are zero. We note that, because of the continuity equation, the Lorentz condition is automatically satisfied for the part of A_μ deriving from the current j_μ.

To be sure that the Lorentz condition is completely satisfied, we must look at the last part $\sum_\mu (\partial A_\mu^0 / \partial x_\mu)$. A_μ^0 is a superposition of plane waves:

$$A_\mu^0(\mathbf{x}, t) = \alpha_\mu \exp[i(\mathbf{k} \cdot \mathbf{x} - \omega t)] = \alpha_\mu \exp \left(i \sum_\lambda k_\lambda x_\lambda \right) \qquad (7.5.15)$$

where k_λ is a four-vector:

$$k_\lambda \equiv \left(\mathbf{k}, \frac{i\omega}{c} \right) \qquad (7.5.16)$$

The Lorentz condition gives

$$\sum_\mu \alpha_\mu k_\mu = 0 \qquad (7.5.17)$$

It is left as an exercise at the end of this chapter to show that, when such a condition is valid,

$$|\mathbf{E}| = |\mathbf{B}|$$

$$\mathbf{E} \perp \mathbf{B}, \mathbf{E} \perp \mathbf{k}, \mathbf{B} \perp \mathbf{k}$$

$$\mathbf{B} = \frac{\mathbf{k}}{|\mathbf{k}|} \times \mathbf{E} \qquad (7.5.18)$$

We have derived

$$A_\mu(x) = \frac{1}{c} \int d^4x' G(x - x') j_\mu(x') + A_\mu^0(x) \qquad (7.5.19)$$

A_μ is a four-vector and j_μ is a four-vector. d^4x is an invariant scalar. In fact $(x_4 = ix_0)$,

$$
\begin{aligned}
id^4x &= dx_1\, dx_2\, dx_3\, dx_4 \\
&= \frac{\partial(x_1, x_2, x_3, x_4)}{\partial(x_1', x_2', x_3', x_4')} dx_1'\, dx_2'\, dx_3'\, dx_4' \\
&= dx_1'\, dx_2'\, dx_3'\, dx_4' = id^4x'
\end{aligned}
\tag{7.5.20}
$$

because the Jacobian

$$
J = \frac{\partial(x_1, x_2, x_3, x_4)}{\partial(x_1', x_2', x_3', x_4')} = 1
\tag{7.5.21}
$$

Therefore, $G(x, x') = G(x - x')$ must be an invariant scalar:

$$
A_\mu''(x'') = \frac{1}{c} d^4x' G''(x'' - x') j_\mu(x') + A_\mu^{0''}(x'')
\tag{7.5.22}
$$

In summary, we must have:

(1) G a function of $(x - x')$. This assures that the Lorentz condition is respected.

(2) G invariant under a Lorentz transformation.

(3) G should reduce conveniently in the electrostatic and magnetostatic limit.

CLAIM (see Fig. 7.7)

$$
\begin{aligned}
G(\mathbf{x} - \mathbf{x}'; t - t') &= \frac{\delta[(x_0 - x_0') - |\mathbf{x} - \mathbf{x}'|]}{|\mathbf{x} - \mathbf{x}'|} \\
&= \frac{\delta[c(t - t') - |\mathbf{x} - \mathbf{x}'|]}{|\mathbf{x} - \mathbf{x}'|} = \frac{\delta(\xi_0 - r)}{r}
\end{aligned}
\tag{7.5.23}
$$

where

$$
|\mathbf{x} - \mathbf{x}'| = r
$$

$$
ct - ct' = x_0 - x_0' = \xi_0
$$

PROOF We begin by noting that

$$
\nabla^2\left(\frac{1}{r}\delta\right) = \delta\nabla^2\frac{1}{r} + \frac{1}{r}\nabla^2\delta + 2\boldsymbol{\nabla}\delta \cdot \boldsymbol{\nabla}\frac{1}{r}
$$

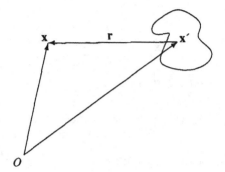

FIGURE 7.7

Letting

$$\frac{\partial \delta}{\partial t} = \delta'$$

we write

$$\frac{\partial \delta}{\partial r} = -\frac{1}{c}\delta'$$

For a function $f(r)$,

$$\nabla^2 f = \frac{1}{r^2}\frac{\partial}{\partial r}\left(r^2\frac{\partial f}{\partial r}\right) = \frac{1}{r^2}2r\frac{\partial f}{\partial r} + \frac{\partial^2 f}{\partial r^2} = \frac{\partial^2 f}{\partial r^2} + \frac{2}{r}\frac{\partial f}{\partial r}$$

$$\nabla f = \frac{\partial f}{\partial r}\frac{\mathbf{r}}{r}$$

Then

$$\nabla^2 \delta = \frac{\partial^2 \delta}{\partial r^2} + \frac{2}{r}\frac{\partial \delta}{\partial r} = \frac{1}{c^2}\delta'' - \frac{2}{rc}\delta'$$

$$\nabla \delta = \frac{\partial \delta}{\partial r}\frac{\mathbf{r}}{r} = -\frac{\mathbf{r}}{rc}\delta'$$

$$\nabla \frac{1}{r} = \frac{\partial(1/r)}{\partial r}\frac{\mathbf{r}}{r} = -\frac{\mathbf{r}}{r^3}$$

We know that

$$\nabla^2 \frac{1}{r} = -4\pi\delta(\mathbf{r}) = -4\pi\delta(\mathbf{x} - \mathbf{x}')$$

Therefore,

$$\nabla^2 \left(\frac{1}{r} \delta \right) = (\delta)[-4\pi\delta(\mathbf{x} - \mathbf{x}')] + \frac{1}{rc^2} \delta'' - \frac{2}{r^2 c} \delta'$$

$$+ 2 \left(-\frac{\mathbf{r}}{rc} \delta' \right) \left(-\frac{\mathbf{r}}{r^3} \right)$$

$$= -4\pi\delta[c(t - t') - r]\delta(\mathbf{x} - \mathbf{x}') + \frac{1}{rc^2} \delta''[c(t - t') - r]$$

$$= -4\pi\delta[c(t - t')]\delta(\mathbf{x} - \mathbf{x}') + \frac{1}{rc^2} \delta''[c(t - t') - r]$$

and

$$\Box^2 \left(\frac{1}{r} \delta \right) = \nabla^2 \left(\frac{1}{r} \delta \right) - \frac{1}{c^2} \frac{\partial^2}{\partial t^2} \left(\frac{1}{r} \delta \right)$$

$$= \nabla^2 \left(\frac{1}{r} \delta \right) - \frac{1}{rc^2} \delta'' = -4\pi\delta[c(t - t')]\delta(\mathbf{x} - \mathbf{x}')$$

$$= -4\pi\delta(x_0 - x_0')\delta(\mathbf{x} - \mathbf{x}') \qquad \text{Q.E.D.}$$

Now the solutions of the potential equations are

$$A_\mu(x) = A_\mu(\mathbf{x}, x_0) = \frac{1}{c} \int d^4 x' G(x - x') j_\mu(x') + A_\mu^0(x)$$

$$= \frac{1}{c} \int d^3 x' dx_0' \frac{\delta[(x_0 - x_0') - |\mathbf{x} - \mathbf{x}'|]}{|\mathbf{x} - \mathbf{x}'|} j_\mu(\mathbf{x}', x_0')$$

$$+ A_\mu^0(\mathbf{x}, x_0)$$

$$= \frac{1}{c} \int d^3 x' \frac{j_\mu(\mathbf{x}', x_0 - |\mathbf{x} - \mathbf{x}'|)}{|\mathbf{x} - \mathbf{x}'|} A_0^\mu(\mathbf{x}, x_0)$$

$$= \frac{1}{c} \int d^3 x' \frac{j_\mu[\mathbf{x}', t - (r/c)]}{r} + A_\mu^0(\mathbf{x}, t) \qquad (7.5.24)$$

Therefore,

$$\mathbf{A}(\mathbf{x}, t) = \mathbf{A}^0(\mathbf{x}, t) + \frac{1}{c} \int d^3 x' \frac{\mathbf{j}[\mathbf{x}', t - (r/c)]}{r}$$

$$\phi(\mathbf{x}, t) = \phi^0(\mathbf{x}, t) + \int d^3 x' \frac{\rho[\mathbf{x}', t - (r/c)]}{r} \qquad (7.5.25)$$

where $r = |\mathbf{x} - \mathbf{x}'|$.

If we have stationary charges and currents, the retardation term does not appear in the brackets:

$$\mathbf{A}(\mathbf{x}) = \frac{1}{c} \int d^3\mathbf{x}' \frac{\mathbf{j}(\mathbf{x}')}{r}$$

$$\phi(\mathbf{x}) = \int d^3\mathbf{x}' \frac{\rho(\mathbf{x}')}{r} \tag{7.5.26}$$

7.6. Retarded, Advanced, and Symmetrical Potentials

Now we investigate the problem of the uniqueness of the Green's function. This function has been defined as a solution of the equation

$$\Box^2 G(\mathbf{x}, x_0; \mathbf{x}', x_0') = -4\pi\delta^{(4)}(x - x') \tag{7.6.1}$$

Solutions can differ depending on the solutions of the homogeneous equation

$$\Box^2 G(\mathbf{x}, x_0; \mathbf{x}', x_0') = 0 \tag{7.6.2}$$

The results of the use of the Green's function are the potentials obtained in Eq. (7.5.25). The time dependence of the potentials shows up in the following way: The current and charge at position x' (see Fig. 7.8) determine the potential at position x. Given a certain *world line*, that is, a trajectory of a point charge in four-dimensional space, we can build different light cones, one for each position.

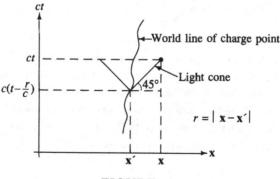

FIGURE 7.8

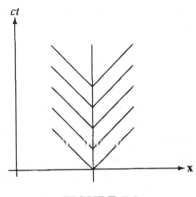

FIGURE 7.9

G contains a δ function whose argument is zero only on the light cone:

$$G_{\mathrm{ret}}(x, x') = \frac{1}{|\mathbf{x} - \mathbf{x}'|} \delta[c(t - t') - |\mathbf{x} - \mathbf{x}'|] \qquad (7.6.3)$$

where the subscript ret = retarded will be soon explained. For a particle at rest, the situation is depicted in Fig. 7.9.

We must now deal with the concept of *causality*. Charges and currents are the sources, that is, the *causes* of the fields. What is produced by something we call cause cannot precede the same cause. The problem is that this concept is not accounted for in the mathematics of our theory. In fact, until now we have dealt with *retarded Green's functions* as written in (7.5.23); these functions give us the potentials at time t as functions of the sources at time $[t - (r/c)]$. Other functions, called *advanced Green's functions*, can be defined as follows:

$$G_{\mathrm{adv}}(x, x') = \frac{1}{|\mathbf{x} - \mathbf{x}'|} \delta[c(t - t') + |\mathbf{x} - \mathbf{x}'|] \qquad (7.6.4)$$

It can also be shown that these functions are solutions of Eq. (7.6.1), so that G_{ret} and G_{adv} differ by a solution of the homogeneous equation Eq. (7.6.2). Since both G_{ret} and G_{adv} are "acceptable" Green's functions, any linear combination of the two is also an acceptable Green's function.

To better examine the implications of an eventual choice, let us consider the situation of a particle that is initially at rest and starts moving at a certain time. In the retarded case, we have, as in Fig. 7.10,

$$\phi(\mathbf{x}, t) = \int d^3x' \frac{\rho[\mathbf{x}', t - (r/c)]}{r} \qquad (7.6.5)$$

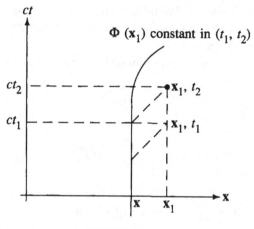

FIGURE 7.10

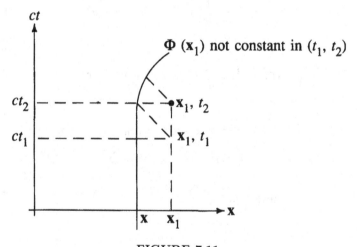

FIGURE 7.11

This potential in the time interval (t_1, t_2) is independent of time. In the advanced case of Fig. 7.11,

$$\phi(\mathbf{x}, t) = \int d^3 x' \frac{\rho[\mathbf{x}', t + (r/c)]}{r} \qquad (7.6.6)$$

This potential in (t_1, t_2) is time dependent.

We have essentially two pictures: one using the retarded potentials and the other using the advanced potentials. The mathematical treatment does not tell us which is correct. We make a choice on the basis of the principle of causality. The fields are determined by the past, not by the future. The option is for *retarded potentials*.

We present now some of the properties of the δ function:

$$\delta(ax) = \frac{1}{|a|}\delta(x) \tag{7.6.7}$$

or

$$\int_{-\infty}^{+\infty} dx f(x)\delta(ax) = \frac{1}{|a|}\int_{-\infty}^{+\infty} dx f(x)\delta(x) = \frac{1}{|a|}f(0) \tag{7.6.8}$$

Also,

$$
\begin{aligned}
\delta(x^2 - a^2) &= \delta[(x+a)(x-a)] \\
&= \delta[2a(x-a)] + \delta[-2a(x+a)] \\
&= \frac{1}{2|a|}[\delta(x-a) + \delta(x+a)]
\end{aligned} \tag{7.6.9}
$$

Therefore,

$$\delta(\xi_0^2 - r^2) = \frac{1}{2r}[\delta(\xi_0 + r) + \delta(\xi_0 - r)] \tag{7.6.10}$$

But

$$
\begin{aligned}
\xi_0^2 - r^2 &= (ct - ct')^2 - |\mathbf{x} - \mathbf{x}'|^2 = (x_0 - x_0')^2 - |\mathbf{x} - \mathbf{x}'|^2 \\
&= -(x_4 - x_4')^2 - |\mathbf{x} - \mathbf{x}'|^2 = -\sum_\mu (x_\mu - x_\mu')^2 \\
&= \text{invariant}
\end{aligned} \tag{7.6.11}
$$

Then

$$
\begin{aligned}
\delta(\xi_0^2 - r^2) &= \frac{1}{2r}[\delta(\xi_0 + r) + \delta(\xi_0 - r)] \\
&= \frac{G_{\text{adv}} + G_{\text{ret}}}{2} = \text{invariant}
\end{aligned} \tag{7.6.12}
$$

where

$$G_{\text{ret}} = \frac{\delta(\xi_0 - r)}{r} = \frac{\delta[c(t - t') - |\mathbf{x} - \mathbf{x}'|]}{|\mathbf{x} - \mathbf{x}'|} \tag{7.6.13}$$

$$G_{\text{adv}} = \frac{\delta(\xi_0 + r)}{r} = \frac{\delta[c(t - t') + |\mathbf{x} - \mathbf{x}'|]}{|\mathbf{x} - \mathbf{x}'|} \tag{7.6.14}$$

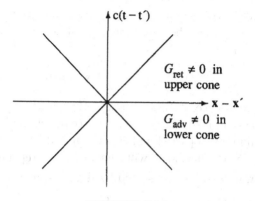

FIGURE 7.12

The sum $(G_{\text{ret}} + G_{\text{adv}})$ is invariant. G_{ret} and G_{adv} are also invariant independently for the following reason. These two functions contain 5δ functions whose arguments are zero only in the retarded and advanced light cones, respectively. These two cones are given in Fig. 7.12. The breaking up of the two cones in invariant with respect to Lorentz transformations. Also, no event can be taken from one cone to the other cone by a Lorentz transformation, since such transformations do not contemplate time reversal.

CHAPTER 7 EXERCISES[1]

7.1. (a) Express the invariants $\sum_{\mu\nu}(F_{\mu\nu})^2$ and $\sum_{\mu\nu}F_{\mu\nu}\hat{F}_{\mu\nu}$ in terms of **E** and **B**.

   (b) Show the invariance of the property $|\mathbf{E}| = |\mathbf{B}|$ of an electromagnetic wave.

   (c) Show the invariance of circular polarization.

7.2. A particle of mass m is subjected to a force **F** that is the gradient of a scalar field S, so that

$$m\frac{d\mathbf{v}}{dt} = -g\boldsymbol{\nabla}S$$

is the nonrelativistic equation of motion of the particle.

---

[1]Some of the exercises of this Chapter are adapted versions of problems that appear in *Space-time Physics* (1st and 2nd editions), by E. F. Taylor and J. A. Wheeler.

(a) Write the correct relativistic equation of motion.

(b) Assuming that the scalar field satisfies the homogeneous field equation

$$(\Box^2 - k^2)S(\mathbf{x}, t) = 0$$

in the absence of matter, how is this equation modified by the presence of a stationary point source of "charge" g?

(c) Assuming two particles of charges g_1 and g_2 at rest a distance r apart, with what force will they attract or repel each other?

7.3. A plane wave can be represented by the four-potential

$$A_\mu = \alpha_\mu \exp\left[i \sum_\lambda k_\lambda x_\lambda\right]$$

α_μ being a *constant* four-vector and $\sum_\lambda k_\lambda^2 = 0$.

(a) Express the Lorentz condition on α_μ.

(b) Find expressions for the fields $F_{\mu\nu}$.

(c) Show that

$$\sum_\nu F_{\mu\nu} k_\nu = 0, \quad \sum_\nu \hat{F}_{\mu\nu} k_\nu = 0$$

(d) Write the two relations above in terms of $\mathbf{E}$ and $\mathbf{B}$ and show that they express the relations

$$\mathbf{E} \cdot \mathbf{k} = \mathbf{B} \cdot \mathbf{k} = 0, \quad \mathbf{E} \times \mathbf{k} = \frac{\omega}{c}\mathbf{E}, \mathbf{k} \times \mathbf{E} = \frac{\omega}{c}\mathbf{B}$$

7.4. A Van de Graaff generator is used to accelerate electrons down a tube 3 m long. The cathode is at a potential of 2.56 million V below the anode potential and the field is uniform. Calculate the following:

(a) The mass of the electron as it reaches the anode.

(b) The transit time.

The rest mass of an electron is 9.1×10^{-28} g. The charge of an electron is 4.8×10^{-10} ESU units.

7.5. A Minkowski force related to a mass point m is defined as the four-vector

$$F_\mu = m\frac{du_\mu}{d\tau}$$

where

$$u_\mu \equiv \left(\frac{\mathbf{v}}{\sqrt{1 - \beta^2}}, \frac{ic}{\sqrt{1 - \beta^2}} \right)$$

τ = proper time

(a) Show that $\sum_\mu F_\mu u_\mu = 0$.

(b) Express F_μ in terms of $\mathbf{F} = d\mathbf{p}/dt$ and $\mathbf{v}$.

7.6. Two particles of equal mass m have momenta $\mathbf{p}_1$ and $\mathbf{p}_2$, which may be noncolinear. Let $E = E_1 + E_2$ and $\mathbf{p} = \mathbf{p}_1 + \mathbf{p}_2$ be the total energy and the total momentum, respectively. Show that the following relation exists between E and p:

$$\frac{E^2}{c^2} - p^2 = M^2 c^2$$

Determine the value of M and show that it is an invariant.

7.7. Consider a relativistic particle moving with a constant acceleration a in a system instantaneously at rest with respect to the particle. At time $t = 0$ the particle has zero velocity with respect to some observer who is unaccelerated and remains so. What is the velocity of the particle with respect to this observer after a time T (measured in the observer's Lorentz frame)?

7.8. The following table lists the space and time coordinates of three events plus the reference event (event 0) as observed in a Lorentz frame S.

(a) Indicate the nature of the four-vector distance (timelike, lightlike, or space-like) between all the couples of events 01, 02, 03, 12, 13, and 23.

(b) Find the speed, with respect to the frame S, of a frame S' moving along the x axis in which events 1 and 2 occur at the same place.

(c) Find the speed, with respect to the frame S, of a frame S'' moving along the x axis in which events 2 and 3 occur at the same time.

Event	x Coordinate (light years)	y Coordinate (light years)	Time (years)
0	0	0	0
1	8	0	5
2	9	0	9
3	5	3	7

7.9. An electron of total energy E is incident upon an electron at rest. Find the energy of each electron in the reference frame in which the electrons have equal and opposite momenta.

7.10. We wish to test experimentally the basis relations of relativistic mechanics by accelerating an electron through a potential difference and bending it by 180° by means of a magnetic field. Discuss quantitatively the pertinent parameters of this problem, in particular the source, the voltage, the magnetic field, the linear dimensions, and the resolution of the detector. Assume that a photographic film is used as the detector. Discuss the health hazards connected with this experiment.

7.11. A train of electromagnetic waves propagates in vacuum, after having been emitted in a previous instant by a source that has been left behind, so that no charges are present in the region occupied by the radiation. Show that the total energy and the momentum of the field form a four-vector.

7.12. Consider an electron in the Coulomb field of a proton. In the rest system of the proton, the force on the electron is

$$\mathbf{F} = -\frac{e^2}{r^3}\mathbf{r}$$

Derive its orbit using the mechanisms of special relativity, but not quantum theory. In particular, obtain the rate of precession of an elliptical orbit. Keep in mind that the total energy and the total angular momentum are conserved. Take into account only the electrostatic Coulomb interaction.

7.13. A particle of mass m_0 has an excited state of energy ΔE, which can be reached by γ-ray absorption. It is assumed that $\Delta E/c^2$ is not small compared to m_0. Find the resonant γ-ray energy $E\gamma$ necessary to excite the particle initially at rest.

7.14. Consider a body of mass m under the action of a force $\mathbf{F}$. According to the theory of special relativity, the relation between force and momentum is

$$\mathbf{F} = \frac{d\mathbf{p}}{dt}$$

Show that the relativistic relation between force and acceleration is given by

$$\mathbf{a} = \frac{\mathbf{F} - [\mathbf{F} \cdot (\mathbf{v}/c)](\mathbf{v}/c)}{m\gamma}$$

where

$$\gamma = \frac{1}{\sqrt{1 - (v^2/c^2)}}$$

7.15. The Compton effect consists of an encounter between a single photon and a single electron. The electron may be considered free because its binding energy to the parent atom (a few electron volts) is much smaller than the energy of the photon. Assume that a photon with energy 1.022 eV (= 2 electron masses) is backscattered by an electron at rest. Calculate the momentum acquired by the electron and the momentum of the backscattered photon. Present your results in an energymomentum diagram.

7.16. If a photon with energy of four electron masses hits an electron at rest, it is generally scattered via the Compton process. Occasionally, it produces a positron-electron pair. This couple and the original electron form a *polyelectron*, a structure similar to the hydrogen molecule ion. The encounter is described in Fig. P7.16. (In this figure the masses are encircled; also, since *c* is set equal to 1, energy and momentum are expressed in the same units.) Redo this diagram in an inertial frame of reference that is going to the right with speed 0.8*c* with respect to the inertial frame in which the encounter is described in the diagram.

7.17. A positron with kinetic energy equal to twice its rest mass is incident on an electron at rest. The positron and the electron annihilate,

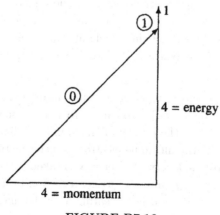

FIGURE P7.16

creating two gamma photons. One photon goes off at an angle of 30°
with respect to the direction of the incoming positron. Calculate the
energies of the two resulting gamma photons, and find the direction
in which the second photon travels.

7.18. Calculate the threshold energy for each of the following two reactions,
where a pion is incident on a proton at rest. The rest energy for each
particle is given in the following table.

(a) $\pi + p \to \Lambda + K$
(b) $\pi + p \to p + K + K$

Particle	Rest energy (MeV)
π	140
p	940
Λ	495
K	1115

7.19. A pi-plus meson has a mean lifetime (= proper time) of 2.6×10^{-8} s
and a rest energy of 140 MeV. Calculate how far such a particle, once
created, will go in its lifetime if its kinetic energy is (a) 0.7 MeV,
(b) 30 MeV, and (c) 700 MeV.

7.20. *Bremsstrahlung*, or *braking radiation*, is produced when a highly
energetic electron experiences the sudden acceleration due to the
attraction exercised on it by a nucleus of a target, such as a tungsten
(W) nucleus. Because of this acceleration, the electron emits gamma
radiation at the expense of its kinetic energy. Assume the following
conditions:

$E = pc$ for the incoming electron and outgoing electron, but not for
the nucleus, Ratio of electron mass to nuclear mass $\ll 1$ Derive a
simple formula for the energy of the gamma radiation as a function
of the energy of the incoming electron and the energy of the outgoing
electron. Show that very little energy is taken by the target nucleus.

7.21. A highly energetic gamma ray strikes a carbon 12 nucleus, converting
a proton into a neutron and creating a pi-plus meson, which moves

away from the nucleus according to the reaction

$$\gamma + {}^{12}_{6}C \rightarrow \pi^+ + {}^{12}_{5}B$$

It may be of interest to determine the relative reaction rate for different direction of motion of the outgoing pi-plus meson. Consider the following data:

mass of the carbon-12 nucleus $= 11{,}175\,\text{MeV}$,

energy of the gamma ray: $800\,\text{MeV}$,

mass of the boron-12 nucleus $= 11{,}188\,\text{MeV}$ in the ground state,

mass of the pi-plus meson $= 139.57\,\text{MeV}$.

Consider in particular the case in which the detector for the pi-plus meson is placed in a direction perpendicular to the direction of the incoming gamma ray. Calculate the energy of the outgoing pi-plus meson, assuming that the outgoing boron-12 nucleus remains in the ground state.

7.22. A pi-plus meson (mass $= 139.57\,\text{MeV}$) decays into a positive muon (μ^+; mass $= 105.66\,\text{MeV}$) and a mu neutrino (ν_μ; mass $= 0$). What must the energy of the pi-plus meson be in order that, following the decay, the positive muon is at rest?

7.23. A positive muon (μ^+; mass $= 105.66\,\text{MeV}$) is at rest and it decays into a positron (mass $= 0.511\,\text{MeV}$) and two neutrinos of different kinds (ν_e and $\bar{\nu}_\mu$; both have zero mass). Calculate the maximum energy of the product positron, taking into account the fact that this maximum energy occurs when the two neutrinos move off together.

7.24. Evaluate and put in a table the three components of the momentum and the value of the energy for the following particles:

(a) Particle a, which moves in the $+x$ direction with kinetic energy $=$ three times its rest energy, as measured by an observer in the laboratory frame.

(b) Particle a above, as observed in a frame in which kinetic energy $=$ rest energy.

(c) Particle c, which moves in the $+y$ direction with momentum $=$ twice its rest energy, as measured by an observer in the laboratory frame.

(d) Particle d, which moves in the $-x$ direction with total energy $=$ four times its rest energy, as measured by an observer in the laboratory frame.

(e) Particle d above, in the rest frame of the particle.

(f) Particle f, which moves with the three components of its momentum all equal and with kinetic energy = four times its rest energy, as measured by an observer in the laboratory frame.

Use the same units for energy, momentum, and mass.

7.25. Two trains of equal masses of 10^6 kg travel on the same track in opposite directions at the speed of 130 mph. Calculate the increase of tho rest mass of the track-trains system following a head-on collision.

7.26. Protons of various total energies emit flashes of light at times that follow each other at a constant rate of proper time. The proper time interval between two sequential flashes is equal to the time it takes light to travel 1 m, that is, "a meter of time." The distance traveled by the protons between flashes is reported in the following table.

(a) Complete the table.

(b) Estimate the proper time it would take a proton with 10^{11} GeV energy to cross the disk of our galaxy.

Take $1\,\mathrm{GeV} = 10^9\,\mathrm{eV}$ as the rest energy of the proton, and use the same units for energy and momentum.

Lab distance travelled between flashes (m)	Momentum	Energy (GeV)	γ	Lab time between flashes
0				
0.1				
1				
5				
10				
10^3				
10^6				
10^{11}				

7.27. Evaluate the rest masses of the following two-particle systems (c is set equal to 1):

(a) Particle with mass m and $\mathbf{p} = 0$, and particle with mass m and energy $E = 3m$.

(b) Particle with mass $3m$ and energy $7m$, and particle with mass m and $\mathbf{p} = 0$.

(c) Particle with mass m and $\mathbf{p} = 0$, and photon with energy $3m$.

(d) Two photons, one with energy E and the other with energy $3E$ traveling in opposite directions.

(e) Two photons, one with energy E and the other with energy $2E$ traveling in same direction.

(f) Two particles, each of mass m, with equal energy $E = 2m$, traveling in perpendicular directions.

(g) Two photons, one with energy E and the other with energy $3E$, traveling in perpendicular directions.

7.28. A charge λ per unit length is deposited on two parallel, thin, and infinitely long straight wires that are a distance a apart, are made of a nonconducting material, and are lying along the x direction. The wires move together in the laboratory frame S' in the x direction with a velocity $v = \beta c$. Calculate the force per unit length between the two wires in a frame of reference S in which they are at rest and compare it with the force calculated in the laboratory frame S'.

8

Radiation from a Moving Point Charge

8.1. Liénard–Wiechert Potentials of a Moving Point Charge

Consider a particle of charge e that travels along a certain trajectory. At time $t = t_p$, it is at $\mathbf{x} = \mathbf{x}_p$ and has velocity v (see Fig. 8.1). What potentials does it produce? We shall approach the solution of this problem in a succession of steps.

STEP 1 We can write

$$j_\mu(x) = ce \int d\tau u_\mu \delta^{(4)}[x - x_p(\tau)]$$

$$= ce \int d\tau u_\mu \delta^{(3)}[\mathbf{x} - \mathbf{x}_p(\tau)]\delta[c(t - t_p(\tau))] \quad (8.1.1)$$

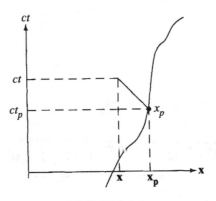

FIGURE 8.1

In fact,

$$\mathbf{j}(\mathbf{x},t) = ce \int dt \sqrt{1-\beta^2} \frac{\mathbf{v}}{\sqrt{1-\beta^2}} \delta^{(3)}(\mathbf{x}-\mathbf{x})_p \delta[c(t-t_p)]$$

$$= e\mathbf{v}\delta^{(3)}(\mathbf{x}-\mathbf{x}_p(t_p)), \quad \mathbf{v} \text{ at time } t=t_p \qquad (8.1.2)$$

$$ic\rho(\mathbf{x},t) = ce \int dt \sqrt{1-\beta^2} \frac{\mathbf{v}}{\sqrt{1-\beta^2}} \delta^{(3)}(\mathbf{x}-\mathbf{x}_p) \delta[c(t-t_p)]$$

$$= ice\delta^{(3)}(\mathbf{x}-\mathbf{x}_p(t_p)) \qquad (8.1.3)$$

STEP 2 We shall cite the formula for $A_\mu(x)$:

$$A_\mu(x) = \frac{1}{c} \int d^4x' G_{\text{ret}}(x-x') j_\mu(x') \qquad (8.1.4)$$

STEP 3 Let us prepare G_{ret}:

$$G_{\text{ret}} = \frac{\delta(\xi_0 - r)}{r} \qquad (8.1.5)$$

But, according to Eq. (7.6.12),

$$\delta(\xi_0^2 - r^2) = \frac{G_{\text{ret}} + G_{\text{adv}}}{2} \qquad (8.1.6)$$

Therefore,

$$G_{\text{ret}}(x - x') = 2\delta(\xi_0^2 - r^2)\eta(x_0 - x'_0)$$

$$= 2\delta\left(\sum_\mu (x_\mu - x'_\mu)^2\right)\eta(x_0 - x'_0) \quad (8.1.7)$$

where

$$\eta(x_0 - x'_0) = \begin{cases} 1, & \text{if } x_0 > x'_0 \\ 0, & \text{if } x_0 < x'_0 \end{cases} \quad (8.1.8)$$

STEP 4 We now substitute the expressions (8.1.1) and (8.1.7) for j_μ and G_{ret}, respectively, into the expression for $A_\mu(x)$:

$$A_\mu(x) = \frac{1}{c}\int d^4x' G_{\text{ret}}(x - x')j_\mu(x')$$

$$= \frac{1}{c}\int d^4x'\, 2\delta\left(\sum_\mu (x_\mu - x'_\mu)^2\right)$$

$$\times \eta(x_0 - x'_0)ce\int d\tau u_\mu \delta^{(4)}(x' - x_p)$$

$$= 2e\int d\tau \delta\left(\sum_\mu (x_\mu - x_{\mu p})^2\right)\eta(x_0 - x_{0p})u_\mu(x_p)$$

$$= 2e\int d\tau \delta\left(\sum_\mu R_\mu^2\right)\eta(x_0 - x_{0p})u_\mu(x_p) \quad (8.1.9)$$

where $R_\mu = x_\mu - x_{\mu p}$. u_μ is a function of τ because it is a function of position $x_p = x_p(\tau)$. Formula (8.1.9) implies that $x_0 > x_{0p}$; that is, $t > t_p$.

STEP 5 Let us keep in mind that we have always to take into account the fact that $x_0 > x_{0p}$ or $t > t_p$. We have

$$\int ds\, \delta[f(s)]g(s) = \int ds \frac{df}{ds}\frac{\delta[f(s)]}{df/ds}g(s)$$

$$= \int df \delta(f)\frac{g(s)}{df/ds} = \left[\frac{g(s)}{df/ds}\right]_{f(s)=0} \quad (8.1.10)$$

Therefore,

$$\int d\tau \, \delta\left(\sum_\mu R_\mu^2\right) u_\mu = \frac{u_\mu}{d\left(\sum_\mu R_\mu^2\right)\Big/ d\tau}\Bigg|_{\sum_\lambda R_\lambda^2 = 0}$$

$$= -\frac{u_\mu}{2\sum_\lambda R_\lambda u_\lambda}\Bigg|_{\sum_\gamma R_\lambda^2 = 0} \qquad (8.1.11)$$

because

$$\frac{dR_\lambda}{d\tau} = \frac{d(x_\lambda - x_{\lambda p})}{d\tau} = -\frac{dx_{\lambda p}}{d\tau} = -\mu_\lambda \qquad (8.1.12)$$

STEP 6 Let us go back to $A_\mu(x)$:

$$A_\mu(x) = 2e \int d\tau \, \delta\left(\sum_\mu R_\mu^2\right) \eta(x_0 - x_{0p}) u_\mu(\tau)$$

$$= \frac{2e u_\mu}{-2\sum_\lambda R_\lambda u_\lambda}\Bigg|_{\sum_\lambda R_\lambda^2 = 0}$$

$$= -\frac{e u_\mu}{\sum_\lambda R_\lambda u_\lambda}\Bigg|_{\sum_\lambda R_\lambda^2 = 0} \qquad (8.1.13)$$

Now

$$u_\mu \equiv \left(\frac{\mathbf{v}}{\sqrt{1 - (v^2/c^2)}}, \frac{ic}{\sqrt{1 - (v^2/c^2)}}\right) \qquad (8.1.14)$$

$$R_\lambda \equiv (\mathbf{x} - \mathbf{x}_p, i(x_0 - x_{0p})) \qquad (8.1.15)$$

$$R_\lambda\Big|_{\sum_\lambda R_\lambda^2 = 0} \equiv (\mathbf{x} - \mathbf{x}_p, i|\mathbf{x} - \mathbf{x}_p|) \equiv (\mathbf{r}, ir) \qquad (8.1.16)$$

where $r = |\mathbf{x} - \mathbf{x}_p|$. In Eq. (8.1.16) we have made

$$x_0 - x_{0p} = r \qquad (8.1.17)$$

(If we had used G_{adv}, we would have used at this point $x_0 - x_{0p} = -r$.)

Therefore,

$$\sum_\lambda R_\lambda u_\lambda \Bigg|_{\sum_\lambda R_\lambda^2 = 0} = \frac{\mathbf{v} \cdot \mathbf{r}}{\sqrt{1 - (v^2/c^2)}} - \frac{cr}{\sqrt{1 - (v^2/c^2)}}$$

$$= -\frac{cr}{\sqrt{1 - \beta^2}} \left(1 - \frac{\mathbf{v}}{c} \cdot \frac{\mathbf{r}}{r} \right)$$

$$= -\frac{cr}{\sqrt{1 - \beta^2}} \left(1 - \mathbf{n} \cdot \frac{\mathbf{v}}{c} \right) \tag{8.1.18}$$

where

$$\mathbf{n} = \frac{\mathbf{r}}{r} \tag{8.1.19}$$

Now

$$A_\mu(x) = -\frac{e u_\mu}{\displaystyle\sum_\lambda R_\lambda u_\lambda} \Bigg|_{\sum_\lambda R_\lambda^2 = 0}$$

$$= -\frac{e u_\mu}{-\dfrac{cr}{\sqrt{1 - \beta^2}} \left(1 - \dfrac{\mathbf{n} \cdot \mathbf{v}}{c} \right)}$$

$$= \frac{e u_\mu}{\dfrac{cr}{\sqrt{1 - \beta^2}} \left(1 - \dfrac{\mathbf{n} \cdot \mathbf{v}}{c} \right)} \tag{8.1.20}$$

We get then

$$\mathbf{A}(\mathbf{x}, t) = \frac{e\mathbf{v}/\sqrt{1 - \beta^2}}{(cr/\sqrt{1 - \beta^2})[1 - (\mathbf{n} \cdot \mathbf{v})/c]}$$

$$= \frac{e}{r} \frac{\mathbf{v}/c}{1 - (\mathbf{n} \cdot \mathbf{v})/c} \tag{8.1.21}$$

For $v/c \ll 1$, this reduces to

$$\mathbf{A}(\mathbf{x}, t) = \frac{e}{r} \frac{\mathbf{v}}{c} \tag{8.1.22}$$

We have also

$$\rho(\mathbf{x}, t) = e\delta^{(3)}(\mathbf{x} - \mathbf{x}_p)$$
$$\mathbf{j}(\mathbf{x}, t) = e\mathbf{v}\delta^{(3)}(\mathbf{x} - \mathbf{x}_p) \tag{8.1.23}$$

and for $v \ll c$,

$$\mathbf{A}(\mathbf{x}, t) = \frac{1}{c} \int \frac{\mathbf{j}(\mathbf{x}', t)}{|\mathbf{x} - \mathbf{x}'|} d^3\mathbf{x}'$$

$$= \frac{1}{c} \int \frac{e\mathbf{v}\delta^{(3)}(\mathbf{x}' - \mathbf{x}_p)}{|\mathbf{x} - \mathbf{x}'|} d^3\mathbf{x}'$$

$$= \frac{e\mathbf{v}}{c|\mathbf{x} - \mathbf{x}_p|} = \frac{e\mathbf{v}}{cr} \tag{8.1.24}$$

in accordance with Eq. (8.1.22).

Analogously,

$$i\phi(\mathbf{x}, t) = \frac{eu_4}{\frac{cr}{\sqrt{1 - \beta^2}}\left(1 - \frac{\mathbf{n} \cdot \mathbf{v}}{c}\right)} = \frac{\dfrac{eic}{\sqrt{1 - \beta^2}}}{\dfrac{cr}{\sqrt{1 - \beta^2}}\left(1 - \dfrac{\mathbf{n} \cdot \mathbf{v}}{c}\right)}$$

and

$$\phi(\mathbf{x}, t) = \frac{e}{r}\frac{1}{1 - (\mathbf{n} \cdot \mathbf{v})/c} \tag{8.1.25}$$

For $v/c \ll 1$, then

$$\phi(x, t) = \frac{e}{r} \tag{8.1.26}$$

We obtain again this result by inserting the expression (8.1.23) $\rho(\mathbf{x}, t)$ in the expression of ϕ for $v \ll c$:

$$\phi(\mathbf{x}, t) = \int \frac{\rho(\mathbf{x}', t)d^3\mathbf{x}'}{|\mathbf{x} - \mathbf{x}'|}$$

$$= \int \frac{e\delta^{(3)}(\mathbf{x}' - \mathbf{x}_p)}{|\mathbf{x} - \mathbf{x}'|} d^3\mathbf{x}'$$

$$= \frac{e}{|\mathbf{x} - \mathbf{x}_p|} = \frac{e}{r} \tag{8.1.27}$$

STEP 7 We now have to take *time* into account. We let

$$x_0 - x_{0p} = r$$

or

$$ct - ct_p = r$$

$$t_p = t - \frac{r}{c} \tag{8.1.28}$$

Then we can write the potentials for a moving point charge:

$$\mathbf{A}(\mathbf{x}, t) = \frac{e}{r} \frac{\mathbf{v}/c}{1 - \mathbf{n} \cdot (\mathbf{v}/c)}\bigg|_{t_p = t - (r/c)}$$

$$\phi(\mathbf{x}, t) = \frac{e}{r} \frac{1}{1 - \mathbf{n} \cdot (\mathbf{v}/c)}\bigg|_{t_p = t - (r/c)}$$

(8.1.29)

These potentials are called the *Liénard-Wiechert potentials*.

In evaluating these potentials at point $\mathbf{x}$ at time t, we have to use in expressions (8.1.29) the velocity, the unit vector $\mathbf{n}$, and the value of $r = |\mathbf{x} - \mathbf{x}_p|$ at the time $t - (r/c)$. A signal that leaves $\mathbf{x}_p$ at time t_p reaches $\mathbf{x}$ at time t (see Fig. 8.1).

8.2. Fields of a Moving Point Charge

We have found the potentials of a moving point charge in Eq. (8.1.29). We could derive the fields by using the expressions

$$\mathbf{B} = \boldsymbol{\nabla} \times \mathbf{A}$$

$$\mathbf{E} = -\frac{1}{c}\frac{\partial \mathbf{A}}{\partial t} - \boldsymbol{\nabla}\phi$$

(8.2.1)

We shall instead follow a different procedure. Let us start with

$$A_\mu(\mathbf{x}, t) = 2e \int d\tau \, u_\mu \, \delta\left(\sum_\lambda R_\lambda^2\right) = 2e \int d\tau \, u_\mu \delta(s) \qquad (8.2.2)$$

where $s = \sum_\lambda R_\lambda^2$. We can then write

$$F_{\mu\nu} = \frac{\partial A_\nu}{\partial x_\mu} - \frac{\partial A_\mu}{\partial x_\nu} = 2e \int d\tau \left[u_\nu \frac{\partial \delta(S)}{\partial x_\mu} - u_\mu \frac{\partial \delta(S)}{\partial x_p} \right]$$

$$= 2e \int d\tau \frac{\partial \delta(s)}{\partial s} \left(u_\nu \frac{ds}{dx_\mu} - u_\mu \frac{ds}{dx_p} \right) \qquad (8.2.3)$$

But

$$\frac{ds}{dx_\mu} = \frac{d\left(\sum_\lambda R_\lambda^2\right)}{dx_\mu} = \frac{d\left[\sum_\lambda (x_\lambda - x_{\lambda p})^2\right]}{dx_\mu}$$

$$= 2(x_\mu - x_{\mu p}) = 2R_\mu \qquad (8.2.4)$$

Then

$$F_{\mu\nu} = 4e \int d\tau \frac{\partial \delta(s)}{\partial s}(u_\nu R_\mu - u_\mu R_\nu)$$

$$= 4e \int d\tau \frac{\partial \delta(s)}{\partial s}\frac{d\tau}{ds}(u_\nu R_\mu - u_\mu R_\nu) \qquad (8.2.5)$$

Also

$$\frac{ds}{d\tau} = \frac{d}{d\tau}\left(\sum_\lambda R_\lambda^2\right) = \sum_\lambda 2R_\lambda \frac{dR_\lambda}{d\tau}$$

$$= \sum_\lambda R_\lambda \frac{d}{d\tau}(x_\lambda - x_{\lambda p})$$

$$= -\sum_\lambda 2R_\lambda \frac{dx_{\lambda p}}{d\tau} = -2\sum_\lambda R_\lambda u_\lambda \qquad (8.2.6)$$

$$d\tau = \frac{ds}{-2\sum_\lambda R_\lambda u_\lambda}$$

We can now write

$$F_{\mu\nu} = 4e \int d\tau \frac{\partial \delta(s)}{\partial \tau}\frac{u_\nu R_\mu - u_\mu R_\nu}{-2\sum_\lambda R_\lambda u_\lambda}$$

$$= -2e \int d\tau \frac{\partial \delta(s)}{\partial \tau}\frac{u_\nu R_\mu - u_\mu R_\nu}{\sum_\lambda R_\lambda u_\lambda}$$

$$= 2e \int d\tau \delta(s)\frac{d}{d\tau}\left[\frac{u_\nu R_\mu - u_\mu R_\nu}{\sum_\lambda R_\lambda u_\lambda}\right]$$

$$= 2e \int \frac{ds}{-2\sum_\lambda R_\lambda u_\lambda}\delta(s)\frac{d}{d\tau}\left[\frac{u_\nu R_\mu - u_\mu R_\nu}{\sum_\lambda R_\lambda u_\lambda}\right]$$

$$= \int ds\delta(s)\frac{e}{\sum_\lambda R_\lambda u_\lambda}\frac{d}{d\tau}\left[\frac{u_\mu R_\nu - u_\nu R_\mu}{\sum_\lambda R_\lambda u_\lambda}\right] \qquad (8.2.7)$$

Then

$$F_{\mu\nu} = \left[\frac{e}{\sum_\lambda R_\lambda u_\lambda} \frac{d}{d\tau} \left(\frac{u_\mu R_\nu - u_\nu R_\mu}{\sum_\lambda R_\lambda u_\lambda} \right) \right]_{s = \sum_\lambda R_\lambda^2 = 0} \tag{8.2.8}$$

Now we have

$$\frac{du_\mu}{d\tau} = \dot{u}_\mu$$

$$\frac{dR_\mu}{d\tau} = \dot{R}_\mu = u_\mu$$

$$-\sum_\lambda u_\lambda^2 = c^2 \tag{8.2.9}$$

$$\frac{d}{d\tau} \left(\sum_\lambda R_\lambda u_\lambda \right) = \sum_\lambda R_\lambda \dot{u}_\lambda - \sum_\lambda u_\lambda^2 = \sum_\lambda R_\lambda \dot{u}_\lambda + c^2$$

Then

$$F_{\mu\nu} = \left\{ \frac{e}{\sum_\lambda R_\lambda u_\lambda} \left[\frac{\dot{u}_\mu R_\nu + u_\mu \dot{R}_\nu - \dot{u}_\nu R_\mu - u_\nu \dot{R}_\mu}{\sum_\lambda R_\lambda u_\lambda} \right. \right.$$

$$\left. \left. - \frac{(u_\mu R_\nu - u_\nu R_\mu)\left(c^2 + \sum_\lambda R_\lambda \dot{u}_\lambda \right)}{\left(\sum_\lambda R_\lambda u_\lambda \right)^2} \right] \right\}_{\sum_\lambda R_\lambda^2 = 0}$$

$$= e \left\{ \frac{\dot{u}_\mu R_\nu - \dot{u}_\nu R_\mu}{\left(\sum_\lambda R_\lambda u_\lambda \right)^2} - (u_\mu R_\nu - u_\nu R_\mu) \frac{c^2 + \sum_\lambda r_\lambda \dot{u}_\lambda}{\left(\sum_\lambda R_\lambda u_\lambda \right)^3} \right\}_{\sum_\lambda R_\lambda^2 = 0} \tag{8.2.10}$$

We can divide $F_{\mu\nu}$ in two parts, the *near field* and the *radiation field*.

$$F_{\mu\nu} = E_{\mu\nu}^{\text{near}} + F_{\mu\nu}^{\text{rad}} \tag{8.2.11}$$

where

$$
F_{\mu\nu}^{\text{near}} = - \left[\frac{ec^2 (u_\mu R_\nu - u_\nu R_\mu)}{\left(\sum_\lambda R_\lambda u_\lambda \right)} \right]_{\sum_\lambda R_\lambda^2 = 0}
\tag{8.2.12}
$$

$$
F_{\mu\nu}^{\text{rad}} = e \left[\frac{\dot{u}_\mu R_\nu - \dot{u}_\nu R_\mu}{\left(\sum_\lambda R_\lambda u_\lambda \right)^2} - (u_\mu R_\nu - u_\nu R_\mu) \frac{\sum_\lambda R_\lambda \dot{u}_\lambda}{\left(\sum_\lambda R_\lambda u_\lambda \right)^3} \right]_{\sum_\lambda R_\lambda^2 = 0}
$$
$$
\tag{8.2.13}
$$

$F_{\mu\nu}^{\text{near}}$ is independent of the acceleration, and goes as $1/r^2$. $F_{\mu\nu}^{\text{rad}}$ depends on the acceleration and goes as $1/r$. This decomposition is independent of the frame of reference.

Near Fields. The near fields are given by

$$
F_{\mu\nu}^{\text{near}} = - \left[\frac{ec^2 (u_\mu R_\nu - u_\nu R_\mu)}{\left(\sum_\lambda R_\lambda u_\lambda \right)^3} \right]_{\sum_\lambda R_\lambda^2 = 0}
\tag{8.2.14}
$$

$\sum_\lambda R_\lambda^2 = 0$ means

$$
x_0 - x_{0p} = |\mathbf{x} - \mathbf{x}_p| = r
$$
$$
t_p = t - \frac{r}{c}
$$

Now from Eq. (8.1.18)

$$
\sum_\lambda R_\lambda u_\lambda \Big|_{\sum_\lambda R_\lambda^2 = 0} = - \frac{cr}{\sqrt{1 - \beta^2}} \left(1 - \mathbf{n} \cdot \frac{\mathbf{v}}{c} \right)
\tag{8.2.15}
$$

Then

$$-\frac{ec^2}{\left(\sum_\lambda R_\lambda u_\lambda\right)^3}\Bigg|_{\sum_\lambda R_\lambda^2=0} = \frac{ec^2(1-\beta^2)^{3/2}}{c^3 r^3[1-\mathbf{n}\cdot(\mathbf{v}/c)]^3}$$

$$= \frac{e(1-\beta^2)^{3/2}}{cr^3[1-\mathbf{n}\cdot(\mathbf{v}/c)]^3} \qquad (8.2.16)$$

and

$$F_{\mu\nu}^{\text{near}} = \frac{e(1-\beta^2)^{3/2}}{cr^3\{1-[(\mathbf{n}\cdot\mathbf{v})/c]\}^3}(u_\mu R_\nu - u_\nu R_\mu)\Bigg|_{\sum_\lambda R_\lambda^2=0} \qquad (8.2.17)$$

We can write

$$B_1^{\text{near}} = F_{23}^{\text{near}}$$

$$= \left[\frac{e(1-\beta^2)^{3/2}}{cr^3\{1-[(\mathbf{n}\cdot\mathbf{v})/c]\}^3}\frac{v_2 R_3 - v_3 R_2}{(1-\beta^2)^{1/2}}\right]_{\sum_\lambda R_\lambda^2=0}$$

$$= \frac{e(1-\beta^2)}{cr^3\{1-[(\mathbf{n}\cdot\mathbf{v})/c]\}^3}(v_2 r_3 - v_3 r_2)\Bigg|_{t_p=t-(r/c)}$$

$$= \frac{e(1-\beta^2)}{cr^3\{1-[(\mathbf{n}\cdot\mathbf{v})/c]\}^3}(\mathbf{v}\times\mathbf{r})_1\Bigg|_{t_p=t-(r/c)} \qquad (8.2.18)$$

because

$$R_\lambda\big|_{\sum_\lambda R_\lambda^2=0} \equiv (\mathbf{r}, ir) \qquad (8.2.19)$$

Analogously

$$B_2^{\text{near}} = F_{12}^{\text{near}} = \frac{e(1-\beta^2)}{cr^3\{1-[(\mathbf{n}\cdot\mathbf{v})/c]\}^3}(\mathbf{v}\times\mathbf{r})_2\Bigg|_{t_p=t-(r/c)} \qquad (8.2.20)$$

and

$$B_3^{\text{near}} = F_{12}^{\text{near}} = \frac{e(1-\beta^2)}{cr^3\{1-[(\mathbf{n}\cdot\mathbf{v})/c]\}^3}(\mathbf{v}\times\mathbf{r})_3\Bigg|_{t_p=t-(r/c)} \qquad (8.2.21)$$

or

$$\mathbf{B}^{\text{near}}(\mathbf{x}, t) = \frac{e(1-\beta^2)}{cr^3\{1-[(\mathbf{n}\cdot\mathbf{v})/c]\}^3}(\mathbf{v}\times\mathbf{r})\Bigg|_{t_p=t-(r/c)} \qquad (8.2.22)$$

If we consider the limiting case $v \ll c$, we obtain

$$\mathbf{B}^{\text{near}} = \frac{e}{r^2} \left(\frac{\mathbf{v}}{c} \times \mathbf{n} \right) \tag{8.2.23}$$

The same result can be obtained in the following way by using for $\mathbf{A}$ an expression valid only for $v \ll c$:

$$\mathbf{B} = \boldsymbol{\nabla} \times \mathbf{A} = \boldsymbol{\nabla} \times \frac{1}{c} \int \frac{\mathbf{j}(\mathbf{x}', t)}{|\mathbf{x} - \mathbf{x}'|} d^3 \mathbf{x}'$$

$$= -\frac{1}{c} \int \mathbf{j}(\mathbf{x}', t) \times \boldsymbol{\nabla}_x \frac{1}{|\mathbf{x} - \mathbf{x}'|} d^3 \mathbf{x}'$$

$$= -\frac{1}{c} \int e \mathbf{v} \delta^{(3)} (\mathbf{x}' - \mathbf{x}_p) \times \boldsymbol{\nabla}_x \frac{1}{|\mathbf{x} - \mathbf{x}'|} d^3 \mathbf{x}'$$

$$= -\frac{1}{c} e \mathbf{v} \times \boldsymbol{\nabla}_x \frac{1}{|\mathbf{x} - \mathbf{x}_p|} = \frac{e}{r^2 c} \mathbf{v} \times \mathbf{n} \tag{8.2.24}$$

where

$$\boldsymbol{\nabla} \frac{1}{|\mathbf{x} - \mathbf{x}_p|} = \boldsymbol{\nabla} \frac{1}{r} = -\frac{1}{r^2} \frac{\mathbf{r}}{r} = -\frac{\mathbf{n}}{r^2} \tag{8.2.25}$$

$$\mathbf{n} = \frac{\mathbf{x} - \mathbf{x}_p}{|\mathbf{x} - \mathbf{x}_p|} \tag{8.2.26}$$

Analogously,

$$E_1^{\text{near}} = i F_{14}^{\text{near}}$$

$$= \left[\frac{ie(1 - \beta^2)^{3/2}}{cr^3 \{1 - [(\mathbf{n} \cdot \mathbf{v})/c]\}^3} \frac{v_1 R_4 - i R_1 c}{\sqrt{1 - \beta^2}} \right]_{\sum_\lambda R_\lambda^2 = 0} \tag{8.2.27}$$

and

$$\mathbf{E}^{\text{near}}(\mathbf{x}, t) = \frac{ie(1 - \beta^2)}{cr^3 \{1 - [(\mathbf{n} \cdot \mathbf{v})/c]\}} (\mathbf{v} R_4 - i c \mathbf{r}) \Big|_{\sum_\lambda R_\lambda^2 = 0}$$

$$= \frac{e(1 - \beta^2)}{cr^3 \{1 - [(\mathbf{n} \cdot \mathbf{v})/c]\}} (-\mathbf{v} r + \mathbf{r} c) \Big|_{t_p = t - (r/c)}$$

$$= \frac{e}{r^2} \left(\mathbf{n} - \frac{\mathbf{v}}{c} \right) \frac{1 - \beta^2}{\{1 - [(\mathbf{n} \cdot \mathbf{v})/c]\}^3} \Big|_{t_p = t - (r/c)} \tag{8.2.28}$$

We find, for $v/c \ll 1$,

$$\mathbf{E}^{\text{near}} = \frac{e}{r^2} \mathbf{n} \tag{8.2.29}$$

In summary, we obtain

$$\mathbf{B}^{\text{near}}(\mathbf{x}, t) = \left\{ \frac{e}{r^2} \left(\frac{\mathbf{v}}{c} \times \mathbf{n} \right) \frac{1 - \beta^2}{\{1 - [(\mathbf{n} \cdot \mathbf{v})/c]\}^3} \right\}_{t_p = t - (r/c)}$$

$$\mathbf{E}^{\text{near}}(\mathbf{x}, t) = \left\{ \frac{e}{r^2} \left(\mathbf{n} - \frac{\mathbf{v}}{c} \right) \frac{1 - \beta^2}{\{1 - [(\mathbf{n} \cdot \mathbf{v})/c]\}^3} \right\}_{t_p = t - (r/c)} \tag{8.2.30}$$

These fields are also called *velocity fields*.

Radiation Fields. We start by reminding ourselves that

$$R_\lambda \Big|_{\sum_\lambda R_\lambda^2 = 0} \equiv (\mathbf{r}, ir) \tag{8.2.31}$$

$$u_\mu \equiv \left(\frac{\mathbf{v}}{\sqrt{1 - \beta^2}}, \frac{ic}{\sqrt{1 - \beta^2}} \right) \tag{8.2.32}$$

$$\dot{u}_\mu = \frac{du_\mu}{d\tau}, \quad \mathbf{a} = \frac{d\mathbf{v}}{dt} \tag{8.2.33}$$

$$d\tau = dt \sqrt{1 - \beta^2} \tag{8.2.34}$$

Then

$$\dot{u}_1 = \left(\frac{d}{dt} \frac{v_1}{\sqrt{1 - \beta^2}} \right) \frac{d}{d\tau}$$

$$= \frac{\dot{v}_1}{\sqrt{1 - \beta^2}} \frac{1}{\sqrt{1 - \beta^2}} + \frac{v_1}{\sqrt{1 - \beta^2}} \frac{1}{1 - \beta^2} \frac{2[(\mathbf{v} \cdot \dot{\mathbf{v}})/c^2]}{2\sqrt{1 - \beta^2}}$$

$$= \frac{a_1}{1 - \beta^2} + \frac{v_1[(\mathbf{v} \cdot \mathbf{a})/c^2]}{(1 - \beta^2)^2}$$

and

$$\dot{\mathbf{u}} = \frac{1}{1 - \beta^2} \left[\mathbf{a} + \mathbf{v} \frac{(\mathbf{v} \cdot \mathbf{a})/c^2}{1 - \beta^2} \right] \tag{8.2.35}$$

$$\dot{u}_4 = \left(\frac{d}{dt} \frac{ic}{\sqrt{1 - \beta^2}} \right) \frac{dt}{d\tau}$$

$$= \frac{1}{\sqrt{1 - \beta^2}} \left[ic \frac{1}{\sqrt{1 - \beta^2}} \frac{(\mathbf{v} \cdot \dot{\mathbf{v}})/c^2}{\sqrt{1 - \beta^2}} \right]$$

$$= \frac{ic}{1 - \beta^2} \frac{(\mathbf{v} \cdot \mathbf{a})/c^2}{1 - \beta^2} \tag{8.2.36}$$

Therefore,

$$\sum_\lambda R_\lambda \dot{u}_\lambda \bigg|_{\sum_\lambda R_\lambda^2 = 0} = \frac{1}{1-\beta^2} \left\{ \left[\mathbf{r} \cdot \mathbf{a} + \mathbf{r} \cdot \mathbf{v} \frac{(\mathbf{v} \cdot \mathbf{a})/c^2}{1-\beta^2} \right] - rc \frac{(\mathbf{v} \cdot \mathbf{a})/c^2}{1-\beta^2} \right\}$$

$$= \frac{1}{1-\beta^2} \left\{ \mathbf{r} \cdot \mathbf{a} - rc \left(1 - \frac{\mathbf{n} \cdot \mathbf{v}}{c} \right) \left[\frac{(\mathbf{v} \cdot \mathbf{a})/c^2}{1-\beta^2} \right] \right\}$$

$$(8.2.37)$$

$$\dot{\mathbf{u}} \times \mathbf{r} = \frac{1}{1-\beta^2} \times \left[\mathbf{a} \times \mathbf{r} + \mathbf{v} \times \mathbf{r} \frac{(\mathbf{v} \cdot \mathbf{a})/c^2}{1-\beta^2} \right] \qquad (8.2.38)$$

$$\dot{\mathbf{u}} R_4 - \mathbf{r} \dot{u}_4 = \frac{1}{1-\beta^2} \left[ir\mathbf{a} + ir\mathbf{v} \frac{(\mathbf{v} \cdot \mathbf{a})/c^2}{1-\beta^2} \right] - \frac{icr}{1-\beta^2} \left[\frac{(\mathbf{v} \cdot \mathbf{a})/c^2}{1-\beta^2} \right]$$

$$= \frac{1}{1-\beta^2} \left[ir\mathbf{a} - icr \left(\mathbf{n} - \frac{\mathbf{v}}{c} \right) \frac{(\mathbf{v} \cdot \mathbf{a})/c^2}{1-\beta^2} \right] \qquad (8.2.39)$$

We also know that

$$\sum_\lambda R_\lambda u_\lambda \bigg|_{\sum_\lambda R_\lambda^2 = 0} = -\frac{cr}{\sqrt{1-\beta^2}} \left(1 - \frac{\mathbf{n} \cdot \mathbf{v}}{c} \right) \qquad (8.2.40)$$

We are now ready to consider the fields. We find

$$B_1^{\text{rad}} = F_{23}^{\text{rad}}$$

$$= e \left[\frac{\dot{u}_2 R_3 - \dot{u}_3 R_2}{\left(\displaystyle\sum_\lambda R_\lambda u_\lambda \right)^2} - (u_2 R_3 - u_3 R_2) \frac{\displaystyle\sum_\lambda R_\lambda \dot{u}_\lambda}{\left(\displaystyle\sum_\lambda R_\lambda u_\lambda \right)^3} \right]_{\sum_\lambda R_\lambda^2 = 0}$$

$$(8.2.41)$$

and

$$\mathbf{B}^{\text{rad}} = e \left\{ \dot{\mathbf{u}} \times \mathbf{r} \frac{1-\beta^2}{c^2 r^2 \{1 - [(\mathbf{n} \cdot \mathbf{v})/c]\}^2} \right.$$

$$\left. - (\mathbf{u} \times \mathbf{r}) \left(\sum_\lambda R_\lambda \dot{u}_\lambda \right) \frac{(1-\beta^2)^{3/2}}{-c^3 r^3 \{1 - [(\mathbf{n} \cdot \mathbf{v})/c]\}^3} \right\}_{\sum_\lambda R_\lambda^2 = 0}$$

$$(8.2.42)$$

But

$$\frac{1-\beta^2}{c^2 r^2 \{1 - [(\mathbf{n} \cdot \mathbf{v})/c]\}^2} - (\dot{\mathbf{u}} \times \mathbf{r})$$

$$= \frac{1-\beta^2}{c^2 r^2 \{1 - [(\mathbf{n} \cdot \mathbf{v})/c]\}^2} \frac{1}{1-\beta^2} \left[\mathbf{a} \times \mathbf{r} + \mathbf{v} \times \mathbf{r} \frac{(\mathbf{v} \cdot \mathbf{a})/c^2}{1-\beta^2} \right]$$

$$= \frac{1}{c^2 r \{1 - [(\mathbf{n} \cdot \mathbf{v})/c]\}^2} (\mathbf{a} \times \mathbf{n})$$

$$+ \frac{1}{c^2 r \{1 - [(\mathbf{n} \cdot \mathbf{v})/c]\}^2} (\mathbf{v} \times \mathbf{n}) \left[\frac{(\mathbf{v} \cdot \mathbf{a})/c^2}{1-\beta^2} \right]$$

and

$$-(\mathbf{u} \times \mathbf{r}) \left(\sum_\lambda R_\lambda \dot{u}_\lambda \right) \frac{(1-\beta^2)^{3/2}}{-c^3 r^3 \left(1 - \dfrac{\mathbf{n} \cdot \mathbf{v}}{c} \right)^3}$$

$$= \frac{(1-\beta^2)^{3/2}}{c^3 r^3 \left(1 - \dfrac{\mathbf{n} \cdot \mathbf{v}}{c} \right)^3} \frac{\mathbf{v} \times \mathbf{r}}{\sqrt{1-\beta^2}}$$

$$\times \left\{ \frac{1}{1-\beta^2} \left[\mathbf{r} \cdot \mathbf{a} - rc \left(1 - \frac{\mathbf{n} \cdot \mathbf{v}}{c} \right) \left(\frac{(\mathbf{v} \cdot \mathbf{a})/c^2}{1-\beta^2} \right) \right] \right\}$$

$$= \frac{1}{c^2 r \left(1 - \dfrac{\mathbf{n} \cdot \mathbf{v}}{c} \right)^3} \left(\frac{\mathbf{v}}{c} \times \mathbf{n} \right) (\mathbf{n} \cdot \mathbf{a})$$

$$- \frac{1}{c_r^2 \left(1 - \dfrac{\mathbf{n} \cdot \mathbf{v}}{c} \right)^3} (\mathbf{v} \times \mathbf{n}) \left(\frac{(\mathbf{v} \cdot \mathbf{a})/c^2}{1-\beta^2} \right)$$

Therefore,

$$\mathbf{B}^{\text{rad}} = \frac{e}{c^2 r} \left[\frac{\mathbf{a} \times \mathbf{n}}{\left(1 - \dfrac{\mathbf{n} \cdot \mathbf{v}}{c} \right)^2} + \frac{(\mathbf{n} \cdot \mathbf{a})[(\mathbf{v}/c) \times \mathbf{n}]}{1 - (\mathbf{n} \cdot \mathbf{v}/c)^3} \right]_{\sum_\lambda R_\lambda^2 = 0} \qquad (8.2.43)$$

Analogously,

$$E_1^{\text{rad}} = iF_{14}^{\text{rad}}$$

$$= ie \left[\frac{\dot{u}_1 R_4 - \dot{u}_4 R_1}{\left(\sum_\lambda R_\lambda u_\lambda\right)^2} - (u_1 R_4 - u_4 R_1) \frac{\sum_\lambda R_\lambda \dot{u}\lambda}{\left(\sum_\lambda R_\lambda u_\lambda\right)^3} \right]_{\sum_\lambda R_\lambda^2 = 0}$$

$$(8.2.44)$$

and

$$\mathbf{E} = ie \left[\frac{\dot{\mathbf{u}} R_4 - \mathbf{r}\dot{u}_4}{\left(\sum_\lambda R_\lambda u_\lambda\right)^2} - (\mathbf{u} R_4 - u_4 \mathbf{r}) \frac{\sum_\lambda R_\lambda \dot{u}\lambda}{\left(\sum_\lambda R_\lambda u_\lambda\right)^3} \right]_{\sum_\lambda R_\lambda^2 = 0}$$

$$(8.2.45)$$

But

$$\frac{(\dot{\mathbf{u}} R_4 - \mathbf{r}\dot{u}_4)}{\left(\sum_\lambda R_\lambda u_\lambda\right)^2} = \frac{(1 - \beta^2)}{c^2 r^2 \left(1 - \dfrac{\mathbf{n} \cdot \mathbf{v}}{c}\right)^2} \frac{1}{1 - \beta^2}$$

$$\times \left\{ i r \mathbf{a} - i c r \left(\mathbf{n} - \frac{\mathbf{v}}{c}\right) \left[\frac{(\mathbf{v} \cdot \mathbf{a}/c^2)}{1 - \beta^2}\right] \right\}$$

$$= \frac{i \mathbf{a}}{c^2 r \left(1 - \dfrac{\mathbf{n} \cdot \mathbf{v}}{c}\right)^2}$$

$$- \frac{i}{c r \left(1 - \dfrac{\mathbf{n} \cdot \mathbf{v}}{c}\right)^2} \frac{(\mathbf{v} \cdot \mathbf{a})/c^2}{1 - \beta^2} \left(\mathbf{n} - \frac{\mathbf{v}}{c}\right)$$

and

$$-(\mathbf{u} R_4 - u_4 \mathbf{r}) \frac{\sum_\lambda R_\lambda \dot{u}\lambda}{\left(\sum_\lambda R_\lambda u_\lambda\right)^3}$$

$$= -\left(\frac{ivr}{\sqrt{1-\beta^2}} - \frac{icr}{\sqrt{1-\beta^2}}\right)\left[-\frac{(1-\beta^2)^{3/2}}{c^2 r^3 \left(1 - \dfrac{\mathbf{n}\cdot\mathbf{v}}{c}\right)^3}\right]$$

$$\times \frac{1}{1-\beta^2}\left\{\mathbf{r}\cdot\mathbf{a} - rc\left(1 - \frac{\mathbf{n}\cdot\mathbf{v}}{c}\right)\left[\frac{(\mathbf{v}\cdot\mathbf{a})/c^2}{1-\beta^2}\right]\right\}$$

$$= \frac{1}{c^3 r^3 \left(1 - \dfrac{\mathbf{n}\cdot\mathbf{v}}{c}\right)^3} icr\left(\frac{\mathbf{v}}{c} - \mathbf{n}\right)$$

$$\times \left\{\mathbf{r}\cdot\mathbf{a} - rc\left(1 - \frac{\mathbf{n}\cdot\mathbf{v}}{c}\right)\frac{(\mathbf{v}\cdot\mathbf{a})/c^2}{1-\beta^2}\right\}$$

$$= \frac{-i}{c^2 r \left(1 - \dfrac{\mathbf{n}\cdot\mathbf{v}}{C}\right)^2}\left(\mathbf{n} - \frac{\mathbf{v}}{c}\right)(\mathbf{n}\cdot\mathbf{a})$$

$$+ \frac{i}{cr\left(1 - \dfrac{\mathbf{n}\cdot\mathbf{v}}{c}\right)^2}\frac{(\mathbf{v}\cdot\mathbf{a})/c^2}{1-\beta^2}\left(\mathbf{n} - \frac{\mathbf{v}}{c}\right)$$

Therefore,

$$\mathbf{E}^{\mathrm{rad}} = \left[-\frac{e\mathbf{a}}{c^2 r\left(1 - \dfrac{\mathbf{n}\cdot\mathbf{v}}{c}\right)^2} + \frac{e[\mathbf{n} - (\mathbf{v}/c)](\mathbf{n}\cdot\mathbf{a})}{c^2 r\left(1 - \dfrac{\mathbf{n}\cdot\mathbf{v}}{c}\right)^3}\right]_{\sum_\lambda R_\lambda^2 = 0}$$

$$= -\left\{\frac{e}{c^2 r}\left[\frac{\mathbf{a}}{\left(1 - \dfrac{\mathbf{n}\cdot\mathbf{v}}{c}\right)^2} - \frac{(\mathbf{n}\cdot\mathbf{a})[\mathbf{n} - (\mathbf{v}/c)]}{\left(1 - \dfrac{\mathbf{n}\cdot\mathbf{v}}{c}\right)^3}\right]\right\}_{\sum_\lambda R_\lambda^2 = 0}$$

$$(8.2.46)$$

Summarizing the result,

$$\mathbf{B}^{\mathrm{rad}}(\mathbf{x}, t)$$

$$= \left\{\frac{e}{c^2 r}\left[\frac{\mathbf{a}\times\mathbf{n}}{\left(1 - \dfrac{\mathbf{n}\cdot\mathbf{v}}{c}\right)^2} - \frac{(\mathbf{n}\cdot\mathbf{a})[(\mathbf{v}/c)]\times\mathbf{n}}{\left(1 - \dfrac{\mathbf{n}\cdot\mathbf{v}}{c}\right)^3}\right]\right\}_{t_p = t - (r/c)}$$

$$\mathbf{E}^{\text{rad}}(\mathbf{x}, t)$$

$$= -\left\{ \frac{e}{c^2 r} \left[\frac{\mathbf{a}}{\left(1 - \dfrac{\mathbf{n} \cdot \mathbf{v}}{c}\right)^2} - \frac{(\mathbf{n} \cdot \mathbf{a})[\mathbf{n} - (\mathbf{v}/c)]}{\left(1 - \dfrac{\mathbf{n} \cdot \mathbf{v}}{c}\right)^3} \right] \right\}_{t_p = t - (r/c)}$$

$$(8.2.47)$$

These fields are also called *acceleration fields*.

Let us examine some properties of these fields:

(1) The fields are proportional to $1/r$ and are linear in $\mathbf{a}$. At great distances from the source, the radiation fields dominate.

(2) $\mathbf{B}^{\text{rad}} \cdot \mathbf{n} = 0$

$$\mathbf{E}^{\text{rad}} \cdot \mathbf{n}$$

$$= -\left\{ \frac{e}{c^2 r} \left[\frac{\mathbf{a} \cdot \mathbf{n}}{\left(1 - \dfrac{\mathbf{n} \cdot \mathbf{v}}{c}\right)^2} - \frac{(\mathbf{n} \cdot \mathbf{a})\left(1 - \dfrac{\mathbf{n} \cdot \mathbf{v}}{c}\right)}{\left(1 - \dfrac{\mathbf{n} \cdot \mathbf{v}}{c}\right)^3} \right] \right\}_{t_p = t - (r/c)}$$

$$= 0$$

$$\mathbf{n} \times \mathbf{E}^{\text{rad}}$$

$$= \left\{ \frac{e}{c^2 r} \left[\frac{\mathbf{a} \times \mathbf{n}}{\left(1 - \dfrac{\mathbf{n} \cdot \mathbf{v}}{c}\right)^2} = \frac{(\mathbf{n} \cdot \mathbf{a})\left(\dfrac{\mathbf{v}}{c} \times \mathbf{n}\right)}{\left(1 - \dfrac{\mathbf{n} \cdot \mathbf{v}}{c}\right)^3} \right] \right\}_{t_p = t - (r/c)}$$

$$= \mathbf{B}^{\text{rad}}$$

or

$$\mathbf{E}^{\text{rad}} \perp \mathbf{B}^{\text{rad}}$$

and

$$|\mathbf{E}| = |\mathbf{B}|$$

The relative orientations of $\mathbf{E}$, $\mathbf{B}$ and $\mathbf{n}$ are given in Fig. 8.2.

We can make the following observations:

(1) $E^{\text{near}}, B^{\text{near}} \propto \frac{1}{r^2}$, $E^{\text{rad}}, B^{\text{rad}} \propto \frac{1}{r}$

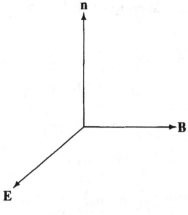

FIGURE 8.2

(2) If we calculate the Poynting vectors for the near and radiation fields, we get

$$N^{\text{near}} \propto \frac{1}{r^4}$$

$$N^{\text{rad}} \propto \frac{1}{r^2}$$

(3) If we integrate $\mathbf{N}$ over the surface of a sphere of radius r, we find that:

(a) Integral containing $N^{\text{near}} \propto 1/r^2$
(b) Integral containing N^{rad} remains finite even for very large r.

(4) The "complete" Poynting vector

$$\mathbf{N} = \frac{c}{4\pi}(\mathbf{E}^{\text{near}} + \mathbf{E}^{\text{rad}}) \times (\mathbf{B}^{\text{near}} + \mathbf{B}^{\text{rad}})$$

contains near–rad cross terms like $(\mathbf{E}^{\text{near}} \times \mathbf{B}^{\text{rad}})$ and $(\mathbf{E}^{\text{rad}} \times \mathbf{B}^{\text{near}})$. These terms vanish as $1/r^3$. The integral of these terms decreases as $1/r$, with increasing r.

(5) The energy of the near fields is always in the vicinity of the source. Therefore, the radiation of a point charge can be evaluated while neglecting the near fields.

8.3. Fields of a Slow-Moving Point Charge

8.3.1. *General Case*

We shall assume $v/c \ll 1$. Then

$$\mathbf{E}^{\mathrm{rad}} = -\left\{\frac{e}{c^2 r}[\mathbf{a} - \mathbf{n}(\mathbf{n} \cdot \mathbf{a})]\right\}_{t_p = t - (r/c)}$$

$$= \left\{\frac{e}{c^2 r}[-\mathbf{a} + \mathbf{n}(\mathbf{n} \cdot \mathbf{a})]\right\}_{t_p = t - (r/c)}$$

$$= \left\{\frac{e}{c^2 r}[(\mathbf{a} \times \mathbf{n}) \times \mathbf{n}]\right\}_{t_p = t - (r/c)} \tag{8.3.1}$$

$$\mathbf{B}^{\mathrm{rad}} = \left\{\frac{e}{c^2 r}(\mathbf{a} \times \mathbf{n})\right\}_{t_p = t - (r/c)} \tag{8.3.2}$$

We get immediately

$$\mathbf{B}^{\mathrm{rad}} \times \mathbf{n} = \mathbf{E}^{\mathrm{rad}} \tag{8.3.3}$$

We drop now the subscript rad and proceed to calculate the Poynting vector:

$$\mathbf{N} = \frac{c}{4\pi}(\mathbf{E} \times \mathbf{B}) = \frac{c}{4\pi}\frac{e^2}{c^4 r^2}[(\mathbf{a} \times \mathbf{n}) \times \mathbf{n}] \times (\mathbf{a} \times \mathbf{n})$$

$$= \frac{e^2}{4\pi c^3 r^2}\{(\mathbf{a} \times \mathbf{n})^2 \mathbf{n} - [\mathbf{n} \cdot (\mathbf{a} \times \mathbf{n})](\mathbf{a} \times \mathbf{n})\}$$

$$= \frac{e^2}{4\pi c^3 r^2}(\mathbf{a} \times \mathbf{n})^2 \mathbf{n} \tag{8.3.4}$$

In Fig. 8.3, we see that the Poynting vector at position $\mathbf{x}$ is directed as

$$\mathbf{n} = \frac{\mathbf{x} - \mathbf{x}_p}{|\mathbf{x} - \mathbf{x}_p|} = \frac{\mathbf{r}}{r} \tag{8.3.5}$$

and is due to

$$(\mathbf{a} \times \mathbf{n})^2 = a^2 \sin^2 \theta \tag{8.3.6}$$

$\mathbf{x}_p$ is the position of the moving charge and $\mathbf{a}$ is the acceleration at point $\mathbf{x}_p$ where the charge finds itself at a time $t - (|\mathbf{x} - \mathbf{x}_p|/c)$.
Therefore,

$$\mathbf{N} = \frac{c}{4\pi}(\mathbf{E} \times \mathbf{B}) = \frac{e^2}{4\pi c^3 r^2}(\mathbf{a} \times \mathbf{n})^2 \mathbf{n}$$

$$= \frac{\mathbf{n}}{r^2}\frac{\sin^2 \theta}{4\pi}\frac{e^2 a^2}{c^3} \tag{8.3.7}$$

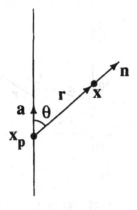

FIGURE 8.3

The total power emitted is

$$S = \iint N r^2 \sin\theta \, d\theta \, d\phi = \frac{e^2 a^2}{4\pi c^3} \int_0^\pi \sin^3\theta \, d\theta \int_0^{2\pi} d\phi$$

$$= \frac{e^2 a^2}{4\pi c^3} \frac{4}{3} 2\pi = \frac{2}{3} \frac{e^2 a^2}{c^3} \tag{8.3.8}$$

To find the magnitude of S, we shall examine a particular case.

8.3.2. Small Periodic Oscillations in One Dimension

The motions of nuclei in molecules or of electrons in atoms can be reduced to harmonic motions. The charged particles undergo motion and radiate energy. The question is, what portion of their kinetic energy do they radiate?

Let x_0 be the maximum excursion of a charged particle, as in Fig. 8.4. If we examine the field at great distance from the origin (0), $\mathbf{r}$ and $\mathbf{r}_0$ are almost parallel and we can use one instead of the other.

Let us assume that the motion of the particle is described by

$$x(t_p) = x_0 \cos \omega t_p \tag{8.3.9}$$

We note that

$$|\mathbf{r}_0 - \mathbf{r}| \le x_0 \tag{8.3.10}$$

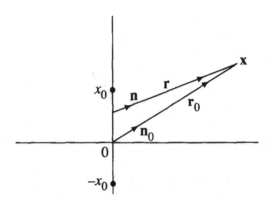

FIGURE 8.4

We have

$$v(t_p) = -\omega x_0 \sin \omega \, t_p \tag{8.3.11}$$

$$a(t_p) = -\omega^2 x_0 \cos \omega \, t_p \tag{8.3.12}$$

Then, if we indicate by O the *order of magnitude*,

$$O\left(\frac{\omega x_0}{c}\right) = O\left(\frac{v}{c}\right) \ll 1 \tag{8.3.13}$$

Let us evaluate the Poynting vector

$$\mathbf{N} = \frac{\mathbf{n}}{r^2} \frac{\sin^2 \theta}{4\pi} \frac{e^2 a^2}{c^3} \simeq \frac{\mathbf{n}_0}{r_0^2} \frac{\sin^2 \theta}{4\pi} \frac{e^2 \omega^4 x_0^2}{c^3} \cos^2 \omega \left(t - \frac{r_0}{c}\right) \tag{8.3.14}$$

The error in the argument is

$$\frac{\omega r_0}{c} - \frac{\omega r}{c} \simeq \frac{\omega x_0}{c} \ll 1 \tag{8.3.15}$$

Let us evaluate the average flow of energy into the solid angle $d\Omega$:

$$dS = \sin^2 \theta \frac{d\Omega}{4\pi} \frac{e^2 \omega^4 x_0^2}{c^3} \frac{1}{T} \int_0^T \cos^2 \omega \left(t - \frac{r_0}{c}\right) dt$$

$$= \sin^2 \theta \frac{d\Omega}{4\pi} \frac{e^2 \omega^4 x_0^2}{c^3} \frac{1}{2} = \frac{e^2 \omega^4 x_0^2}{8\pi c^3} \sin^2 \theta \, d\Omega \tag{8.3.16}$$

where $T = 2\pi/\omega$. The average energy loss per unit time is given by

$$\overline{S} = \frac{e^2\omega^4 x_0^2}{8\pi c^3} \iint \sin^2\theta \, d\Omega = \frac{e^2\omega^4 x_0^2}{8\pi c^3} \frac{8\pi}{3} = \frac{1}{3}\frac{e^2\omega^4 x_0^2}{c^3} \qquad (8.3.17)$$

The energy lost in a period is given by

$$\overline{S}\frac{2\pi}{\omega} = \frac{2\pi}{3}e^2 \left(\frac{\omega}{c}\right)^3 x_0^2 \qquad (8.3.18)$$

The mechanical energy of the harmonic oscillator is

$$\varepsilon = \frac{1}{2}m\omega^2 x_0^2 \qquad (8.3.19)$$

The fractional loss of energy per period is then

$$\frac{(2\pi/3)e^2(\omega/c)^3 x_0^2}{\frac{1}{2}m\omega^2 x_0^2} = \frac{4\pi}{3}\frac{e^2}{mc^2}\frac{\omega}{c} \qquad (8.3.20)$$

EXAMPLE: Molecule of HC

In a molecule of HC, since the Cl atom is much more massive than the **H** atom, it is the latter that vibrates and the relevant quantity is

$$\frac{e^2}{m_{\mathrm{H}}c^2} = \frac{e^2}{2000 m_{\mathrm{el}}c^2} \simeq 10^{-16}\mathrm{cm}$$

On the other hand,

$$\frac{\omega}{c} = \frac{2\pi}{\lambda} = \frac{2\pi}{10^{-4}}\mathrm{cm}^{-1}$$

Since $\lambda = 10^{-4}\,\mathrm{cm}$ (infrared radiation), the fractional loss per period is

$$\frac{4\pi}{3}\frac{\omega}{c}\frac{e^2}{m_{\mathrm{H}}c^2} = \frac{4\pi}{3}\frac{2\pi}{10^{-4}}10^{-16} = 2.6 \times 10^{-11} \simeq 10^{-11}$$

How long does it take for the molecule to lose all its vibrational energy? The frequency of vibration is

$$\nu = \frac{c}{\lambda} = \frac{3 \times 10^{10}}{10^{-4}} = 3 \times 10^{14}\mathrm{s}^{-1}$$

This is the number of periods of vibration occurring in a second. The fractional loss in a second is then

$$3 \times 10^{14} \times 10^{-11} = 3000$$

In about

$$\frac{1}{3000}S = 3.3 \times 10^{-4} = 0.33\,\text{ms}$$

the molecule will lose all its energy.

EXAMPLE: Atom

In the case of an atom the relevant quantity is

$$\frac{e^2}{m_{\text{el}}c^2} = 2.8 \times 10^{-13}\,\text{cm}$$

$$\lambda = 4000\,\text{Å} = 4 \times 10^{-5}\,\text{cm}$$

$$\frac{\omega}{c} = \frac{2\pi}{\lambda} = \frac{2\pi}{4 \times 10^{-5}} \simeq 10^5\,\text{cm}^{-1}$$

$$\frac{4\pi}{3}\frac{\omega}{c}\frac{e^2}{m_{\text{el}}c^2} = \frac{4\pi}{3} \times 10^5 \times 2.8 \times 10^{-13} = 11.7 \times 10^{-8}$$

$$\nu = \frac{c}{\lambda} = \frac{3 \times 10^{10}}{4 \times 10^{-5}} = 0.75 \times 10^{15}\,\text{S}^{-1}$$

The fractional loss per second is $11.7 \times 10^{-8} \times 0.75 \times 10^{15} = 8.77 \times 10^7 \text{s}^{-1}$. In about

$$\frac{1}{8.77 \times 10^7} = 1.14 \times 10^{-8} \simeq 10\,\text{ns}$$

the atom will lose all its energy.

We have seen that the energy loss per unit time is given by

$$\overline{S} = \frac{1}{3}\frac{e^2\omega^4 x_0^2}{c^3}\frac{\text{erg}}{\text{s}}$$

If we call τ the time the system takes to emit a quantum $\hbar\omega$, we have

$$\overline{S}\tau = \hbar\omega \tag{8.3.21}$$

Then

$$\tau^{-1} = \frac{\overline{S}}{\hbar\omega} = \frac{1}{3}\frac{e^2}{\hbar c}\left(\frac{\omega}{c}\right)^2 x_0^2\omega = \frac{e^2\omega^3}{3\hbar c^3}x_0^2$$

$$\simeq \frac{e^2}{\hbar c}\omega\left(\frac{x_0}{\lambda}\right)^2 \simeq \frac{1}{137}\omega\left(\frac{x_0}{\lambda}\right)^2 \qquad (8.3.22)$$

This formula gives the correct quantum mechanical result if we make the substitution $x_0 \to x_{mn}$, where x_{mn} is the matrix element between the two states m and n involved in the transition.

Let us now consider again the cases of a molecule and an atom.

EXAMPLE: Molecule (infrared)

For a molecule we can write

$$x_0 = 10^{-8}\,\text{cm}$$

$$\lambda = 10^{-4}\,\text{cm}$$

$$\omega = 2\pi \times 3 \times 10^{14} = 1.88 \times 10^{15}\,\text{s}^{-1}$$

$$\tau = \frac{137}{\omega}\left(\frac{\lambda}{x_0}\right)^2 = \frac{137}{1.9 \times 10^{15}}\left(\frac{10^{-4}}{10^{-8}}\right)^2 = 7.2 \times 10^{-6}\,\text{s}$$

EXAMPLE: Atom (optical)

For an atom we can write

$$x_0 = 10^{-8}\,\text{cm}$$

$$\lambda = 4 \times 10^{-5}\,\text{cm}$$

$$\omega = 2\pi \times 7.5 \times 10^{14} = 4.7 \times 10^{15}\,\text{s}^{-1}$$

$$\tau = \frac{137}{\omega}\left(\frac{\lambda}{x_0}\right)^2 = \frac{137}{4.7 \times 10^{15}}\left(\frac{4 \times 10^{-5}}{10^{-8}}\right)^2 = 4.7 \times 10^{-7}\,\text{s}$$

The following additional remarks can be made:

(1) In the limit $v/c \ll 1$, which implies $\omega x_0/c \ll 1$, the frequency of the emitted radiation is equal to the frequency of the harmonic motion; $\mathbf{E}$ and $\mathbf{B}$ are $\propto \cos\omega[t - (r_0/c)]$.

(2) If the motion is not simple harmonic, it is still possible to make a Fourier analysis of it:

$$x(t) = \sum_n x_n \cos(n\omega t + \phi_n)$$

If, for every n,

$$\frac{n x_n \omega}{c} \ll 1$$

We can apply to this case the same conclusions found in the simple harmonic case.

(3) A systematic investigation of the angular distribution of the radiation emitted by a time-varying source reveals the "type" of this radiation (dipole, quadrupole, and so on). In nuclear physics most of the radiation is *multipole radiation*. In atomic physics the radiation is chiefly *dipole radiation*. This is related to the value of the *radiative lifetime*, that is, the time it takes the system to get rid of its excitation energy via radiation.

Excited atoms can emit radiation or can give their excitation energy to other atoms via collisions. The dipole radiation is fast enough to occur before a collision takes place. The quadrupole radiation, instead, is slower and cannot compete with collisions.

It is extremely difficult to observe quadrupole radiation from terrestrial light sources, for, in these systems, it is practically impossible to reach pressures low enough to eliminate the effects of collisions; in any case, even at very low pressures, collisions with the walls cause the loss of excitation energy by the atoms or molecules. Conditions of extremely low pressure, suitable for the observation of quadrupole radiation, may be present in cosmic sources, such as intergalactic dust or gases.

8.4. Radiation from a Moving Charged Particle

Let us go back to the general expressions for $\mathbf{E}^{\mathrm{rad}}$ and $\mathbf{B}^{\mathrm{rad}}$:

$$\mathbf{E}^{\mathrm{rad}}(\mathbf{x}\ t) = -\left\{ \frac{e}{c^2 r} \left[\frac{\mathbf{a}}{\left(1 - \frac{\mathbf{n}\cdot\mathbf{v}}{c}\right)^2} - \frac{\mathbf{n}\cdot\mathbf{a}\left(\mathbf{n} - \frac{\mathbf{v}}{c}\right)}{\left(1 - \frac{\mathbf{n}\cdot\mathbf{v}}{c}\right)^3} \right] \right\}_{t_p = t - (r/c)}$$

$$\mathbf{B}^{\mathrm{rad}}(\mathbf{x}\,t) = \left\{ \frac{e}{c^2 r} \left[\frac{\mathbf{a} \times \mathbf{n}}{\left(1 - \dfrac{\mathbf{n} \cdot \mathbf{v}}{c}\right)^2} + \frac{(\mathbf{n} \cdot \mathbf{a})\left(\dfrac{\mathbf{v}}{c} \times \mathbf{n}\right)}{\left(1 - \dfrac{\mathbf{n} \cdot \mathbf{v}}{c}\right)^3} \right] \right\}_{t_p = t - (r/c)}$$

$$(8.4.1)$$

The Poynting vector in the general case is given by

$$\mathbf{N} = \frac{c}{4\pi} [\mathbf{E}^{\mathrm{rad}} \times \mathbf{B}^{\mathrm{rad}}]_{t_p = t - (r/c)}$$

$$= \left\{ \frac{e^2}{4\pi c^3 r^2} \left[-\frac{\mathbf{a}}{A^2} + \frac{\mathbf{n} \cdot \mathbf{a}\left(\mathbf{n} - \dfrac{\mathbf{v}}{c}\right)}{A^3} \right] \right.$$

$$\left. \times \left[\frac{\mathbf{a} \times \mathbf{n}}{A^2} + \frac{(\mathbf{n} \cdot \mathbf{a})\left(\dfrac{\mathbf{v}}{c} \times \mathbf{n}\right)}{A^3} \right] \right\}_{t_p = t - (r/c)}$$

$$(8.4.2)$$

where

$$A = 1 - \frac{\mathbf{n} \cdot \mathbf{v}}{c}$$

The calculation to be performed to find $\mathbf{N}$ is a bit laborious. We start from the relations

$$\mathbf{c} \times (\mathbf{a} \times \mathbf{b}) = (\mathbf{b} \cdot \mathbf{c})\mathbf{a} - (\mathbf{a} \cdot \mathbf{c})\mathbf{b}$$

which produces the following relations:

$$\mathbf{a} \times (\mathbf{a} \times \mathbf{n}) = (\mathbf{a} \cdot \mathbf{n})\mathbf{a} - a^2 \mathbf{n}$$

$$\mathbf{n} \times (\mathbf{a} \times \mathbf{n}) = \mathbf{a} - (\mathbf{a} \cdot \mathbf{n})\mathbf{n}$$

$$\frac{\mathbf{v}}{c} \times (\mathbf{a} \times \mathbf{n}) = \left(\frac{\mathbf{v}}{c} \cdot \mathbf{n}\right) \mathbf{a} - \left(\mathbf{a} \cdot \frac{\mathbf{v}}{c}\right) \mathbf{n}$$

$$\mathbf{a} \times \left(\frac{\mathbf{v}}{c} \times \mathbf{n}\right) = (\mathbf{a} \cdot \mathbf{n})\frac{\mathbf{v}}{c} - \left(\frac{\mathbf{v}}{c} \cdot \mathbf{a}\right) \mathbf{n}$$

$$\mathbf{n} \times \left(\frac{\mathbf{v}}{c} \times \mathbf{n}\right) = \frac{\mathbf{v}}{c} - \left(\frac{\mathbf{v}}{c} \cdot \mathbf{n}\right) \mathbf{n}$$

$$\frac{\mathbf{v}}{c} \times \left(\frac{\mathbf{v}}{c} \times \mathbf{n}\right) = \left(\frac{\mathbf{v}}{c} \cdot \mathbf{n}\right) \frac{\mathbf{v}}{c} - \frac{v^2}{c^2}\mathbf{n}$$

We have then

$$\left[-\frac{\mathbf{a}}{A^2} + \frac{(\mathbf{n}\cdot\mathbf{a})\left(\mathbf{n}-\frac{\mathbf{v}}{c}\right)}{A^3}\right] \times \left[\frac{\mathbf{a}\times\mathbf{n}}{A^2} + \frac{(\mathbf{n}\cdot\mathbf{a})\left(\frac{\mathbf{v}}{c}\times\mathbf{n}\right)}{A^3}\right]$$

$$= -\frac{\mathbf{a}\times(\mathbf{a}\times\mathbf{n})}{A^4} + \frac{(\mathbf{n}\cdot\mathbf{a})\left[\mathbf{n}\times(\mathbf{a}\times\mathbf{n}) - \frac{\mathbf{v}}{c}\times(\mathbf{a}\times\mathbf{n})\right]}{A^5}$$

$$-\frac{(\mathbf{n}\cdot\mathbf{a})\left[\mathbf{a}\times\left(\frac{\mathbf{v}}{c}\times\mathbf{n}\right)\right]}{A^5}$$

$$+\frac{(\mathbf{n}\cdot\mathbf{a})^2\left[\mathbf{n}\times\left(\frac{\mathbf{v}}{c}\times\mathbf{n}\right) - \frac{\mathbf{v}}{c}\times\left(\frac{\mathbf{v}}{c}\times\mathbf{n}\right)\right]}{A^6}$$

$$= -\frac{\mathbf{a}(\mathbf{a}\cdot\mathbf{n}) - \mathbf{n}a^2}{A^4}$$

$$= \frac{(\mathbf{n}\cdot\mathbf{a})\left[\mathbf{a} - \mathbf{n}(\mathbf{a}\cdot\mathbf{n}) - \mathbf{a}\left(\frac{\mathbf{v}}{c}\cdot\mathbf{n}\right) + \mathbf{n}\left(\mathbf{a}\cdot\frac{\mathbf{v}}{c}\right)\right]}{A^5}$$

$$-\frac{(\mathbf{n}\cdot\mathbf{a})\left[\frac{\mathbf{v}}{c}(\mathbf{a}\cdot\mathbf{n}) - \mathbf{n}\left(\frac{\mathbf{v}}{c}\cdot\mathbf{a}\right)\right]}{A^5}$$

$$+\frac{(\mathbf{n}\cdot\mathbf{a})^2\left[\frac{\mathbf{v}}{c} - \mathbf{n}\left(\frac{\mathbf{v}}{c}\cdot\mathbf{n}\right) - \frac{\mathbf{v}}{c}\left(\frac{\mathbf{v}}{c}\cdot\mathbf{n}\right) + \mathbf{n}\left(\frac{\mathbf{v}}{c}\right)^2\right]}{A^6}$$

$$= \frac{1}{A^6}\left\{\mathbf{n}a^2A^2 - \mathbf{a}(\mathbf{a}\cdot\mathbf{n})A^2 + (\mathbf{n}\cdot\mathbf{a})A\right.$$

$$\times\left[\mathbf{a} - \mathbf{n}(\mathbf{a}\cdot\mathbf{n}) - \mathbf{a}\left(\frac{\mathbf{v}}{c}\cdot\mathbf{n}\right) + \mathbf{n}\left(\mathbf{a}\cdot\frac{\mathbf{v}}{c}\right)\right.$$

$$\left.+\mathbf{n}\left(\frac{\mathbf{v}}{c}\cdot\mathbf{a}\right) - \frac{\mathbf{v}}{c}(\mathbf{a}\cdot\mathbf{n})\right]$$

$$\left.+(\mathbf{n}\cdot\mathbf{a})^2\left[\frac{\mathbf{v}}{c} - \mathbf{n}\left(\frac{\mathbf{v}}{c}\cdot\mathbf{n}\right) - \frac{\mathbf{v}}{c}\left(\frac{\mathbf{v}}{c}\cdot\mathbf{n}\right) + \mathbf{n}\left(\frac{\mathbf{v}}{c}\right)^2\right]\right\}$$

$$= \frac{1}{A^6}\left\{\mathbf{n}a^2A^2 - \mathbf{a}(\mathbf{a}\cdot\mathbf{n})\left[1 + \left(\frac{\mathbf{n}\cdot\mathbf{v}}{c}\right)^2 - 2\frac{\mathbf{n}\cdot\mathbf{v}}{c}\right]\right.$$

$$+(\mathbf{n}\cdot\mathbf{a})\left(1 - \frac{\mathbf{n}\cdot\mathbf{v}}{c}\right)[\mathbf{a} - \mathbf{n}(\mathbf{a}\cdot\mathbf{n})$$

$$-\mathbf{a}\left(\frac{\mathbf{v}}{c}\cdot\mathbf{n}\right)+\mathbf{n}\left(\mathbf{a}\cdot\frac{\mathbf{v}}{c}\right)$$

$$+\mathbf{n}\left(\frac{\mathbf{v}}{c}\cdot\mathbf{a}\right)-\frac{\mathbf{v}}{c}(\mathbf{a}\cdot\mathbf{n})\bigg]$$

$$+(\mathbf{n}\cdot\mathbf{a})^2\left[\frac{\mathbf{v}}{c}-\mathbf{n}\left(\frac{\mathbf{v}}{c}\cdot\mathbf{n}\right)-\frac{\mathbf{v}}{c}\left(\frac{\mathbf{v}}{c}\cdot\mathbf{n}\right)+\mathbf{n}\left(\frac{\mathbf{v}}{c}\right)^2\right]\bigg\}$$

$$=\frac{1}{A^6}\bigg\{\mathbf{n}a^2A^2-\mathbf{a}(\mathbf{a}\cdot\mathbf{n})-\mathbf{a}(\mathbf{a}\cdot\mathbf{n})\left(\frac{\mathbf{n}\cdot\mathbf{v}}{c}\right)^2$$

$$+2\mathbf{a}(\mathbf{a}\cdot\mathbf{n})\left(\frac{\mathbf{n}\cdot\mathbf{v}}{c}\right)$$

$$+(\mathbf{n}\cdot\mathbf{a})\mathbf{a}-(\mathbf{n}\cdot\mathbf{a})^2\mathbf{n}-\mathbf{a}(\mathbf{a}\cdot\mathbf{n})\left(\frac{\mathbf{v}}{c}\cdot\mathbf{n}\right)$$

$$+\mathbf{n}(\mathbf{n}\cdot\mathbf{a})\left(\mathbf{a}\cdot\frac{\mathbf{v}}{c}\right)$$

$$+\mathbf{n}(\mathbf{n}\cdot\mathbf{a})\left(\frac{\mathbf{v}}{c}\cdot\mathbf{a}\right)-\frac{\mathbf{v}}{c}(\mathbf{a}\cdot\mathbf{n})^2-(\mathbf{n}\cdot\mathbf{a})\left(\frac{\mathbf{n}\cdot\mathbf{v}}{c}\right)\mathbf{a}$$

$$+(\mathbf{n}\cdot\mathbf{a})^2\left(\mathbf{n}\cdot\frac{\mathbf{v}}{c}\right)\mathbf{n}+(\mathbf{n}\cdot\mathbf{a})\left(\mathbf{n}\cdot\frac{\mathbf{v}}{c}\right)^2\mathbf{a}$$

$$-(\mathbf{n}\cdot\mathbf{a})\left(\mathbf{n}\cdot\frac{\mathbf{v}}{c}\right)\left(\mathbf{a}\cdot\frac{\mathbf{v}}{c}\right)\mathbf{n}$$

$$-(\mathbf{n}\cdot\mathbf{a})\left(\frac{\mathbf{n}\cdot\mathbf{v}}{c}\right)\left(\frac{\mathbf{v}}{c}\cdot\mathbf{a}\right)\mathbf{n}+(\mathbf{n}\cdot\mathbf{a})^2\left(\frac{\mathbf{n}\cdot\mathbf{v}}{c}\right)\frac{\mathbf{v}}{c}$$

$$+(\mathbf{n}\cdot\mathbf{a})^2\frac{\mathbf{v}}{c}-\mathbf{n}(\mathbf{n}\cdot\mathbf{a})^2\left(\frac{\mathbf{v}}{c}\cdot\mathbf{n}\right)-(\mathbf{n}\cdot\mathbf{a})^2\left(\frac{\mathbf{v}}{c}\cdot\mathbf{n}\right)\frac{\mathbf{v}}{c}$$

$$+(\mathbf{n}\cdot\mathbf{a})^2\left(\frac{\mathbf{v}^2}{c^2}\right)\mathbf{n}\bigg\}$$

$$=\frac{1}{A^6}\bigg\{\mathbf{n}a^2A^2-(\mathbf{a}\cdot\mathbf{n})^2\mathbf{n}+\mathbf{n}(\mathbf{n}\cdot\mathbf{a})^2\frac{v^2}{c^2}$$

$$+2\mathbf{n}(\mathbf{n}\cdot\mathbf{a})\left(\frac{\mathbf{v}}{c}\cdot\mathbf{a}\right)-2\mathbf{n}(\mathbf{n}\cdot\mathbf{a})\left(\frac{\mathbf{n}\cdot\mathbf{v}}{c}\right)\left(\frac{\mathbf{v}}{c}\cdot\mathbf{a}\right)\bigg\}$$

$$=\frac{1}{A^6}\mathbf{n}\bigg\{a^2\left(1-\frac{\mathbf{n}\cdot\mathbf{v}}{c}\right)^2-(\mathbf{a}\cdot\mathbf{n})^2(1-\beta^2)$$

$$+2(\mathbf{a}\cdot\mathbf{n})\left(\mathbf{a}\cdot\frac{\mathbf{v}}{c}\right)\left(1-\frac{\mathbf{n}\cdot\mathbf{v}}{c}\right)\bigg\}$$

The result of these calculations is

$$
\mathbf{N} = \mathbf{n} \frac{e^2}{4\pi c^3 r^2} \left\{ a^2 \left(1 - \frac{\mathbf{n} \cdot \mathbf{v}}{c} \right)^2 - (\mathbf{a} \cdot \mathbf{n})^2 (1 - \beta^2) \right.
$$

$$
\left. + 2(\mathbf{a} \cdot \mathbf{n}) \left(\mathbf{a} \cdot \frac{\mathbf{v}}{c} \right) \left(1 - \frac{\mathbf{n} \cdot \mathbf{v}}{c} \right) \right\} \frac{1}{\left(1 - \dfrac{\mathbf{n} \cdot \mathbf{v}}{c} \right)^6} \Bigg|_{t_p} \qquad (8.4.3)
$$

We shall now examine some particular cases.

8.4.1. *Parallel Velocity and Acceleration*

In this case

$$
(\mathbf{a} \cdot \mathbf{n}) \left(\mathbf{a} \cdot \frac{\mathbf{v}}{c} \right) = \left(\frac{\mathbf{v}}{c} \cdot \mathbf{n} \right) a^2 \qquad (8.4.4)
$$

and

$$
\mathbf{N} = \mathbf{n} \frac{e^2}{4\pi c^3 r^2} \left\{ a^2 \left[1 - \frac{2\mathbf{n} \cdot \mathbf{v}}{c} + \left(\frac{\mathbf{n} \cdot \mathbf{v}}{c} \right)^2 \right] - (\mathbf{a} \cdot \mathbf{n})^2 (1 - \beta^2) \right.
$$

$$
\left. + 2 \left(\frac{\mathbf{v}}{c} \cdot \mathbf{n} \right) a^2 \left(1 - \frac{\mathbf{n} \cdot \mathbf{v}}{c} \right) \right\} \frac{1}{\left(1 - \dfrac{\mathbf{n} \cdot \mathbf{v}}{c} \right)^6}
$$

$$
= \mathbf{n} \frac{e^2}{4\pi c^3 r^2} \left\{ a^2 \left[1 - \frac{2\mathbf{n} \cdot \mathbf{v}}{c} + \left(\frac{\mathbf{n} \cdot \mathbf{v}}{c} \right)^2 + \frac{2\mathbf{n} \cdot \mathbf{v}}{c} \left(1 - \frac{\mathbf{n} \cdot \mathbf{v}}{c} \right) \right] \right.
$$

$$
\left. - (\mathbf{a} \cdot \mathbf{n})^2 (1 - \beta^2) \right\} \frac{1}{\left(1 - \dfrac{\mathbf{n} \cdot \mathbf{v}}{c} \right)^6}
$$

$$
= \mathbf{n} \frac{e^2}{4\pi c^3 r^2} \left\{ a^2 \left[1 + \left(\frac{\mathbf{n} \cdot \mathbf{v}}{c} \right)^2 - 2 \left(\frac{\mathbf{n} \cdot \mathbf{v}}{c} \right)^2 \right] \right.
$$

$$
\left. - (\mathbf{a} \cdot \mathbf{n})^2 + (\mathbf{a} \cdot \mathbf{n})^2 \frac{v^2}{c^2} \right\} \frac{1}{\left(1 - \dfrac{\mathbf{n} \cdot \mathbf{v}}{c} \right)^6}
$$

$$
= \mathbf{n} \frac{e^2}{4\pi c^3 r^2} \left\{ a^2 \left[1 - \left(\frac{\mathbf{n} \cdot \mathbf{v}}{c} \right)^2 \right] \right.
$$

$$+(\mathbf{a} \cdot \mathbf{n})^2 \frac{v^2}{c^2} - (\mathbf{a} \cdot \mathbf{n})^2 \bigg\} \frac{1}{\left(1 - \dfrac{\mathbf{n} \cdot \mathbf{v}}{c}\right)^6}$$

$$= \mathbf{n} \frac{e^2}{4\pi c^3 r^2} \bigg\{ a^2 - a^2 \left(\frac{\mathbf{n} \cdot \mathbf{v}}{c}\right)^2$$

$$+(\mathbf{a} \cdot \mathbf{n})^2 \frac{v^2}{c^2} - (\mathbf{a} \cdot \mathbf{n})^2 \bigg\} \frac{1}{\left(1 - \dfrac{\mathbf{n} \cdot \mathbf{v}}{c}\right)^6}$$

$$= \mathbf{n} \frac{e^2}{4\pi c^3 r^2} \{ a^2 - (\mathbf{a} \cdot \mathbf{n})^2 \} \frac{1}{\left(1 - \dfrac{\mathbf{n} \cdot \mathbf{v}}{c}\right)^6} \tag{8.4.5}$$

If θ is the angle between $\mathbf{a}$ (and $\mathbf{v}$) and $\mathbf{n}$ (see Fig. 8.5), then

$$\mathbf{N} = \mathbf{n} \frac{e^2}{4\pi c^3 r^2} \left\{ a^2 (1 - \cos^2 \theta) \frac{1}{(1 - \beta \cos \theta)^6} \right\}$$

$$= \mathbf{n} \frac{e^2}{4\pi c^3 r^2} \frac{a^2 \sin^2 \theta}{(1 - \beta \cos \theta)^6} \tag{8.4.6}$$

Let us set

$$\zeta = \frac{1}{(1 - \beta \cos \theta)^6} \tag{8.4.7}$$

Then

$$\mathbf{N} = \mathbf{n} \frac{e^2 a^2 \sin^2 \theta}{4\pi c^3 r^2} \zeta \bigg|_{t_p = t - (r/c)} \tag{8.4.8}$$

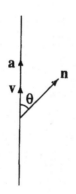

FIGURE 8.5

This formula for $v/c \ll (\zeta \simeq 1)$ reduces to the value given in Eq. (8.3.7). The factor ζ changes with the direction of **n**:

$$\theta = 90°, \quad \zeta = 1$$
$$\theta < 90°, \quad \zeta > 1$$
$$\theta > 90°, \quad \zeta < 1$$

The θ dependence of the ζ factor distorts the $\sin^2\theta$ pattern. The value of θ at which N takes its largest magnitude for a certain β is given by

$$\frac{\cos\theta}{1 + 2\sin^2\theta} = \beta \tag{8.4.9}$$

The energy measured by an observer in an interval of time dt is not equal to the energy emitted by the particle in the same interval. If the observer is in the direction of the velocity (Fig. 8.6), a signal emitted by the source at the time t_p arrives at **x** (position of the observer) at the time

$$t = t_p + \frac{|\mathbf{x} - \mathbf{x}_p|}{c} = t_p + \frac{r}{c} \tag{8.4.10}$$

A signal emitted by the source at the time $t_p + \Delta t_p$ arrives at **x** at the time

$$t' = t_p + \Delta t_p + \frac{r - v\Delta t_p}{c}$$

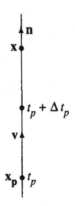

FIGURE 8.6

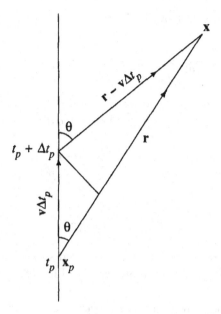

FIGURE 8.7

Then

$$\Delta t = t' - t = \left(t_p + \Delta t_p + \frac{r - v\Delta t_p}{c} \right) - \left(t_p + \frac{r}{c} \right)$$

$$= \Delta t_p \left(1 - \frac{v}{c} \right) \tag{8.4.11}$$

The interval Δt_p is compressed if the direction of the movement of the particle is also the direction of observation.

If that is not the case (see Fig. 8.7), in general, the signal emitted at the time t_p arrives at $\mathbf{x}$ at the time

$$t = t_p + \frac{r}{c} \tag{8.4.12}$$

and the signal emitted at the time tp arrives at $\mathbf{x}$ at the time

$$t' \simeq t_p + \Delta t_p + \frac{r - v \cos \theta \Delta t_p}{c}$$

It is then

$$\Delta t = t' - t = \Delta t_p \left(1 - \frac{v}{c} \cos \theta \right) = \Delta t_p (1 - \beta \cos \theta) \tag{8.4.13}$$

Therefore, the power lost by the particle by radiation into the solid angle $d\Omega$ is given by

$$Nr^2 d\Omega \frac{dt}{dt_p} = Nr^2 d\Omega (1 - \beta \cos\theta)$$

$$= \frac{d\Omega}{4\pi} \frac{e^2}{c^3} \frac{a^2 \sin^2\theta}{(1 - \beta\cos\theta)^5}$$

$$= d\Omega \frac{e^2 a^2 \sin^2\theta}{4\pi c^3} \eta \bigg|_{t_p = t - (r/c)} \qquad (8.4.14)$$

where

$$\eta = \frac{1}{(1 - \beta\cos\theta)^5} \qquad (8.4.15)$$

For $\theta = 90°$, $< 90°$, $> 90°$, η is 1, > 1, and < 1, respectively. The two patterns $\sin^2\theta$ and $\eta\sin^2\theta$ are represented in Fig. 8.8; the value $\beta = 0.25$ is used.

If we integrate expression (8.4.15) over all angles, we obtain the energy lost by the particle per unit time. The value of θ at which the power in Eq. (8.4.14) has its largest magnitude for a certain β is given by

$$\frac{2\cos\theta}{2 + 3\sin^2\theta} = \beta \qquad (8.4.16)$$

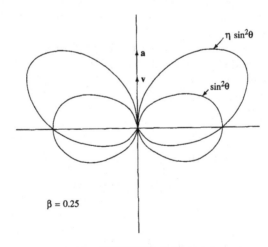

$\beta = 0.25$

FIGURE 8.8

We list now the values of β corresponding to some typical values of θ that are solutions of Eqs. (8.4.9) and (8.4.16):

$\theta(°)$	$\beta = \dfrac{\cos\theta}{1 + 2\sin^2\theta}$	$\beta = \dfrac{2\cos\theta}{2 + 3\sin^2\theta}$
0	1	1
30	0.5773	0.6298
60	0.2	0.2353
90	0	0

8.4.2. *Perpendicular Velocity and Acceleration*

This case is represented in Fig. 8.9. Let us rewrite the general expression for the Poynting vector:

$$\mathbf{N} = \mathbf{n}\frac{e^2}{4\pi c^3 r^2}\left\{a^2\left(1 - \frac{\mathbf{n}\cdot\mathbf{v}}{c}\right)^2 - (\mathbf{a}\cdot\mathbf{n})(1 - \beta^2)\right.$$

$$\left. + 2(\mathbf{a}\cdot\mathbf{n})\left(\mathbf{a}\cdot\frac{\mathbf{v}}{c}\right)\left(1 - \frac{\mathbf{n}\cdot\mathbf{v}}{c}\right)\right\}\frac{1}{\left(1 - \dfrac{\mathbf{n}\cdot\mathbf{v}}{c}\right)^6}\Bigg|_{t_p = t - (r/c)}$$

$$(8.4.17)$$

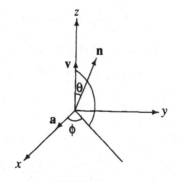

FIGURE 8.9

We have in this case

$$\frac{\mathbf{v} \cdot \mathbf{n}}{c} = \beta \cos \theta$$

$$\mathbf{a} \cdot \mathbf{n} = a \sin \theta \cos \phi$$

$$\mathbf{a} \cdot \mathbf{v} = 0$$

Then the expression in braces becomes

$$a^2 \left(1 - \frac{\mathbf{n} \cdot \mathbf{v}}{c}\right)^2 - (\mathbf{a} \cdot \mathbf{n})^2 (1 - \beta^2)$$

$$= a^2 (1 - \beta \cos \theta)^2 - (1 - \beta^2)(a^2 \sin^2 \theta \cos^2 \phi)$$

$$= a^2 [(1 - \beta \cos \theta)^2 - (1 - \beta^2) \sin^2 \theta \cos^2 \phi]$$

and

$$\mathbf{N} = \mathbf{n} \frac{e^2 a^2}{4\pi c^3 r^2} \frac{(1 - \beta \cos \theta)^2 - (1 - \beta^2) \sin^2 \theta \cos^2 \phi}{(1 - \beta \cos \theta)^6} \bigg|_{t_p} \qquad (8.4.18)$$

The power lost in a solid angle $d\Omega$ is given by

$$N r^2 d\Omega \frac{dt}{dt_p} = \frac{e^2}{4\pi c^3} a^2$$

$$\times \left[\frac{(1 - \beta \cos \theta)^2 - (1 - \beta^2) \sin^2 \theta \cos^2 \phi}{(1 - \beta \cos \theta)^5}\right] d\Omega \bigg|_{t_p} \qquad (8.4.19)$$

Assume $\phi = 0$; that is, $\mathbf{n}$ is in the same plane of $\mathbf{a}$ and $\mathbf{v}$. This could be the case of a circular orbit as in Fig. 8.10. Since $\cos \phi = 1$,

$$(1 - \beta \cos \theta)^2 - (1 - \beta^2) \sin^2 \theta \cos^2 \phi$$

$$= 1 + \beta^2 \cos^2 \theta - 2\beta \cos \theta - \sin^2 \theta + \beta^2 \sin^2 \theta$$

$$= \cos^2 \theta - 2\beta \cos \theta + \beta^2$$

$$= (\beta - \cos \theta)^2$$

Then, for angles $\phi \approx 0$,

$$N r^2 d\Omega \frac{dt}{dt_p} = \frac{e^2 a^2}{4\pi c^3} \frac{(\beta - \cos \theta)^2}{(1 - \beta \cos \theta)^5} d\Omega \bigg|_{t_p} \qquad (8.4.20)$$

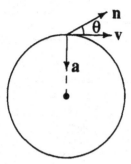

FIGURE 8.10

The emission is zero for θ such that $\cos\theta = \beta$. Let us call $\bar{\theta}$ this particular value of θ, and let us consider the case of a highly relativistic particle ($\beta \simeq 1$):

$$\cos\bar{\theta} = \beta$$

$$1 - \cos\bar{\theta} = 2\sin^2\frac{\bar{\theta}}{2} \simeq 2\frac{\bar{\theta}^2}{4} = 1 - \beta$$

$$\bar{\theta}^2 = 2(1 - \beta)$$

$$\bar{\theta} = \sqrt{2(1 - \beta)} \simeq \sqrt{2\frac{1}{2}(1 - \beta)(1 + \beta)}$$

$$= \sqrt{1 - \beta^2} = \frac{mc^2}{E_{\text{tot}}} \simeq \frac{mc^2}{E_p}$$

$$(8.4.21)$$

The angular width of the beam is of the order $\sqrt{1 - \beta^2} = 1/\gamma$.

EXAMPLE

If

$$E_p = 500\,\text{MeV} \simeq E_{\text{tot}}$$

since

$$mc^2 = 0.51\,\text{MeV}$$

Table 8.1. Velocities and Energies of Electrons $(mc^2 = 0.51\,\text{MeV})$.

E_p (MeV)	β	$\dfrac{E}{mc^2} = \dfrac{1}{\sqrt{1 - \beta^2}}$
0	0	1
0.01	0.195	1.02
0.1	0.5486	1.196
1	0.9412	2.96
10	0.9988	20.42
100	0.999987	196.1
1000 (1 GeV)	0.99999987	1,961.2
2,000	0.9999999675	3,922.6
10,000	0.9999999987	19,649.4
20,000	0.9999999997	41,169.3

We find

$$\frac{mc^2}{E_p} = \bar{\theta} \simeq 10^{-3}$$

In this case the radiation is highly peaked in the direction of motion.

In Table 8.1, we report some relevant data on fast relativistic electrons. The total energy is given by

$$E = \frac{mc^2}{\sqrt{1 - \beta^2}} \tag{8.4.22}$$

The kinetic energy is given by

$$E_p = E - mc^2 = mc^2 \left(\frac{1}{\sqrt{1 - \beta^2}} - 1 \right) \tag{8.4.23}$$

from which we derive

$$\beta = \frac{v}{c} = \sqrt{1 - \frac{1}{[(E_p/mc^2) + 1]^2}} \tag{8.4.24}$$

For electrons, $mc^2 = 0.51\,\text{MeV}$; for protons $mc^2 = 936\,\text{MeV}$.

8.5. Synchrotron Radiation

Synchrotron radiation is emitted by relativistic electrons whose paths are bent by a magnetic field. The general properties of the system are the following:

(1) The energies of the electrons are several hundred million electron volts or a few billion electron volts; in any case

$$E_p \gg mc^2 \tag{8.5.1}$$

(2) The directional pattern of the radiation does not have the ordinary dipole pattern, but it is peaked forward.

(3) The fields in the direction $\mathbf{n} \perp \mathbf{a}$ can be obtained from Eq. (8.4.1) by putting $\mathbf{n} \cdot \mathbf{a} = 0$:

$$
\begin{aligned}
\mathbf{E}^{\text{rad}}(\mathbf{x}, t) &= -\frac{e}{c^2 r} \frac{\mathbf{a}}{\{1 - [(\mathbf{n} \cdot \mathbf{v})/c]\}^2}\bigg|_{t_p} \\
\mathbf{B}^{\text{rad}}(\mathbf{x}, t) &= -\frac{e}{c^2 r} \frac{\mathbf{a} \times \mathbf{n}}{\{1 - [(\mathbf{n} \cdot \mathbf{v})/c]\}^2}\bigg|_{t_p}
\end{aligned}
\tag{8.5.2}
$$

For $\mathbf{n} \perp \mathbf{a}$, the polarization of the radiation is parallel to the radius. The Poynting vector is given by

$$
\begin{aligned}
\mathbf{N} &= \frac{c}{4\pi}(\mathbf{E} \times \mathbf{B}) = \frac{c}{4\pi}\left(-\frac{e^2}{c^4 r^2}\right) \frac{\mathbf{a} \times (\mathbf{a} \times \mathbf{n})}{\{1 - [(\mathbf{n} \cdot \mathbf{v})/c]\}^4} \\
&= \frac{e^2 a^2}{4\pi c^3 r^2} \frac{1}{\{1 - [(\mathbf{n} \cdot \mathbf{v})/c]\}^4}\mathbf{n}
\end{aligned}
\tag{8.5.3}
$$

which is the value $\mathbf{N}$ in Eq. (8.4.18) when we put $\sin\theta\cos\phi = 0$.

(4) The power emitted in a solid angle $d\Omega$ is given by Eq. (8.4.20), that is, by

$$Nr^2 d\Omega \frac{dt}{dt_p} = \frac{e^2 a^2}{4\pi c^3} \frac{(\beta - \cos\theta)^2}{(1 - \beta\cos\theta)^5} d\Omega \tag{8.5.4}$$

The pattern of the radiation is sketchily represented in Fig. 8.11. In Table 8.2, we report the value of the θ-dependent factor for $\beta = 0.95$ and 0.99. We see that the higher β is, the narrower and more peaked is the beam; also, the beam has two additional lobes whose maxima occur at

$$\cos\theta = \frac{5\beta^2 - 2}{3\beta} \tag{8.5.5}$$

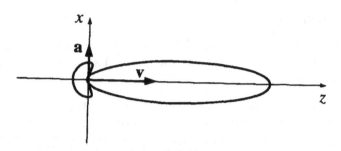

FIGURE 8.11

Table 8.2.

$\beta = 0.95$		$\beta = 0.99$	
$\theta°$	$\dfrac{(\beta - \cos\theta)^2}{(1 - \beta\cos\theta)^5}$	$\theta°$	$\dfrac{(\beta - \cos\theta)^2}{(1 - \beta\cos\theta)^5}$
0	8000	0	1,000,000
18.19 =	0	8.11 =	0
$\cos^{-1} 0.95$		$\cos^{-1} 0.99$	0
20	7.47	10	2738.49
25	36.78	12.41945 =	4474.5
		$\cos^{-1}[(5\beta^2 - 2)/3\beta]$	
28.166 =	41.35	20	1538.03
$\cos^{-1}[(5\beta^2 - 2)/3\beta]$			
30	40.276	30	260.3
45	15.482	45	32.95
60	5.077	60	7.31
90	0.9025	90	0.9801
120	0.3011	120	0.297
135	0.21	135	0.2028
150	0.1639	150	0.1558
180	0.13486	180	0.1269

We shall consider now the total power emitted.[1] The argument by which this quantity is calculated is very simple. We consider the rest system of the particle and calculate the loss of energy in this system. We call $W^{(0)}$

---

[1] See J. Schwinger, *Phys. Rev. 75*, 1912 (1949).

the energy in the rest system and consider the energy lost in the unit time

$$\frac{dW^{(0)}}{d\tau}$$

The energy enters the four-vector P_μ in the following way:

$$P_\mu \equiv \left(\mathbf{p}, \frac{iE}{c} \right)$$

In the rest system, $\mathbf{p} = 0$; the three spatial components of P_μ are zero, and only the fourth component is different from zero. How does this component transform if we consider another coordinate system?

$$P'_4 = a_{44} P_4 = \frac{1}{\sqrt{1 - \beta^2}} P_4 \tag{8.5.6}$$

or

$$dW = \frac{dW^{(0)}}{\sqrt{1 - \beta^2}} \tag{8.5.7}$$

But

$$dt = \frac{d\tau}{\sqrt{1 - \beta^2}} \tag{8.5.8}$$

Then

$$\frac{dW}{dt} = \frac{dW^{(0)}}{\sqrt{1 - \beta^2}} \frac{\sqrt{1 - \beta^2}}{d\tau} = \frac{dW^{(0)}}{d\tau} = \text{invariant} \tag{8.5.9}$$

But, on the other hand [see Eq. (8.3.8)],

$$\frac{dW^{(0)}}{d\tau} = \int N r^2 d\Omega = \frac{2}{3} \frac{e^2}{c^3} |\mathbf{a}|^2_{\text{rest}} = \frac{2e^2}{2c^3} \left(\frac{d\mathbf{v}}{d\tau} \right)^2_{\text{rest}} \tag{8.5.10}$$

From Eqs. (8.2.35) and (8.2.36), we derive

$$\frac{du_\mu}{d\tau} \equiv \left[\frac{1}{1 - \beta^2} \left(\mathbf{a} + \mathbf{v} \frac{(\mathbf{v} \cdot \mathbf{a})/c^2}{1 - \beta^2} \right), \frac{ic}{1 - \beta^2} \frac{(\mathbf{v} \cdot \mathbf{a})/c^2}{1 - \beta^2} \right]$$

$$\xrightarrow[\mathbf{v}=0]{} [\mathbf{a}, 0] = \left[\left(\frac{d\mathbf{v}}{d\tau} \right)_{\text{rest}}, 0 \right] \tag{8.5.11}$$

Then

$$\frac{dW^{(0)}}{d\tau} = \frac{2e^2}{3c^3} \sum_\mu \left(\frac{du_\mu}{d\tau}\right)^2_{\text{rest}} \tag{8.5.12}$$

But $\sum_\mu (du_\mu/d\tau)^2$ is invariant and, as such, valid in any system. Therefore,

$$\left(\frac{dW}{dt}\right)_{\substack{\text{any} \\ \text{system}}} = \frac{2e^2}{3c^3} \sum_\mu \left(\frac{du_\mu}{d\tau}\right)^2_{\substack{\text{any} \\ \text{system}}} \tag{8.5.13}$$

We find

$$\sum_\mu \left(\frac{du_\mu}{d\tau}\right)^2 = \frac{a^2}{(1-\beta^2)^2} + \frac{\beta^2[(\mathbf{v}/c) \cdot \mathbf{a}]^2}{(1-\beta^2)^4}$$

$$+ \frac{2[(\mathbf{v}/c) \cdot \mathbf{a}]^2}{(1-\beta^2)^3} - \frac{[(\mathbf{v}/c) \cdot \mathbf{a}]^2}{(1-\beta^2)^4}$$

$$= \frac{a^2}{(1-\beta^2)^2} + \frac{2[(\mathbf{v}/c) \cdot \mathbf{a}]^2}{(1-\beta^2)^3} - (1-\beta^2)\frac{[(\mathbf{v}/c) \cdot \mathbf{a}]^2}{(1-\beta^2)^4}$$

$$= \frac{a^2}{(1-\beta^2)^2} + \frac{2[(\mathbf{v}/c) \cdot \mathbf{a}]^2}{(1-\beta^2)^3} - \frac{[(\mathbf{v}/c) \cdot \mathbf{a}]^2}{(1-\beta^2)^3}$$

$$= \frac{a^2}{(1-\beta^2)^2} + \frac{[(\mathbf{v}/c) \cdot \mathbf{a}]^2}{(1-\beta^2)^3}$$

$$= \frac{a^2(1-\beta^2) + [(\mathbf{v}/c) \cdot \mathbf{a}]^2}{(1-\beta^2)^3}$$

$$= \frac{a^2 - \{a^2\beta^2 - [(\mathbf{v}/c) \cdot \mathbf{a}]^2\}}{(1-\beta^2)^3}$$

$$= \frac{a^2 - [a^2\beta^2 - a^2\beta^2\cos^2\alpha]}{(1-\beta^2)^3}$$

$$= \frac{a^2 - a^2\beta^2\sin^2\alpha}{(1-\beta^2)^3} = \frac{a^2 - [\mathbf{a} \times (\mathbf{v}/c)]^2}{(1-\beta^2)^3} \tag{8.5.14}$$

Therefore,

$$\frac{dW}{dt_p} = \frac{2e^2}{3c^3}\frac{a^2 - [\mathbf{a} \times (\mathbf{v}/c)]^2}{(1-\beta^2)^3} \tag{8.5.15}$$

Let us apply these considerations to the case of the synchrotron, Fig. 8.10. If R = radius of orbits,

$$a = \frac{v^2}{R} \tag{8.5.16}$$

$$a^2 - \left(a \times \frac{\mathbf{v}}{c}\right)^2 = \frac{v^4}{R^2} - \left(\frac{av}{c}\right)^2$$

$$= \frac{v^4}{R^2} - \frac{v^4}{R^2}\beta^2 = \frac{v^4}{R^2}(1 - \beta^2) \tag{8.5.17}$$

$$\frac{dW}{dt_p} = \frac{2e^2}{3c^3}\frac{1}{R^2}\frac{v^4}{(1 - \beta^2)^2} = \frac{2e^2c}{3R^2}\frac{\beta^4}{(1 - \beta^2)^2} \tag{8.5.18}$$

Taking into account the fact that

$$1 - \beta^2 = \left(\frac{mc^2}{E}\right)^2 \tag{8.5.19}$$

assuming $E_p \gg mc^2$ or $\beta \simeq 1$ (kinetic energy $\gg$ rest energy), we obtain

$$\frac{dW}{dt_p} \simeq \frac{2}{3}\frac{e^2}{R}\frac{c}{R}\left(\frac{E_p}{mc^2}\right)^4 \tag{8.5.20}$$

The energy lost in one turn is then given by

$$\frac{dW/dt_p}{\text{frequency}} = \frac{dW}{dt_p}\frac{2\pi R}{c} = \frac{4\pi}{3}\frac{e^2}{R}\left(\frac{E_p}{mc^2}\right)^4 \tag{8.5.21}$$

EXAMPLE

If

$$E_p = 500\,\text{MeV}$$

and

$$R = 100\,\text{cm}$$

we obtain

$$\frac{dW}{dt_p}\frac{2\pi R}{c} = \frac{4\pi}{3}\frac{e^2}{R}\left(\frac{E_p}{mc^2}\right)^4 = 8.916 \times 10^{-9}\,\text{ergs} = 5.57\,\text{keV}$$

Electrons are given high velocities as they are accelerated in synchrotrons, while they are confined by the magnetic field present to undergo motion in circular orbits. The electrons radiate energy while they are brought to their high velocities because they are being accelerated and because they move in circular orbits. There will be a point at which the energy radiated is practically equal to that supplied. On the other hand, in linear accelerators larger energies may be feasible.

We shall consider now the spectrum of synchrotron radiation. This is a very elaborate problem and Schwinger's paper should be consulted.

Many frequencies will be present in the beam as seen by a detector, due to the fact that the passage of the lobe over the detector produces a pulse. We can assume that each electron radiates independently as if no other electron were present and sum over the contributions due to all electrons. This is not a correct procedure; coherence effects are neglected. If two electrons are within a fraction of a wavelength of the radiation they emit, each electron will affect the radiation of the other. We shall not go into any calculations. We report Schwinger's conclusions.

In the spectrum there are two characteristic frequencies:

$$\omega_0 = \frac{c}{R} \tag{8.5.22}$$

$$\omega_c = \frac{3}{2}\omega_0 \left(\frac{E_p}{mc^2}\right)^3 \gg \omega_0 \tag{8.5.23}$$

In our example,

$$\frac{E_p}{mc^2} = 10^3, \quad \left(\frac{E_p}{mc^2}\right)^3 = 10^9$$

$$\omega_0 = \frac{c}{R} = \frac{3 \times 10^{10}}{100} = 3 \times 10^8 \text{s}^{-1}$$

$$\omega_c = \frac{3}{2} \times 3 \times 10^8 \times 10^9 = 4.5 \times 10^{17}\text{s}^{-1}$$

$$\lambda_c = \frac{2\pi c}{4.5 \times 10^{17}} = 4 \times 10^{-7}\,\text{cm} = 40\,\text{Å} \quad \text{(soft x-rays)}$$

We can write λ_c in general as follows:

$$\lambda_c = 5.56\frac{R}{E_p^3}\,\text{Å} \tag{8.5.24}$$

where R is expressed in meters and E_p in GeV.[2]

---

[2]R. E. Watson and M. L. Perlman, *Science* **199**, 1295 (1978).

A derivation of the approximate formula (8.5.23) follows:

(1) The angular width of the beam is [see Eq. (8.4.20)]

$$\frac{1}{\gamma} = \sqrt{1 - \beta^2} = \frac{mc^2}{E_p} \qquad (8.5.25)$$

(2) The time it takes the electron to go around once is

$$T \simeq \frac{2\pi R}{c} \qquad (8.5.26)$$

(3) The time the synchrotron radiation illuminates a fixed observer is given by

$$\frac{T}{2\pi} = \frac{\Delta t'}{1/\gamma}$$

$$\Delta t' = \frac{T}{2\pi\gamma} = \frac{R}{c\gamma} \qquad (8.5.27)$$

(4) The time interval during which the observer sees the radiation is

$$\Delta t = \frac{dt}{dt'}\Delta t' = (1 - \beta)\Delta t' \simeq \frac{\Delta t'}{\gamma^2} = \frac{R}{c\gamma^3} \qquad (8.5.28)$$

because

$$1 - \beta \simeq \frac{1}{2}(1 - \beta)(1 + \beta) = \frac{1}{2}(1 - \beta^2) = \frac{1}{2\gamma^2} \qquad (8.5.29)$$

(5) The highest frequency of the spectrum is

$$\omega_c \simeq \frac{1}{\Delta t} = \frac{c\gamma^3}{R} = \frac{c}{R}\left(\frac{E_p}{mc^2}\right)^3 = \omega_0 \left(\frac{E_p}{mc^2}\right) \qquad (8.5.30)$$

CHAPTER 8 EXERCISES

8.1. A particle is at rest at $t = 0$. At time $t = 0$, a *constant* force $F = dp/dt$ in the x direction is applied to this particle. Show that, if the particle is at $x = 0$ at the time $t = 0$, its world line is a hyperbola

$$\left(x + \frac{mc^2}{F}\right)^2 - (ct)^2 = \left(\frac{mc^2}{F}\right)^2$$

Sketch the world line in a $x \leftrightarrow ct$ plane.

8.2. A proton travels on a circular path perpendicular to a homogeneous magnetic induction field B. Show that the number of revolutions per second, as measured by an observer in the proton's frame of reference, is independent of the proton's energy, even at relativistic energies.

8.3. Two electrons ejected from a filament at rest in a certain frame S move off with equal speeds of magnitude $0.6c$ (c = velocity of light), one toward, say, x and the other toward $-x$. Their speed relative to each other, as measured in S, is $1.2c$, which exceeds c. Calculate the speed of the electron moving towards x measured in a frame S' that moves with the electron going towards $-x$.

8.4. A proton (charge e, mass m) of asymptotic velocity $v_0(v_0 \ll c)$ is approaching a nucleus of charge Ze head on.

(a) Find the velocity $v(r)$ and the acceleration $a(r)$ of the proton as a function of the distance r from the nucleus.

(b) Find the minimum distance r_0 in terms of v_0.

(c) The particle radiates at the rate

$$S = \frac{2e^2}{3c^3}a^2(t)$$

Calculate the total loss of energy during the collision by estimating

$$\int a^2(t)dt = 2 \int_{r_0}^{\infty} dr \frac{a^2(r)}{v(r)}$$

(d) Indicate the angular distribution of the emitted radiation.

(e) Estimate the range of the frequencies present in the emitted radiation.

Note that S is given m Gaussian units; in such units, the Coulomb potential is Ze/r.

8.5. J. J. Thomson, having identified the electron as a constituent of all atoms, proceeded to construct a model in which the atom consists of a ball of positively charged matter representing most of the atomic mass with the electrons embedded in it. In this model, spectral lines are produced by the periodic motion of the electrons inside the atom. Consider the atom of hydrogen. Assume that the positive charge has uniform density up to the radius R of the atom, and calculate the frequency of the electron's oscillations that we may expect in such an atom and the wavelength of the corresponding electromagnetic radiation.

8.6. (a) Show that the four-velocity $u_\mu = dx_\mu/d\tau$ satisfies the relation

$$\sum_\mu u_\mu^2 = -c^2$$

(b) Introduce the energy–momentum four-vector

$$p_\mu = \left(\mathbf{p}, \frac{iE}{c}\right)$$

and show that

$$\sum_\mu p_\mu dp_\mu = 0, \quad \mathbf{v} = \frac{\mathbf{p}c^2}{E}, \quad dE = \mathbf{v} \cdot d\mathbf{p}$$

(c) Define an angular momentum tensor of a mass point

$$L_{\mu\nu} = x_\mu p_\nu - x_\nu p_\mu$$

Show that $dL_{\mu\nu}/dt = 0$ for a *free* particle. This means that $L_{\mu\nu} = $ constant. $L_{\mu\nu} = $ constant gives us six relations; three of them express the constancy of the angular momentum. Find the meaning of the other three relations.

8.7. A point charge e located at $\mathbf{x}_p(\tau)$ and moving with velocity $\mathbf{v}$ has a charge and current density given by

$$\rho(\mathbf{x}, t) = e\delta^{(3)}(\mathbf{x} - \mathbf{x}_p(t))$$
$$\mathbf{j}(\mathbf{x}, t) = e\mathbf{v}\delta^{(3)}(\mathbf{x} - \mathbf{x}_p(t))$$

Prove that $\mathbf{j}$ and $j_4 = icp$ form a four-vector. To carry out this proof, start from the expression

$$eu_\mu \int d\tau \delta^{(3)}(\mathbf{x} - \mathbf{x}_p(\tau))\delta(t - t_p(\tau))$$

$\mathbf{x}_p(\tau)$ and $t_p(\tau)$ are a parametric representation of the particle's world line in terms of the proper time, and u_μ is the four-velocity of the particle.

(a) Show that this expression is a four-vector.
(b) Reduce it by carrying out the τ integration.

8.8. A point charge e moves with constant velocity and produces the potentials

$$\phi(\mathbf{x}, t) = \frac{e}{r(t')} \left\{ \frac{1}{1 - [(\mathbf{n} \cdot \mathbf{v})/c]} \right\} \Bigg|_{t' = t - (r/c)}$$

$$\mathbf{A}(\mathbf{x}, t) = \frac{e\mathbf{v}/c}{r(t')} \left\{ \frac{1}{1 - [(\mathbf{n} \cdot \mathbf{v})/c]} \right\} \Bigg|_{t' = t - (r/c)}$$

where $r = |\mathbf{x} - \mathbf{x}_p(t')|$

$\mathbf{x}_p(t)$ = position of the point charge at the time t

$$\mathbf{n} = \frac{\mathbf{r}}{r}$$

Derive these potentials from the static Coulomb potential (the potential in the frame of reference in which the particle is at rest):

$$\phi^{(0)} = \frac{e}{r}, \quad \mathbf{A}^{(0)} = 0$$

by the use of a Lorentz transformation. To write $1/r$ in the new variables, use the fact that

$$\frac{1}{r} = \frac{-c}{\displaystyle\sum_\lambda R_\lambda u_\lambda \big|_{\sum_\lambda R_\lambda^2 = 0}}$$

8.9. Obtain the fields of a point charge in uniform motion by means of a Lorentz transformation from the potentials of a charge at rest.

8.10. The relations of the retarded position and velocity to the present position and velocity of a charged particle are not known in general. For this reason the Liénard–Wiechert potentials allow the evaluation of the fields in terms of the *retarded* positions and velocities of the charges, and not in terms of the *present* positions and velocities. However, if a charge is in uniform motion, it is possible to express the potentials (and the fields) in terms of the present position of the charge. Prove the above for the case in which a point charge is at position $\mathbf{x}_0 = 0$ at the time $t = 0$ and moves with velocity $\mathbf{v}$ in the z direction (see Fig. P8.10). Make your proof by showing that

$$\phi(\mathbf{x}, t) = \frac{e}{\sqrt{(x^2 + y^2)(1 - \beta^2) + (z - vt)^2}}$$

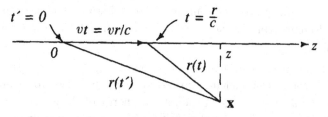

FIGURE P8.10

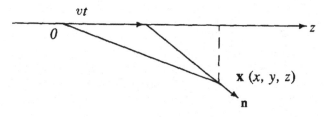

FIGURE P8.11

8.11. An electron with velocity $\mathbf{v}$ moves along the z axis (see Fig. P8.11). Its potentials at $\mathbf{x}$ are

$$\phi(\mathbf{x}, t) = \frac{e}{s}$$

$$\mathbf{A}(\mathbf{x}, t) = \frac{v}{c}\phi\mathbf{k}$$

where

$$s = \sqrt{(1 - \beta^2)(x^2 + y^2) + (z - vt)^2}$$

$\mathbf{k} =$ unit vector in the z direction

(a) Show that ϕ and $\mathbf{A}$ satisfy the Lorentz condition.
(b) Show also that

$$\mathbf{B} = \frac{\mathbf{v}}{c} \times \mathbf{E}$$

(c) Calculate $\mathbf{E}$ and $\mathbf{B}$ explicitly and show that $\mathbf{E}$ is parallel to $\mathbf{n}$.

$$Note : \frac{\partial f(\mathbf{x} - \mathbf{v}t)}{\partial t} = -(\mathbf{v} \cdot \boldsymbol{\nabla})f$$

8.12. Consider a hydrogen atom in its ground state. The electron runs in a circular orbit of radius $r = 0.529\,\text{Å}$ in the Coulomb field of the nucleus.

(a) Calculate the energy lost in one complete turn of the electron around the nucleus. Express your result in electron volts.

(b) What percentage of the kinetic energy of the electron is equal to the energy lost by the electron in one complete turn?

Disregard throughout quantum-mechanical effects.

9

Radiation Damping
and Electromagnetic Mass

9.1. Introduction

Given a certain distribution of charges and currents, the force density is expressed by

$$f_\mu = \frac{1}{c} \sum_\nu F_{\mu\nu} j_\nu \tag{9.1.1}$$

where $F_{\mu\nu}$ represents the fields and

$$j_\nu \equiv (\mathbf{j}, ic\rho) \tag{9.1.2}$$

The force acting on all charges and currents in a volume V is given by

$$\int_V d^3\mathbf{x}\, \mathbf{f} = \frac{d\mathbf{p}}{dt} \tag{9.1.3}$$

On the other hand, the inhomogeneous Maxwell equations in covariant form,

$$\sum_{\sigma} \frac{\partial F_{\nu\sigma}}{\partial x_{\sigma}} = \frac{4\pi}{c} j_{\nu} \tag{9.1.4}$$

give us

$$j_{\nu} = \frac{c}{4\pi} \sum_{\sigma} \frac{\partial F_{\nu\sigma}}{\partial x_{\sigma}} \tag{9.1.5}$$

This formalism has led us to assume a momentum of density $\mathbf{G}$, given by Eq. (7.4.18), and an energy density W, given by Eq. (7.4.15), for the electromagnetic field. The validity of these assumptions rests on the fact that, when we deal with the conservation laws for a closed system that contains matter and radiation, we consider these laws correct only if the momentum and the energy of the electromagnetic field are included. If the system is not closed, in the most general circumstances, the conservation laws will require that, in setting up the expressions for the changes of the "total momentum" and of "total energy" of the system, we keep into account the energy and momentum that flow across the boundaries of the volume in which the system resides.

The conservation laws require that the electromagnetic field that enters the calculation of the force acting on an elementary charge include not only the contributions due to the other charges, but also the field produced by the same charge, the *self-field!* This can be easily understood by the following argument: we shall consider, for simplicity, a system consisting of a point charge in motion.

If a point charge is in uniform motion, the self-field of the charge does not act on the charge itself, but contributes to the total momentum of the system; in fact a small, virtual change of the velocity of the charge results in simultaneous changes of both the momentum of the charge and the momentum of the fields.

If an external force acts on the point charge, the charge will accelerate and radiate. This external force will have to provide the energy and the momentum that accompany the changes in the electromagnetic field. For this to be possible, it will be necessary for the radiation produced by the accelerating charge to act back on the charge by means of a reaction force. The only way for the radiation field "to communicate" with the external force is via the point charge!

In the following section we shall calculate this reaction force, called *self-force*, for the case of an electron, following the treatment by Abraham and Lorentz.[1]

9.2. Evaluation of the Self-Force and Radiation Damping

Consider a system consisting of a single a electron. Let $j_\mu(\mathbf{x}, t)$ be the four-current vector associated with it. The potentials are given by

$$A_\mu(\mathbf{x}, t) = \frac{1}{c} \int d^3\xi \frac{j_\mu[\mathbf{x} - \boldsymbol{\xi}, t - (|\boldsymbol{\xi}|/c)]}{|\boldsymbol{\xi}|} \tag{9.2.1}$$

where

$$\boldsymbol{\xi} = \mathbf{x} - \mathbf{x}' \tag{9.2.2}$$

The notation is explained in Fig. 9.1. We can write

$$F_{\mu\nu}^{\text{self}}(\mathbf{x}, t) = \frac{1}{c} \int d^3\xi \frac{1}{|\boldsymbol{\xi}|}$$

$$\times \left(\frac{\partial j_\nu(\mathbf{x} - \boldsymbol{\xi}, t')}{\partial x_\mu} - \frac{\partial j_\mu(\mathbf{x} - \boldsymbol{\xi}, t')}{\partial x_\nu} \right)_{t' = t - (|\boldsymbol{\xi}|/c)} \tag{9.2.3}$$

The self-force density is given by

$$F_\mu^{\text{self}}(\mathbf{x}, t) = \frac{1}{c} \sum_\nu F_{\mu\nu}^{\text{self}}(\mathbf{x}, t) j_\nu(\mathbf{x}, t) \tag{9.2.4}$$

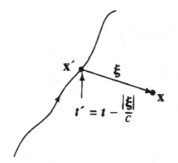

FIGURE 9.1

---

[1]H. A. Lorentz, *Theory of Electrons*, 2nd ed. (1915), Dover, New York, 1952.

The kth component of the total self-force is

$$F_k^{\text{self}}(t) = \int d^3\mathbf{x}\, f_k^{\text{self}}(\mathbf{x}, t)$$

$$= \int d^3\mathbf{x} \frac{1}{c} \sum_\nu f_{k\nu}^{\text{self}}(\mathbf{x}, t) j_\nu(\mathbf{x}, t)$$

$$= \int d^3\mathbf{x} \frac{1}{c^2} \sum_\nu \int d^3\boldsymbol{\xi} \frac{1}{|\boldsymbol{\xi}|}$$

$$\times \left(\frac{\partial j_\nu(\mathbf{x} - \boldsymbol{\xi}, t')}{\partial x_k} - \frac{\partial j_k(\mathbf{x} - \boldsymbol{\xi}, t')}{\partial x_\nu} \right)_{t' = t - (|\boldsymbol{\xi}|/c)} j_\nu(\mathbf{x}, t) \qquad (9.2.5)$$

Let us calculate this force for an electron that at time t has $\mathbf{v} = 0$ (but $d\mathbf{v}/dt \neq 0$). The four-current is then

$$j_\mu \equiv (0, ic\rho) \qquad (9.2.6)$$

We shall assume the charge symmetric in this rest system. Then

$$F_k^{\text{self}}(t) = \int d^3\mathbf{x} \frac{1}{c^2} \int \sum_\nu d^3\boldsymbol{\xi} \frac{1}{|\boldsymbol{\xi}|}$$

$$\times \left(\frac{\partial j_\nu(\mathbf{x} - \boldsymbol{\xi}, t')}{\partial x_k} - \frac{\partial j_k(\mathbf{x} - \boldsymbol{\xi}, t')}{\partial x_\nu} \right)_{t' = t - (|\boldsymbol{\xi}|/c)} j_\nu(\mathbf{x}, t)$$

$$= \int d^3\mathbf{x} \frac{1}{c^2} \int d^3\boldsymbol{\xi} \frac{1}{|\boldsymbol{\xi}|}$$

$$\times \left(\frac{\partial ic\rho(\mathbf{x} - \boldsymbol{\xi}, t')}{\partial x_k} - \frac{\partial j_k(\mathbf{x} - \boldsymbol{\xi}, t')}{\partial ict} \right)_{t' = t - (|\boldsymbol{\xi}|/c)} ic\rho(x, t)$$

$$= - \int d^3\mathbf{x}\rho(\mathbf{x}, t) \int d^3\boldsymbol{\xi} \frac{1}{|\boldsymbol{\xi}|}$$

$$\times \left(\frac{\partial \rho(\mathbf{x} - \boldsymbol{\xi}, t')}{\partial x_k} + \frac{1}{c^2} \frac{\partial j_k(\mathbf{x} - \boldsymbol{\xi}, t')}{\partial t} \right)_{t' = t - (|\boldsymbol{\xi}|/c)}$$

$$= \mathrm{I} + \mathrm{II} \qquad (9.2.7)$$

where

$$\mathrm{I} = - \int d^3\mathbf{x}\rho(\mathbf{x}, t)$$

$$\times \int d^3\boldsymbol{\xi} \frac{1}{|\boldsymbol{\xi}|} \left[\frac{\partial \rho(\mathbf{x} - \boldsymbol{\xi}, t')}{\partial x_k} \right]_{t' = t - (|\boldsymbol{\xi}|/c)} \qquad (9.2.8)$$

$$\mathrm{II} = -\frac{1}{c^2} \int d^3\mathbf{x}\rho(\mathbf{x}, t)$$

$$\times \int d^3\boldsymbol{\xi} \frac{1}{|\boldsymbol{\xi}|} \left[\frac{\partial j_k(\mathbf{x} - \boldsymbol{\xi}, t')}{\partial t} \right]_{t'=t-(|\boldsymbol{\xi}|/c)} \tag{9.2.9}$$

Let us now consider separately the two integrals I and II.

Integral I. We can expand ρ as follows:

$$\rho(\mathbf{x} - \boldsymbol{\xi}, t') = \rho\left(\mathbf{x} - \boldsymbol{\xi}, t - \frac{|\boldsymbol{\xi}|}{c}\right)$$

$$= \rho(\mathbf{x} - \boldsymbol{\xi}, t) - \frac{|\boldsymbol{\xi}|}{c} \frac{\partial \rho(\mathbf{x} - \boldsymbol{\xi}, t)}{\partial t}$$

$$+ \frac{1}{2} \frac{|\boldsymbol{\xi}|^2}{c^2} \frac{\partial^2 \rho(\boldsymbol{x} - \boldsymbol{\xi}, t)}{\partial t^2}$$

$$- \frac{1}{6} \frac{|\boldsymbol{\xi}|^3}{c^3} \frac{\partial^3 \rho(\boldsymbol{x} - \boldsymbol{\xi}, t)}{\partial t^3} + \cdots \tag{9.2.10}$$

Now we have, neglecting terms with power greater than 3,

$$\mathrm{I} = -\int d^3\mathbf{x}\rho(\mathbf{x}, t) \int d^3\boldsymbol{\xi} \frac{1}{|\boldsymbol{\xi}|} \left[\frac{\partial \rho(\mathbf{x} - \boldsymbol{\xi}, t')}{\partial x_k} \right]_{t'=t-(|\boldsymbol{\xi}|/c)}$$

$$= -\int d^3\mathbf{x}\rho(\mathbf{x}, t) \int d^3\boldsymbol{\xi} \frac{1}{|\boldsymbol{\xi}|}$$

$$\times \frac{\partial}{\partial x_k} \left[\rho(\mathbf{x} - \boldsymbol{\xi}, t) - \frac{|\boldsymbol{\xi}|}{c} \frac{\partial \rho(\mathbf{x} - \boldsymbol{\xi}, t)}{\partial t} \right.$$

$$\left. + \frac{1}{2} \frac{|\boldsymbol{\xi}|^2}{c^2} \frac{\partial^2 \rho(\mathbf{x} - \boldsymbol{\xi}, t)}{\partial t^2} - \frac{1}{6} \frac{|\boldsymbol{\xi}|^3}{c^3} \frac{\partial^3 \rho(\mathbf{x} - \boldsymbol{\xi}, t)}{\partial t^3} \right]$$

$$= -\iint d^3\mathbf{x} d^3\boldsymbol{\xi} \rho(\mathbf{x}, t) \frac{1}{|\boldsymbol{\xi}|} \frac{\partial}{\partial x_k} \rho(\mathbf{x} - \boldsymbol{\xi}, t)$$

$$+ \iint d^3\mathbf{x} d^3\boldsymbol{\xi} \rho(\mathbf{x}, t) \frac{1}{|\boldsymbol{\xi}|} \frac{\partial}{\partial x_k} \left[\frac{|\boldsymbol{\xi}|}{c} \frac{\partial \rho(\mathbf{x} - \boldsymbol{\xi}, t)}{\partial t} \right]$$

$$- \frac{1}{2} \iint d^3\mathbf{x} d^3\boldsymbol{\xi} \rho(\mathbf{x}, t) \frac{1}{|\boldsymbol{\xi}|} \frac{\partial}{\partial x_k} \left[\frac{|\boldsymbol{\xi}|^2}{c^2} \frac{\partial^2 \rho(\mathbf{x} - \boldsymbol{\xi}, t)}{\partial t^2} \right]$$

$$+ \frac{1}{6} \iint d^3x d^3\xi \rho(\mathbf{x}, t) \frac{1}{|\boldsymbol{\xi}|} \frac{\partial}{\partial x_k} \left[\frac{|\boldsymbol{\xi}|^3}{c^3} \frac{\partial^3 \rho(\mathbf{x} - \boldsymbol{\xi}, t)}{\partial t^3} \right]$$

$$= \textcircled{1} + \textcircled{2} + \textcircled{3} + \textcircled{4} \qquad (9.2.11)$$

Now

$$\textcircled{1} = -\iint d^3x d^3\xi \rho(\mathbf{x}, t) \frac{1}{|\boldsymbol{\xi}|} \frac{\partial}{\partial x_k} \rho(\mathbf{x} - \boldsymbol{\xi}, t)$$

$$= \iint d^3x d^3\xi \rho(\mathbf{x}, t) \frac{1}{|\boldsymbol{\xi}|} \frac{\partial}{\partial \xi_k} \rho(\mathbf{x} - \boldsymbol{\xi}, t)$$

$$\overset{\text{IP}}{=} -\iint d^3x d^3\xi \rho(\mathbf{x}, t) \rho(\mathbf{x} - \boldsymbol{\xi}, t) \frac{\partial}{\partial \xi_k} \frac{1}{|\boldsymbol{\xi}|}$$

$$= \iint d^3x d^3\xi \rho(\mathbf{x}, t) \rho(\mathbf{x} - \boldsymbol{\xi}, t) \frac{\xi_k}{|\boldsymbol{\xi}|^3} = 0 \qquad (9.2.12)$$

by symmetry. IP indicates an integration by parts.

$$\textcircled{2} = \iint d^3x d^3\xi \rho(\mathbf{x}, t) \frac{1}{|\boldsymbol{\xi}|} \frac{\partial}{\partial x_k} \left[\frac{|\boldsymbol{\xi}|}{c} \frac{\partial \rho(\mathbf{x} - \boldsymbol{\xi}, t)}{\partial t} \right]$$

But

$$\frac{\partial \rho}{\partial t} = -\boldsymbol{\nabla} \cdot \mathbf{j} = -\boldsymbol{\nabla} \cdot (\rho \mathbf{v}) = -\mathbf{v} \cdot \boldsymbol{\nabla}_\rho = 0$$

since $\mathbf{v} = 0$. Thus, the second term in Eq. (9.2.11) is zero:

$$\textcircled{2} = 0$$

$$\textcircled{3} = -\frac{1}{2} \iint d^3x d^3\xi \rho(\mathbf{x}, t) \frac{1}{|\boldsymbol{\xi}|} \frac{\partial}{\partial x_k} \left[\frac{|\boldsymbol{\xi}|^2}{c^2} \frac{\partial^2 \rho(\mathbf{x} - \boldsymbol{\xi}, t)}{\partial t^2} \right] \qquad (9.2.13)$$

But

$$\frac{\partial^2 \rho(\mathbf{x} - \boldsymbol{\xi}, t)}{\partial t^2} = -\frac{\partial}{\partial t} [\mathbf{v} \cdot \boldsymbol{\nabla}_x \rho(\mathbf{x} - \boldsymbol{\xi}, t)]$$

$$= -\frac{\partial \mathbf{v}}{\partial t} \cdot \boldsymbol{\nabla}_x \rho(\mathbf{x} - \boldsymbol{\xi}, t) - \mathbf{v} \cdot \boldsymbol{\nabla}_x \frac{\partial}{\partial t} \rho(\mathbf{x} - \boldsymbol{\xi}, t)$$

$$= -\frac{\partial \mathbf{v}}{\partial t} \cdot \boldsymbol{\nabla}_x \rho(\mathbf{x} - \boldsymbol{\xi}, t) = \frac{\partial \mathbf{v}}{\partial t} \cdot \boldsymbol{\nabla}_\xi \rho(\mathbf{x} - \boldsymbol{\xi}, t)$$

Then

$$\text{③} = -\frac{1}{2c^2} \iint d^3\mathbf{x}\, d^3\boldsymbol{\xi}\, \rho(\mathbf{x},t)|\boldsymbol{\xi}| \frac{\partial}{\partial x_k}\left[\frac{\partial^2 \rho(\mathbf{x}-\boldsymbol{\xi},t)}{\partial t^2}\right]$$

$$= \frac{1}{2c^2} \int d^3\mathbf{x}\, d^3\boldsymbol{\xi}\, \rho(\mathbf{x},t)|\boldsymbol{\xi}| \frac{\partial}{\partial \xi_k}\left[\frac{\partial^2 \rho(\mathbf{x}-\boldsymbol{\xi},t)}{\partial t^2}\right]$$

$$\overset{\text{IP}}{=} -\frac{1}{2c^2} \iint d^3\mathbf{x}\, d^3\boldsymbol{\xi}\, \rho(\mathbf{x},t)\frac{\partial^2 \rho(\mathbf{x}-\boldsymbol{\xi},t)}{\partial t^2}\frac{\partial}{\partial \xi_k}|\boldsymbol{\xi}|$$

$$= -\frac{1}{2c^2} \iint d^3\mathbf{x}\, d^3\boldsymbol{\xi}\, \rho(\mathbf{x},t)\frac{\partial^2 \rho(\mathbf{x}-\boldsymbol{\xi},t)}{\partial t^2}\frac{\xi_k}{|\boldsymbol{\xi}|}$$

$$= -\frac{1}{2c^2} \iint d^3\mathbf{x}\, d^3\boldsymbol{\xi}\, \rho(\mathbf{x},t)\left[\left(\frac{\partial \mathbf{v}}{\partial t}\cdot\boldsymbol{\nabla}_\xi\right)\rho(\mathbf{x}-\boldsymbol{\xi},t)\right]\frac{\xi_k}{|\boldsymbol{\xi}|}$$

$$\overset{\text{IP}}{=} \frac{1}{2c^2} \iint d^3\mathbf{x}\, d^3\boldsymbol{\xi}\, \rho(\mathbf{x},t)(\mathbf{x}-\boldsymbol{\xi},t)\left(\frac{\partial \mathbf{v}}{\partial t}\cdot\boldsymbol{\nabla}_\xi\right)\frac{\xi_k}{|\boldsymbol{\xi}|}$$

Now

$$\frac{\partial \mathbf{v}}{\partial t}\cdot\boldsymbol{\nabla}_\xi = \frac{\partial v_x}{\partial t}\frac{\partial}{\partial \xi_x} + \frac{\partial v_y}{\partial t}\frac{\partial}{\partial \xi_y} + \frac{\partial v_z}{\partial t}\frac{\partial}{\partial \xi_z}$$

$$\left(\frac{\partial \mathbf{v}}{\partial t}\cdot\boldsymbol{\nabla}_\xi\right)\frac{\xi_x}{|\boldsymbol{\xi}|} = \frac{\partial v_x}{\partial t}\frac{\partial}{\partial \xi_x}\frac{\xi_x}{\sqrt{\xi_x^2+\xi_y^2+\xi_z^2}}$$

$$+ \frac{\partial v_y}{\partial t}\frac{\partial}{\partial \xi_y}\frac{\xi_x}{\sqrt{\xi_x^2+\xi_y^2+\xi_z^2}}$$

$$+ \frac{\partial v_z}{\partial t}\frac{\partial}{\partial \xi_z}\frac{\xi_x}{\sqrt{\xi_x^2+\xi_y^2+\xi_z^2}}$$

$$= \frac{\partial v_x}{\partial t}\frac{|\boldsymbol{\xi}|-\xi_x(\xi_x/|\boldsymbol{\xi}|)}{|\boldsymbol{\xi}|^2} - \frac{\partial v_y}{\partial t}\frac{\xi_x\xi_y}{|\boldsymbol{\xi}|^3} - \frac{\partial v_z}{\partial t}\frac{\xi_x\xi_z}{|\boldsymbol{\xi}|^3}$$

$$= -\frac{\xi_x[(\partial\mathbf{v}/\partial t)\cdot\boldsymbol{\xi}]}{|\boldsymbol{\xi}|^3} + \frac{\partial v_x}{\partial t}\frac{1}{|\boldsymbol{\xi}|}$$

Then

$$\text{③} = \frac{1}{2c^2} \iint d^3\mathbf{x}\, d^3\boldsymbol{\xi}\, \rho(\mathbf{x},t)\rho(\mathbf{x}-\boldsymbol{\xi},t)\left(\frac{\partial \mathbf{v}}{\partial t}\cdot\boldsymbol{\nabla}_\xi\right)\frac{\xi_k}{|\boldsymbol{\xi}|}$$

$$= \frac{1}{2c^2} \iint d^3\mathbf{x}\, d^3\boldsymbol{\xi}\, \rho(\mathbf{x},t)\rho(\mathbf{x}-\boldsymbol{\xi},t)$$

$$\times \left\{ -\frac{\xi_k[(\partial \mathbf{v}/\partial t) \cdot \boldsymbol{\xi}]}{|\boldsymbol{\xi}|^3} + \frac{\partial v_k}{\partial t}\frac{1}{|\boldsymbol{\xi}|} \right\}$$

$$= \iint d^3\mathbf{x} d^3\boldsymbol{\xi}\, \rho(\mathbf{x},t)\rho(\mathbf{x}-\boldsymbol{\xi},t)$$

$$\times \left\{ \frac{1}{2c^2}\left[\frac{\partial v_x}{\partial t}\frac{1}{|\boldsymbol{\xi}|} - \frac{\xi_k[(\partial \mathbf{v}/\partial t) \cdot \boldsymbol{\xi}]}{|\boldsymbol{\xi}|^3} \right] \right\}$$

But

$$\int d^3\boldsymbol{\xi}\,\frac{\xi_k \xi_i}{|\boldsymbol{\xi}|^3} = 0, \quad \text{if } k \neq i$$

$$= \frac{1}{3}\int \frac{d^3\boldsymbol{\xi}}{|\boldsymbol{\xi}|}, \quad \text{if } k = i$$

Therefore,

$$③ = \frac{\partial v_k}{\partial t}\iint d^3\mathbf{x} d^3\boldsymbol{\xi}\, \rho(\mathbf{x},t)\rho(\mathbf{x}-\boldsymbol{\xi},t)\frac{1}{2c^2}\left[\frac{1}{|\boldsymbol{\xi}|} - \frac{1}{3|\boldsymbol{\xi}|} \right]$$

$$= \frac{1}{3c^2}\frac{\partial v_k}{\partial t}\int d^3\mathbf{x} d^3\boldsymbol{\xi}\, \frac{\rho(\mathbf{x},t)\rho(\mathbf{x}-\boldsymbol{\xi},t)}{|\boldsymbol{\xi}|} \tag{9.2.14}$$

Consider now the last integral in I:

$$④ = \iint d^3\mathbf{x} d^3\boldsymbol{\xi}\, \rho(\mathbf{x},t)\frac{1}{|\boldsymbol{\xi}|}\frac{\partial}{\partial x_k}\left[\frac{1}{6c^3}|\boldsymbol{\xi}|^3\frac{\partial^3\rho(\mathbf{x}-\boldsymbol{\xi},t)}{\partial t^3} \right]$$

$$= \frac{1}{6c^3}\iint d^3\mathbf{x} d^3\boldsymbol{\xi}\, \rho(\mathbf{x},t)|\boldsymbol{\xi}|^2\frac{\partial}{\partial x_k}\left[\frac{\partial^3\rho(\mathbf{x}-\boldsymbol{\xi},t)}{\partial t^3} \right]$$

$$= -\frac{1}{6c^3}\iint d^3\mathbf{x} d^3\boldsymbol{\xi}\, \rho(\mathbf{x},t)|\boldsymbol{\xi}|^2\frac{\partial}{\partial \xi_k}\left[\frac{\partial^3\rho(\mathbf{x}-\boldsymbol{\xi},t)}{\partial t^3} \right]$$

$$\overset{\text{IP}}{=} \frac{1}{6c^3}\iint d^3\mathbf{x} d^3\boldsymbol{\xi}\, \rho(\mathbf{x},t)\frac{\partial^3\rho(\mathbf{x}-\boldsymbol{\xi},t)}{\partial t^3}\frac{\partial}{\partial \xi_k}|\boldsymbol{\xi}|^2$$

$$= \frac{1}{3c^3}\iint d^3\mathbf{x} d^3\boldsymbol{\xi}\, \rho(\mathbf{x},t)\frac{\partial^3\rho(\mathbf{x}-\boldsymbol{\xi},t)}{\partial t^3}\xi_k$$

$$= \frac{1}{3c^3}\iint d^3\mathbf{x} d^3\boldsymbol{\xi}\, \rho(\mathbf{x},t)\left[\left(\frac{\partial^2\mathbf{v}}{\partial t^2} \cdot \boldsymbol{\nabla}_\xi \right)\rho(\mathbf{x}-\boldsymbol{\xi},t) \right]\xi_k$$

$$\stackrel{\text{IP}}{=} -\frac{1}{3c^3}\iint d^3\mathbf{x}\,d^3\boldsymbol{\xi}\,\rho(\mathbf{x},t)\rho(\mathbf{x}-\boldsymbol{\xi},t)\frac{\partial^2 v_k}{\partial t^2}$$

$$= -\frac{1}{3c^3}\frac{\partial v_k}{\partial t^2}\iint d^3\mathbf{x}\,d^3\boldsymbol{\xi}\,\rho(\mathbf{x},t)\rho(\mathbf{x}-\boldsymbol{\xi},t) = -\frac{e^3}{3c^3}\frac{\partial^2 v_k}{\partial t^2} \qquad (9.2.15)$$

Therefore, we can write

$$I = \frac{1}{3c^2}\frac{\partial v_k}{\partial t}\iint d^3\mathbf{x}\,d^3\boldsymbol{\xi}\,\frac{\rho(\mathbf{x},t)\rho(\mathbf{x}-\boldsymbol{\xi},t)}{|\boldsymbol{\xi}|} - \frac{1}{3}\frac{\partial^2 v_k}{\partial t^2}\frac{e^2}{c^3} \qquad (9.2.16)$$

Integral II. Let us now consider the integral

$$II = -\iint d^3\mathbf{x}\,d^3\boldsymbol{\xi}\,\rho(\mathbf{x},t)\frac{1}{|\boldsymbol{\xi}|}\frac{1}{c^2}\left[\frac{\partial j_k(\mathbf{x}-\boldsymbol{\xi},t')}{\partial t}\right]_{t'=t-(|\boldsymbol{\xi}|/c)}$$

We have

$$\left[\frac{\partial j_k(\mathbf{x}-\boldsymbol{\xi},t')}{\partial t}\right]_{t'=t-(|\boldsymbol{\xi}|/c)}$$

$$= \frac{\partial}{\partial t}\left[v_k\left(t-\frac{|\boldsymbol{\xi}|}{c}\right)\rho\left(\mathbf{x}-\boldsymbol{\xi},t-\frac{|\boldsymbol{\xi}|}{c}\right)\right]$$

$$= \frac{\partial}{\partial t}\left\{\left[v_k(t)-\frac{|\boldsymbol{\xi}|}{c}\frac{\partial v_k(t)}{\partial t}+\cdots\right]\right.$$

$$\left.\times\left[\rho(\mathbf{x}-\boldsymbol{\xi},t)-\frac{|\boldsymbol{\xi}|}{c}\frac{\partial\rho(\mathbf{x}-\boldsymbol{\xi},t)}{\partial t}+\cdots\right]\right\}$$

$$= \frac{\partial v_k(t)}{\partial t}\rho(\mathbf{x}-\boldsymbol{\xi},t)-\frac{|\boldsymbol{\xi}|}{c}\frac{\partial^2 v_k(t)}{\partial t^2}\rho(\mathbf{x}-\boldsymbol{\xi},t)+\cdots$$

Then

$$II = -\iint d^3\mathbf{x}\,d^3\boldsymbol{\xi}\,\frac{1}{|\boldsymbol{\xi}|}\frac{1}{c_2}\left[\frac{\partial v_k}{\partial t}\rho(\mathbf{x}-\boldsymbol{\xi},t)\right.$$

$$\left.-\frac{|\boldsymbol{\xi}|}{c}\frac{\partial^2 v_k}{\partial t^2}\rho(\mathbf{x}-\boldsymbol{\xi},t)\right]\rho(\mathbf{x},t)$$

$$= -\iint d^3\mathbf{x}\,d^3\boldsymbol{\xi}\,\rho(\mathbf{x},t)\frac{1}{c^2}\left[\frac{1}{|\boldsymbol{\xi}|}\rho(\mathbf{x}-\boldsymbol{\xi},t)\frac{\partial v_k}{\partial t}\right.$$

$$- \frac{1}{c}\rho(\mathbf{x} - \boldsymbol{\xi}, t)\frac{\partial^2 v_k}{\partial t^2}\Bigg]$$

$$= -\frac{1}{c^2}\frac{\partial v_k}{\partial t} \iint d^3x d^3\boldsymbol{\xi} \frac{\rho(\mathbf{x}, t)\rho(\mathbf{x} - \boldsymbol{\xi}, t)}{|\boldsymbol{\xi}|}$$

$$+ \frac{1}{c^3}\frac{\partial^2 v_k}{\partial t^2} \iint d^3x d^3\boldsymbol{\xi} \rho(\mathbf{x}, t)\rho(\mathbf{x} - \boldsymbol{\xi}, t)$$

$$= -\frac{1}{c^2}\frac{\partial v_k}{\partial t} \iint d^3x d^3\boldsymbol{\xi} \frac{\rho(\mathbf{x}, t)\rho(\mathbf{x} - \boldsymbol{\xi}, t)}{|\boldsymbol{\xi}|} + \frac{e^2}{c^3}\frac{\partial^2 v_k}{\partial t^2} \qquad (9.2.17)$$

Therefore,

$$F_k^{\text{self}}(t) = \text{I} + \text{II}$$

$$= \frac{1}{3c^2}\frac{\partial v_k}{\partial t} \iint d^3x d^3\boldsymbol{\xi} \frac{\rho(\mathbf{x}, t)\rho(\mathbf{x} - \boldsymbol{\xi}, t)}{|\boldsymbol{\xi}|} - \frac{e^2}{3c^3}\frac{\partial^2 v_k}{\partial t^2}$$

$$- \frac{1}{c^2}\frac{\partial v_k}{\partial t} \iint d^3x d^3\boldsymbol{\xi} \frac{\rho(\mathbf{x}, t)\rho(\mathbf{x} - \boldsymbol{\xi}, t)}{|\boldsymbol{\xi}|} + \frac{e^2}{c^3}\frac{\partial^2 v_k}{\partial t^2}$$

$$= \left(\frac{1}{3c^2} - \frac{1}{c^2}\right)\frac{\partial v_k}{\partial t} \iint d^3x d^3\boldsymbol{\xi} \frac{\rho(\mathbf{x}, t)\rho(\mathbf{x} - \boldsymbol{\xi}, t)}{|\boldsymbol{\xi}|}$$

$$+ \left(1 - \frac{1}{3}\right)\frac{e^2}{c^3}\frac{\partial^2 v_k}{\partial t^2}$$

$$= -\frac{4}{3}\frac{dv_k}{dt}\frac{U_{\text{el}}}{c^2} + \frac{2}{3}\frac{e^2}{c^3}\frac{d^2 v_k}{dt^2}$$

or

$$F_k^{\text{self}}(t) = -\frac{4}{3}\frac{dv_k}{dt}\frac{U_{\text{el}}}{c^2} + \frac{2}{3}\frac{e^2}{c^3}\frac{d^2 v_k}{dt^2} \qquad (9.2.18)$$

Where

$$U_{\text{el}} = \frac{1}{2} \iint \frac{\rho(\mathbf{x}, t)\rho(\mathbf{x} - \boldsymbol{\xi}, t)}{|\boldsymbol{\xi}|} d^3x d^3\boldsymbol{\xi} \qquad (9.2.19)$$

is the *electrostatic self-energy*.

In the expression for the self-force we have:

(1) A term proportional to the acceleration,

$$-\frac{4}{3}\frac{U_{\text{el}}}{c^2}\frac{dv_k}{dt}$$

(2) A term proportional to the derivative of the acceleration with respect to time.

Note that U_{el}/c^2 has the dimension of a mass; we shall call it the *electromagnetic mass* of the particle.

Equation of Motion in the Case $v \ll c$. This equation is given by

$$m_0 \frac{d\mathbf{v}}{dt} = e\left(\mathbf{E}^{ext} + \frac{\mathbf{v}}{c} \times \mathbf{B}^{ext}\right) + \mathbf{F}_{self}$$

$$= e\left(\mathbf{E}^{ext} + \frac{\mathbf{v}}{c} \times \mathbf{B}^{ext}\right) - \frac{4}{3}\frac{U_{el}}{c^2}\frac{d\mathbf{v}}{dt} + \frac{2e^2}{3c^3}\frac{d^2\mathbf{v}}{dt^2} \qquad (9.2.20)$$

where m_0 = rest mass of the electron. We can write

$$\left(m_0 + \frac{4}{3}\frac{U_{el}}{c^2}\right)\frac{d\mathbf{v}}{dt} = e\left(\mathbf{E}^{ext} + \frac{\mathbf{v}}{c} \times \mathbf{B}^{ext}\right) + \frac{2e^2}{3c^3}\ddot{\mathbf{v}} \qquad (9.2.21)$$

We call

$$m_{exp} = m_0 + \frac{4}{3}\frac{U_{el}}{c^2}$$

the *observed* or *experimental mass*. Then

$$m_{exp}\frac{d\mathbf{v}}{dt} = e\left(\mathbf{E}^{ext} + \frac{\mathbf{v}}{c} \times \mathbf{B}^{ext}\right) + \frac{2e^2}{3c^3}\ddot{\mathbf{v}} \qquad (9.2.22)$$

The term $(2e^2/3c^3)\ddot{\mathbf{v}}$ is called the *radiation damping term*.

We note here that, since we have assumed the particle instantaneously at rest, and the charge distribution rigid and spherically symmetrical, the above results are applicable only to the nonrelativistic motion of charged particles and do not have the proper Lorentz-transformation properties.

It is of interest to consider the form that Eq. (9.2.22) takes in the absence of an external force:

$$\dot{\mathbf{v}} = \frac{2}{3}\frac{e^2}{mc^3}\ddot{\mathbf{v}} \qquad (9.2.23)$$

where m stands for m_{exp}, or

$$\dot{\mathbf{v}} = \tau\ddot{\mathbf{v}} \qquad (9.2.24)$$

where

$$\tau = \frac{2e^2}{3mc^3} = 6.25 \times 10^{-24}\text{s} \qquad (9.2.25)$$

The solution of (9.2.24) is

$$\mathbf{v} = \mathbf{v}_0 + \mathbf{v}_1 e^{t/\tau} \qquad (9.2.26)$$

where $\mathbf{v}_0$ and $\mathbf{v}_1$ are constant vectors and Eq. (9.2.26) is called the *runaway solution*. It predicts that, in the absence of an applied external force, the charged particle, starting with a finite velocity, self-accelerates to infinite velocity. This paradox can be resolved by assuming that the expression for the self-force applies only when there is an external force acting on the particle that is large compared to the self-force, and the energy lost by the particle, because of the self-force, is small in comparison to the total energy of the particle.

Let us make some observations regarding the self-force:

(1) The damping force is not complete as it is written in Eq. (9.2.22). If the velocity of the particle is great, other terms must be added. This can be seen by looking at Figs. 9.2a and 9.2b. In the first case we have, in the various time intervals,

$0 - t_0$: radiation damping and radiation

$t_0 - t_1$: no radiation damping, but radiation present

$t_1 - t_2$: radiation damping and radiation

In the second case, we have radiation damping and radiation in the interval $0 - t_0$, and radiation, but *not* radiation damping, for $t \geq t_0$.

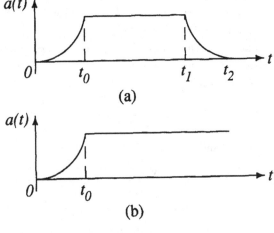

(a)

(b)

FIGURE 9.2

Clearly, if the acceleration is constant after t_0, as the velocity grows, the neglected terms in the radiation damping expression become relevant.

(2) To have an idea of the order of magnitude of the radiation damping, let us make the following calculation:

$$\frac{2}{3}\frac{e^2}{c^3}\frac{d^2}{dt^2}v = \frac{2}{3}\frac{e^2}{mc^2}\frac{1}{c}\frac{d}{dt}m\frac{dv}{dt}$$

$$= \frac{2}{3}\frac{e^2}{mc^2}\frac{1}{c}\frac{d\,\text{force}}{dt}$$

$$= \frac{2 \times (4.8 \times 10^{-10})^2}{3 \times 9.1 \times 10^{-28} \times (3 \times 10^{10})^3}\frac{dF}{dt}$$

$$= 6.25 \times 10^{-24}\frac{dF}{dt} \tag{9.2.27}$$

The radiation damping term indeed seems very small. We shall see that there is a situation in which it is not negligible: the case of forced oscillations.

(3) Radiation damping, dependent on the third derivative of the position, makes a distinction between the two directions of time.

Newton's law of motion does not change if we change t into $-t$, and all the laws of motion do not change if we change t into $-t$.

The cause of the dependence of the radiation damping on the sign of time is the use we have made of retarded potentials.

If we had used a symmetrical combination of retarded and advanced potentials, we would have obtained a t dependence such that a change of t into $-t$ would not have affected the radiation damping.

(4) The energy U_{el} can be easily calculated if we assume that the charge is distributed over a sphere of radius r_0. In such a case

$$U_{\text{el}} = \frac{1}{8\pi}\int E^2 d^3\mathbf{x} = \frac{1}{8\pi}\int_{r_0}^{\infty}\frac{e^2}{r^4}4\pi r^2 dr = \frac{1}{2}\frac{e^2}{r_0} \tag{9.2.28}$$

The electromagnetic mass of the electron according to this model is given by

$$m_{\text{el}} = \frac{4}{3}\frac{U_{\text{el}}}{c^2} = \frac{4}{3}\frac{e^2}{2r_0}\frac{1}{c^2} = \frac{2}{3}\frac{e^2}{r_0 c^2} \tag{9.2.29}$$

The electromagnetic mass is the greater, the smaller the radius r_0. By properly choosing r_0, we could explain the mass of any charged particle as an electromagnetic mass.

Formula (9.2.29) allows us to calculate the radius the electron would have if its total mass, as deduced from experiments on the deviation of the electron's trajectory under the action of a field, were entirely of electromagnetic origin:

$$r_0 = \frac{2}{3}\frac{e^2}{mc^2} = \frac{2 \times (4.8 \times 10^{-10})^2}{3 \times 9.1 \times 10^{-28} \times 9 \times 10^{20}}$$

$$= 1.9 \times 10^{-13}\,\text{cm} \qquad (9.2.30)$$

This quantity is called the *classical radius of the electron.*

On the other hand, if the mass of the proton were entirely of electromagnetic origin, the proton would have a radius *smaller* than that of the electron and equal to about

$$\frac{1}{2000}r_0 = \frac{1.9 \times 10^{-13}}{2000} \simeq 10^{-16}\text{cm!} \qquad (9.2.31)$$

The subject of the structure of the electron is open to investigation. A unique mode of finding the structure of a particle is by shooting at it other high-energy probing particles and studying the deviations from pure Coulomb scattering. The closer the probing particle goes to the particle under investigation, the greater is the amount of information that becomes available regarding the latter particle. What counts is the energy in the center of mass system, and this energy has not been enough to give evidence of the structure of the electron. However, the present experimental limit for the radius of the electron is below the value of r_0 found in Eq. (9.2.30).

There is evidence of a structure of the proton to which a radius of 0.8×10^{-13} cm has been given.

9.3. Energy Loss by Radiation. Application to Periodic Motion

Consider the case of an electron that oscillates between two position $\pm z_0$ with a periodic motion of frequency ω:

$$
\begin{aligned}
z &= z_0 \cos \omega t \\
\dot{z} &= -z_0\omega \sin \omega t = v \\
\ddot{z} &= -z_0\omega^2 \cos \omega t = a = \dot{v} \\
\dddot{z} &= z_0\omega^3 \sin \omega t = \dot{a} = \ddot{v}
\end{aligned}
\qquad (9.3.1)
$$

Then

$$F^{\text{self}} = \frac{2e^2}{3c^3}\,\dddot{z} = \frac{2e^2}{3c^3}\omega^3 z_0 \sin \omega t \qquad (9.3.2)$$

The work done by this force in the unit time is

$$F^{\text{self}} \cdot v = \frac{2e^2}{3c^3}\omega^3 z_0 \sin \omega t \cdot (-z_0\omega \sin \omega t)$$

$$= -\frac{2e^2}{3c^3}\omega^4 z_0^2 \sin^2 \omega t \qquad (9.3.3)$$

The average work done by F^{self} in the unit time is

$$-\frac{2e^2}{3c^3}\omega^4 z_0^2 \overline{\sin^2 \omega t} = -\frac{e^2}{3c^3}\omega^4 z_0^2 \qquad (9.3.4)$$

This quantity corresponds to the energy radiated by the particle in the unit time, as can be seen comparing Eq. (9.3.4) with Eq. (8.3.17).

9.4. Forced Vibrations

Let us consider now a particle of charge e, elastically bound to an equilibrium position. Let

$$\omega_0 = \sqrt{\frac{K}{m}} \qquad (9.4.1)$$

be the frequency of free oscillations, with K = force constant and m = mass of the particle.

Under the action of a radiation field, the particle will be displaced from equilibrium; a consequence of this displacement is the creation of a *dipole moment*. This moment may not be in the direction of the **E** field because of some anisotropy, but we shall neglect this possibility.

The equation of motion of the particle will be

$$\ddot{\mathbf{z}} = -\frac{K}{m}\mathbf{z} + \frac{\mathbf{F}}{m} + \frac{\mathbf{F}^{\text{self}}}{m} \qquad (9.4.2)$$

where **F** is the force due to the **E** field. We may write

$$\ddot{\mathbf{z}} + \omega_0^2\mathbf{z} = \frac{e}{m}\mathbf{E}(t) + \frac{2e^2}{3mc^3}\dddot{\mathbf{z}} \qquad (9.4.3)$$

We shall set

$$\mathbf{E}(t) = \mathbf{E}_0 e^{i\omega t} \tag{9.4.4}$$

$$\mathbf{z}(t) = \mathbf{z}_0 e^{i\omega t} \tag{9.4.5}$$

Then

$$\ddot{\mathbf{z}} = -\omega^2 \mathbf{z}_0 e^{i\omega t} \tag{9.4.6}$$

$$\dddot{\mathbf{z}} = -i\omega^3 \mathbf{z}_0 e^{i\omega t} \tag{9.4.7}$$

and Eq. (9.4.3) gives

$$(-\omega^2 + \omega_0^2)\mathbf{z}_0 e^{i\omega t} = \frac{e}{m}\mathbf{E}_0 e^{i\omega t} - \frac{2e^2}{3mc^3}i\omega^3 \mathbf{z}_0 e^{i\omega t} \tag{9.4.8}$$

$$\left(\omega_0^2 - \omega^2 + i\frac{2e^2}{3mc^3}\omega^3 \right) \mathbf{z}_0 = \frac{e}{m}\mathbf{E}_0 \tag{9.4.9}$$

Then if

$$\mathbf{p}_0 e^{i\omega t} = e\mathbf{z} = e\mathbf{z}_0 e^{i\omega t} \tag{9.4.10}$$

$$\mathbf{p}_0 = e\mathbf{z}_0 = \frac{(e^2/m)\mathbf{E}_0}{\omega_0^2 - \omega^2 + i(2e^2/3mc^3)\omega^3} = \alpha(\omega)\mathbf{E}_0 \tag{9.4.11}$$

where we define

$$\alpha(\omega) = polarizability = \frac{e^2/m}{\omega_0^2 - \omega^2 + i(2e^2/3mc^3)\omega^3} \tag{9.4.12}$$

The damping force makes $\alpha(\omega)$ complex.

How about the magnetic force? The magnetic field in a plane wave has (in the gaussian system of units) the same value as the $\mathbf{E}$ field. The order of magnitude of the magnetic force is v/c times that of the electric force. But the velocity is

$$v \simeq \omega z_0$$

and

$$\frac{v}{c} = \frac{\omega z_0}{c} = \frac{2\pi z_0}{\lambda}$$

Using $z_0 = 10^{-8}$ cm and $\lambda = 1000 \times 10^{-8}$ cm, we have

$$\frac{v}{c} = 10^{-3} \tag{9.4.13}$$

and the magnetic force is negligible. If we use x rays, the situation is different.

Let us consider now the ω-dependent factor in $\alpha(\omega)$. Let $A = (2e^2/3mc^3)w^3$; it is

$$\frac{1}{\omega_0^2 - \omega^2 + iA} = \frac{\omega_0^2 - \omega^2}{(\omega_0^2 - \omega^2)^2 + A^2} - i\frac{A}{(\omega_0^2 - \omega^2)^2 + A^2}$$

$$= \frac{1}{\sqrt{(\omega_0^2 - \omega^2)^2 + A^2}}\left[\frac{\omega_0^2 - \omega^2}{\sqrt{(\omega_0^2 - \omega^2)^2 + A^2}}\right.$$

$$\left. - i\frac{A}{\sqrt{(\omega_0^2 - \omega^2)^2 + A^2}}\right] \tag{9.4.14}$$

Therefore,

$$\alpha(\omega)\mathbf{E}_0 e^{i\omega t} = \frac{e^2}{m}\frac{\mathbf{E}_0 e^{i\omega t}}{\sqrt{(\omega_0^2 - \omega^2)^2 + A^2}}\left[\frac{\omega_0^2 - \omega^2}{\sqrt{(\omega_0^2 - \omega^2)^2 + A^2}}\right.$$

$$\left. - i\frac{A}{\sqrt{(\omega_0^2 - \omega^2)^2 + A^2}}\right] \tag{9.4.15}$$

Taking the real part of expression (9.4.15), we obtain the induced electric dipole:

$$\mathbf{p}(t) = \text{Re}[\alpha(\omega)\mathbf{E}_0 e^{i\omega t}] = \text{Re}\left\{\frac{e^2}{m}\frac{\mathbf{E}_0}{\sqrt{(\omega_0^2 - \omega^2)^2 + A^2}}\right.$$

$$\times(\cos\omega t + i\sin\omega t)\left[\frac{\omega_0^2 - \omega^2}{\sqrt{(\omega_0^2 - \omega^2)^2 + A^2}}\right.$$

$$\left.\left. - i\frac{A}{\sqrt{(\omega_0^2 - \omega^2)^2 + A^2}}\right]\right\}$$

$$= \frac{e^2}{m}\frac{\mathbf{E}_0}{\sqrt{(\omega_0^2 - \omega^2)^2 + A^2}}\left[\cos\omega t\frac{\omega_0^2 - \omega^2}{\sqrt{(\omega_0^2 - \omega^2)^2 + A^2}}\right.$$

$$\left. + \sin\omega t\frac{A}{\sqrt{(\omega_0^2 - \omega^2)^2 + A^2}}\right]$$

$$= \frac{e^2}{m} \frac{\mathbf{E}_0}{\sqrt{(\omega_0^2 - \omega^2)^2 + A^2}} [\cos \omega t \cos \phi + \sin \omega t \sin \phi]$$

$$= \frac{e^2}{m} \frac{\mathbf{E}_0}{\sqrt{(\omega_0^2 - \omega^2)^2 + A^2}} \cos(\omega t - \phi) \qquad (9.4.16)$$

where

$$\cos \psi - \frac{\omega_0^2 - \omega^2}{\sqrt{(\omega_0^2 - \omega^2)^2 + A^2}} \qquad (9.4.17)$$

$$\sin \phi = \frac{A}{\sqrt{(\omega_0^2 - \omega^2)^2 + A^2}} \qquad (9.4.18)$$

$$\tan \phi = \frac{A}{\omega_0^2 - \omega^2} = \frac{(2/3)(e^2/mc^3)\omega^3}{\omega_0^2 - \omega^2} \qquad (9.4.19)$$

Therefore,

$$\begin{aligned} \phi &= 0, \quad \text{for } \omega = 0 \\ \phi &= \frac{\pi}{2}, \quad \text{for } \omega = \omega_0 \\ \phi &= \pi, \quad \text{for } \omega \gg \omega_0 \text{ (if } A \text{ remains small)} \end{aligned} \qquad (9.4.20)$$

Note that

$$\frac{2}{3} \frac{e^2}{mc^3} = \frac{2 \times (4.8 \times 10^{-10})^2}{3 \times 9.1 \times 10^{-28} \times (3 \times 10^{10})^3} = 6.25 \times 10^{-24}$$

9.5. Scattering of Radiation

A sinusoidal electric field $\mathbf{E}_0 e^{i\omega t}$ acting on a polarizable system, consisting of a charged particle elastically bound to a point, induces in the system a dipole moment that is proportional to the field and depends on the frequencies ω of the field and ω_0 of the natural oscillations of the particle. This dipole is given by

$$\mathbf{p}(t) = \text{Re}[\alpha(\omega)\mathbf{E}(t)] = \text{Re}\left[\frac{e^2}{m} \frac{1}{\omega_0^2 - \omega^2 + iA} \mathbf{E}_0 e^{i\omega t}\right]$$

$$= \frac{(e^2/m)\mathbf{E}_0}{\sqrt{(\omega_0^2 - \omega^2)^2 + A^2}} \cos(\omega t - \phi) \qquad (9.5.1)$$

where

$$A = \frac{2}{3}\frac{e^2\omega^3}{mc^3} \qquad (9.5.2)$$

$$\tan\phi = \frac{(2e^3/3mc^3)\omega^3}{\omega_0^2 - \omega^2} \qquad (9.5.3)$$

$$\alpha(\omega) = \text{polarizability} = \frac{e^2/m}{\omega_0^2 - \omega^2 + i(2e^2\omega^3)/(3mc^3)} \qquad (9.5.4)$$

The vibrating dipole induced by the field will produce radiation. We shall calculate the amount of radiation emitted by the dipole in the elementary solid angle $d\omega$ (see Fig. 9.3). We shall use for the fields the formulas derived in Sec. 8.3:

$$\mathbf{E}^{\text{rad}}(\mathbf{x}, t) = -\left\{\frac{e}{c^2 r}[\mathbf{a} - \mathbf{n}(\mathbf{n} \cdot \mathbf{a})]\right\}_{t_p=t-(r/c)}$$

$$= \left\{\frac{e}{c^2 r}[(\mathbf{a} \times \mathbf{n}) \times \mathbf{n}]\right\}_{t_p=t-(r/c)} \qquad (9.5.5)$$

$$\mathbf{B}^{\text{rad}}(\mathbf{x}, t) = \left\{\frac{e}{c^2 r}[\mathbf{a} \times \mathbf{n}]\right\}_{t_p=t-(r/c)}$$

Dropping the superscript rad, we write

$$\mathbf{n} \times \mathbf{E} = -\frac{e}{c^2 r}\mathbf{n} \times \mathbf{a} = \frac{e}{c^2 r}\mathbf{a} \times \mathbf{n} = \mathbf{B} \qquad (9.5.6)$$

$$\mathbf{B} \times \mathbf{n} = \frac{e}{c^2 r}[(\mathbf{a} \times \mathbf{n}) \times \mathbf{n}] = \mathbf{E} \qquad (9.5.7)$$

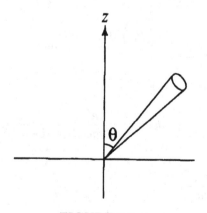

FIGURE 9.3

Also, from Eq. (8.3.7),

$$\mathbf{N} = \frac{c}{4\pi}\mathbf{E} \times \mathbf{B} = \mathbf{n}\frac{e^2}{4\pi c^3 r^2}(\mathbf{a} \times \mathbf{n})^2$$

$$= \mathbf{n}\frac{1}{r^2}\frac{\sin^2\theta}{4\pi}\frac{e^2 a^2}{c^3} \tag{9.5.8}$$

The total power emitted in the solid angle $d\Omega$ is

$$Nr^2 d\Omega = \frac{1}{c^3}\frac{d\Omega}{4\pi}\sin^2\theta(ea)^2 = \frac{1}{c^3}\frac{d\Omega}{4\pi}\sin^2\theta(\ddot{p})^2 \tag{9.5.9}$$

where

$$p(t) = ez(t) \tag{9.5.10}$$

But

$$p(t) = \frac{e^2}{m}\frac{E_0}{\sqrt{(\omega_0^2 - \omega^2)^2 + A^2}}\cos(\omega t - \phi) \tag{9.5.11}$$

$$\ddot{p}(t) = -\frac{e^2}{m}\frac{\omega^2 E_0}{\sqrt{(\omega_0^2 - \omega^2)^2 + A^2}}\cos(\omega t - \phi) \tag{9.5.12}$$

Taking the time average of the square of $\ddot{p}(t)$,

$$\overline{[\ddot{p}(t)]^2} = \frac{e^4}{m^2}\frac{\omega^4 E_0^2}{\sqrt{(\omega_0^2 - \omega^2)^2 + A^2}}\frac{1}{2} \tag{9.5.13}$$

Therefore,

$$\overline{N}r^2 d\Omega = \frac{1}{c^3}\frac{d\Omega}{4\pi}\sin^2\theta\frac{1}{2m^2}\frac{e^4\omega^4 E_0^2}{(\omega_0^2 - \omega^2)^2 + A^2}$$

$$= \left(\frac{c}{4\pi}\frac{1}{2}E_0^2\right)d\Omega\sin^2\theta\left(\frac{e^2}{mc^2}\right)^2$$

$$\times \frac{\omega^4}{(\omega_0^2 - \omega^2)^2 + A^2}$$

$$= \overline{N}_0 d\Omega\sin^2\theta\left(\frac{e^2}{mc^2}\right)^2\frac{\omega^4}{(\omega_0^2 - \omega^2)^2 + A^2} \tag{9.5.14}$$

where

$$A = \frac{2e^2\omega^3}{3mc^3} \tag{9.5.15}$$

$$\overline{N}_0 = \text{average flux per unit area of incident radiation} \tag{9.5.16}$$

We define the *differential cross section*:

$$d\sigma = \frac{\text{flux into } d\Omega}{\text{flux per unit area of incident radiation}}$$

$$= \frac{\overline{N} r^2 d\Omega}{\overline{N}_0} = d\Omega \sin^2 \theta r_e^2 f(\omega) \qquad (9.5.17)$$

where

$$r_e = \frac{e^2}{mc^2} = \frac{(4.8 \times 10^{-10})^2}{(9.1 \times 10^{-28})(9 \times 10^{20})} = 2.8 \times 10^{-13}\,\text{cm} \qquad (9.5.18)$$

$$f(\omega) = \frac{\omega^4}{(\omega_0^2 - \omega^2)^2 + \left(\dfrac{2e^2\omega^3}{3mc^3}\right)^2} \qquad (9.5.19)$$

The total cross section is given by

$$\sigma(\omega) = \int \frac{d\sigma}{d\Omega} d\Omega = r_e^2 f(\omega) \int d\Omega \sin^2 \theta = \frac{8\pi}{3} r_e^2 f(\omega) \qquad (9.5.20)$$

Let us examine the magnitude of the radiation damping term:

$$R_d = A = \frac{2e^2\omega^2}{3mc^3} = \frac{2}{3}\frac{e^2}{mc^2}\frac{\omega}{c}\omega^2 = \frac{2}{3}r_e\frac{2\pi}{\lambda}\omega^2 = \frac{4\pi}{3}\frac{r_e}{\lambda}\omega^2 \qquad (9.5.21)$$

This term may become great even if λ is large with respect to r_e. For λ to be on the order of r_e, we must have γ radiation.

Let us now consider three different cases for scattering based on the frequency range.

9.5.1. *Rayleigh Scattering*

$$R_d \le \omega^2 \ll \omega_0^2 \qquad (9.5.22)$$

ω is in the visible range

$$\lambda = 5000\,\text{Å} = 5 \times 10^{-5}\,\text{cm}$$

$$\frac{r_e}{\lambda} = \frac{2.8 \times 10^{-13}}{5 \times 10^{-5}} = 5.6 \times 10^{-9}$$

Some substances have many strong resonance frequencies in the ultraviolet; when visible light strikes these substances, the condition (9.5.22) may be verified. We have in this case

$$f(\omega) \simeq \frac{\omega^4}{\omega_0^4} = \left(\frac{\omega}{\omega_0}\right)^4 \qquad (9.5.23)$$

and

$$d\sigma = r_e^2 \sin^2 \theta \left(\frac{\omega}{\omega_0} \right)^4 d\Omega \qquad (9.5.24)$$

The total cross section is given by

$$\sigma(\omega) = \int \frac{d\sigma}{d\Omega} d\Omega = \frac{8\pi}{3} r_e^2 \left(\frac{\omega}{\omega_0} \right)^4 \qquad (9.5.25)$$

This type of scattering is called *Rayleigh scattering*. The sky is blue because of the preferential scattering of light in the blue region.

9.5.2. *Thomson Scattering*

For Thomson scattering

$$\omega \gg \omega_0 = 0 \qquad (9.5.26)$$

This is the case of free electrons, that is, electrons not subjected to any restoring force. We can, therefore, disregard R_d for any electromagnetic radiation for which $\lambda \gg r_e$. We have then

$$f(\omega) \simeq 1 \qquad (9.5.27)$$

and

$$d\sigma = r_e^2 \sin^2 \theta d\Omega \qquad (9.5.28)$$

Also,

$$\sigma(\omega) = \int \frac{d\sigma}{d\Omega} d\Omega = \int r_e^2 \sin^2 \theta d\Omega = \frac{8\pi}{3} r_e^2 \qquad (9.5.29)$$

This type of scattering is independent of frequency and is called *Thomson scattering*.

The total cross section is on the order of

$$\frac{8\pi}{3} \times (2.8 \times 10^{-13})^2 = 6.6 \times 10^{-25} \, \text{cm}^2$$

These considerations are valid for electrons that are almost free. In atoms, electrons are quantum mechanically bound. However, if the energy $h\nu$ of the incident radiation is great with respect to the binding energy of the electron, the electron behaves as if it were free. At very high frequencies, the quantity $h\nu(h = \text{Planck's constant}, \nu = \omega/2\pi)$ may approach the value

$$mc^2 = 9.1 \times 10^{-28} \times 9 \times 10^{20} = 8.19 \times 10^{-7} \text{erg}$$

and the wavelength the value

$$\lambda = \frac{c}{\nu} = \frac{hc}{h\nu} = \frac{hc}{mc^2} = \frac{6.625 \times 10^{-27} \times 3 \times 10^{10}}{8.19 \times 10^{-7}}$$

$$= 2.43 \times 10^{-10} \, \text{cm} = 0.024 \text{Å}$$

When such is the case, we are in the region of *Compton scattering*, where quantum mechanical modifications become important. Thomson scattering is a classical (nonquantistic) effect and is valid at lower frequencies.

9.5.3. *Resonance Scattering*

For resonance scattering,

$$\omega \simeq \omega_0 \qquad\qquad (9.5.30)$$

and

$$f(\omega) = \frac{\omega^4}{(\omega_0^2 - \omega^2)^2 + \left(\dfrac{2e^2\omega^3}{3mc^3}\right)^2}$$

$$= \frac{\omega_0^4}{(\omega_0 - \omega)^2(\omega_0 + \omega)^2 + \left(\dfrac{2e^2\omega^3}{3mc^3}\right)^2}$$

$$\simeq \frac{\omega_0^2}{4(\omega_0 - \omega)^2 + \left(\dfrac{2e^2\omega_0^2}{3mc^3}\right)^2} = \frac{\omega_0^2/4}{(\omega_0 - \omega)^2 + \frac{1}{4}\gamma^2} \qquad (9.5.31)$$

where

$$\gamma = \frac{2e^2\omega_0^2}{3mc^3} = \frac{2}{3}\left(\frac{e^2}{mc^2}\right)\frac{\omega}{c}\omega_0 = \frac{4\pi}{3}\frac{r_e}{\lambda}\omega_0 \qquad (9.5.32)$$

Then

$$d\sigma = r_e^2 \sin^2\theta d\Omega f(\omega) = r_e^2 \sin^2\theta d\Omega \frac{\omega_0^2/4}{(\omega_0 - \omega)^2 + \frac{1}{4}\gamma^2}$$

$$= \frac{1}{4}r_e^2 \sin^2\theta d\Omega \frac{\omega_0^2/(\gamma/2)^2}{1 + [(\omega - \omega_0)/(\gamma/2)]^2} \qquad (9.5.33)$$

The relative intensity falls from its maximum value at ω_0 symmetrically to half-value at the frequency

$$\omega = \omega_0 \pm \frac{\gamma}{2} \tag{9.5.34}$$

The ratio of the width of the resonance line to the resonance frequency is given by

$$\frac{\gamma}{\omega_0} = \frac{4\pi}{3} \frac{r_e}{\lambda} \simeq \frac{10^{-13}}{10^{-5}} 10^{-8} \tag{9.5.35}$$

Also

$$f(\omega_0) = \left(\frac{\omega_0}{\gamma}\right)^2 = \left(\frac{3}{4\pi} \frac{\lambda}{r_e}\right)^2 \tag{9.5.36}$$

and

$$d\sigma(\omega = \omega_0) = d\Omega \sin^2 \theta r_e^2 f(\omega_0)$$

The total cross section is given by

$$\sigma(\omega) = \int \frac{d\sigma}{d\Omega} d\Omega = \frac{8\pi}{3} r_e^2 f(\omega) \tag{9.5.37}$$

Then

$$\sigma(\omega_0) = \frac{8\pi}{3} r_e^2 \left(\frac{3}{4\pi} \frac{\lambda}{r_e}\right)^2 = \frac{3}{2\pi} \lambda^2 \tag{9.5.38}$$

$\sigma(\omega_0)$ depends only on the wavelength of the incident radiation. It is $\sim 10^{16}$ times larger than the Thomson scattering area.

Table 9.1 and Fig. 9.4 summarize these results; the sketch in the figure is only indicative of the behavior of $\sigma(\omega)$ versus ω.

Table 9.1.

	$d\sigma$	$\sigma(\omega)$
$\omega \ll \omega_0$: Rayleigh scattering	$r_e^2 \sin^2 \theta d\Omega \left(\dfrac{\omega}{\omega_0}\right)^4$	$\dfrac{8\pi}{3} r_e^2 \left(\dfrac{\omega}{\omega_0}\right)^4$
$\omega \simeq \omega_0$: Resonance scattering $\left(\gamma = \dfrac{4\pi}{3}\dfrac{r_e}{\lambda}\omega_0\right)$	$r_e^2 \sin^2 \theta d\Omega \dfrac{\omega_0^2/4}{(\omega_0 - \omega)^2 + \frac{1}{4}\gamma^2}$	$\dfrac{8\pi}{3} r_e^2 \dfrac{\omega_0^2/4}{(\omega_0 - \omega)^2 + \frac{1}{4}\gamma^2}$ $\sigma(\omega_0) = \dfrac{3}{2\pi}\lambda^2$
$w \gg \omega_0$: Thomson scattering	$r_e^2 \sin^2 \theta d\Omega$	$\dfrac{8\pi}{3} r_e^2$

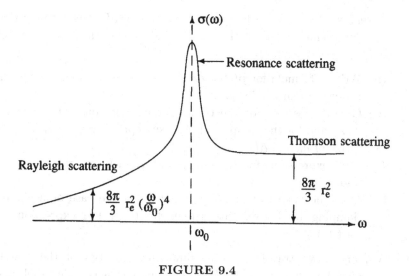

FIGURE 9.4

CHAPTER 9 EXERCISES

9.1. **(a)** Express the Minkowski force F_μ in terms of $\mathbf{F} = d\mathbf{p}/dt$ and $\mathbf{v}$ and show that

$$\sum_\mu F_\mu p_\mu = 0$$

where

$$p_\mu \equiv \left(\frac{m\mathbf{v}}{\sqrt{1-\beta^2}}, \frac{imc}{\sqrt{1-\beta^2}} \right)$$

(b) Show that

$$F_\mu^{\mathrm{rad}} = \frac{2e^2}{3mc^3} \left[\frac{d^2 p_\mu}{d\tau^2} - \frac{p_\mu}{m^2 c^2} \sum_v \left(\frac{dp_v}{d\tau} \right)^2 \right]$$

by verifying that

$$\sum_\mu F_\mu^{\mathrm{rad}} p_\mu = 0$$

and that F_μ^{rad} goes to the proper limit when $\mathbf{v} = 0$.

9.2. Consider a system of N independent oscillating mass points of charge e and mass m. Let $n_0 = N/V$ be the density of these points and w_0 the angular frequency of their natural oscillations. Assume that this system is interacting with a plane wave in which $\mathbf{E} = \mathbf{E}_0 e^{i\omega t}$ and that the displacements $z(t)$ of the charges induced by the field are much smaller than $\lambda = (2\pi c)/w$.

(a) Write a formula for $\mathbf{p}(t) = e\mathbf{z}(t)$ in complex form and define the complex polarizability $\alpha(\omega)$.

(b) Give expressions for the complex dielectric constant $K_c = K_r +iK_i$ and for the complex electric susceptibility $\chi_c = \chi_r + i\chi_i$ in terms of n_0 and $\alpha(\omega)$.

(c) Find expressions for K_r, K_i, χ_r and χ_i in terms of e, m, n_0, ω_0, and ω.

(d) Give the values of K_r, K_i, χ_r and χ_i in the case of negligible damping by finding the proper limits of the expressions in part (c).

9.3. A plane wave travels in a medium that consists of the system considered in Exercise 9.2. This medium has a complex dielectric constant $K_c = K_r + iK_i$ and a complex index of refraction

$$n = n_r + in_i = \sqrt{K_c}$$

The plane wave in the medium has the form

$$\mathbf{E}(\mathbf{x}, t) = \mathbf{E}_0 e^{-i\boldsymbol{\eta}\cdot\mathbf{r}+i\omega t}$$

where $\boldsymbol{\eta}$ = complex wave vector, and $\eta = \frac{\omega}{(c/n)}$.

(a) Find the formulas that relate n_r and n_i to K_r and K_i.

(b) Show that the intensity of the wave that is proportional to $|\mathbf{E}|^2$ drops off as $e^{-\mu(\omega)x}$, where

$$\mu(\omega) = -\frac{2\omega}{c}n_i = -\frac{\omega K_i}{cn_r}$$

and verify that n_i is intrinsically negative.

(c) Assume the damping negligible and show that n_0 plane wave propagation is possible for frequencies ω in the region

$$\omega_0 < \omega < \omega_L$$

where

$$\omega_L = \omega_0\sqrt{1 + \frac{4\pi n_0 e^2}{m\omega_0^2}}$$

9.4. (a) A sphere of radius R and dielectric constant K is placed in a uniform electric field. Find the induced electric dipole moment.

(b) The sphere described in part (a) is placed in a beam of electromagnetic radiation with long wavelength ($\lambda \gg R$). What is the functional dependence of the scattered radiation on the intensity and the frequency of the incident radiation?

9.5. The appearance of the factor $\frac{4}{3}$ in the formula for the electro-magnetic mass of the electron shows that the dynamical behavior of the electron cannot be explained by taking into account only its field. Present some hypothesis that may account for the presence of such a factor.

9.6. X-rays are produced by the electron bombardment of a solid anticathode. Assume an electron energy of less than $500\,\mathrm{keV}$.

(a) What is the ratio of the X-ray intensity at $90°$ to the direction of the electron beam to the intensity at $45°$ to this direction?

(b) How are the X-rays polarized?

9.7. Consider a classical radiating particle of mass m and charge e that is bound with a resonant angular frequency ω_0. A monochromatic electric field $E_0 e^{i\omega t}$ is switched on at time $t = 0$ and acts on this particle. The equation of motion is given by

$$\ddot{x} + \gamma\,\dot{x} + \omega_0^2 x = \frac{eE_0}{m} e^{i\omega t}$$

where $\gamma\dot{x}=$ radiation damping term and $\omega_0 \gg \gamma$. Assume that, at time $t = 0, x = \dot{x} = 0$

(a) Derive the expression

$$\sigma = \frac{3}{2\pi}\lambda^2$$

for the scattering cross section at resonance ($\omega = \omega_0$). λ is the wavelength of the radiation.

(b) Comment on the fact that the expression for σ in part (a) is independent of the charge of the particle and remains the same as $e \to 0$.

10

Radiation from Periodic Charge and Current Distributions

10.1. Multipole Expansion

A current or a charge distribution is periodic if it is of the following form:

$$\mathbf{j}(\mathbf{x}, t) = \mathbf{j}_0(\mathbf{x})e^{-i\omega t}$$
$$\rho(\mathbf{x}, t) = \rho_0(\mathbf{x})e^{-i\omega t}$$

$$(10.1.1)$$

This case is distinct from the case of a charge vibrating harmonically with frequency ω:

$$\mathbf{x}_p(t) = \mathbf{x}_0 \cos \omega t \qquad (10.1.2)$$

In this last case,

$$\mathbf{j}(\mathbf{x}, t) = e\dot{\mathbf{x}}_p \delta(\mathbf{x} - \mathbf{x}_p(t))$$

$$= -e\omega \mathbf{x}_0 \sin \omega t \delta(\mathbf{x} - \mathbf{x}_0 \cos \omega t) \qquad (10.1.3)$$

$$\rho(\mathbf{x}, t) = e\delta(\mathbf{x} - \mathbf{x}_p(t)) = e\delta(\mathbf{x} - \mathbf{x}_0 \cos \omega t)$$

and the current an charge do not depend on time as $e^{-i\omega t}$. Eventually, we can make the following expansion:

$$\mathbf{j}(\mathbf{x},t) = \sum_{n=1}^{\infty} \mathbf{j}_n(\mathbf{x})e^{-in\omega t}$$

$$\rho(\mathbf{x},t) = \sum_{n=1}^{\infty} \rho_n(\mathbf{x})e^{-in\omega t}$$

(10.1.4)

A harmonic oscillator can give dipole, quadrupole,..., radiation. In the present chapter we examine these elementary types of radiation separately.

If we want to be sure that the radiation has a certain frequency, we must assume that the sources are of the type $e^{-i\omega t}$. It is more convenient to describe the situation in terms of $\mathbf{j}$ and ρ rather than in terms of the motion of a particle.

Assume now that our sources are given by the expressions in Eqs. (10.1.1). We can write (see Fig. 10.1)

$$\mathbf{A}(\mathbf{x},t) = \frac{1}{c} \int d^3\mathbf{x}' \frac{\mathbf{j}\left(\mathbf{x}', t - \frac{|\mathbf{x}-\mathbf{x}'|}{c}\right)}{|\mathbf{x}-\mathbf{x}'|}$$

$$= \frac{1}{c} \int d^3\mathbf{x}' \frac{\mathbf{j}_0(\mathbf{x}')}{|\mathbf{x}-\mathbf{x}'|} \exp\left[-i\omega\left(t - \frac{|\mathbf{x}-\mathbf{x}'|}{c}\right)\right] \quad (10.1.5)$$

$$\phi(\mathbf{x},t) = \int d^3\mathbf{x}' \frac{\rho\left(\mathbf{x}', t - \frac{|\mathbf{x}-\mathbf{x}'|}{c}\right)}{|\mathbf{x}-\mathbf{x}'|}$$

$$= \int d^3\mathbf{x}' \frac{\rho_0(\mathbf{x}')}{|\mathbf{x}-\mathbf{x}'|} \exp\left[-i\omega\left(t - \frac{|\mathbf{x}-\mathbf{x}'|}{c}\right)\right] \quad (10.1.6)$$

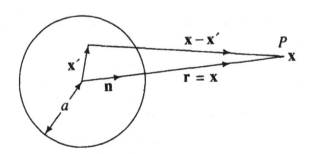

FIGURE 10.1

We shall assume that currents and charges are confined in a finite region of space of radius a, and that

$$r \gg a \qquad (10.1.7)$$

We can then write

$$|\mathbf{x} - \mathbf{x}'| = \sqrt{r^2 + x'^2 - 2rx' \cdot \mathbf{n}}$$

$$= r \left(1 + \frac{x'^2}{r^2} - \frac{2}{r} \mathbf{x}' \cdot \mathbf{n} \right)^{1/2}$$

$$\simeq r \left(1 + \frac{x'^2}{2r^2} - \frac{1}{r} \mathbf{x}' \cdot \mathbf{n} \right) = r - \mathbf{n} \cdot \mathbf{x}' + \frac{x'^2}{2r} \qquad (10.1.8)$$

But

$$O \left(\frac{x'^2}{2r} \right) \simeq 0$$

Then

$$|\mathbf{x} - \mathbf{x}'| \simeq r \left(1 - \frac{\mathbf{n} \cdot \mathbf{x}'}{r} \right) \qquad (10.1.9)$$

Then

$$\mathbf{A}(\mathbf{x}, t) \simeq \frac{1}{c} \int d^3 \mathbf{x}' \frac{\mathbf{j}_0(\mathbf{x}')}{1 - \frac{n \cdot x'}{r}}$$

$$\times \frac{\exp \left[-i\omega \left(t - \frac{r}{c} + \frac{\mathbf{n} \cdot \mathbf{x}}{c} \right) \right]}{r}$$

$$= \frac{1}{c} \frac{\exp \left(-i\omega t + i\omega \frac{r}{c} \right)}{r}$$

$$\times \int d^3 \mathbf{x}' \frac{\mathbf{j}_0(\mathbf{x}') \exp \left[-\frac{i\omega}{c} (\mathbf{n} \cdot \mathbf{x}') \right]}{1 - \frac{n \cdot x'}{r}} \qquad (10.1.10)$$

But

$$\frac{1}{1 - [(\mathbf{n} \cdot \mathbf{x}')/r]} \simeq 1 + \frac{\mathbf{n} \cdot \mathbf{x}'}{r}$$

The term $(\mathbf{n} \cdot \mathbf{x}')/r$ when multiplied by the factor $1/r$, which is outside the integral, gives a contribution proportional to $1/r^2$, which can be neglected.

The factor

$$\exp\left[-\frac{i\omega}{c}(\mathbf{n}\cdot\mathbf{x}')\right] = \exp\left[-2\pi i\left(\frac{\mathbf{n}\cdot\mathbf{x}'}{\lambda}\right)\right] \qquad (10.1.11)$$

is independent of r and therefore of how far we observe the field. It depends on the ratio of the size of the region in which currents and charges are confined to the wavelength of the emitted radiation.

Let us examine now the asymptotic fields (at points like P in Fig. 10.1, very far away).

$$\mathbf{A}(\mathbf{x},t) = \frac{e^{ikr-i\omega t}}{r}\int d^3x' \frac{\mathbf{j}_0(\mathbf{x}')}{c}e^{-i\mathbf{k}\cdot\mathbf{x}'}$$

$$\phi(\mathbf{x},t) = \frac{e^{ikr-i\omega t}}{r}\int d^3x' \rho_0(\mathbf{x}')e^{-i\mathbf{k}\cdot\mathbf{x}'} \qquad (10.1.12)$$

where

$$\frac{\omega}{c} = k = \frac{2\pi}{\lambda}$$

$$\frac{\omega}{c}\mathbf{n} = k\mathbf{n} = \mathbf{k} \qquad (10.1.13)$$

We can also consider the far $\mathbf{E}$ and $\mathbf{B}$ fields. When we consider these asymptotic values, we can disregard terms $\propto 1/r^2$. We have

$$\mathbf{B} = \boldsymbol{\nabla}\times\mathbf{A} = \boldsymbol{\nabla}\times\frac{e^{ikr-i\omega t}}{r}\int d^3x' \frac{\mathbf{j}_0(\mathbf{x}')}{c}e^{-i\mathbf{k}\cdot\mathbf{x}'}$$

$$= \boldsymbol{\nabla}\times\frac{e^{ikr-i\omega t}}{r}\mathbf{q} \qquad (10.1.14)$$

where

$$\mathbf{q}(\mathbf{k}) = \int d^3x' \frac{\mathbf{j}_0(\mathbf{x}')}{c}e^{-i\mathbf{k}\cdot\mathbf{x}'} \qquad (10.1.15)$$

We know that

$$\boldsymbol{\nabla}\times\phi\mathbf{u} = \phi\boldsymbol{\nabla}\times\mathbf{u} + \boldsymbol{\nabla}\phi\times\mathbf{u} = \boldsymbol{\nabla}\phi\times\mathbf{u} \qquad (10.1.16)$$

if $\mathbf{u} = $ constant. Then

$$\mathbf{B} = \boldsymbol{\nabla}\left(\frac{e^{ikr-i\omega t}}{r}\right)\times\mathbf{q} \qquad (10.1.17)$$

If $kr \gg 1$, that is, $r \gg \lambda$ (in addition to the condition $r \gg a$), we say that we are in the *wave zone*. If we are in the wave zone, the derivative of $1/r$

can be neglected:

$$\frac{\partial}{\partial r}\frac{e^{ikr}}{r} = \frac{ikre^{ikr} - e^{ikr}}{r^2} \simeq \frac{ike^{ikr}}{r} \qquad (10.1.18)$$

Therefore,

$$\mathbf{B} = \frac{\boldsymbol{\nabla} e^{ikr - i\omega t}}{r} \times \mathbf{q} = \frac{e^{-i\omega t}}{r}\boldsymbol{\nabla} e^{ikr} \times \mathbf{q}$$

$$= \frac{e^{-i\omega t}}{r}\frac{\mathbf{r}}{r}\frac{\partial}{\partial r}e^{ikr} \times \mathbf{q}$$

$$= \frac{e^{-i\omega t}}{r}\mathbf{n}ike^{ikr} \times \mathbf{q} = \frac{e^{ikr - i\omega t}}{r}(i\mathbf{k} \times \mathbf{q}) \qquad (10.1.19)$$

Analogously,

$$\mathbf{E} = -\frac{1}{c}\frac{\partial \mathbf{A}}{\partial t} - \boldsymbol{\nabla}\phi$$

$$= -\frac{1}{c}(-i\omega)\frac{e^{ikr - i\omega t}}{r}\int d^3\mathbf{x}'\frac{\mathbf{j}_0(\mathbf{x}')}{c}e^{-i\mathbf{k}\cdot\mathbf{x}'}$$

$$-\boldsymbol{\nabla}\frac{e^{ikr - i\omega t}}{r}\int d^3\mathbf{x}'\rho_0(\mathbf{x}')e^{-i\mathbf{k}\cdot\mathbf{x}'}$$

$$= \frac{i\omega}{c}\mathbf{q}\frac{e^{ikr - i\omega t}}{r} - i\mathbf{k}q_0\frac{e^{ikr - i\omega t}}{r}$$

$$= \left(\frac{i\omega}{c}\mathbf{q} - i\mathbf{k}q_0\right)\frac{e^{ikr - i\omega t}}{r} \qquad (10.1.20)$$

where

$$q_0 = \int d^3\mathbf{x}\,\rho_0(\mathbf{x}')e^{-i\mathbf{k}\cdot\mathbf{x}'} \qquad (10.1.21)$$

q_0 and $\mathbf{q}$ are not independent. The equation of continuity

$$\boldsymbol{\nabla}\cdot\mathbf{j} + \frac{\partial\rho}{\partial t} = 0 \qquad (10.1.22)$$

gives us

$$\boldsymbol{\nabla}\cdot\mathbf{j}_0(\mathbf{x}) - i\omega\rho_0(\mathbf{x}) = 0 \qquad (10.1.23)$$

Multiplying by $e^{-i\mathbf{k}\cdot\mathbf{x}}$ and integrating,

$$\int d^3\mathbf{x}[\boldsymbol{\nabla}\cdot\mathbf{j}_0(\mathbf{x})]e^{-i\mathbf{k}\cdot\mathbf{x}} = i\omega\int d^3\mathbf{x}\,\rho_0(\mathbf{x})e^{-i\mathbf{k}\cdot\mathbf{x}}$$

Integrating the left members by parts,

$$-\int d^3\mathbf{x}\,\mathbf{j}_0(\mathbf{x})\cdot\boldsymbol{\nabla}e^{-i\mathbf{k}\cdot\mathbf{x}} = i\omega\int d^3\mathbf{x}\,\rho_0(\mathbf{x})e^{-i\mathbf{k}\cdot\mathbf{x}}$$

$$i\mathbf{k}\cdot\int d^3\mathbf{x}\,\mathbf{j}_0(\mathbf{x})e^{-i\mathbf{k}\cdot\mathbf{x}} = i\omega\int d^3\mathbf{x}\,\rho_0(\mathbf{x})e^{-i\mathbf{k}\cdot\mathbf{x}}$$

$$i\mathbf{k}\cdot\mathbf{q}c = i\omega q_0$$

or

$$i\mathbf{k}\cdot\mathbf{q} = \frac{i\omega}{c}q_0$$

$$q_0 = \frac{\mathbf{k}\cdot\mathbf{q}}{\omega/c} = \frac{\mathbf{k}\cdot\mathbf{q}}{k} = \mathbf{n}\cdot\mathbf{q} \qquad (10.1.24)$$

Therefore,

$$\mathbf{E}(\mathbf{x},t) = (ik\mathbf{q} - ik q_0)\frac{e^{ikr-i\omega t}}{r}$$

$$= (ik\mathbf{q} - ik\mathbf{n}q_0)\frac{e^{ikr-i\omega t}}{r}$$

$$= ik[\mathbf{q} - \mathbf{n}(\mathbf{n}\cdot\mathbf{q})]\frac{e^{ikr-i\omega t}}{r} \qquad (10.1.25)$$

In summary,

$$\mathbf{B}(\mathbf{x},t) = (i\mathbf{k}\times\mathbf{q})\frac{e^{ikr-i\omega t}}{r}$$

$$\mathbf{E}(\mathbf{x},t) = ik[\mathbf{q} - \mathbf{n}(\mathbf{n}\cdot\mathbf{q})]\frac{e^{ikr-i\omega t}}{r} \qquad (10.1.26)$$

We note that

$$\mathbf{E}\cdot\mathbf{n} \propto \mathbf{n}\cdot\mathbf{q} - \mathbf{n}\cdot\mathbf{q} = 0 \qquad (10.1.27)$$

$$\mathbf{B}\cdot\mathbf{n} \propto \mathbf{n}\cdot(i\mathbf{k}\times\mathbf{q}) = \mathbf{n}\cdot k(i\mathbf{n}\times\mathbf{q}) = 0 \qquad (10.1.28)$$

$$\mathbf{n}\times\mathbf{E} = ik(\mathbf{n}\times\mathbf{q})\frac{e^{ikr-i\omega t}}{r}$$

$$= (i\mathbf{k}\times\mathbf{q})\frac{e^{ikr-i\omega t}}{r} = \mathbf{B} \qquad (10.1.29)$$

$$\mathbf{n}\times\mathbf{B} = \mathbf{n}\times(\mathbf{n}\times\mathbf{E}) = -\mathbf{E} \qquad (10.1.30)$$

We shall now discuss the relevant quantity

$$q(\mathbf{k}) = \frac{1}{c} \int d^3\mathbf{x}\, \mathbf{j}_0(\mathbf{x}) e^{-i\mathbf{k}\cdot\mathbf{x}} \qquad (10.1.31)$$

We can expand the exponential inside the integral:

$$q(\mathbf{k}) = \frac{1}{c} \int d^3\mathbf{x}\, \mathbf{j}_0(\mathbf{x})(1 - i\mathbf{k}\cdot\mathbf{x} + \cdots)$$

$$= \frac{1}{c} \int d^3\mathbf{x}\, \mathbf{j}_0(\mathbf{x}) - \frac{i}{c} \int d^3\mathbf{x}\, \mathbf{j}_0(x)(\mathbf{k}\cdot\mathbf{x}) + \cdots \qquad (10.1.32)$$

Let us examine these terms separately.

(1) $(1/c) \int d^3\mathbf{x}\, \mathbf{j}_0(\mathbf{x})$ is essentially an electric dipole moment. We can write

$$\mathbf{j}_0 = -\mathbf{x}(\nabla \cdot \mathbf{j}_0) + \sum_k \frac{\partial}{\partial x_k}(\mathbf{x}j_{0k}) \qquad (10.1.33)$$

In fact,

$$\sum_k \frac{\partial}{\partial x_k}(\mathbf{x}j_{0k}) = \frac{\partial}{\partial x}(\mathbf{x}j_{0x}) + \frac{\partial}{\partial y}(\mathbf{x}j_{0y}) + \frac{\partial}{\partial z}(\mathbf{x}j_{0z})$$

$$= \frac{\partial}{\partial x}[(\hat{\mathbf{i}}x + \hat{\mathbf{j}}y + \hat{\mathbf{k}}z)j_{0x}]$$

$$+ \frac{\partial}{\partial y}[(\hat{\mathbf{i}}x + \hat{\mathbf{j}}y + \hat{\mathbf{k}}z)j_{0y}]$$

$$+ \frac{\partial}{\partial z}[(\hat{\mathbf{i}}x + \hat{\mathbf{j}}y + \hat{\mathbf{k}}z)j_{0z}]$$

$$= (\hat{\mathbf{i}}x + \hat{\mathbf{j}}y + \hat{\mathbf{k}}z)\frac{\partial j_{0x}}{\partial x} + \hat{\mathbf{i}}j_{0x}$$

$$+ (\hat{\mathbf{i}}x + \hat{\mathbf{j}}y + \hat{\mathbf{k}}z)\frac{\partial j_{0y}}{\partial y} + \hat{\mathbf{j}}j_{0y}$$

$$+ (\hat{\mathbf{i}}x + \hat{\mathbf{j}}y + \hat{\mathbf{k}}z)\frac{\partial j_{0z}}{\partial z} + \hat{\mathbf{k}}j_{0z}$$

$$= (\hat{\mathbf{i}}x + \hat{\mathbf{j}}y + \hat{\mathbf{k}}z)(\nabla \cdot \mathbf{j}_0) + \mathbf{j}_0 = \mathbf{x}(\nabla \cdot \mathbf{j}_0) + \mathbf{j}_0$$

where $\hat{\mathbf{i}}, \hat{\mathbf{j}}$, and $\hat{\mathbf{k}}$, are unit vectors in the x, y, and z directions, respectively. Therefore,

$$\frac{1}{c} \int d^3\mathbf{x}\, \mathbf{j}_0(\mathbf{x}) = -\frac{1}{c} \int d^3\mathbf{x}\, \mathbf{x}(\boldsymbol{\nabla} \cdot \mathbf{j}_0)$$

$$+ \frac{1}{c} \int d^3\mathbf{x} \sum_k \frac{\partial}{\partial x_k}(\mathbf{x}j_{0k})$$

$$= -\frac{i\omega}{c} \int \mathbf{x}\rho_0(\mathbf{x})d^3\mathbf{x}$$

$$= -ik \int \mathbf{x}\rho_0(\mathbf{x})d^3\mathbf{x} = -ik\mathbf{D} \qquad (10.1.34)$$

where the second term in the expression for $(1/c) \int d^3\mathbf{x}\, \mathbf{j}_0(\mathbf{x})$ has been dropped because $\mathbf{j}_0$ is confined to a finite region of space, and where

$$\mathbf{D} = \int \mathbf{x}\rho_0(\mathbf{x})d^3\mathbf{x} \qquad (10.1.35)$$

is the *electric dipole moment* of the charge distribution. $\mathbf{D}$ is independent of $\mathbf{k}$, but may have an angular dependence. If $\mathbf{D} \neq 0$, the radiation field due to dipole radiation will predominate as long as $ka \ll 1$.

If $\mathbf{D} = 0$, the most important term in the expansion of $\mathbf{q}(\mathbf{k})$ is the following:

(2) $-(i/c) \int d^3\mathbf{x}\, \mathbf{j}_0(\mathbf{x})(\mathbf{k} \cdot \mathbf{x})$, for which we can write

$$(\mathbf{k} \cdot \mathbf{x})\mathbf{j}_0(\mathbf{x}) = \frac{1}{2}[(\mathbf{k} \cdot \mathbf{x})\mathbf{j}_0(\mathbf{x}) + (\mathbf{k} \cdot \mathbf{j}_0)\mathbf{x}]$$

$$+ \frac{1}{2}[(\mathbf{k} \cdot \mathbf{x})j_0(\mathbf{x}) - (\mathbf{k} \cdot \mathbf{j}_0)\mathbf{x}]$$

$$= \frac{1}{2}[(\mathbf{k} \cdot \mathbf{x})\mathbf{j}_0(\mathbf{x}) + (\mathbf{k} \cdot \mathbf{j}_0)\mathbf{x}]$$

$$- \frac{1}{2}\mathbf{k} \times (\mathbf{x} \times \mathbf{j}_0) \qquad (10.1.36)$$

because

$$\mathbf{k} \times (\mathbf{x} \times \mathbf{j}_0) = \mathbf{x}(\mathbf{k} \cdot \mathbf{j}_0) - \mathbf{j}_0(\mathbf{k} \cdot \mathbf{x}) \qquad (10.1.37)$$

Taking the ith component of the vector $(\mathbf{k} \cdot \mathbf{x})\mathbf{j}_0$,

$$[(\mathbf{k} \cdot \mathbf{x})\mathbf{j}_0(\mathbf{x})]_i = \frac{1}{2}\left[\left(\sum_j k_j\, x_j\right) j_{0i} + \left(\sum_j k_j\, j_{0j}\right) x_i\right]$$

$$-\frac{1}{2}[\mathbf{k} \times (\mathbf{x} \times \mathbf{j}_0)]_i \qquad (10.1.38)$$

Then

$$\left[-\frac{i}{c}\int d^3\mathbf{x}\,\mathbf{j}_0(\mathbf{x})(\mathbf{k} \cdot \mathbf{x})\right]_i$$

$$= -\frac{i}{2c}\sum_j k_j \int d^3\mathbf{x}(x_j j_{0i} + x_i j_{0j})$$

$$+ i\left[\mathbf{k} \times \int d^3\mathbf{x}\,\frac{1}{2c}(\mathbf{x} \times \mathbf{j}_0)\right]_i \qquad (10.1.39)$$

CLAIM:

$$\int d^3\mathbf{x}(x_j j_{0i} + x_i j_{0j}) = -\int d^3\mathbf{x}\,x_i x_j (\boldsymbol{\nabla} \cdot \mathbf{j}_0) \qquad (10.1.40)$$

PROOF: For $x_i = x$, $x_j = y$, relation (10.1.40) gives

$$\int d^3\mathbf{x}(x j_{0y} + y j_{0x}) = -\int d^3\mathbf{x}\,xy\left(\frac{\partial j_{0x}}{\partial x} + \frac{\partial j_{0y}}{\partial y} + \frac{\partial j_{0z}}{\partial z}\right)$$

If we work on the right member, we find, integrating by parts, that

$$\int d^3\mathbf{x}\,xy\frac{\partial j_{0x}}{\partial x} \overset{\mathrm{IP}}{=} -\int d^3\mathbf{x}\,j_{0x}\frac{\partial}{\partial x}(xy) = -\int d^3\mathbf{x}\,j_{0x}y$$

$$\int d^3\mathbf{x}\,xy\frac{\partial j_{0y}}{\partial y} \overset{\mathrm{IP}}{=} -\int d^3\mathbf{x}\,j_{0y}\frac{\partial}{\partial y}(xy) = -\int d^3\mathbf{x}\,j_{0y}x$$

$$\int d^3\mathbf{x}\,xy\frac{\partial j_{0z}}{\partial z} \overset{\mathrm{IP}}{=} -\int d^3\mathbf{x}\,j_{0z}\frac{\partial}{\partial z}(xy) = 0 \qquad \text{Q.E.D.}$$

Therefore,

$$\left[-\frac{i}{c}\int d^3\mathbf{x}\,\mathbf{j}_0(\mathbf{x})(\mathbf{k} \cdot \mathbf{x})\right]_i$$

$$= \frac{i}{2c}\sum_j k_j \int d^3\mathbf{x}\,x_i x_j (\boldsymbol{\nabla} \cdot \mathbf{j}_0)$$

$$+ \left[i\mathbf{k} \times \int d^3\mathbf{x} \frac{1}{2c}(\mathbf{x} \times \mathbf{j}_0) \right]_i$$

$$= \frac{i}{2c} \sum_j k_j \int d^3\mathbf{x}\, x_i x_j (i\omega \rho_0)$$

$$+ i \left[\mathbf{k} \times \int d^3\mathbf{x} \frac{1}{2c}(\mathbf{x} \times \mathbf{j}_0) \right]_i$$

$$= -\frac{\omega}{2c} \sum_j k_j Q_{ij} + i(\mathbf{k} \times \mathbf{M})_i \qquad (10.1.41)$$

The second term in the expansion (10.1.33) consists of two terms, one containing the *electric quadrupole moment*,

$$Q_{ij} = \int d^3\mathbf{x}\, x_i x_j \rho_0(\mathbf{x}) \qquad (10.1.42)$$

and the other containing the *magnetic dipole moment*,

$$\mathbf{M} = \int d^3\mathbf{x} \frac{1}{2c}[\mathbf{x} \times \mathbf{j}_0(\mathbf{x})] \qquad (10.1.43)$$

We could proceed with calculations of additional terms in the expansion of $\mathbf{q}(\mathbf{k})$ and find *electric octupole* and *magnetic quadrupole* contributions, and so on. If we so proceed, we could find terms that contain the charge distribution $\rho_0(\mathbf{x})$ and give electric multipoles and terms that contain the current density $\mathbf{j}_0(\mathbf{x})$ and give magnetic multipoles. This decomposition between electric and magnetic multipoles is general, as we shall see in the next section.

Some observations are in order regarding the relative size of the emitting sources and of the wavelength of the emitted radiation. In atoms, we have typically

$$a = \text{radius of atom} = 10^{-8} \text{ cm}$$

and

$$ka = \frac{2\pi}{\lambda} a \simeq \frac{a}{\lambda} = \frac{10^{-8}}{5 \times 10^{-5}} \simeq 2 \times 10^{-4}$$

which is a number much smaller than 1.

In nuclei, the situation may be different. The radius of a nucleus is given by

$$a = 1.2 \times 10^{-13}\, A^{1/3} \text{ cm}$$

where A = number of nucleons. Then

$$ka = \hbar\omega \frac{a}{\hbar c}$$

But

$$\frac{a}{\hbar c} = \frac{1.2 \times 10^{-13} \, A^{1/3}}{10^{-27} \times 3 \times 10^{10}} = 4 \times 10^3 \, A^{1/3} \, \text{erg}^{-1}$$

$$1 \, \text{MeV} = 1.6 \times 10^{-6} \, \text{erg}$$

$$\frac{a}{\hbar c} = 4 \times 10^3 \, A^{1/3} \times 1.6 \times 10^{-6} = 6.4 \times 10^{-3} \, A^{1/3} \, \text{MeV}^{-1}$$

Then

$$ka = A^{1/3} \times 6.4 \times 10^{-3} \times (\hbar\omega)_{\text{MeV}}$$

EXAMPLE

For silver,

$$A = 100, \quad A^{1/3} \simeq 5$$
$$ka = 5 \times 6.4 \times 10^{-3} (\hbar\omega)_{\text{MeV}} = 0.003 (\hbar\omega)_{\text{MeV}}$$

This number is small with respect to 1, but not very small.

For γ-rays of 0.4 MeV, we find

$$ka \simeq 0.03 \times 0.4 \simeq 1.2 \times 10^{-2}$$

two orders of magnitude bigger than the same quantity in the atomic case. From this we may deduce the relevance of multipoles other than dipoles when dealing with nuclei.

10.2. Electric and Magnetic Multipoles

Let us start with the following expression:

$$\int d^3\mathbf{x} \, \mathbf{x} (\boldsymbol{\nabla} \cdot \mathbf{j}_0) e^{-i\mathbf{k}\cdot\mathbf{x}} = \int d^3\mathbf{x} (\boldsymbol{\nabla} \cdot \mathbf{j}_0) \mathbf{x} e^{-i\mathbf{k}\cdot\mathbf{x}}$$

$$\overset{\text{IP}}{=} -\int d^3\mathbf{x} [\mathbf{j}_0 \cdot \boldsymbol{\nabla}] \mathbf{x} e^{-i\mathbf{k}\cdot\mathbf{x}}$$

$$= -\int d^3\mathbf{x} \, \mathbf{j}_0 e^{-i\mathbf{k}\cdot\mathbf{x}}$$

$$+ \int d^3\mathbf{x} (\mathbf{j}_0 \cdot i\mathbf{k}) \mathbf{x} e^{-i\mathbf{k}\cdot\mathbf{x}} \qquad (10.2.1)$$

But

$$\mathbf{k} \times (\mathbf{x} \times \mathbf{j}_0) = \mathbf{x}(\mathbf{k} \cdot \mathbf{j}_0) - \mathbf{j}_0(\mathbf{k} \cdot \mathbf{x}) \qquad (10.2.2)$$

$$i\mathbf{k} \times (\mathbf{x} \times \mathbf{j}_0) + \mathbf{j}_0(i\mathbf{k} \cdot \mathbf{x}) = (\mathbf{j}_0 \cdot i\mathbf{k})\mathbf{x} \qquad (10.2.3)$$

and

$$\int d^3\mathbf{x}\, \mathbf{x}(\boldsymbol{\nabla} \cdot \mathbf{j}_0)e^{-i\mathbf{k}\cdot\mathbf{x}}$$

$$= -\int d^3\mathbf{x}\, \mathbf{j}_0\, e^{-i\mathbf{k}\cdot\mathbf{x}}$$

$$+ \int d^3\mathbf{x}[i\mathbf{k} \times (\mathbf{x} \times \mathbf{j}_0) + \mathbf{j}_0(i\mathbf{k} \cdot \mathbf{x})]e^{-i\mathbf{k}\cdot\mathbf{x}} \qquad (10.2.4)$$

$$\int d^3\mathbf{x}\, \mathbf{j}_0(1 - i\mathbf{k} \cdot \mathbf{x})e^{-i\mathbf{k}\cdot\mathbf{x}}$$

$$= -\int d^3\mathbf{x}\, \mathbf{x}(\boldsymbol{\nabla} \cdot \mathbf{j}_0)e^{-i\mathbf{k}\cdot\mathbf{x}}$$

$$+ i\mathbf{k} \times \int d^3\mathbf{x}(\mathbf{x} \times \mathbf{j}_0)e^{-i\mathbf{k}\cdot\mathbf{x}}$$

$$= -i\omega \int d^3\mathbf{x}\, \rho_0(\mathbf{x})\mathbf{x}e^{-i\mathbf{k}\cdot\mathbf{x}}$$

$$+ i\mathbf{k} \times \int d^3\mathbf{x}(\mathbf{x} \times \mathbf{j}_0)e^{-i\mathbf{k}\cdot\mathbf{x}} \qquad (10.2.5)$$

Let us call

$$\mathbf{q}(\lambda\mathbf{k}) = \frac{1}{c} \int d^3\mathbf{x}\, \mathbf{j}_0(\mathbf{x})e^{-i\lambda\mathbf{k}\cdot\mathbf{x}} \qquad (10.2.6)$$

Then

$$\frac{\partial[\lambda\mathbf{q}(\lambda\mathbf{k})]}{\partial\lambda} = \mathbf{q}(\lambda\mathbf{k}) + \lambda\frac{\partial\mathbf{q}(\lambda\mathbf{k})}{\partial\lambda}$$

$$= \frac{1}{c} \int d^3\mathbf{x}\, \mathbf{j}_0(\mathbf{x})(1 - i\lambda\mathbf{k} \cdot \mathbf{x})e^{-i\lambda\mathbf{k}\cdot\mathbf{x}} \qquad (10.2.7)$$

Relation (10.2.7), together with Eq. (10.2.5), gives us

$$\frac{1}{c} \int d^3\mathbf{x}\, \mathbf{j}_0(1 - i\lambda\mathbf{k} \cdot \mathbf{x})e^{-i\lambda\mathbf{k}\cdot\mathbf{x}}$$

$$= -\frac{i\omega}{c} \int d^3\mathbf{x}\, \rho_0(\mathbf{x})\mathbf{x}e^{-i\lambda\mathbf{k}\cdot\mathbf{x}}$$

$$+ \frac{i\lambda \mathbf{k}}{c} \times \int d^3 \mathbf{x} (\mathbf{x} \times \mathbf{j}_0) e^{-i\lambda \mathbf{k} \cdot \mathbf{x}}$$

$$= \frac{d[\lambda \mathbf{q}(\lambda \mathbf{k})]}{d\lambda} \tag{10.2.8}$$

We then find

$$\mathbf{q}(\mathbf{k}) = \int_0^1 d\lambda \frac{d}{d\lambda} [\lambda \mathbf{q}(\lambda \mathbf{k})]$$

$$= -\frac{i\omega}{c} \int d^3 \mathbf{x} \, \rho_0(\mathbf{x}) \, \mathbf{x} \int_0^1 e^{-i\lambda \mathbf{k} \cdot \mathbf{x}} d\lambda$$

$$+ \frac{i\mathbf{k}}{c} \times \int d^3 \mathbf{x} (\mathbf{x} \times \mathbf{j}_0) \int_0^1 \lambda e^{-i\lambda \mathbf{k} \cdot \mathbf{x}} d\lambda \tag{10.2.9}$$

We have divided $\mathbf{q}(\mathbf{k})$ into two parts: the electric part depending on $\rho_0(\mathbf{x})$, and the magnetic part depending on $\mathbf{j}_0(\mathbf{x})$.

Let us expand

$$e^{-i\lambda \mathbf{k} \cdot \mathbf{x}} = 1 - i\lambda \mathbf{k} \cdot \mathbf{x} + \cdots \tag{10.2.10}$$

and retain the first term 1:

$$\mathbf{q}(\mathbf{k}) = -\frac{i\omega}{c} \int d^3 \mathbf{x} \, \rho_0(\mathbf{x}) \mathbf{x} \int_0^1 d\lambda$$

$$+ \frac{i\mathbf{k}}{c} \times \int d^3 \mathbf{x} (\mathbf{x} \times \mathbf{j}_0) \int_0^1 \lambda d\lambda$$

$$= -i\mathbf{k} \int d^3 \mathbf{x} \, \rho_0(\mathbf{x}) \mathbf{x} + i\mathbf{k} \times \int d^3 \mathbf{x} \frac{1}{2c} (\mathbf{x} \times \mathbf{j}_0)$$

$$= -i\mathbf{k} \mathbf{D} + i\mathbf{k} \times \mathbf{M} \tag{10.2.11}$$

Consider now the part $-i\lambda \mathbf{k} \cdot \mathbf{x}$; the first term in Eq. (10.2.9),

$$\mathbf{q}(\mathbf{k}) = -\frac{i\omega}{c} \int d^3 \mathbf{x} \, \rho_0(\mathbf{x}) \mathbf{x} \int_0^1 -i\lambda \mathbf{k} \cdot \mathbf{x} d\lambda$$

$$= -\frac{\omega}{2c} \int d^3 \mathbf{x} \, \rho_0(\mathbf{x}) \mathbf{x} (\mathbf{k} \cdot \mathbf{x}) \tag{10.2.12}$$

gives us the electric quadrupole contribution, and so on.

In summary, we can write

$$\mathbf{q}(\mathbf{k}) = \frac{1}{c} \int d^3 \mathbf{x} \, \mathbf{j}_0(\mathbf{x}) e^{-i\mathbf{k} \cdot \mathbf{x}} = \mathbf{q}^{\text{el}}(\mathbf{k}) + \mathbf{q}^{\text{magn}}(\mathbf{k}) \tag{10.2.13}$$

where

$$\mathbf{q}^{\text{el}}(\mathbf{k}) = -\frac{i\omega}{c} \int d^3(\mathbf{x})\rho_0(\mathbf{x})\mathbf{x} \int_0^1 d\lambda\, e^{-i\lambda\mathbf{k}\cdot\mathbf{x}} \qquad (10.2.14)$$

$$\mathbf{q}^{\text{magn}}(\mathbf{k}) = i\mathbf{k} \times \int d^3\mathbf{x}\frac{1}{c}(\mathbf{x} \times \mathbf{j}_0) \int_0^1 \lambda e^{-i\lambda\mathbf{k}\cdot\mathbf{x}}\, d\lambda \qquad (10.2.15)$$

10.3. Multipole Expansion Using Spherical Harmonics

Let us consider a distribution of charges and currents in a certain region of space, as in Fig. 10.2. $\mathbf{k}$ is the wave vector in the direction of observation. We shall give, without demonstration, the following expansion:[1]

$$e^{-i\mathbf{k}\cdot\mathbf{x}} = e^{-ikR\cos\psi} = \sum_{l=0}^{\infty}(2l+1)(-i)^l j_l(kR)P_l(\cos\psi) \qquad (10.3.1)$$

where

$$j_l(kR) = \text{spherical Bessel function}$$
$$P_l(\cos\psi) = \text{Legendre polynomial}$$

We shall assume that

$$ka \ll 1; \quad \text{that is, } \lambda \gg a \qquad (10.3.2)$$

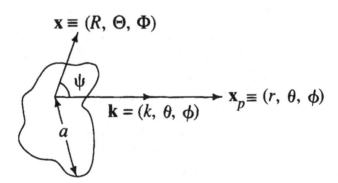

FIGURE 10.2

---

[1]See for example: R. M. Eisberg, *Fundamentals of Modern Physics*, Wiley, New York, London 1964, pp. 537–540.

where a = dimension of the region where charges and currents are located. Then we can use an approximation for $j_l(kr)$. If $z \ll l + 1$,

$$j_l(z) \simeq \frac{z^l}{(2l + 1)!!} \tag{10.3.3}$$

where

$$\frac{1}{(2l + 1)!!} = \frac{1}{1 \cdot 3 \cdot 5 \cdots (2l + 1)} \tag{10.3.4}$$

The Legendre polynomial can be expanded as follows:

$$P_l(\cos \psi) = \frac{4\pi}{(2l + 1)} \sum_{m=-l}^{l} Y_l^{m^*}(\Theta, \Phi) Y_l^m(\theta, \phi) \tag{10.3.5}$$

where we should refer to Fig. 10.2 for the definition of the angles. We can now write

$$e^{-i\mathbf{k}\cdot\mathbf{x}} = e^{-ikR\cos\psi} = \sum_{l=0}^{\infty}(2l + 1)(-i)^l j_l(kR) P_l(\cos\psi)$$

$$\simeq \sum_{l=0}^{\infty}(2l + 1)(-i)^l \frac{(kR)^l}{(2l + 1)!!}\frac{4\pi}{2l + 1}$$

$$\times \sum_m Y_l^{m^*}(\Theta, \Phi) Y_l^m(\theta, \phi)$$

$$= 4\pi \sum_{l=0}^{\infty}(-i)^l \frac{(kR)^l}{(2l + 1)!!} \sum_m Y_l^{m^*}(\Theta, \Phi) Y_l^m(\theta, \phi) \tag{10.3.6}$$

Let us write again, from Eq. (10.2.14),

$$\mathbf{q}^{\text{el}}(\mathbf{k}) = -i\frac{\omega}{c} \int d^3\mathbf{x}\, \rho_0(\mathbf{x})\mathbf{x} \int_0^1 d\lambda\, e^{-i\lambda\mathbf{k}\cdot\mathbf{x}} \tag{10.3.7}$$

On the other hand,

$$\frac{\partial}{\partial\mathbf{k}}e^{-i\lambda\mathbf{k}\cdot\mathbf{x}} = -i\lambda\mathbf{x}e^{-i\lambda\mathbf{k}\cdot\mathbf{x}} \tag{10.3.8}$$

$$\frac{i}{\lambda}\frac{\partial}{\partial\mathbf{k}}e^{-i\lambda\mathbf{k}\cdot\mathbf{x}} = \mathbf{x}e^{-i\lambda\mathbf{k}\cdot\mathbf{x}} \tag{10.3.9}$$

Then

$$\mathbf{q}^{\text{el}}(\mathbf{k}) = -ik \int d^3\mathbf{x}\, \rho_0(\mathbf{x}) \int_0^1 \mathbf{x} e^{-i\lambda \mathbf{k}\cdot\mathbf{x}} d\lambda$$

$$= -ik \int d^3\mathbf{x}\, \rho_0(\mathbf{x}) \frac{\partial}{\partial \mathbf{k}} i \int_0^1 \frac{1}{\lambda} e^{-i\lambda \mathbf{k}\cdot\mathbf{x}} d\lambda$$

$$= k \frac{\partial}{\partial \mathbf{k}} \int d^3\mathbf{x}\, \rho_0(\mathbf{x}) \int_0^1 \frac{e^{-i\lambda \mathbf{k}\cdot\mathbf{x}}}{\lambda} d\lambda$$

$$= k \frac{\partial}{\partial \mathbf{k}} \int d^3\mathbf{x}\, \rho_0(\mathbf{x}) \int_0^1 \frac{d\lambda}{\lambda}$$

$$\times \left[4\pi \sum_{l=0}^{\infty} (-i)^l \frac{(\lambda k R)^l}{(2l+1)!!} \sum_m Y_l^{m^*}(\Theta, \Phi) Y_l^m(\theta, \phi) \right]$$

$$= k \frac{\partial}{\partial \mathbf{k}} 4\pi \sum_{l=0}^{\infty} (-i)^l \frac{k^l}{(2l+1)!!} \sum_m Y_l^m(\theta, \phi) \int d^3\mathbf{x}\, \rho_0(\mathbf{x}) R^l$$

$$\times \left(\int_0^1 \frac{d\lambda}{\lambda} \lambda^l \right) Y_l^{m^*}(\Theta, \Phi) \tag{10.3.10}$$

But

$$\int_0^1 \lambda^{l-1} d\lambda = \frac{1}{l} \tag{10.3.11}$$

Then, disregarding the term with $l = 0$, because of the constant in k,

$$\mathbf{q}^{\text{el}}(\mathbf{k}) = k \frac{\partial}{\partial \mathbf{k}} 4\pi \sum_{l=1}^{\infty} \frac{(-i)^l k^l}{l(2l+1)!!} \sum_m Y_l^m(\theta, \phi)$$

$$\times \int d^3\mathbf{x}\, \rho_0(\mathbf{x}) R^l Y_l^{m^*}(\Theta, \Phi)$$

$$= k \frac{\partial}{\partial \mathbf{k}} 4\pi \sum_{l=1}^{\infty} \frac{(-i)^l k^l}{l(2l+1)!!} \sum_m Y_l^m(\theta, \phi) a_{lm}^{\text{el}} \tag{10.3.12}$$

where

$$a_{lm}^{el} = \int d^3\mathbf{x}\, \rho_0(\mathbf{x}) R^l Y_l^{m^*}(\Theta, \Phi) \tag{10.3.13}$$

Let us now derive the fields

$$\mathbf{B}^{el}(\mathbf{x}_p, t) = i\mathbf{k} \times \mathbf{q}\frac{e^{ikr-i\omega t}}{r}$$

$$= -4\pi\frac{e^{ikr-i\omega t}}{r}\sum_{lm}\left(-i\mathbf{k} \times \frac{\partial}{\partial\mathbf{k}}\right) \tag{10.3.14}$$

$$\times \frac{(-i)^l k^{l+1}}{l(2l+1)!!}a_{lm}^{el}Y_l^m(\theta, \phi)$$

$$\mathbf{E}^{el}(\mathbf{x}_p, t) = -\mathbf{n} \times \mathbf{B}^{el}(\mathbf{x}_p, t)$$

We note that

$$\left(\mathbf{k} \times \frac{\partial}{\partial\mathbf{k}}\right) f(|\mathbf{k}|) = 0 \tag{10.3.15}$$

The contribution to $\mathbf{B}^{el}$ comes from the angles θ and ϕ.

In a perfectly analogous way, we derive for the magnetic case

$$\mathbf{E}^{magn}(\mathbf{x}_p, t) = 4\pi\frac{e^{ikr-i\omega t}}{r}\sum_{lm}\left(-i\mathbf{k} \times \frac{\partial}{\partial\mathbf{k}}\right)$$

$$\times \frac{(-i)^l k^{l+1}}{l(l+1)(2l+1)!!}a_{lm}^{magn}\, Y_l^m(\theta, \phi) \tag{10.3.16}$$

$$\mathbf{B}^{magn}(\mathbf{x}_p, t) = \mathbf{n} \times \mathbf{E}^{magn}(\mathbf{x}_p, t)$$

where

$$a_{lm}^{magn} = \int d^3\mathbf{x}\, \boldsymbol{\nabla} \cdot \left(\mathbf{x} \times \frac{\mathbf{j}_0}{c}\right) R^l Y_l^{m^*}(\Theta, \phi) \tag{10.3.17}$$

The dimensions of a_{lm}^{magn} are the same as those of a_{lm}^{el}.

10.4. Angular Distribution of Multipole Radiation

Let us examine the following operator:

$$\mathcal{L} = -i\left(\mathbf{k} \times \frac{\partial}{\partial\mathbf{k}}\right) \tag{10.4.1}$$

or

$$\mathcal{L} = -i \begin{vmatrix} \hat{\mathbf{i}} & \hat{\mathbf{j}} & \hat{\mathbf{k}} \\ k_x & k_y & k_z \\ \dfrac{\partial}{\partial k_x} & \dfrac{\partial}{\partial k_y} & \dfrac{\partial}{\partial k_z} \end{vmatrix}$$

$$= \hat{\mathbf{i}} \left[-i \left(k_y \frac{\partial}{\partial k_z} - k_z \frac{\partial}{\partial k_y} \right) \right]$$

$$+ \hat{\mathbf{j}} \left[-i \left(k_z \frac{\partial}{\partial k_x} - k_x \frac{\partial}{\partial k_z} \right) \right]$$

$$+ \hat{\mathbf{k}} \left[-i \left(k_x \frac{\partial}{\partial k_y} - k_y \frac{\partial}{\partial k_x} \right) \right] \tag{10.4.2}$$

This is the "dimensionless" angular momentum operator, and we know its component in spherical coordinates:

$$\mathcal{L}_x = i \left(\sin\phi \frac{\partial}{\partial\theta} + \cot\theta \cos\phi \frac{\partial}{\partial\phi} \right)$$

$$\mathcal{L}_y = i \left(-\cos\phi \frac{\partial}{\partial\theta} + \cot\theta \sin\phi \frac{\partial}{\partial\phi} \right) \tag{10.4.3}$$

$$\mathcal{L}_z = -i \frac{\partial}{\partial\phi}$$

Also,

$$\mathcal{L}_+ = \mathcal{L}_x + i\mathcal{L}_y = i \left(\sin\phi \frac{\partial}{\partial\theta} + \cot\theta \cos\phi \frac{\partial}{\partial\phi} \right)$$

$$+ \cos\phi \frac{\partial}{\partial\theta} - \cot\theta \sin\phi \frac{\partial}{\partial\phi}$$

$$= e^{i\phi} \frac{\partial}{\partial\theta} + (i \cot\theta \cos\phi - \cot\theta \sin\phi) \frac{\partial}{\partial\phi}$$

$$= e^{i\phi} \frac{\partial}{\partial\theta} + i \cot\theta (\cos\phi + i\sin\phi) \frac{\partial}{\partial\phi}$$

$$= e^{i\phi} \left(\frac{\partial}{\partial\theta} + i \cot\theta \frac{\partial}{\partial\phi} \right) \tag{10.4.4}$$

$$\mathcal{L}_- = \mathcal{L}_x - i\mathcal{L}_y = i \left(\sin\phi \frac{\partial}{\partial\theta} + \cot\theta \cos\phi \frac{\partial}{\partial\phi} \right)$$

$$+ \left(-\cos\phi \frac{\partial}{\partial\theta} + \cot\theta \sin\phi \frac{\partial}{\partial\phi} \right)$$

$$= -(\cos\phi - i\sin\phi)\frac{\partial}{\partial\theta}$$

$$+ i\cot\theta(\cos\phi - i\sin\phi)\frac{\partial}{\partial\phi}$$

$$= e^{-i\phi} \left(-\frac{\partial}{\partial\theta} + i\cot\theta \frac{\partial}{\partial\phi} \right) \tag{10.4.5}$$

Therefore, in accord with Fig. 10.3,

$$\mathcal{L}_\phi = -\mathcal{L}_x \sin\phi + \mathcal{L}_y \cos\phi$$

$$= -\mathcal{L}_x \frac{e^{i\phi} - e^{-i\phi}}{2i} + \mathcal{L}_y \frac{e^{i\phi} + e^{-i\phi}}{2}$$

$$= \frac{i}{2}\mathcal{L}_x(e^{i\phi} - e^{-i\phi}) + \frac{i}{2}(-i)\mathcal{L}_y(e^{i\phi} + e^{-i\phi})$$

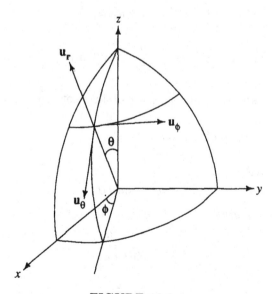

FIGURE 10.3

$$= \frac{i}{2}[(e^{i\phi} - e^{-i\phi})\mathcal{L}_x - i(e^{i\phi} + e^{-i\phi})\mathcal{L}_y]$$

$$= \frac{i}{2}[e^{i\phi}(\mathcal{L}_x - i\mathcal{L}_y) - e^{-i\phi}(\mathcal{L}_x + i\mathcal{L}_y)]$$

$$= \frac{i}{2}[e^{i\phi}\mathcal{L}_- - e^{-i\phi}\mathcal{L}_+]$$

$$= \frac{i}{2}\left(-\frac{\partial}{\partial\theta} + i\cot\theta\frac{\partial}{\partial\phi} - \frac{\partial}{\partial\theta} - i\cot\theta\frac{\partial}{\partial\phi}\right)$$

$$= -i\frac{\partial}{\partial\theta} \tag{10.4.6}$$

Unit vectors:

$$\mathbf{u}_\phi = -\hat{\mathbf{i}}\sin\phi + \hat{\mathbf{j}}\cos\phi$$

$$\mathbf{u}_\theta = \hat{\mathbf{i}}\cos\theta\cos\phi + \hat{\mathbf{j}}\cos\theta\sin\phi - \hat{\mathbf{k}}\sin\theta$$

$$\mathbf{u}_r = \hat{\mathbf{i}}\sin\theta\cos\phi + \hat{\mathbf{j}}\sin\theta\sin\phi + \hat{\mathbf{k}}\cos\theta$$

$\hat{\mathbf{i}}, \hat{\mathbf{j}}$, and $\hat{\mathbf{k}}$ are unit vectors along the x, y, and z axes, respectively.

$$\mathcal{L}_\theta = \mathcal{L}_x\cos\theta\cos\phi + \mathcal{L}_y\cos\theta\sin\phi - \mathcal{L}_z\sin\theta$$

$$= i\cos\theta\cos\phi\left(\sin\phi\frac{\partial}{\partial\theta} + \cot\theta\cos\phi\frac{\partial}{\partial\phi}\right)$$

$$+ i\cos\theta\sin\phi\left(-\cos\phi\frac{\partial}{\partial\theta} + \cot\theta\sin\phi\frac{\partial}{\partial\phi}\right) + i\sin\theta\frac{\partial}{\partial\phi}$$

$$= i\cos\theta\cos^2\phi\cot\theta\frac{\partial}{\partial\phi} + i\cos\theta\sin^2\theta\cot\theta\frac{\partial}{\partial\phi} + i\sin\theta\frac{\partial}{\partial\phi}$$

$$= i\cos\theta\cot\theta\frac{\partial}{\partial\phi} + i\sin\theta\frac{\partial}{\partial\phi}$$

$$= i\frac{\cos^2\theta}{\sin\theta}\frac{\partial}{\partial\phi} + i\sin\theta\frac{\partial}{\partial\phi} = \frac{i}{\sin\theta}\frac{\partial}{\partial\phi} \tag{10.4.7}$$

Therefore, we can write

$$\mathcal{L}_\theta = \frac{i}{\sin\theta}\frac{\partial}{\partial\phi}$$

$$\mathcal{L}_\phi = -i\frac{\partial}{\partial\theta} \tag{10.4.8}$$

We shall now examine several examples:

Electric Multipoles. For the case of electric multipoles, we can write

$$B_\theta^{el} \sim \mathcal{L}_\theta Y_l^m(\theta, \phi) = \frac{i}{\sin\theta} \frac{\partial}{\partial\phi} Y_l^m(\theta, \phi)$$

$$= -\frac{m}{\sin\theta} Y_l^m(\theta, \phi) \qquad (10.4.9)$$

$$B_\phi^{el} \sim \mathcal{L}_\phi Y_l^m(\theta, \phi) = -i \frac{\partial}{\partial\theta} Y_l^m(\theta, \phi)$$

Now, remembering that

$$\mathbf{E} = -\mathbf{n} \times \mathbf{B}$$

we have

$$E_\theta^{el} = B_\phi^{el} \sim -i \frac{\partial}{\partial\theta} Y_l^m(\theta, \phi)$$

$$E_\phi^{el} = -B_\theta^{el} \sim \frac{m}{\sin\theta} Y_l^m(\theta, \phi) \qquad (10.4.10)$$

EXAMPLE: Electric Dipole $l = 1$

The relevant spherical harmonics are

$$Y_1^1 = -\sqrt{\frac{3}{8\pi}} \sin\theta \, e^{i\phi}$$

$$Y_1^0 = \sqrt{\frac{3}{4\pi}} \cos\theta$$

$$Y_1^{-1} = \sqrt{\frac{3}{8\pi}} \sin\theta \, e^{-i\phi}$$

Let us consider first the moment a_{10}:

$$a_{10} = \int d^3\mathbf{x} \, \rho_0(\mathbf{x}) R Y_1^{0*}(\Theta, \Phi) \sim \int d^3\mathbf{x} \, \rho_0(\mathbf{x}) z$$

where $z = R\cos\Theta$. We have then ($m = 0$)

$$E_\theta \sim -i \frac{\partial}{\partial\theta} Y_1^0(\theta, \phi) \sim \sin\theta$$

$$E_\phi \sim \frac{m}{\sin\theta} Y_1^0(\theta, \phi) = 0$$

and

$$B_\theta = -E_\phi = 0$$
$$B_\phi = E_\theta \sim \sin\theta$$

The emitted power is proportional to $\sin^2\theta$.

Let us consider now the moment a_{11}:

$$a_{11} = \int d^3\mathbf{x}\rho_0(\mathbf{x})RY_1^{1*}(\Theta,\Phi) \sim \int d^3\mathbf{x}\,\rho_0(\mathbf{x})R\sin\Theta e^{i\Phi}$$

We have then $(m = 1)$

$$E_\theta \sim -i\frac{\partial}{\partial\theta}Y_1^1(\theta,\phi) \sim -i\cos\theta\,e^{i\phi}$$

$$E_\phi \sim \frac{m}{\sin\theta}Y_1^1(\theta,\phi) \sim \frac{1}{\sin\theta}\sin\theta\,e^{i\phi} = e^{i\phi}$$

The power emitted is proportional to $1 + \cos^2\theta$.

Similar results are obtained when we consider the moment $a_{1,-1}$:

$$a_{1,-1} = \int d^3\mathbf{x}\,\rho_0(\mathbf{x})RY_1^{-1*}(\Theta,\Phi) \sim \int d^3\mathbf{x}\,\rho_0(\mathbf{x})R\sin\Theta e^{-i\Phi}$$

We have then $(m = -1)$

$$E_\theta \sim -i\frac{\partial}{\partial\theta}Y_1^{-1}(\theta,\phi) \sim -i\cos\theta\,e^{-i\phi}$$

$$E_\phi \sim \frac{m}{\sin\theta}Y_1^{-1}(\theta,\phi) \sim -\frac{1}{\sin\theta}\sin\theta\,e^{-i\phi} = -e^{-i\phi}$$

The power emitted is proportional to $1 + \cos^2\theta$.

Note that, if a_{10}, a_{11}, and $a_{1,-1}$ have the same value, the power emitted by the three dipoles together is independent of direction.

EXAMPLE: Electric Quadrupole $l = 2$

The relevant spherical harmonics are

$$Y_2^2 = \sqrt{\frac{15}{32\pi}}\sin^2\theta\,e^{2i\phi}$$

$$Y_2^1 = -\sqrt{\frac{15}{8\pi}}\cos\theta\sin\theta\,e^{i\phi}$$

$$Y_2^0 = \sqrt{\frac{15}{16\pi}}(3\cos^2\theta - 1)$$

$$Y_2^{-1} = \sqrt{\frac{15}{8\pi}}\cos\theta\sin\theta\,e^{-i\phi}$$

$$Y_2^{-2} = \sqrt{\frac{15}{32\pi}}\sin^2\theta\,e^{-2i\phi}$$

Then

$$a_{20} = \int d^3\mathbf{x}\,\rho_0(\mathbf{x})R^2 Y_2^{0*}(\Theta, \Phi)$$

$$\sim \int d^3\mathbf{x}\,\rho_0(\mathbf{x})R^2(3\cos^2\Theta - 1)$$

For $m = 0$,

$$E_\theta \sim -i\frac{\partial}{\partial\theta}Y_2^0(\theta,\phi) \sim -i\frac{\partial}{\partial\theta}(3\cos^2\theta - 1)$$

$$\sim \sin\theta\cos\theta \sim \sin 2\theta$$

$$E_\phi \sim \frac{m}{\sin\theta}Y_2^0(\theta,\phi) = 0$$

The power emitted is proportional to $\sin^2 2\theta$.

Magnetic Multipoles. For the case of magnetic multipoles, we can write

$$E_\theta^{\text{magn}} \sim \mathcal{L}_\theta Y_l^m(\theta,\phi) = \frac{i}{\sin\theta}\frac{\partial}{\partial\phi}Y_l^m(\theta,\phi)$$

$$= -\frac{m}{\sin\theta}Y_l^m(\theta,\phi) \qquad (10.4.11)$$

$$E_\phi^{\text{magn}} \sim \mathcal{L}_\phi Y_l^m(\theta,\phi) = -i\frac{\partial}{\partial\theta}Y_l^m(\theta,\phi)$$

and

$$B_\theta^{\text{magn}} = -E_\phi^{\text{magn}} \sim i\frac{\partial}{\partial\theta}Y_l^m(\theta,\phi)$$

$$B_\phi^{\text{magn}} = E_\theta^{\text{magn}} \sim -\frac{m}{\sin\theta}Y_l^m(\theta,\phi) \qquad (10.4.12)$$

CHAPTER 10 EXERCISES

10.1. **(a)** A point charge oscillates in the z direction with frequency ω (see Fig. P10.1a).

The radiation field for large distances, $r_0 \gg z_0, r_0 \gg (c/\omega)$, gives

$$E, B \sim \cos \omega \left(t - \frac{r(t')}{c} \right)$$

The first approximation, $r(t') = r_0$. Calculate the correction to this approximation, assuming $(\omega z_0)/c \ll 1$. Show that the correction is a wave of frequency 2ω and angular distribution $\sin 2\theta$.

(b) The frequency 2ω-component is called *quadrupole radiation.* Show that this component is the only radiation in the lowest order in $\omega z_0/c$ emitted by the oscillating quadrupole in Fig. P10.1b.

10.2. A square plate of side a is in the $y - z$ plane, with the center at the origin, and carries a uniform surface charge. It oscillates harmonically with small amplitude in the z direction, thereby effectively producing a sheet of currents in the z direction of surface density

$$J_z(y, z) \cos \omega t$$

where

$$J_z(y, z) = \text{const} = J_0 \quad \text{if}$$
$$-a/2 < y < a/2$$
$$-a/2 < z < a/2$$

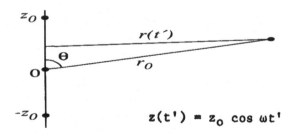

FIGURE P10.1a

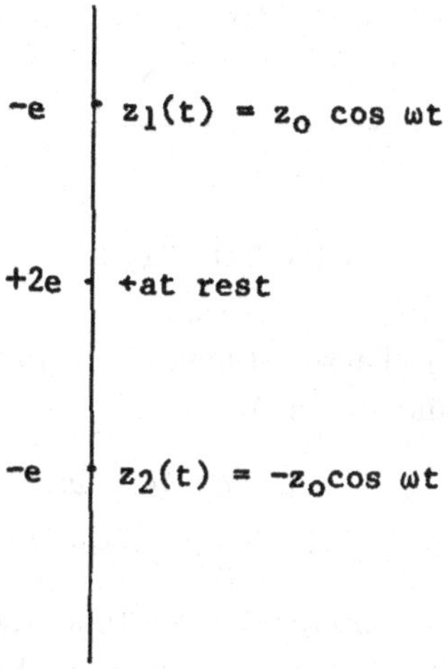

FIGURE P10.1b

and

$$J_z(y, z) = 0 \text{ otherwise}$$

The wavelength of the radiation $(2\pi c)/\omega$ is not large compared to a. Find the pattern of the radiation, that is, the Poynting vector, produced by this current sheet at distance $r \gg \lambda$ as a function of the two spherical angular coordinates θ and φ.

10.3. A gas of identical molecules emits electrical multipole radiation of order l. Assume that all the multipole moments

$$a_{lm} = \int \rho_0(\mathbf{x}) r^l Y_l^{m^*}(\theta, \phi) d^3\mathbf{x}$$

have equal value, independent of m. Show that in this case the emitted radiation is *isotropic*.

Note: You may use the relations

$$L_z Y_l^m = m Y_l^m$$
$$L_\pm Y_l^m = (L_x \pm iL_y)Y_l^m = \sqrt{l(l+1) - m(m \pm l)} Y_l^{m \pm 1}$$

and

$$\sum_m Y_l^m(\theta\phi)Y_l^{m^*}(\theta\phi) = \frac{2l+1}{4\pi}$$

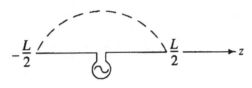

FIGURE P10.5

10.4. Two positive point charges of magnitude q are located on the x axis at the points x_1 and x_2, where

$$x_1 = d - a \sin \omega t$$
$$x_2 = -d + a \sin \omega t$$

Find the energy radiated per unit time and per unit solid angle. Assume that $a \ll d$, and $(a\omega)/c \ll 1$, but do *not* assume that $(d\omega/c) \ll 1$.

10.5. Consider a linear antenna of length L (Fig. P10.5) in the z direction with a current

$$J(z,t) = J_0 \cos \frac{\pi z}{L} \cos \omega t$$

(a) Show that the antenna carries a charge per unit length of

$$Q(z,t) = \frac{\pi J_0}{L\omega} \sin \frac{\pi z}{L} \sin \omega t$$

(b) Calculate the vector

$$q(\mathbf{k}) = \frac{1}{c} \int d^3\mathbf{x}\, \mathbf{j}_0(\mathbf{x}) e^{-\mathbf{k}\cdot\mathbf{x}}$$

and show that, for $(kL)/\pi = 1$,

$$q_z(\mathbf{k}) \sim \frac{\cos[(\pi/2)\cos\theta]}{\sin^2\theta}$$

where $\cos\theta = k_z/k$.

(c) Calculate $q_z(\mathbf{k})$ in the electric dipole approximation.
Note:

$$\int e^{ax} \cos px\, dx = \frac{e^{ax}(a \cos px + p \sin px)}{a^2 + p^2}$$

10.6. (a) What is an electric quadrupole?

 (b) Show how many independent numbers are necessary to specify an electric quadrupole in the most general case.

 (c) Explain why we may talk of the quadrupole moment of a nucleus in its ground state as though it were a single number.

10.7. A metallic medium of conductivity σ contains a small region A of conductivity $\sigma_1 \neq \sigma$ that perturbs an electric field that in its absence would be constant. Find how the effect of this perturbation on the value of the electric field depends on the distance from A in steady-state conditions.

11

Lagrangian and Hamiltonian Formulations of Electrodynamics

11.1. Outline of Classical Mechanics

The aim of this chapter is to give the Lagrangian and the Hamiltonian formulations of the following systems:

(1) Charged particles interacting with external fields
(2) Fields and their sources (charges and currents)
(3) Charged particles and their fields, plus their interactions

These systems are discussed separately and independently. Systems 1 and 2 contain both free terms (either free particle or free fields) and interaction terms, the latter being the same in both cases. System 3 is essentially a combination of 1 and 2, taking into account the fact that the interaction terms have to be included only once.

Before proceeding with the tasks described above, we shall review briefly the Lagrangian and Hamiltonian formulation of classical mechanics.

441

Consider a classical system with f degrees of freedom. For N point particles, $f = 3N$. Assume that we have a set of *generalized coordinates* for the system

$$q_1, q_2, \ldots, q_f \tag{11.1.1}$$

These coordinates may be Cartesian, polar, or of some other convenient type. The *generalized velocities* associated with these coordinates are

$$\dot{q}_1, \dot{q}_2, \ldots, \dot{q}_f \tag{11.1.2}$$

The time evolution of the generalized coordinates is expressed by the *Lagrange equations* (f in number):

$$\frac{d}{dt}\left(\frac{\partial L}{\partial \dot{q}_i}\right) - \frac{\partial L}{\partial q_i} = 0 \tag{11.1.3}$$

where for a conservative and nonrelativistic system the *Lagrangian L* is given by

$$L(q_i, \dot{q}_i) = T - V \tag{11.1.4}$$

T and V are the kinetic and potential energy, respectively. In general, $L = L(q_i, \dot{q}_i, t)$.

Equation (11.1.3) is easily verified if the q_i's are Cartesian coordinates. In this case,

$$L = \frac{1}{2}\sum_j M_j \, \dot{q}_j^2 - V \tag{11.1.5}$$

and, letting $q_j = x$, Eq. (11.1.3) gives

$$M_j \, \ddot{x} = -\frac{\partial}{\partial x} = F_x \tag{11.1.6}$$

In most cases it is sufficient to define L as $T - V$. However, Eq. (11.1.3) provides a more general definition of L, and the Lagrangian L can be considered to be that expression, which, when substituted in Eq. (11.1.3), gives the correct equations of motion.

The Hamiltonian form of the equations of motion replaces the f second-order differential Lagrange equations with $2f$ first-order differential equations. We define the *generalized momenta* as follows:

$$p_i = \frac{\partial L}{\partial \dot{q}_i} \tag{11.1.7}$$

The Hamiltonian H is defined as follows:

$$H(p_i, q_i, t) = \sum_i p_i \dot{q}_i - L(q_i, \dot{q}_i, t) \qquad (11.1.8)$$

Then

$$dH = \sum_i \left(\frac{\partial H}{\partial p_i} dp_i + \frac{\partial H}{\partial q_i} dq_i \right) + \frac{\partial H}{\partial t} dt$$

$$= \sum_i \left(p_i d\dot{q}_i + \dot{q}_i dp_i - \frac{\partial L}{\partial q_i} dq_i - \frac{\partial L}{\partial \dot{q}_i} d\dot{q}_i \right) - \frac{\partial L}{\partial t} dt$$

$$= \sum_i \left(p_i d\dot{q}_i + \dot{q}_i dp_i - \frac{\partial L}{\partial q_i} dq_i - p_i d\dot{q}_i \right) - \frac{\partial L}{\partial t} dt$$

$$= \sum_i (\dot{q}_i dp_i - \dot{p}_i dq_i) - \frac{\partial L}{\partial t} dt \qquad (11.1.9)$$

because

$$\frac{\partial L}{\partial q_i} = \frac{d}{dt} \left(\frac{\partial L}{\partial \dot{q}_i} \right) = \frac{d}{dt} p_i = \dot{p}_i \qquad (11.1.10)$$

Then we must have

$$\frac{\partial H}{\partial p_i} = \dot{q}_i$$

$$\frac{\partial H}{\partial q_i} = -\dot{p}_i \qquad (11.1.11)$$

$$\frac{\partial H}{\partial t} = -\frac{\partial L}{\partial t}$$

These equations are called *Hamilton's* or *canonical equations*.

11.2. Lagrangian Formulation of the Motion of a Charged Particle in Given Fields

We start from the equation

$$\frac{d}{dt} \frac{m\mathbf{v}}{\sqrt{1 - \beta^2}} = e \left(\mathbf{E} + \frac{\mathbf{v}}{c} \times \mathbf{B} \right) \qquad (11.2.1)$$

This equation is not written in covariant form, but it contains all the information we need.

Let us introduce the potentials:

$$\mathbf{E} = -\boldsymbol{\nabla}\phi - \frac{1}{c}\frac{\partial \mathbf{A}}{\partial t}$$
$$\mathbf{B} = \boldsymbol{\nabla} \times \mathbf{A}$$

(11.2.2)

We have then

$$\mathbf{E} + \frac{\mathbf{v}}{c} \times \mathbf{B} = -\boldsymbol{\nabla}\phi - \frac{1}{c}\dot{\mathbf{A}} + \frac{\mathbf{v}}{c} \times (\boldsymbol{\nabla} \times \mathbf{A})$$

(11.2.3)

But

$$\mathbf{v} \times (\boldsymbol{\nabla} \times \mathbf{A}) = \boldsymbol{\nabla}(\mathbf{v} \cdot \mathbf{A}) - (\mathbf{A} \cdot \boldsymbol{\nabla})\mathbf{v}$$
$$- (\mathbf{v} \cdot \boldsymbol{\nabla})\mathbf{A} - \mathbf{A} \times (\boldsymbol{\nabla} \times \mathbf{v})$$
$$= \boldsymbol{\nabla}(\mathbf{v} \cdot \mathbf{A}) - (\mathbf{v} \cdot \boldsymbol{\nabla})\mathbf{A}$$

(11.2.4)

because $\mathbf{v}$ is function of t only. Then

$$\mathbf{E} + \frac{\mathbf{v}}{c} \times \mathbf{B} = -\boldsymbol{\nabla}\phi - \frac{1}{c}\dot{\mathbf{A}} + \boldsymbol{\nabla}\left(\frac{\mathbf{v}}{c} \cdot \mathbf{A}\right) - \left(\frac{\mathbf{v}}{c} \cdot \boldsymbol{\nabla}\right)\mathbf{A}$$
$$= -\boldsymbol{\nabla}\left(\phi - \frac{\mathbf{v}}{c} \cdot \mathbf{A}\right) - \frac{1}{c}\left[\frac{\partial \mathbf{A}}{\partial t} + (\mathbf{v} \cdot \boldsymbol{\nabla})\mathbf{A}\right]$$

(11.2.5)

The fields we are considering are $\mathbf{E}(\mathbf{x}_p, t)$ and $\mathbf{B}(\mathbf{x}_p, t)$ at the position $\mathbf{x}_p$ of the particle. We have two dependencies of the fields on time: one due to the change in the position of the charged particle, and the other due to the explicit time variation of the fields:

$$\frac{d\mathbf{A}(\mathbf{x}_p, t)}{dt} = \frac{\partial \mathbf{A}(\mathbf{x}_p, t)}{\partial t} + \left(\frac{\partial \mathbf{x}_p}{\partial t} \cdot \boldsymbol{\nabla}\right)\mathbf{A}(\mathbf{x}_p, t)$$
$$= \frac{\partial \mathbf{A}(\mathbf{x}_p, t)}{\partial t} + (\mathbf{v} \cdot \boldsymbol{\nabla})\mathbf{A}(\mathbf{x}_p, t)$$

(11.2.6)

Then

$$\mathbf{E} + \frac{\mathbf{v}}{c} \times \mathbf{B} = -\boldsymbol{\nabla}\left(\phi - \frac{\mathbf{v}}{c} \cdot \mathbf{A}\right) - \frac{1}{c}\left[\frac{\partial \mathbf{A}}{\partial t} + (\mathbf{v} \cdot \boldsymbol{\nabla})\mathbf{A}\right]$$
$$= -\boldsymbol{\nabla}\left(\phi - \frac{\mathbf{v}}{c} \cdot \mathbf{A}\right) - \frac{1}{c}\frac{d\mathbf{A}}{dt}$$

(11.2.7)

and

$$\frac{d}{dt}\frac{m\mathbf{v}}{\sqrt{1-\beta^2}} = -e\mathbf{\nabla}\left(\phi - \frac{\mathbf{v}}{c}\cdot\mathbf{A}\right) - \frac{e}{c}\frac{d\mathbf{A}}{dt} \tag{11.2.8}$$

or

$$\frac{d}{dt}\left[\frac{m\mathbf{v(t)}}{\sqrt{1-\beta^2}} + \frac{e}{c}\mathbf{A}(\mathbf{x}_p,t)\right]$$

$$= -\mathbf{\nabla}_{x_p}\left[e\phi(\mathbf{x}_p,t) - \frac{e\mathbf{v}(t)}{c}\cdot\mathbf{A}(\mathbf{x}_p,t)\right] \tag{11.2.9}$$

But

$$(-mc^2)\frac{d}{d\mathbf{v}}\sqrt{1-\beta^2} = (-mc^2)\frac{-2(\mathbf{v}/c^2)}{2\sqrt{1-\beta^2}} = \frac{m\mathbf{v}}{\sqrt{1-\beta^2}} \tag{11.2.10}$$

and

$$\frac{d}{d\mathbf{v}}\left(\frac{e}{c}\mathbf{v}\cdot\mathbf{A}\right) = \frac{e}{c}\mathbf{A} \tag{11.2.11}$$

Then we can write

$$\frac{d}{dt}\left\{(-mc^2)\frac{d}{d\mathbf{v}}\sqrt{1-\beta^2} + \frac{d}{d\mathbf{v}}\left[\frac{e}{c}\mathbf{v}\cdot\mathbf{A}(\mathbf{x}_p,t)\right]\right\}$$

$$-\mathbf{\nabla}_{x_p}\left[e\phi(\mathbf{x}_p,t) - \frac{e\mathbf{v}}{c}\cdot\mathbf{A}(\mathbf{x}_p,t)\right] \tag{11.2.12}$$

CLAIM:

$$L = -mc^2\sqrt{1-\beta^2} + \frac{e}{c}(\mathbf{v}\cdot\mathbf{A}) - e\phi \tag{11.2.13}$$

Proof.

$$\frac{\partial L}{\partial v_k} = -mc^2\frac{\partial}{\partial v_k}\sqrt{1-\beta^2} + \frac{e}{c}\frac{\partial}{\partial v_k}(\mathbf{v}\cdot\mathbf{A})$$

$$= \frac{mv_k}{\sqrt{1-\beta^2}} + \frac{e}{c}A_k \tag{11.2.14}$$

$$\frac{\partial L}{\partial x_k} = \frac{\partial}{\partial x_k}\left(\frac{e}{c}\mathbf{v}\cdot\mathbf{A} - e\phi\right) \tag{11.2.15}$$

where x_k stands for the kth component of $\mathbf{x}_p$. Then, because of Eq. (11.1.9) or Eq. (11.2.12),

$$\frac{d}{dt}\left(\frac{\partial L}{\partial \dot{x}_k}\right) - \frac{\partial L}{\partial x_k} = 0 \qquad\qquad \text{Q.E.D.}$$

11.3. Hamiltonian Formulation of the Motion of a Charged Particle in Given Fields

Let

$$p_k = \frac{\partial L}{\partial v_k} = \textit{canonical momentum} \qquad (11.3.1)$$

Since

$$L = -mc^2\sqrt{1 - \beta^2} + \frac{e}{c}(\mathbf{v} \cdot \mathbf{A}) - e\phi \qquad (11.3.2)$$

the canonical momentum is given by

$$p_k = \frac{\partial L}{\partial v_k} = \frac{mv_k}{\sqrt{1 - \beta^2}} + \frac{e}{c}A_k \qquad (11.3.3)$$

On the other hand,

$$\Pi_k = \frac{mv_k}{\sqrt{1 - \beta^2}} = \textit{kinetic momentum} \qquad (11.3.4)$$

There is a gauge invariance in the Lagrangian and Hamiltonian formulations in the sense that the trajectory of the particle is determined only by the fields.

The Hamiltonian is given by

$$H = \sum_k p_k v_k - L = \sum_k \left(\frac{mv_k^2}{\sqrt{1 - \beta^2}} + \frac{e}{c}v_k A_k\right)$$
$$+ mc^2\sqrt{1 - \beta^2} - \frac{e}{c}(\mathbf{v} \cdot \mathbf{A}) + e\phi$$
$$= \frac{mv^2}{\sqrt{1 - \beta^2}} + \frac{e}{c}(\mathbf{v} \cdot \mathbf{A}) + mc^2\sqrt{1 - \beta^2} - \frac{e}{c}(\mathbf{v} \cdot \mathbf{A}) + e\phi$$
$$= e\phi + \frac{mc^2}{\sqrt{1 - \beta^2}}(\beta^2 + 1 - \beta^2) = e\phi + \frac{mc^2}{\sqrt{1 - \beta^2}}$$

or

$$H = e\phi + \frac{mc^2}{\sqrt{1 - \beta^2}} \tag{11.3.5}$$

where

$$e\phi = \text{potential energy} \tag{11.3.6}$$

$$\frac{mc^2}{\sqrt{1 - \beta^2}} = \text{rest energy} + \text{kinetic energy} \tag{11.3.7}$$

We want to express H as a function of p_k and x_k. Because of Eq. (11.3.3),

$$\mathbf{p} = \frac{m\mathbf{v}}{\sqrt{1 - \beta^2}} + \frac{e}{c}\mathbf{A} \tag{11.3.8}$$

and

$$\left(\mathbf{p} - \frac{e}{c}\mathbf{A}\right)^2 = \frac{m^2 v^2}{1 - \beta^2} = \frac{(mc)^2 \beta^2}{1 - \beta^2} = (mc)^2 \left(\frac{1}{1 - \beta^2} - 1\right)$$

$$= \frac{(mc)^2}{1 - \beta^2} - (mc)^2 \tag{11.3.9}$$

or

$$(mc)^2 + \left(\mathbf{p} - \frac{e}{c}\mathbf{A}\right)^2 = \frac{(mc)^2}{1 - \beta^2} \tag{11.3.10}$$

and

$$\frac{mc^2}{\sqrt{1 - \beta^2}} = c\sqrt{(mc)^2 + \left(\mathbf{p} - \frac{e}{c}\mathbf{A}\right)^2} \tag{11.3.11}$$

Then we can write

$$H(\mathbf{p}, \mathbf{x}_p) = c\sqrt{\left(\mathbf{p} - \frac{e}{c}\mathbf{A}\right)^2 + (mc)^2} + e\phi \tag{11.3.12}$$

From this expression, we can derive the equations

$$\frac{dx_k}{dt} = \frac{\partial H}{\partial p_k}$$

$$\frac{dp_k}{dt} = -\frac{\partial H}{\partial x_k} \tag{11.3.13}$$

The reader can verify that these two equations give

$$\mathbf{p} = \frac{m\mathbf{v}}{\sqrt{1 - \beta^2}} + \frac{e}{c}\mathbf{A}$$

$$\frac{d}{dt}\frac{m\mathbf{v}}{\sqrt{1 - \beta^2}} = e\left(\mathbf{E} + \frac{\mathbf{v}}{c} \times \mathbf{B}\right) \qquad (11.3.14)$$

We can make the following observations:

(1) In the nonrelativistic limit ($v/c \ll 1$),

$$\left(\mathbf{p} - \frac{e}{c}\mathbf{A}\right)^2 \ll m^2 c^2 \qquad (11.3.15)$$

and

$$H(\mathbf{p}, \mathbf{x}_p) = c\sqrt{\left(\mathbf{p} - \frac{e}{c}\mathbf{A}\right)^2 + (mc)^2} + e\phi$$

$$= mc^2\sqrt{1 + \frac{[\mathbf{p} - (e/c)\mathbf{A}]^2}{(mc)^2}} + e\phi$$

$$= mc^2\left[1 + \frac{1}{2}\frac{[\mathbf{p} - (e/c)\mathbf{A}]^2}{(mc)^2} + \cdots\right] + e\phi \qquad (11.3.16)$$

Then

$$H(\mathbf{p}, \mathbf{x}_p) - mc^2 = \frac{[\mathbf{p} - (e/c)\mathbf{A}]^2}{2m} + e\phi \qquad (11.3.17)$$

where

$$\frac{[\mathbf{p} - (e/c)\mathbf{A}]^2}{2m} = \text{kinetic energy} \qquad (11.3.18)$$

$$e\phi = \text{potential energy} \qquad (11.3.19)$$

(2) If the particle moves in time-independent fields, the energy of the particle is conserved:

$$\frac{dH}{dt} = \frac{\partial H}{\partial t} + \sum_k \left(\frac{\partial H}{\partial p_k}\frac{dp_k}{dt} + \frac{\partial H}{\partial x_k}\frac{dx_k}{dt}\right)$$

$$= \frac{\partial H}{\partial t} + \sum_k \left(-\frac{\partial H}{\partial p_k}\frac{\partial H}{\partial x_k} + \frac{\partial H}{\partial x_k}\frac{\partial H}{\partial p_k}\right)$$

$$= \frac{\partial H}{\partial t} = 0 \qquad (11.3.20)$$

where use has been made of Eq. (11.3.13). In this case, the sum of kinetic and potential energy is constant.

If the fields are time-dependent, there is an exchange of energy between particle and fields.

11.4. Lagrangian Formulation of the Maxwell Equations

We want now to give a Lagrangian and Hamiltonian formulation of the electromagnetic field. It is advantageous not to use the Lorentz gauge that is expressed by

$$\sum_{\mu} \frac{\partial A_{\mu}}{\partial x_{\mu}} = 0 \qquad (11.4.1)$$

The Lorentz gauge is useful for the covariant formulation, but for the present formulation it is better to use the *radiation gauge*, called also the *Coulomb gauge*

$$\boldsymbol{\nabla} \cdot \mathbf{A} = 0 \qquad (11.4.2)$$

We shall now write the four Maxwell equations:

$$\boldsymbol{\nabla} \cdot \mathbf{E} = 4\pi\rho \qquad (11.4.3a)$$

$$\boldsymbol{\nabla} \cdot \mathbf{B} = 0 \qquad (11.4.3b)$$

$$\boldsymbol{\nabla} \times \mathbf{E} + \frac{1}{c} \frac{\partial \mathbf{B}}{\partial t} = 0 \qquad (11.4.3c)$$

$$\boldsymbol{\nabla} \times \mathbf{B} - \frac{1}{c} \frac{\partial \mathbf{E}}{\partial t} = \frac{4\pi}{c} \mathbf{j} \qquad (11.4.3d)$$

By the use of Eqs. (11.4.3b) and (11.4.3c), we get

$$\mathbf{B} = \boldsymbol{\nabla} \times \mathbf{A}$$
$$\mathbf{E} = -\boldsymbol{\nabla}\phi - \frac{1}{c} \dot{\mathbf{A}} \qquad (11.4.4)$$

Replacing these expressions in Eq. (11.4.3a), we get

$$\boldsymbol{\nabla} \cdot \left(-\boldsymbol{\nabla}\phi - \frac{1}{c} \dot{\mathbf{A}} \right) = 4\pi\rho$$

$$\nabla^2\phi + \frac{1}{c} \frac{\partial}{\partial t} \boldsymbol{\nabla} \cdot \mathbf{A} = -4\pi\rho \qquad (11.4.5)$$

$$\left(\nabla^2\phi - \frac{1}{c^2} \frac{\partial^2\phi}{\partial t^2} \right) + \frac{1}{c} \frac{\partial}{\partial t} \left(\boldsymbol{\nabla} \cdot \mathbf{A} + \frac{1}{c} \frac{\partial\phi}{\partial t} \right) = -4\pi\rho$$

Using the same expressions in Eq. (11.4.3d), we find

$$\nabla \times (\nabla \times \mathbf{A}) - \frac{1}{c}\frac{\partial}{\partial t}\left(-\nabla\phi - \frac{1}{c}\dot{\mathbf{A}}\right) = \frac{4\pi}{c}\mathbf{j}$$

$$\nabla(\nabla \cdot \mathbf{A}) - \nabla^2\mathbf{A} + \frac{1}{c}\frac{\partial}{\partial t}\left(\nabla\phi + \frac{1}{c}\dot{\mathbf{A}}\right) = \frac{4\pi}{c}\mathbf{j} \qquad (11.4.6)$$

$$\left(\nabla^2\mathbf{A} - \frac{1}{c^2}\frac{\partial^2\mathbf{A}}{\partial t^2}\right) - \nabla\left(\nabla \cdot \mathbf{A} + \frac{1}{c}\frac{\partial\phi}{\partial t}\right) = -\frac{4\pi}{c}\mathbf{j}$$

Taking into account the radiation gauge ($\nabla \cdot \mathbf{A} = 0$), we obtain

$$\nabla^2\phi = -4\pi\rho$$

$$\Box^2\mathbf{A} = -\frac{4\pi}{c}\mathbf{j} + \frac{1}{c}\nabla\frac{\partial\phi}{\partial t} \qquad (11.4.7)$$

Then

$$\phi(\mathbf{x},t) = \int d^3x' \frac{\rho(\mathbf{x}',t)}{|\mathbf{x}-\mathbf{x}'|} \qquad (11.4.8)$$

In the present case, ϕ derives its existence from ρ; if $\rho = 0$, also $\phi = 0$.

The situation is different for $\mathbf{A}$, which depends on the currents, but has existence (velocity of propagation, polarization, and so on) even in the absence of currents and charges.

We have, from Eq. (11.4.8),

$$\nabla\dot{\phi}(\mathbf{x},t) = \nabla\int d^3x' \frac{\dot{\rho}(\mathbf{x}',t)}{|\mathbf{x}-\mathbf{x}'|} = \nabla\int d^3x' \frac{-\nabla_{x'} \cdot \mathbf{j}(\mathbf{x}',t)}{|\mathbf{x}-\mathbf{x}'|}$$

$$\stackrel{\text{IP}}{=} \nabla\int d^3x'\mathbf{j}(\mathbf{x}',t) \cdot \nabla_{x'}\frac{1}{|\mathbf{x}-\mathbf{x}'|}$$

$$= -\nabla\left[\nabla \cdot \int d^3x' \frac{\mathbf{j}(\mathbf{x}',t)}{|\mathbf{x}-\mathbf{x}'|}\right] \qquad (11.4.9)$$

Then

$$\Box^2\mathbf{A}(x,t) = -\frac{4\pi}{c}\mathbf{j}(\mathbf{x},t) - \frac{1}{c}\nabla\left[\nabla \cdot \int d^3x' \frac{\mathbf{j}(\mathbf{x}',t)}{|\mathbf{x}-\mathbf{x}'|}\right]$$

$$= -\frac{4\pi}{c}\mathbf{j}_t(\mathbf{x},t) \qquad (11.4.10)$$

where

$$\mathbf{j}_t(\mathbf{x}, t) = \mathbf{j}(\mathbf{x}, t) + \frac{1}{4\pi}\boldsymbol{\nabla}\left[\boldsymbol{\nabla}\cdot\int d^3\mathbf{x}'\frac{\mathbf{j}(\mathbf{x}', t)}{|\mathbf{x} - \mathbf{x}'|}\right]$$

$$= \mathbf{j}(\mathbf{x}, t) + \frac{1}{4\pi}\boldsymbol{\nabla}\int d^3\mathbf{x}'\frac{\boldsymbol{\nabla}_{x'}\cdot\mathbf{j}(\mathbf{x}', t)}{|\mathbf{x} - \mathbf{x}'|} \tag{11.4.11}$$

Note that

$$\boldsymbol{\nabla}\cdot\mathbf{j}_t(\mathbf{x}, t) = 0 \tag{11.4.12}$$

In fact,

$$\boldsymbol{\nabla}\cdot\mathbf{j}_t = \boldsymbol{\nabla}\cdot\mathbf{j} + \frac{1}{4\pi}\nabla^2\int d^3\mathbf{x}'\frac{\boldsymbol{\nabla}_{x'}\cdot\mathbf{j}(\mathbf{x}', t)}{|\mathbf{x} - \mathbf{x}'|}$$

$$= \boldsymbol{\nabla}\cdot\mathbf{j} + \frac{1}{4\pi}\int d^3\mathbf{x}'[-4\pi\delta(\mathbf{x} - \mathbf{x}')\boldsymbol{\nabla}_{x'}\cdot\mathbf{j}(\mathbf{x}', t)]$$

$$= \boldsymbol{\nabla}\cdot\mathbf{j} - \boldsymbol{\nabla}\cdot\mathbf{j} = 0$$

A vector whose divergence is zero is called *transverse* or *solenoidal*; it can be expressed as a curl of another vector. In the present case, we make the following claim.

CLAIM:

$$\mathbf{j}_t(\mathbf{x}, t) = \boldsymbol{\nabla}\times\int d^3\mathbf{x}'\frac{\boldsymbol{\nabla}_{x'}\times\mathbf{j}(\mathbf{x}', t)}{4\pi|\mathbf{x} - \mathbf{x}'|} \tag{11.4.13}$$

Proof:

$$\boldsymbol{\nabla}\times\int d^3\mathbf{x}'\frac{\boldsymbol{\nabla}_{x'}\times\mathbf{j}(\mathbf{x}', t)}{|\mathbf{x} - \mathbf{x}'|} = \boldsymbol{\nabla}\times\left(\boldsymbol{\nabla}\times\int d^3\mathbf{x}'\frac{\mathbf{j}(\mathbf{x}', t)}{|\mathbf{x} - \mathbf{x}'|}\right)$$

$$= \boldsymbol{\nabla}\left[\boldsymbol{\nabla}\cdot\int d^3\mathbf{x}'\frac{\mathbf{j}(\mathbf{x}', t)}{|\mathbf{x} - \mathbf{x}'|}\right]$$

$$- \nabla^2\int d^3\mathbf{x}'\frac{\mathbf{j}(\mathbf{x}', t)}{|\mathbf{x} - \mathbf{x}'|}$$

$$= \boldsymbol{\nabla}\left[\boldsymbol{\nabla}\cdot\int d^3\mathbf{x}'\frac{\mathbf{j}(\mathbf{x}', t)}{|\mathbf{x} - \mathbf{x}'|}\right]$$

$$+ \int d^3\mathbf{x}'\mathbf{j}(\mathbf{x}', t)4\pi\delta(\mathbf{x} - \mathbf{x}')$$

$$= \nabla \left[\nabla \cdot \int d^3x' \frac{\mathbf{j}(\mathbf{x}',t)}{|\mathbf{x}-\mathbf{x}'|} \right] + 4\pi \, \mathbf{j}(\mathbf{x},t)$$

$$\mathbf{j}(\mathbf{x},t) = -\nabla \left[\nabla \cdot \int d^3x' \frac{\mathbf{j}(\mathbf{x}',t)}{4\pi|\mathbf{x}-\mathbf{x}'|} \right]$$

$$+ \nabla \times \int d^3x' \frac{\nabla_{x'} \times \mathbf{j}(\mathbf{x}',t)}{4\pi|\mathbf{x}-\mathbf{x}'|}$$

Comparing this last relation with Eq. (11.4.11), we see that Eq. (11.4.13) is proved.

We can write

$$\mathbf{j}(\mathbf{x},t) = \mathbf{j}_t(\mathbf{x},t) - \nabla \left[\nabla \cdot \int d^3x' \frac{\mathbf{j}(\mathbf{x}',t)}{4\pi|\mathbf{x}-\mathbf{x}'|} \right]$$

$$= \mathbf{j}_t(\mathbf{x},t) - \nabla \int d^3x' \frac{\nabla_{x'} \cdot \mathbf{j}(\mathbf{x}',t)}{4\pi|\mathbf{x}-\mathbf{x}'|} = \mathbf{j}_t + \mathbf{j}_l \qquad (11.4.14)$$

where

$$\mathbf{j}_l(\mathbf{x},t) = -\nabla \left[\nabla \cdot \int d^3x' \frac{\mathbf{j}(\mathbf{x}',t)}{4\pi|\mathbf{x}-\mathbf{x}'|} \right]$$

$$= -\nabla \int d^3x' \frac{\nabla_{x'} \cdot \mathbf{j}(\mathbf{x}',t)}{4\pi|\mathbf{x}-\mathbf{x}'|} = \frac{1}{4\pi} \nabla \, \dot{\phi}\,(\mathbf{x},t) \qquad (11.4.15)$$

It is easy to see that

$$\nabla \times \mathbf{j}_l(\mathbf{x},t) = 0 \qquad (11.4.16)$$

A vector whose curl is zero is called *longitudinal* or *irrotational*; it can be expressed as the gradient of a scalar function.

From Eq. (11.4.14) we see that the current $\mathbf{j}(x,t)$ is the sum of a transverse vector $\mathbf{j}_t(\mathbf{x},t)$ and of a longitudinal vector $\mathbf{j}_l(\mathbf{x},t)$. Any vector that is a function of position can be expressed as a sum of a transversal vector and a longitudinal vector.

The differential equation for the vector potential $\mathbf{A}(\mathbf{x},t)$ is

$$\Box^2 \mathbf{A}(\mathbf{x},t) = -\frac{4\pi}{c} \mathbf{j}_t(\mathbf{x},t) \qquad (11.4.17)$$

where

$$\mathbf{j}_t(\mathbf{x},t) = \mathbf{j}(\mathbf{x},t) + \frac{1}{4\pi} \nabla \int d^3x' \frac{\nabla_{x'} \cdot \mathbf{j}(\mathbf{x}' \cdot t)}{|\mathbf{x}-\mathbf{x}'|} \qquad (11.4.18)$$

The homogeneous equation

$$\Box^2 \mathbf{A}(\mathbf{x}, t) = \nabla^2 \mathbf{A}(\mathbf{x}, t) - \frac{1}{c^2} \frac{\partial^2}{\partial t^2} \mathbf{A}(\mathbf{x}, t) = 0 \qquad (11.4.19)$$

represents the free electromagnetic waves that can exist independently of the sources.

Let us consider the set function

$$\frac{1}{\sqrt{V}} e^{i\mathbf{k} \cdot \mathbf{x}} \qquad (11.4.20)$$

where V = volume of a large cube = L^3. These functions have the following properties:

(1) They are eigenfunctions of the linear momentum operator of quantum mechanics with "box normalization".
(2) The values that $\mathbf{k}$ can take are determined by the boundary conditions. If we specify these boundary conditions to be periodic

$$k_x = \frac{2\pi n_x}{L}$$

$$k_y = \frac{2\pi n_y}{L} \qquad (11.4.21)$$

$$k_z = \frac{2\pi n_z}{L}$$

with n_x, n_y, and n_z integer positive or negative or zero. Then the number of allowed $\mathbf{k}$ vectors with k_x in $(k_x, k_x + dk_x), k_y$ in $(k_y, k_y + dk_y)$, and k_z in $(k_z, k_z + dk_z)$ is given by

$$dn_x = dn_y = dn_z = \frac{dk_x L}{2\pi} \frac{dk_y L}{2\pi} \frac{dk_z L}{2\pi} - \frac{V}{8\pi^3} d^3\mathbf{k} \qquad (11.4.22)$$

If L is large enough, the allowed $\mathbf{k}$ vectors are closely spaced and we can replace the sum over $\mathbf{k}$ with an integral:

$$\frac{1}{V} \sum_{\mathbf{k}} \to \frac{1}{8\pi^3} \int d^3\mathbf{k} \qquad (11.4.23)$$

(3) They are orthonormal:

$$\int_V d^3\mathbf{x} \frac{e^{i\mathbf{k} \cdot \mathbf{x}}}{\sqrt{V}} \cdot \left(\frac{e^{i\mathbf{k'} \cdot \mathbf{x}}}{\sqrt{V}} \right)^* = \delta_{\mathbf{k} \cdot \mathbf{k'}} \qquad (11.4.24)$$

(4) They present the closure property:

$$\frac{1}{V} \sum_{\mathbf{k}} e^{i\mathbf{k}\cdot(\mathbf{x}-\mathbf{x}')} = \delta(\mathbf{x} - \mathbf{x}') \qquad (11.4.25)$$

(5) Given a certain function $f(\mathbf{x})$ defined in V, we can expand it as follows:

$$f(\mathbf{x}) = \frac{1}{\sqrt{V}} \sum_{\mathbf{k}} f_{\mathbf{k}} e^{i\mathbf{k}\cdot\mathbf{x}} \qquad (11.4.26)$$

where $f_{\mathbf{k}}$, the Fourier transform, is given by

$$f_{\mathbf{k}} = \frac{1}{\sqrt{V}} \int_V f(\mathbf{x}) e^{i\mathbf{k}\cdot\mathbf{x}} d^3\mathbf{x} \qquad (11.4.27)$$

In fact,

$$\frac{1}{\sqrt{v}} \int_V f(\mathbf{x}) e^{i\mathbf{k}\cdot\mathbf{x}} d^3\mathbf{x} = \frac{1}{\sqrt{V}} \int_V \left(\frac{1}{\sqrt{v}} \sum_{\mathbf{k}'} f_{\mathbf{k}'} e^{-i\mathbf{k}'\cdot\mathbf{x}} \right) e^{-i\mathbf{k}\cdot\mathbf{x}} d^3\mathbf{x}$$

$$= \sum_{\mathbf{k}'} \frac{f_{\mathbf{k}'}}{V} \int_V d^3\mathbf{x} e^{i(\mathbf{k}'-\mathbf{k})\cdot\mathbf{x}} = f_{\mathbf{k}}$$

(6) Given a function $\mathbf{j}(\mathbf{x}, t)$, we can write

$$\mathbf{j}(\mathbf{x}, t) = \frac{1}{\sqrt{V}} \sum_{\mathbf{k}} \mathbf{j}_{\mathbf{k}}(t) e^{i\mathbf{k}\cdot\mathbf{x}} \qquad (11.4.28)$$

$$\mathbf{j}_{\mathbf{k}}(t) = \frac{1}{\sqrt{V}} \int \mathbf{j}(\mathbf{x}, t) e^{-i\mathbf{k}\cdot\mathbf{x}} d^3\mathbf{x} \qquad (11.4.29)$$

For any $\mathbf{k}$ vector, we introduce three unit vectors $\varepsilon_{\mathbf{k}s}$ $(s = 1, 2, 3)$, as in Fig. 11.1 $\varepsilon_{\mathbf{k}1}$ and $\varepsilon_{\mathbf{k}2}$ are perpendicular to $\varepsilon_{\mathbf{k}3}$ and perpendicular to each other.

We make the following expansion

$$\mathbf{A}(\mathbf{x}, t) = \frac{1}{\sqrt{V}} \sum_{\mathbf{k}} \sum_{s=1,2,3} \varepsilon_{\mathbf{k}s} Q_{\mathbf{k}s}(t) e^{i\mathbf{k}\cdot\mathbf{x}} \qquad (11.4.30)$$

We note the following:

(1) A must be real. Therefore,

$$Q_{\mathbf{k}s} = Q_{-\mathbf{k}s}^8, \quad \varepsilon_{\mathbf{k}s} = \varepsilon_{-\mathbf{k}s} \qquad (11.4.31)$$

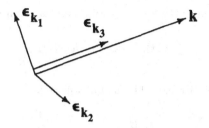

FIGURE 11.1

(2) Since $\boldsymbol{\nabla} \cdot \mathbf{A} = 0$,

$$\sum_{\mathbf{k}s}(i\mathbf{k} \cdot \varepsilon_{\mathbf{k}s})Q_{\mathbf{k}s}(t)e^{i\mathbf{k}\cdot\mathbf{x}} = 0$$

or

$$i\mathbf{k} \cdot \varepsilon_{\mathbf{k}s} = 0 \tag{11.4.32}$$

Therefore, $s = 3$ is out and

$$\mathbf{A}(\mathbf{x}, t) = \frac{1}{\sqrt{V}} \sum_{\mathbf{k}} \sum_{s=1,2} \varepsilon_{\mathbf{k}s}Q_{\mathbf{k}s}(t)e^{i\mathbf{k}\cdot\mathbf{x}} \tag{11.4.33}$$

Let us expand $\mathbf{j}(\mathbf{x}, t)$:

$$\mathbf{j}(\mathbf{x}, t) = \frac{1}{\sqrt{V}} \sum_{\mathbf{k}} \sum_{s=1,2,3} \varepsilon_{\mathbf{k}s}j_{\mathbf{k}s}(t)e^{i\mathbf{k}\cdot\mathbf{x}} \tag{11.4.34}$$

where

$$j_{\mathbf{k}s}(t) = \mathbf{j}_{\mathbf{k}}(t) \cdot \varepsilon_{\mathbf{k}s}, \quad s = 1, 2, 3 \tag{11.4.35}$$

Let us expand $\mathbf{j}_t(\mathbf{x}, t)$:

$$\mathbf{j}_t(\mathbf{x}, t) = \frac{1}{\sqrt{V}} \sum_{\mathbf{k}} \sum_{s=1,2} j_{\mathbf{k}s}(t)\varepsilon_{\mathbf{k}s}e^{i\mathbf{k}\cdot\mathbf{x}} \tag{11.4.36}$$

where

$$j_{\mathbf{k}s}(t) = \mathbf{j}_{\mathbf{k}}(t) \cdot \varepsilon_{\mathbf{k}s}, \quad s = 1, 2 \tag{11.4.37}$$

On the other hand, the $\mathbf{j}_l(\mathbf{x}, t)$ component is given by

$$\mathbf{j}_l(\mathbf{x}, t) = \frac{1}{\sqrt{V}} \sum_{\mathbf{k}} j_{\mathbf{k}3}(t)\varepsilon_{\mathbf{k}3}e^{i\mathbf{k}\cdot\mathbf{x}} \tag{11.4.38}$$

We can now rewrite Eq. (11.4.17) as follows:

$$\Box^2 \frac{1}{\sqrt{V}} \sum_{\mathbf{k}} \sum_{s=1,2} \varepsilon_{\mathbf{k}s} \mathbf{Q}_{\mathbf{k}s}(t)e^{i\mathbf{k}\cdot\mathbf{x}} = -\frac{4\pi}{c}\frac{1}{\sqrt{V}} \sum_{\mathbf{k}}$$

$$\times \sum_{s=1,2} j_{\mathbf{k}s}(t)\varepsilon_{\mathbf{k}s}e^{i\mathbf{k}\cdot\mathbf{x}}$$

$$\left(\nabla^2 - \frac{1}{c^2}\frac{\partial^2}{\partial t^2}\right) \mathbf{Q}_{\mathbf{k}s}(t)e^{i\mathbf{k}\cdot\mathbf{x}} = -\frac{4\pi}{c}j_{\mathbf{k}s}e^{i\mathbf{k}\cdot\mathbf{x}} \tag{11.4.39}$$

$$-k^2\mathbf{Q}_{\mathbf{k}s}(t) - \frac{1}{c^2}\ddot{\mathbf{Q}}_{\mathbf{k}s}(t) = -\frac{4\pi}{c}j_{\mathbf{k}s}(t)$$

If we call $k_c = w_{\mathbf{k}}$, we get

$$\ddot{\mathbf{Q}}_{\mathbf{k}s}(t) + \omega_{\mathbf{k}}^2\mathbf{Q}_{\mathbf{k}s}(t) = 4\pi c j_{\mathbf{k}s}(t), \quad s = 1, 2 \tag{11.4.40}$$

This is essentially the equation of a driven harmonic oscillator. The general solution of differential equation (11.4.40) is equal to the general solution of the homogeneous equation plus a particular solution of the inhomogeneous equation.

Assume that we have the following expression:

$$F(x) = \int_{u(x)}^{v(x)} f(x, y)\,dy \tag{11.4.41}$$

Let

$$f_x(x, y) = \frac{\partial f(x, y)}{\partial x},$$

$$f_{x^2}(x, y) = \frac{\partial^2 f(x, y)}{\partial x^2} \tag{11.4.42}$$

$$f'(x, v) = \left[\frac{\partial}{\partial x}f(x, y)\right]_{y=v},$$

$$f'(x, u) = \left[\frac{\partial}{\partial x}f(x, y)\right]_{y=u} \tag{11.4.43}$$

Then

$$F'(x) = \int_{u(x)}^{v(x)} dy\, f_x(x,y) + v'f(x,v) - u'f(x,u) \qquad (11.4.44)$$

$$F''(x) = \int_{u(x)}^{v(x)} dy\, f_{x^2}(x,y) + v'f_x(x,v) - u'f_x(x,u)$$

$$+ v'f'(x,v) + v''f(x,v)$$

$$- u'f'(x,u) - u''f(x,u) \qquad (11.4.45)$$

CLAIM:

$$F(t) = Q_{\mathbf{ks}}(t) = \int_{-\infty}^{t} 4\pi c \frac{\sin \omega(t-t')}{\omega} j_{\mathbf{ks}}(t')dt' \qquad (11.4.46)$$

is a particular solution of Eq. (11.4.40).

Proof: Comparing Eq. (11.4.46) with Eq. (11.4.41), we have

$$x \to t, \quad y \to t'$$

$$v \to t, \quad v' = 1, \quad v'' = 0$$

$$u \to -\infty$$

Then

$$F''(t) = \int_{-\infty}^{t} f t_2(t,t')dt' + f_t(t,t) - u'f_t(t,-\infty)$$

$$+ f'(t,t) - u'f'(t,-\infty) - u''f(t,-\infty)$$

where

$$f(t,t') = 4\pi c \frac{\sin \omega(t-t')}{\omega} j_{\mathbf{ks}}(t')$$

Thus

$$f(t,t) = 0$$

$$f_t(t,t) = \frac{\partial}{\partial t} f(t,t) = 0$$

$$f'(t,t) = \left[\frac{\partial}{\partial t} f(t,t') \right]_{t'-t} = [4\pi c \cos \omega(t-t') j_{\mathbf{ks}}(t')]_{t'=t}$$

$$= 4\pi c j_{\mathbf{ks}}(t)$$

$$f(t, -\infty) = 0$$

$$f(t, -\infty) = \frac{\partial}{\partial t} f(t, -\infty) = 0$$

$$f'(t, -\infty) = \left[\frac{\partial}{\partial t} f(t, t') \right]_{t'=-\infty}$$

$$= [4\pi \, cj_{\mathbf{ks}}(t') \cos \omega(t - t')]_{t'=-\infty} = 0$$

Also

$$f_t(t, t') = \frac{\partial}{\partial t} f(t, t') = 4\pi c \cos \omega(t - t') j_{\mathbf{ks}}(t')$$

$$f_{t^2}(t, t') = \frac{\partial}{\partial t} f_t(t, t') = -4\pi c\omega \sin \omega(t - t') j_{\mathbf{ks}}(t')$$

$$= -\omega^2 f(t, t')$$

Therefore, we have

$$\ddot{F}(t) = \ddot{Q}_{\mathbf{ks}}(t) = -\omega^2 \int_{-\infty}^{t} 4\pi c \frac{\sin \omega(t - t')}{\omega} j_{\mathbf{ks}}(t') + 4\pi cj_{\mathbf{ks}}(t)$$

or

$$\ddot{Q}_{\mathbf{ks}}(t) + \omega_{\mathbf{k}}^2 Q_{\mathbf{ks}}^{(t)} = 4\pi cj_{\mathbf{ks}}(t) \qquad\qquad \text{Q.E.D.}$$

Therefore, the solution of Eq. (11.4.40) is given by the particular solution (11.4.46) of the inhomogeneous equation plus the general solution of the homogeneous equation. This last quantity is

$$Q_{\mathbf{ks}}(t) = Q_{\mathbf{ks}}^{(0)} \cos(\omega t + \phi) \qquad\qquad (11.4.47)$$

CLAIM: The Lagrangian of the **A** field is given by

$$L = \frac{1}{4\pi c^2} \left[\frac{1}{2} \sum_{ks} \dot{Q}_{\mathbf{ks}} \dot{Q}_{\mathbf{ks}}^* - \frac{1}{2} \sum_{ks} \omega_{\mathbf{k}}^2 Q_{\mathbf{ks}} Q_{\mathbf{ks}}^* + 4\pi c \sum_{ks} j_{\mathbf{ks}} Q_{\mathbf{ks}}^* \right] \quad (11.4.48)$$

Proof:

$$\frac{\partial L}{\partial \dot{Q}_{\mathbf{k}s}^{*}} = \frac{1}{4\pi c^2}\frac{1}{2}2\,\dot{Q}_{\mathbf{k}s} = \frac{\dot{Q}_{\mathbf{k}s}}{4\pi c^2}$$

$$\frac{\partial L}{\partial Q_{\mathbf{k}s}^{*}} = -\frac{\omega_{\mathbf{k}}^2 Q_{\mathbf{k}s}}{4\pi c^2} = \frac{1}{c} - j_{\mathbf{k}s}$$

$$\frac{d}{dt}\left(\frac{\partial L}{\partial \dot{Q}_{\mathbf{k}s}^{*}}\right) - \frac{\partial L}{\partial Q_{\mathbf{k}s}^{*}} = \frac{\ddot{Q}_{\mathbf{k}s}}{4\pi c^2} + \frac{\omega_{\mathbf{k}}^2 Q_{\mathbf{k}s}}{4\pi c^2} - \frac{1}{c}j_{\mathbf{k}s} = 0$$

or

$$\ddot{Q}_{\mathbf{k}s} + \omega_k^2 Q_{\mathbf{k}s} - 4\pi c j_{\mathbf{k}s} = 0 \qquad \text{Q.E.D.}$$

Formula (11.4.48) gives us the Lagrangian L of the fields in terms of Fourier transforms; that is, the fields are described as a system of harmonic oscillators. However, we should not get the impression that the Lagrangian formulation of the Maxwell equations is possible only in terms of harmonic oscillator expansion. The following treatment of the Hamiltonian of the fields will include a formulation for continuous systems that can be easily extended to the Lagrangian L.

11.5. Hamiltonian Formulation of the Maxwell Equations

The Hamiltonian of the $\mathbf{A}$ field is given by

$$H = \sum_{\mathbf{k}s} P_{\mathbf{k}s}\,\dot{Q}_{\mathbf{k}s} - L \qquad (11.5.1)$$

where

$$L = \frac{1}{4\pi c^2}\left[\frac{1}{2}\sum_{\mathbf{k}s} \dot{Q}_{\mathbf{k}s}\dot{Q}_{\mathbf{k}s}^{*} - \frac{1}{2}\sum_{\mathbf{k}s}\omega_k^2\,Q_{\mathbf{k}s}Q_{\mathbf{k}s}^{*} + 4\pi c\sum_{\mathbf{k}s} j_{\mathbf{k}s}Q_{\mathbf{k}s}^{*}\right] \qquad (11.5.2)$$

and

$$P_{\mathbf{k}s} = \frac{\partial L}{\partial \dot{Q}_{\mathbf{k}s}} = \frac{1}{4\pi c^2}\,\dot{Q}_{\mathbf{k}s}^{*} \qquad (11.5.3)$$

Then

$$H = \frac{1}{4\pi c^2}\left[\sum_{ks} \dot{Q}_{ks}\,\dot{Q}^*_{ks} - \frac{1}{2}\sum_{ks}\dot{Q}_{ks}\,\dot{Q}^*_{ks}\right.$$

$$\left. + \frac{1}{2}\sum_{ks}\omega_k^2 Q_{ks}\,Q^*_{ks} - 4\pi c\sum_{ks}j_{ks}\,Q^*_{ks}\right]$$

$$= \frac{1}{8\pi c^2}\sum_{ks}[\dot{Q}_{ks}\,\dot{Q}^*_{ks} + \omega_{ks}^2\,Q_{ks}\,Q^*_{ks}]\quad \frac{1}{c}\sum_{ks}j_{ks}\,Q^*_{ks}$$

$$= \frac{1}{2}\sum_{ks}\left[4\pi c^2 P_{ks}P^*_{ks} + \frac{\omega_k^2}{4\pi c^2}Q_{ks}\,Q^*_{ks}\right] - \frac{1}{c}\sum_{ks}j_{ks}Q^*_{ks} \quad (11.5.4)$$

Hamilton's equations give

$$\dot{Q}_{ks} = \frac{\partial H}{\partial P_{ks}} = 2\frac{1}{2}4\pi c^2 P^*_{ks} = 4\pi c^2 P^*_{ks} \quad\quad\quad (11.5.5)$$

which is in accord with Eq. (11.5.3), and

$$\dot{P}_{ks} = -\frac{\partial H}{\partial Q_{ks}} = -\frac{\omega_k^2}{4\pi c^2}Q^*_{ks} + \frac{1}{c}j_{-ks} \quad\quad (11.5.6)$$

or

$$\frac{1}{4\pi c^2}\ddot{Q}^*_{ks} + \frac{\omega_k^2}{4\pi c^2}Q^*_{ks} = \frac{1}{c}j - ks$$

Changing $\mathbf{k}$ into $-\mathbf{k}$, we obtain

$$\ddot{Q}_{ks} + \omega_k^2 Q_{ks} = 4\pi c j_{ks} \quad\quad\quad (11.5.7)$$

which gives the known result.

We recall that

$$\mathbf{A}(\mathbf{x}, t) = \frac{1}{\sqrt{V}}\sum_{ks}\varepsilon_{ks}Q_{ks}e^{i\mathbf{k}\cdot\mathbf{x}}, \quad s = 1, 2 \quad\quad (11.5.8)$$

and

$$\mathbf{j}(\mathbf{x}, t) = \frac{1}{\sqrt{V}}\sum_{ks}\varepsilon_{ks}j_{ks}e^{i\mathbf{k}\cdot\mathbf{x}}, \quad s = 1, 2, 3 \quad\quad (11.5.9)$$

Then

$$\frac{1}{c}\int d^3\mathbf{x}\,\mathbf{j}(\mathbf{x},t)\cdot\mathbf{A}(\mathbf{x},t) = \frac{1}{c}\int d^3\mathbf{x}\frac{1}{\sqrt{V}}\sum_{\mathbf{k}s}\varepsilon_{\mathbf{k}s}j_{\mathbf{k}s}e^{i\mathbf{k}\cdot\mathbf{x}}$$

$$\cdot\frac{1}{\sqrt{V}}\sum_{\mathbf{k}'s'}\varepsilon_{\mathbf{k}'s'}Q_{\mathbf{k}'s'}e^{i\mathbf{k}'\cdot\mathbf{x}}$$

$$=\frac{1}{c}\sum_{\mathbf{k}s}j_{\mathbf{k}s}\varepsilon_{\mathbf{k}s}\sum_{\mathbf{k}'s}Q_{\mathbf{k}'s'}\varepsilon_{\mathbf{k}'s}\delta_{\mathbf{k},-\mathbf{k}'}$$

$$=\frac{1}{c}\sum_{\mathbf{k}s}j_{\mathbf{k}s}Q_{\mathbf{k}s}^{*} \tag{11.5.10}$$

Also, calling $\mathbf{E}^{\mathrm{rad}}(\mathbf{x},t)$ the component of the electric field associated with $\mathbf{A}$,

$$\mathbf{E}^{\mathrm{rad}}(\mathbf{x},t) = -\frac{1}{c}\,\dot{\mathbf{A}} = -\frac{1}{c}\frac{1}{\sqrt{V}}\sum_{\mathbf{k}s}\varepsilon_{\mathbf{k}s}\,\dot{Q}_{\mathbf{k}s}\,e^{i\mathbf{k}\cdot\mathbf{x}} \tag{11.5.11}$$

and

$$\int[\mathbf{E}^{\mathrm{rad}}(\mathbf{x},t)]^2 d^3\mathbf{x} = \frac{1}{c^2}\frac{1}{V}\sum_{\mathbf{k}s}\sum_{\mathbf{k}'s'}\varepsilon_{\mathbf{k}s}\cdot\varepsilon_{\mathbf{k}'s'}\,\dot{Q}_{\mathbf{k}s}\dot{Q}_{\mathbf{k}'s'}$$

$$\times\int e^{i(\mathbf{k}+\mathbf{k}')\cdot\mathbf{x}}d^3\mathbf{x}$$

$$=\frac{1}{c^2}\sum_{\mathbf{k}s}\sum_{\mathbf{k}'s'}\varepsilon_{\mathbf{k}s}\cdot\varepsilon_{\mathbf{k}'s'}\,\dot{Q}_{\mathbf{k}s}\dot{Q}_{\mathbf{k}'s'}\,\delta_{\mathbf{k},-\mathbf{k}'}$$

$$=\frac{1}{c^2}\sum_{\mathbf{k}s}\sum_{s'}\varepsilon_{\mathbf{k}s}\cdot\varepsilon_{-\mathbf{k}s'}\,\dot{Q}_{\mathbf{k}s}\dot{Q}_{-\mathbf{k}s'}$$

$$=\frac{1}{c^2}\sum_{\mathbf{k}s}\dot{Q}_{\mathbf{k}s}\dot{Q}_{\mathbf{k}s'}^{*}$$

$$\frac{1}{8\pi}\int[\mathbf{E}^{\mathrm{rad}}(\mathbf{x},t)]^2 d^3\mathbf{x} = \frac{1}{8\pi c^2}\sum_{\mathbf{k}s}\dot{Q}_{\mathbf{k}s}\dot{Q}_{\mathbf{k}s}^{*}$$

$$=\frac{1}{2}\sum_{\mathbf{k}s}4\pi c^2 P_{\mathbf{k}s}P_{\mathbf{k}s}^{*} \tag{11.5.12}$$

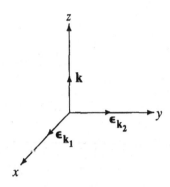

FIGURE 11.2

Note that the field may also have a longitudinal component, so the total field is given by

$$\mathbf{E}_{\text{tot}} = \mathbf{E}^{\text{long}} + \mathbf{E}^{\text{rad}} = -\nabla\phi - \frac{1}{c}\dot{\mathbf{A}} \qquad (11.5.13)$$

This separation of the $\mathbf{E}$ field into longitudinal and transverse (radiative) parts is possible only when using the present $\nabla \cdot \mathbf{A} = 0$ gauge.

Finally, assume $\mathbf{k}$ to be in the z direction, as in Fig. 11.2. The $\mathbf{k}$ component of $\mathbf{A}$ would then be

$$\frac{1}{\sqrt{V}}\varepsilon_1 Q_1 e^{ikz} + \frac{1}{\sqrt{V}}\varepsilon_2 Q_2 e^{ikz}$$

and

$$
\begin{aligned}
(\nabla \times \mathbf{A})_{\text{kcomp}} &= \frac{1}{\sqrt{V}}\left[\hat{\mathbf{j}}\frac{\partial}{\partial z}Q_1 e^{ikz} - \hat{\mathbf{k}}\frac{\partial}{\partial y}Q_1 e^{ikz}\right.\\
&\qquad\left. -\hat{\mathbf{i}}\frac{\partial}{\partial z}Q_2 e^{ikz} + \hat{\mathbf{k}}\frac{\partial}{\partial z}Q_2 e^{ikz}\right]\\
&= \frac{1}{\sqrt{V}}[\hat{\mathbf{j}}(ik)Q_1 e^{ikz} - \hat{\mathbf{i}}(ik)Q_2 e^{ikz}] \qquad (11.5.14)
\end{aligned}
$$

where $\hat{\mathbf{i}}, \hat{\mathbf{j}}$, and $\hat{\mathbf{k}}$ are the unit vectors in the x, y, and z directions, respectively. Since

$$\frac{1}{V}\int d^3\mathbf{x}e^{i(\mathbf{k}+\mathbf{k}')\cdot\mathbf{x}} = \delta_{\mathbf{k},-\mathbf{k}'} \qquad (11.5.15)$$

when calculating the integral of $(\nabla \times \mathbf{A})^2$, we shall have, for each $\mathbf{k}$ component,

$$[\hat{\mathbf{j}}(ik)Q_1 e^{ikz} - \hat{\mathbf{i}}(ik)Q_2 e^{ikz}] \cdot [\hat{\mathbf{j}}(-ik)Q_1^* e^{-ikz} - \hat{\mathbf{i}}(-ik)Q_2^* e^{-ikz}]$$

$$= k^2 Q_1 Q_1^* + k^2 Q_2 Q_2^* \tag{11.5.16}$$

Then

$$\frac{1}{8\pi} \int [\mathbf{B}^{\text{rad}}(\mathbf{x}, t)]^2 d^3\mathbf{x} = \sum_{\mathbf{k}s} \frac{\omega_{\mathbf{k}}^2}{8\pi c^2} Q_{\mathbf{k}s} Q_{\mathbf{k}s}^* \tag{11.5.17}$$

Because of Eqs. (11.5.10), (11.5.12), and (11.5.17), we can write

$$H = \frac{1}{8\pi} \int (\mathbf{E}^{\text{rad}})^2 d^3\mathbf{x} + \frac{1}{8\pi} \int (\mathbf{B}^{\text{rad}})^2 d^3\mathbf{x}$$

$$- \frac{1}{c} \int \mathbf{j}(\mathbf{x}, t) \cdot \mathbf{A}(\mathbf{x}, t) d^3\mathbf{x} \tag{11.5.18}$$

An expression for the Lagrangian in terms of the fields can be easily derived:

$$L = \frac{1}{8\pi} \int (\mathbf{E}^{\text{rad}})^2 d^3\mathbf{x} - \frac{1}{8\pi} \int (\mathbf{B}^{\text{rad}})^2 d^3\mathbf{x}$$

$$+ \frac{1}{c} \int \mathbf{j}(\mathbf{x}, t) \cdot \mathbf{A}(\mathbf{x}, t) d^3\mathbf{x} \tag{11.5.19}$$

11.6. Poisson Bracket Method

In this section we give a detailed treatment of the fields based on the use of the *Poisson brackets*. The main motivation for the presentation of this method, widely used in classical mechanics, is that it provides the starting point for the quantization of the fields and the quantum theory of radiation.

Consider a set of generalized coordinates q_s and momenta p_s. Let

$$H = H(p_s, q_s, t) \tag{11.6.1}$$

be the Hamiltonian of the system. A Poisson bracket of two quantities F and G is defined as follows:

$$\{F, G\} = \sum_s \left(\frac{\partial F}{\partial p_s} \frac{\partial G}{\partial q_s} - \frac{\partial F}{\partial q_s} \frac{\partial G}{\partial p_s} \right) \tag{11.6.2}$$

The Poisson brackets have the following properties

(1)

$$\{p_s, q_t\} = \delta_{st} \tag{11.6.3}$$

$$\{p_s, p_t\} = \{q_s, q_t\} = 0 \tag{11.6.4}$$

(2)

$$\{F_1 F_2, G\} = F_1\{F_2, G\} + F_2\{F_1, G\} \tag{11.6.5}$$

This follows from the fact that we are dealing with linear differential operations.

(3)

$$\{F(p_s, q_s), q_t\}$$
$$= \sum_s \left(\frac{\partial F}{\partial p_s} \frac{\partial q_t}{\partial q_s} - \frac{\partial F}{\partial q_s} \frac{\partial q_t}{\partial p_s} \right) = \frac{\partial F}{\partial p_t} \tag{11.6.6}$$

$$\{F(p_s, q_s), p_t\}$$
$$= \sum_s \left(\frac{\partial F}{\partial p_s} \frac{\partial p_t}{\partial q_s} - \frac{\partial F}{\partial q_s} \frac{\partial p_t}{\partial p_s} \right) = -\frac{\partial F}{\partial q_t} \tag{11.6.7}$$

$$\{H(p_s, q_s, t), F(p_s, q_s)\}$$
$$= \sum_s \left(\frac{\partial H}{\partial p_s} \frac{\partial F}{\partial q_s} - \frac{\partial H}{\partial q_s} \frac{\partial F}{\partial p_s} \right)$$
$$= \sum_s \left(\frac{dq_s}{dt} \frac{\partial F}{\partial q_s} + \frac{dp_s}{dt} \frac{\partial F}{\partial p_s} \right) = \frac{dF}{dt} \tag{11.6.8}$$

In our case, the generalized coordinates are $Q_{\mathbf{k}s}$, and the generalized momenta are

$$P_{\mathbf{k}s} = \frac{1}{4\pi c^2} \dot{Q}_{\mathbf{k}s}^* \tag{11.6.9}$$

We have

$$\{Q_{\mathbf{k}s}, Q_{\mathbf{k}'s'}\} = \{P_{\mathbf{k}s}, P_{\mathbf{k}'s'}\} = 0 \tag{11.6.10}$$

$$\{\dot{Q}_{\mathbf{k}s}, Q_{\mathbf{k}'s'}\} = 4\pi c^2 \{P_{\mathbf{k}s}^*, Q_{\mathbf{k}'s'}\}$$
$$= 4\pi c^2 \{P_{-\mathbf{k}s}, Q_{\mathbf{k}'s'}\} = 4\pi c^2 \delta_{\mathbf{k}, -\mathbf{k}'} \delta_{ss'} \tag{11.6.11}$$

We find

$$\{A_i(\mathbf{x}, t), A_j(\mathbf{x}', t)\} = 0 \tag{11.6.12}$$

because the $A_i(\mathbf{x}, t)$ components depend only on the generalized coordinates and are independent of the generalized momenta.

On the other hand,

$$\{\dot{A}_i(\mathbf{x}, t), A_j(\mathbf{x}', t)\}$$

$$= \frac{1}{V} \sum_{\mathbf{k}s} \sum_{\mathbf{k}'s'} \{\varepsilon_{\mathbf{k}si} \dot{Q}_{\mathbf{k}s}(t) e^{i\mathbf{k}\cdot\mathbf{x}}, \varepsilon_{\mathbf{k}'s'j} Q_{\mathbf{k}'s'}(t) e^{i\mathbf{k}'\cdot\mathbf{x}'}\}$$

$$= \frac{1}{V} \sum_{\mathbf{k}s} \sum_{\mathbf{k}'s'} \varepsilon_{\mathbf{k}si} \varepsilon_{\mathbf{k}'s'j} e^{i\mathbf{k}\cdot\mathbf{x}} e^{i\mathbf{k}'\cdot\mathbf{x}'} \{\dot{Q}_{\mathbf{k}s}, Q_{\mathbf{k}'s'}\}$$

$$= \frac{1}{V} \sum_{\mathbf{k}s} \sum_{\mathbf{k}'s'} \varepsilon_{\mathbf{k}si} \varepsilon_{\mathbf{k}'s'j} e^{i\mathbf{k}\cdot\mathbf{x}} e^{i\mathbf{k}'\cdot\mathbf{x}'} 4\pi c^2 \delta_{\mathbf{k}, -\mathbf{k}'} \delta_{ss'}$$

$$= \sum_{\mathbf{k}s} \varepsilon_{\mathbf{k}si} \varepsilon_{\mathbf{k}sj} e^{i\mathbf{k}\cdot(\mathbf{x}-\mathbf{x}')} \frac{4\pi c^2}{V}$$

$$= 4\pi c^2 \sum_{\mathbf{k}} \frac{1}{V} e^{i\mathbf{k}\cdot(\mathbf{x}-\mathbf{x}')} \sum_s \varepsilon_{\mathbf{k}si} \varepsilon_{\mathbf{k}sj} \qquad (11.6.13)$$

But

$$\sum_{s=1}^{3} \varepsilon_{\mathbf{k}si} \varepsilon_{\mathbf{k}sj} = \delta_{ij} \qquad (11.6.14)$$

$$\sum_{s=1}^{2} \varepsilon_{\mathbf{k}si} \varepsilon_{\mathbf{k}sj} = \delta_{ij} - \frac{k_i k_j}{k^2} \qquad (11.6.15)$$

Therefore,

$$\{\dot{A}_i(\mathbf{x}, t), A_j(\mathbf{x}', t)\}$$

$$= 4\pi c^2 \sum_{\mathbf{k}} \frac{1}{V} e^{i\mathbf{k}\cdot(\mathbf{x}-\mathbf{x}')} \left(\delta_{ij} - \frac{k_i k_j}{k^2}\right)$$

$$= 4\pi c^2 \delta_{ij} \sum_{\mathbf{k}} \frac{1}{V} e^{i\mathbf{k}\cdot(\mathbf{x}-\mathbf{x}')} - 4\pi c^2 \sum_{\mathbf{k}} \frac{1}{V} e^{i\mathbf{k}\cdot(\mathbf{x}-\mathbf{x}')} \frac{k_i k_j}{k^2}$$

$$= 4\pi c^2 \left[\delta_{ij} \delta(\mathbf{x} - \mathbf{x}') + \frac{\partial}{\partial x_i} \frac{\partial}{\partial x_j} \sum_{\mathbf{k}} \frac{1}{V} \frac{e^{i\mathbf{k}\cdot(\mathbf{x}-\mathbf{x}')}}{k^2}\right] \qquad (11.6.16)$$

We present now the following claim.

CLAIM:

$$\frac{1}{V} \sum_{\mathbf{k}} \frac{e^{i\mathbf{k} \cdot (\mathbf{x} - \mathbf{x}')}}{k^2} = \frac{1}{4\pi |\mathbf{x} - \mathbf{x}'|} \qquad (11.6.17)$$

Proof:

$$\nabla^2 \frac{1}{V} \sum_{\mathbf{k}} \frac{e^{i\mathbf{k} \cdot (\mathbf{x} - \mathbf{x}')}}{k^2} = -\frac{1}{V} \sum_{\mathbf{k}} e^{i\mathbf{k} \cdot (\mathbf{x} - \mathbf{x}')} = -\delta(\mathbf{x} - \mathbf{x}')$$

$$\nabla^2 \frac{1}{4\pi |\mathbf{x} - \mathbf{x}'|} = -\frac{4\pi \delta(\mathbf{x} - \mathbf{x}')}{4\pi} = -\delta(\mathbf{x} - \mathbf{x}')$$

Considerations similar to those of Sec. 2.7 are in order. The left and right members of Eq. (11.6.17) may differ only by a function whose Laplacian is zero. This function has to go to zero at infinity, and therefore is identically zero.   Q.E.D.

Therefore,

$$\{\dot{A}_i(\mathbf{x}, t), A_j(\mathbf{x}', t)\} = 4\pi c^2 \left\{ \delta_{ij}\delta(\mathbf{x} - \mathbf{x}') + \frac{\partial^2}{\partial x_i \partial x_j} \frac{1}{4\pi |\mathbf{x} - \mathbf{x}'|} \right\}$$

$$= 4\pi c^2 \delta_{ij}^{\text{tr}}(\mathbf{x} - \mathbf{x}') \qquad (11.6.18)$$

where

$$\delta_{ij}^{\text{tr}}(\mathbf{x} - \mathbf{x}') = \delta_{ij}\delta(\mathbf{x} - \mathbf{x}') + \frac{\partial^2}{\partial x_i \partial x_j} \frac{1}{4\pi |\mathbf{x} - \mathbf{x}'|} \qquad (11.6.19)$$

The Poisson brackets consist of differential operations over the P's and Q's, but not over the $\mathbf{x}$'s, which are simply parameters.

The Poisson brackets must be consistent with the fact that $\nabla \cdot \mathbf{A} = 0$. This can be verified.

Note also

$$\{E_i(\mathbf{x}, t), A_j(\mathbf{x}', t)\} = \left\{ -\frac{\partial \phi(\mathbf{x}, t)}{\partial x_i} - \frac{1}{c} \dot{A}_i(\mathbf{x}, t), A_j(\mathbf{x}', t) \right\}$$

$$= -\frac{1}{c} \{\dot{A}_i(\mathbf{x}, t), A_j(\mathbf{x}', t)\}$$

$$= -4\pi c \delta_{ij}^{\text{tr}}(\mathbf{x} - \mathbf{x}') \qquad (11.6.20)$$

because ϕ is function only of x.

CLAIM: Given a certain vector field $\mathbf{F}(\mathbf{x}, t)$

$$\int \delta_{ij}^{\mathrm{tr}}(\mathbf{x} - \mathbf{x}') F_j(\mathbf{x}', t) d^3 x' = G_i^{\mathrm{tr}}(\mathbf{x}, t) \qquad (11.6.21)$$

where $G^{\mathrm{tr}}(\mathbf{x}, t)$ is a vector that is transverse:

$$\boldsymbol{\nabla} \cdot \mathbf{G}^{\mathrm{tr}}(\mathbf{x}, t) = 0 \qquad (11.6.22)$$

but is *not* the transverse part of $\mathbf{F}(\mathbf{x}, t)$.

Proof:

$$\delta_{ij}^{\mathrm{tr}}(\mathbf{x} - \mathbf{x}') = \delta_{ij} \delta(\mathbf{x} - \mathbf{x}') + \frac{\partial^2}{\partial x_i \partial x_j} \frac{1}{4\pi |\mathbf{x} - \mathbf{x}'|}$$

Set $j = 1$. Then we get

$$\delta_{11}^{\mathrm{tr}}(\mathbf{x} - \mathbf{x}') = \delta(\mathbf{x} - \mathbf{x}') + \frac{\partial^2}{\partial x^2} \frac{1}{4\pi |\mathbf{x} - \mathbf{x}'|}$$

$$\delta_{21}^{\mathrm{tr}}(\mathbf{x} - \mathbf{x}') = \frac{\partial^2}{\partial x \partial y} \frac{1}{4\pi |\mathbf{x} - \mathbf{x}'|}$$

$$\delta_{31}^{\mathrm{tr}}(\mathbf{x} - \mathbf{x}') = \frac{\partial^2}{\partial x \partial z} \frac{1}{4\pi |\mathbf{x} - \mathbf{x}'|}$$

and

$$\int \delta_{11}^{\mathrm{tr}}(\mathbf{x} - \mathbf{x}') F_x(\mathbf{x}', t) d^3 x' = \int \delta(\mathbf{x} - \mathbf{x}') F_x(\mathbf{x}', t) d^3 x'$$

$$+ \frac{\partial^2}{\partial x^2} \int \frac{F_x(\mathbf{x}', t)}{4\pi |\mathbf{x} - \mathbf{x}'|} d^3 x'$$

$$= F_x(\mathbf{x}, t) + \frac{\partial^2}{\partial x^2} \int \frac{F_x(\mathbf{x}', t)}{4\pi |\mathbf{x} - \mathbf{x}'|} d^3 x'$$

$$\int \delta_{21}^{\mathrm{tr}}(\mathbf{x} - \mathbf{x}') F_x(\mathbf{x}', t) d^3 x' = \frac{\partial^2}{\partial x \partial y} \int \frac{F_x(\mathbf{x}', t)}{4\pi |\mathbf{x} - \mathbf{x}'|} d^3 x'$$

$$\int \delta_{31}^{\mathrm{tr}}(\mathbf{x} - \mathbf{x}') F_x(\mathbf{x}', t) d^3 x' = \frac{\partial^2}{\partial x \partial z} \int \frac{F_x(\mathbf{x}', t)}{4\pi |\mathbf{x} - \mathbf{x}'|} d^3 x'$$

We have now a vector with three components:

$$G_x^{\text{tr}}(\mathbf{x},t) = F_x(\mathbf{x},t) + \frac{\partial^2}{\partial x^2}\int \frac{F_x(\mathbf{x}',t)}{4\pi|\mathbf{x}-\mathbf{x}'|}d^3\mathbf{x}'$$

$$G_y^{\text{tr}}(\mathbf{x},t) = \frac{\partial^2}{\partial x\partial y}\int \frac{F_x(\mathbf{x}',t)}{4\pi|\mathbf{x}-\mathbf{x}'|}d^3\mathbf{x}'$$

$$G_z^{\text{tr}}(\mathbf{x},t) = \frac{\partial^2}{\partial x\partial z}\int \frac{F_x(\mathbf{x}',t)}{4\pi|\mathbf{x}-\mathbf{x}'|}d^3\mathbf{x}'$$

The divergence of the vector G^{tr} is zero:

$$\boldsymbol{\nabla}\cdot\mathbf{G}^{\text{tr}}(\mathbf{x},t) = \frac{\partial F_x}{\partial x} + \frac{\partial}{\partial x}\left[\frac{\partial^2}{\partial x^2}\int \frac{F_x(\mathbf{x}',t)}{4\pi|\mathbf{x}-\mathbf{x}'|}d^3\mathbf{x}'\right]$$

$$+ \frac{\partial}{\partial y}\left[\frac{\partial^2}{\partial x\partial y}\int \frac{F_x(\mathbf{x}',t)}{4\pi|\mathbf{x}-\mathbf{x}'|}d^3\mathbf{x}'\right]$$

$$+ \frac{\partial}{\partial z}\left[\frac{\partial^2}{\partial x\partial z}\int \frac{F_x(\mathbf{x}',t)}{4\pi|\mathbf{x}-\mathbf{x}'|}d^3\mathbf{x}'\right]$$

$$= \frac{\partial F_x}{\partial x} + \frac{\partial}{\partial x}\left[\left(\frac{\partial^2}{\partial x^2}+\frac{\partial^2}{\partial y^2}+\frac{\partial^2}{\partial z^2}\right)\int \frac{F_x(\mathbf{x}',t)}{4\pi|\mathbf{x}-\mathbf{x}'|}d^3\mathbf{x}'\right]$$

$$= \frac{\partial F_x}{\partial x} + \frac{\partial}{\partial x}\nabla^2\int \frac{F_x(\mathbf{x}',t)}{4\pi|\mathbf{x}-\mathbf{x}'|}d^3\mathbf{x}'$$

$$= \frac{\partial F_x}{\partial x} + \frac{\partial}{\partial x}(-4\pi)\int \frac{F_x(\mathbf{x}',t)}{4\pi}\delta(\mathbf{x}-\mathbf{x}')d^3\mathbf{x}' = 0 \quad \text{Q.E.D.}$$

CLAIM: Given a certain field $\mathbf{F}(\mathbf{x},t)$,

$$\sum_j \int \delta_{ij}^{\text{tr}}(\mathbf{x}-\mathbf{x}')F_j(\mathbf{x}',t)d^3\mathbf{x}' = F_i^{\text{tr}}(\mathbf{x},t) \tag{11.6.23}$$

where $F^{\text{tr}}(\mathbf{x},t)$ is the transverse part of $\mathbf{F}(\mathbf{x},t)$.

Proof: For $j=2$,

$$\int \delta_{12}^{\text{tr}}(\mathbf{x}-\mathbf{x}')F_y(\mathbf{x}',t)d^3\mathbf{x}' = \frac{\partial^2}{\partial x\partial y}\int \frac{F_y(\mathbf{x}',t)}{4\pi|\mathbf{x}-\mathbf{x}'|}d^3\mathbf{x}'$$

$$\int \delta_{22}^{\text{tr}}(\mathbf{x}-\mathbf{x}')F_y(\mathbf{x}',t)d^3\mathbf{x}' = F_y(\mathbf{x},t) + \frac{\partial^2}{\partial y^2}\int \frac{F_y(\mathbf{x}',t)}{4\pi|\mathbf{x}-\mathbf{x}'|}d^3\mathbf{x}'$$

$$\int \delta_{32}^{\text{tr}}(\mathbf{x} - \mathbf{x}')F_y(\mathbf{x}',t)d^3\mathbf{x}' = \frac{\partial^2}{\partial x \partial z} \int \frac{F_y(\mathbf{x}',t)}{4\pi|\mathbf{x} - \mathbf{x}'|}d^3\mathbf{x}'$$

For $j = 3$,

$$\int \delta_{13}^{\text{tr}}(\mathbf{x} - \mathbf{x}')F_z(\mathbf{x}',t)d^3\mathbf{x}' = \frac{\partial^2}{\partial x \partial z} \int \frac{F_z(\mathbf{x}',t)}{4\pi|\mathbf{x} - \mathbf{x}'|}d^3\mathbf{x}'$$

$$\int \delta_{23}^{\text{tr}}(\mathbf{x} - \mathbf{x}')F_z(\mathbf{x}',t)d^3\mathbf{x}' = \frac{\partial^2}{\partial y \partial z} \int \frac{F_z(\mathbf{x}',t)}{4\pi|\mathbf{x} - \mathbf{x}'|}d^3\mathbf{x}'$$

$$\int \delta_{33}^{\text{tr}}(\mathbf{x} - \mathbf{x}')F_z(\mathbf{x}',t)d^3\mathbf{x}' = F_z(\mathbf{x},t) + \frac{\partial^2}{\partial z^2} \int \frac{F_z(\mathbf{x}',t)}{4\pi|\mathbf{x} - \mathbf{x}'|}d^3\mathbf{x}'$$

Now, for $i = 1$,

$$\sum_j \int \delta_{1j}^{\text{tr}}(\mathbf{x} - \mathbf{x}')F_j(\mathbf{x}',t)d^3\mathbf{x}' = \int \delta_{11}^{\text{tr}}(\mathbf{x} - \mathbf{x}')F_x(\mathbf{x}',t)d^3\mathbf{x}'$$

$$+ \int \delta_{12}^{\text{tr}}(\mathbf{x} - \mathbf{x}')F_y(\mathbf{x}',t)d^3\mathbf{x}'$$

$$+ \int \delta_{13}^{\text{tr}}(\mathbf{x} - \mathbf{x}')F_z(\mathbf{x}',t)d^3\mathbf{x}'$$

$$= F_x(\mathbf{x},t) + \frac{\partial^2}{\partial x^2} \int \frac{F_x(\mathbf{x}',t)}{4\pi|\mathbf{x} - \mathbf{x}'|}d^3\mathbf{x}'$$

$$+ \frac{\partial^2}{\partial x \partial y} \int \frac{F_y(\mathbf{x}',t)}{4\pi|\mathbf{x} - \mathbf{x}'|}d^3\mathbf{x}'$$

$$+ \frac{\partial^2}{\partial x \partial z} \int \frac{F_z(\mathbf{x}',t)}{4\pi|\mathbf{x} - \mathbf{x}'|}d^3\mathbf{x}'$$

$$= F_x + \frac{\partial}{\partial x} \left[\frac{1}{4\pi} \boldsymbol{\nabla} \cdot \int d^3\mathbf{x}' \frac{F(\mathbf{x}',t)}{|\mathbf{x} - \mathbf{x}'|} \right]$$

Q.E.D.

We can now summarize some result:

$$\{E_i(\mathbf{x},t), E_j(\mathbf{x}',t)\} = 0 \qquad (11.6.24)$$

$$\{B_i(\mathbf{x},t), B_j(\mathbf{x}',t)\} = 0 \qquad (11.6.25)$$

$$\{E_i(\mathbf{x},t), A_j(\mathbf{x}',t)\} = -4\pi c \delta_{ij}^{\text{tr}}(\mathbf{x} - \mathbf{x}') \qquad (11.6.26)$$

$$\{B_i(\mathbf{x},t), A_j(\mathbf{x}',t)\} = 0 \qquad (11.6.27)$$

$$\{E_i(\mathbf{x},t), B_j(\mathbf{x}',t)\} \neq 0 \qquad (11.6.28)$$

We can verify that the following relation gives a correct result:

$$\dot{\mathbf{A}}(\mathbf{x}, t) = \{H, \mathbf{A}(\mathbf{x}, t)\} \tag{11.6.29}$$

If we use the result (11.6.26), we obtain

$$\dot{A}_i(\mathbf{x}, t) = \frac{1}{8\pi} \int d^3x' \sum_s \{E_s^2(\mathbf{x}', t), A_i(\mathbf{x}, t)\}$$

$$- \frac{1}{4\pi} \int d^3x' \sum_s E_s(\mathbf{x}', t)\{E_s(\mathbf{x}', t), A_i(\mathbf{x}, t)\}$$

$$= \frac{1}{4\pi} \int d^3x' \sum_s E_s(\mathbf{x}', t)[-4\pi c\delta_{si}^{\text{tr}}(\mathbf{x} - \mathbf{x}')]$$

$$= -cE_i^{\text{tr}}(\mathbf{x}, t)$$

which is consistent with the Maxwell equations.

CLAIM: If H is the field Hamiltonian,

$$\ddot{\mathbf{A}} = \{\mathbf{H}, \dot{\mathbf{A}}\} = c^2 \nabla^2 \mathbf{A} + 4\pi c\mathbf{j_t} \tag{11.6.30}$$

consistent with

$$\nabla^2 \mathbf{A} - \frac{1}{c^2} \frac{\partial^2 \mathbf{A}}{\partial t^2} = -4\pi c\mathbf{j_t} \tag{11.6.31}$$

Proof: We can write

$$\mathbf{A}(\mathbf{x}, t) = \frac{1}{\sqrt{V}} \sum_{\mathbf{k}s} \varepsilon_{\mathbf{k}s} Q_{\mathbf{k}s}(t) e^{i\mathbf{k} \cdot \mathbf{x}}$$

$$\dot{\mathbf{A}}(\mathbf{x}, t) = \frac{1}{\sqrt{V}} \sum_{\mathbf{k}s} \varepsilon_{\mathbf{k}s} \dot{Q}_{\mathbf{k}s}(t) e^{i\mathbf{k} \cdot \mathbf{x}}$$

Also

$$\{P_{\mathbf{k}s}, Q_{\mathbf{k}s}\} = \frac{1}{4\pi c^2} \{\dot{Q}_{\mathbf{k}s}^*, Q_{\mathbf{k}s}\} = 1$$

$$\{Q_{\mathbf{k}s}, \dot{Q}_{\mathbf{k}s}^*\} = -4\pi c^2$$

$$\{Q_{\mathbf{k}s}^*, \dot{Q}_{\mathbf{k}s}\} = -4\pi c^2$$

The Hamiltonian can be written as follows:

$$H = \frac{1}{8\pi c^2} \sum_{\mathbf{k}s} [\dot{Q}_{\mathbf{k}s}\dot{Q}_{\mathbf{k}s}^* + \omega_{\mathbf{k}}^2 Q_{\mathbf{k}s}Q_{\mathbf{k}s}^*] - \frac{1}{c} \sum_{\mathbf{k}s} j_{\mathbf{k}s}Q_{\mathbf{k}s}^*$$

Then

$$\ddot{\mathbf{A}} = \{H, \dot{\mathbf{A}}\} = \left\{ \frac{1}{8\pi c^2} \sum_{\mathbf{k}s} \omega_{\mathbf{k}}^2 Q_{\mathbf{k}s} Q_{\mathbf{k}s}^* \right.$$

$$\left. - \frac{1}{c} \sum_{\mathbf{k}s} j_{\mathbf{k}s} Q_{\mathbf{k}s}^*, \frac{1}{\sqrt{V}} \sum_{\mathbf{k}'s'} \varepsilon_{\mathbf{k}'s'} \dot{Q}_{\mathbf{k}'s'} e^{i\mathbf{k}\cdot\mathbf{x}} \right\}$$

$$= \frac{2}{8\pi c^2} \sum_{\mathbf{k}s} \omega_{\mathbf{k}}^2 Q_{\mathbf{k}s} \frac{e^{i\mathbf{k}\cdot\mathbf{x}}}{\sqrt{V}} \varepsilon_{\mathbf{k}s} \{Q_{\mathbf{k}s}^*, \dot{Q}_{\mathbf{k}s}\}$$

$$- \frac{1}{c} \sum_{\mathbf{k}s} j_{\mathbf{k}s} \frac{e^{i\mathbf{k}\cdot\mathbf{x}}}{\sqrt{V}} \varepsilon_{\mathbf{k}s} \{Q_{\mathbf{k}s}^*, \dot{Q}_{\mathbf{k}s}\}$$

$$= \frac{2}{8\pi c^2} \sum_{\mathbf{k}s} \omega_{\mathbf{k}}^2 Q_{\mathbf{k}s} \frac{e^{i\mathbf{k}\cdot\mathbf{x}}}{\sqrt{V}} (-4\pi c^2) \varepsilon_{\mathbf{k}s}$$

$$- \frac{1}{c} \sum_{\mathbf{k}s} j_{\mathbf{k}s} \frac{e^{i\mathbf{k}\cdot\mathbf{x}}}{\sqrt{V}} \varepsilon_{\mathbf{k}s} (-4\pi c^2)$$

$$= - \sum_{\mathbf{k}s} \omega_{\mathbf{k}}^2 Q_{\mathbf{k}s} \frac{e^{i\mathbf{k}\cdot\mathbf{x}}}{\sqrt{V}} \varepsilon_{\mathbf{k}s} + 4\pi c \sum_{\mathbf{k}s} j_{\mathbf{k}s} \frac{e^{i\mathbf{k}\cdot\mathbf{x}}}{\sqrt{V}} \varepsilon_{\mathbf{k}s}$$

$$= c^2 \nabla^2 \mathbf{A} + 4\pi c \mathbf{j}_t$$

or

$$\nabla^2 \mathbf{A} - \frac{1}{c^2} \frac{\partial^2 \mathbf{A}}{\partial t^2} = -\frac{4\pi}{c} \mathbf{j}_t \qquad\qquad \text{Q.E.D.}$$

11.7. Hamiltonian of a Closed System

We have discussed until now *open* systems, covering:

(1) The motion of particles in given fields
(2) The Hamiltonian of the **A** field.

We want now to write the Hamiltonian that is applicable to be *closed* system. A closed system is a set of charged particles (generic mass $= m_s$, generic charge $= e_s$) plus the fields that they produce.

We can go from a closed system to an open system under certain circumstances. Consider the closed system $A + B$ in Fig. 11.3. The part B is an open system if we can take into account the presence of A only by

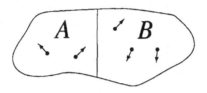

FIGURE 11.3

means of a field produced by A on B, this can be done if the reactions of B on A are negligible.

When we use a power supply to energize a system, we can consider the system "open" if we can disregard its reaction on the power supply. A single atom or a single molecule in the absence of any external field is an example of a closed system.

CLAIM: The Hamiltonian of a closed system is

$$H = \sum_s \sqrt{(m_s c^2)^2 + [c\mathbf{p}_s - e_s \mathbf{A}(\mathbf{x}_s, t)]^2}$$

$$+ \frac{1}{2} \sum_{\substack{ss' \\ s \neq s'}} \frac{e_s e_s}{|\mathbf{x}_s(t) - \mathbf{x}_{s'}(t)|}$$

$$+ \frac{1}{8\pi} \int d^3\mathbf{x} [(\boldsymbol{\nabla} \times \mathbf{A})^2 + (\mathbf{E}^{\mathrm{rad}})^2] \tag{11.7.1}$$

In this description, $\mathbf{V} \cdot \mathbf{A} = 0$. The term

$$-\frac{1}{c} \int \mathbf{j}(\mathbf{x}, t) \cdot \mathbf{A}(\mathbf{x}, t) d^3\mathbf{x} \tag{11.7.2}$$

is not present. We claim that it is contained in

$$\sum_s \sqrt{(m_s c^2)^2 + [c\mathbf{p}_s - e\mathbf{A}(\mathbf{x}_s, t)]^2} \tag{11.7.3}$$

Proof: The basic Poisson brackets are

$$\{p_{is}, p_{ks'}\} = \{x_{is}, x_{ks'}\} = 0 \tag{11.7.4}$$

$$\{p_{is}, x_{ks'}\} = \delta_{ik}\delta_{ss'} \tag{11.7.5}$$

$$\{E_i^{\mathrm{rad}}(\mathbf{x}, t), A_j(\mathbf{x}', t)\} = -4\pi c \delta_{ij}^{\mathrm{tr}}(\mathbf{x} - \mathbf{x}') \tag{11.7.6}$$

$$\{A_i(\mathbf{x},t), A_j(\mathbf{x}',t)\} = 0 \tag{11.7.7}$$

$$\{\dot{A}_i(\mathbf{x},t), A_j(\mathbf{x}',t)\} = 4\pi c^2 \delta_{ij}^{\mathrm{tr}}(\mathbf{x} - \mathbf{x}') \tag{11.7.8}$$

$$\{E_i(\mathbf{x},t), E_j(\mathbf{x}',t)\} = \{B_i(\mathbf{x},t), B_j(\mathbf{x}',t)\} = 0 \tag{11.7.9}$$

The equations of motion of the particles and the Maxwell equations are condensed in the following expressions:

$$\dot{\mathbf{P}}_s = \{H, \mathbf{p}_s\} \tag{11.7.10}$$

$$\dot{\mathbf{x}}_s = \{H, \mathbf{x}_s\} \tag{11.7.11}$$

$$\dot{\mathbf{A}}_i = \{H, A_i\} \tag{11.7.12}$$

$$\dot{E}_i^{\mathrm{rad}} = \{H, E_i^{\mathrm{rad}}\} \tag{11.7.13}$$

$\mathbf{x}_s$ and $\mathbf{p}_s$ are conjugate variables, and so are $\mathbf{E}^{\mathrm{rad}}$ and $\mathbf{A}$. The Hamiltonian H in Eq. (11.7.1) must be equivalent to

$$H = \sum_s \frac{m_s c^2}{\sqrt{1 - \beta_s^2}} + \frac{1}{8\pi} \int d^3\mathbf{x}(\mathbf{E}_{\mathrm{tot}}^2 + \mathbf{B}^2) \tag{11.7.14}$$

We have

$$\nabla^2 \phi = -4\pi\rho \tag{11.7.15}$$

and

$$\mathbf{E}_{\mathrm{tot}} = -\nabla\phi - \frac{1}{c}\dot{\mathbf{A}} = \mathbf{E}^{\mathrm{long}} + \mathbf{E}^{\mathrm{rad}} \tag{11.7.16}$$

In the present gauge ($\nabla \cdot \mathbf{A} = 0$), this separation is possible; in other gauges it is not.

We have now to prove that the Hamiltonian H is correct. Let us examine first the particles' equations of motion:

$$\dot{x}_{si} = \frac{\partial H}{\partial p_{si}} = c \frac{cp_{si} - e_s A_i}{\sqrt{(m_s c^2)^2 + (c\mathbf{p}_s - e_s \mathbf{A})^2}} \tag{11.7.17}$$

Then, because of Eq. (11.3.11),

$$\frac{\dot{x}_{si}}{c^2} = \frac{p_{si} - e_s(A_i/c)}{\sqrt{(m_s c^2)^2 + (c\mathbf{p}_s - e_s \mathbf{A})^2}} = \frac{p_{si} - e_s(A_i/c)}{m_s c^2 / \sqrt{1 - \beta_s^2}} \tag{11.7.18}$$

and

$$p_{si} = e_s \frac{A_i}{c} = \frac{m_s \dot{x}_{si}}{\sqrt{1 - \beta_s^2}}$$

$$p_{si} = \frac{m_s \dot{x}_{si}}{\sqrt{1 - \beta_s^2}} + e_s \frac{A_i}{c} \tag{11.7.19}$$

in accord with Eq. (11.3.8).

We can also write

$$\dot{p}_{si} = -\frac{\partial H}{\partial x_{si}} = \sum_j \frac{cp_{sj} - e_s A_j}{\sqrt{(m_s c^2)^2 + (c\mathbf{p}_s - e_s \mathbf{A})^2}} \left(e_s \frac{\partial A_j}{\partial x_{si}} \right)$$

$$- e_s \frac{\partial}{\partial x_{si}} \sum_t \frac{e_t}{|\mathbf{x}_s - \mathbf{x}_t|}$$

$$= \sum_j \frac{\dot{x}_{sj}}{c} e_s \frac{\partial A_j}{\partial x_{si}} - e_s \frac{\partial}{\partial x_{si}} \phi(\mathbf{x}_s)$$

$$= e_s \frac{\partial}{\partial x_{si}} \sum_j \left(\frac{\dot{x}_{sj}}{c} A_j \right) - e_s \frac{\partial}{\partial x_{si}} \phi(\mathbf{x}_s)$$

$$= e_s \frac{\partial}{\partial x_{si}} \left(\frac{\mathbf{v}_s}{c} \cdot \mathbf{A} \right) - e_s \frac{\partial}{\partial x_{si}} \phi(\mathbf{x}_s)$$

$$= \frac{d}{dt} \left(\frac{m_s \dot{x}_{sj}}{\sqrt{1 - \beta_s^2}} + e_s \frac{A_i}{c} \right) \tag{11.7.20}$$

Therefore,

$$\frac{d}{dt} \left(\frac{m_s \mathbf{v}_s}{\sqrt{1 - \beta_s^2}} + e_s \frac{\mathbf{A}}{c} \right) = e_s \mathbf{\nabla}_s \left(\frac{\mathbf{v}_s}{c} \cdot \mathbf{A} \right) - e_s \mathbf{\nabla}_s \phi(\mathbf{x}_s) \tag{11.7.21}$$

or

$$\frac{d}{dt} \frac{m_s \mathbf{v}_s}{\sqrt{1 - \beta_s^2}} = e_s \left[-\mathbf{\nabla}\phi - \frac{1}{c} \frac{d\mathbf{A}}{dt} + \mathbf{\nabla} \left(\frac{\mathbf{v}_s}{c} \cdot \mathbf{A} \right) \right] \tag{11.7.22}$$

But, according to Eq. (11.2.4),

$$\mathbf{\nabla} \left(\frac{\mathbf{v}_s}{c} \cdot \mathbf{A} \right) = \left(\frac{\mathbf{v}_s}{c} \cdot \mathbf{\nabla} \right) \mathbf{A} + \frac{\mathbf{v}_s}{c} \times (\mathbf{\nabla} \times \mathbf{A}) \tag{11.7.23}$$

and, according to Eq. (11.2.6),

$$\frac{d\mathbf{A}}{dt} = \frac{\partial \mathbf{A}}{dt} + (\mathbf{v}_s \cdot \boldsymbol{\nabla})\mathbf{A} \tag{11.7.24}$$

Then Eq. (11.7.22) becomes

$$\frac{d}{dt} \frac{m_s \mathbf{v}_s}{\sqrt{1 - \beta_s^2}} = e_s \left[-\boldsymbol{\nabla}\phi - \frac{1}{c}\frac{d\mathbf{A}}{dt} + \left(\frac{\mathbf{v}_s}{c} \cdot \boldsymbol{\nabla}\right)\mathbf{A} + \frac{\mathbf{v}_s}{c} \times (\nabla \times \mathbf{A}) \right]$$

$$= e_s \left\{ -\boldsymbol{\nabla}\phi - \frac{1}{c}\left[\frac{d\mathbf{A}}{dt} - (\mathbf{v}_s \cdot \boldsymbol{\nabla})\mathbf{A}\right] + \frac{\mathbf{v}_s}{c} \times (\nabla \times \mathbf{A}) \right\}$$

$$= e_s \left\{ -\boldsymbol{\nabla}\phi - \frac{1}{c}\frac{\partial \mathbf{A}}{dt} + \frac{\mathbf{v}_s}{c} \times (\nabla \times \mathbf{A}) \right\}$$

$$= e_s \left\{ \mathbf{E}_{\text{tot}} + \frac{\mathbf{v}_s}{c} \times \mathbf{B} \right\} \tag{11.7.25}$$

which gives us a correct result.

Let us examine now the fields equations. We have

$$\dot{A}_i(\mathbf{x}, t) = \{H, A_i\}$$

$$= \frac{1}{8\pi} \int d^3 \mathbf{x}' \sum_j \{[E_j^{\text{rad}}(\mathbf{x}', t)]^2, A_i(\mathbf{x}, t)\}$$

$$= \frac{1}{4\pi} \int d^3 \mathbf{x}' \sum_j E_j^{\text{rad}}(\mathbf{x}', t), \{E_j^{\text{rad}}(\mathbf{x}', t), A_i(\mathbf{x}, t)\}$$

$$= \frac{1}{4\pi} \int d^3 \mathbf{x}' \sum_j \{E_j^{\text{rad}}(\mathbf{x}', t)[-4\pi c \delta_{ij}^{\text{tr}}(\mathbf{x} - \mathbf{x}')]\}$$

$$= -c \sum_i \int d^3 \mathbf{x}' \sum_j \{E_j^{\text{rad}}(\mathbf{x}', t)\delta_{ij}^{\text{tr}}(\mathbf{x} - \mathbf{x}')$$

$$= -c E_i^{\text{rad}}(\mathbf{x}, t) \tag{11.7.26}$$

Also, we expect (note that in what follows the superscript rad has been dropped)

$$\dot{E}(\mathbf{x}, t) = -\frac{1}{c}\ddot{\mathbf{A}}(\mathbf{x}, t) = \{H, \mathbf{E}\}$$

$$= \frac{1}{8\pi} \int d^3 \mathbf{x}' \{[\boldsymbol{\nabla} \times \mathbf{A}(\mathbf{x}', t)]^2, \mathbf{E}(\mathbf{x}, t)\}$$

$$+ \sum_s \{\sqrt{(m_s c^2)^2 + [c\mathbf{p}_s - e_s \mathbf{A}(\mathbf{x}_s, t)]^2}, \mathbf{E}(\mathbf{x}, t)\}$$

$$= \left\{ \frac{1}{8\pi c^2} \sum_{\mathbf{k}s} \omega_{\mathbf{k}}^3 Q_{\mathbf{k}s} Q_{\mathbf{k}s}^*, \mathbf{E} \right\}$$

$$+ \sum_s \{\sqrt{(m_s c^2)^2 + [c\mathbf{p}_s - e_s \mathbf{A}(\mathbf{x}_s, t)]^2}, \mathbf{E}(\mathbf{x}, t)\} \quad (11.7.27)$$

where use has been made of Eq. (11.5.17). We have

$$\mathbf{A}(\mathbf{x}, t) = \frac{1}{\sqrt{V}} \sum_{\mathbf{k}s} Q_{\mathbf{k}s}(t) e^{i\mathbf{k}\cdot\mathbf{x}} \quad (11.7.28)$$

and

$$-c\nabla^2 \mathbf{A}(\mathbf{x}, t) = \frac{c}{\sqrt{V}} \sum_{\mathbf{k}s} k^2 \varepsilon_{\mathbf{k}s} Q_{\mathbf{k}s} e^{i\mathbf{k}\cdot\mathbf{x}}$$

$$= \frac{1}{c\sqrt{V}} \sum_{\mathbf{k}s} \omega_{\mathbf{k}}^2 Q_{\mathbf{k}s} \varepsilon_{\mathbf{k}s} e^{i\mathbf{k}\cdot\mathbf{x}} \quad (11.7.29)$$

Also

$$\left\{ \frac{1}{8\pi c^2} \sum_{\mathbf{k}s} \omega_{\mathbf{k}}^2 Q_{\mathbf{k}s} Q_{\mathbf{k}s}^*, \frac{\partial \mathbf{A}}{\partial t} \right\}$$

$$= \frac{1}{8\pi c^2} \left\{ \sum_{\mathbf{k}s} \omega_{\mathbf{k}}^2 Q_{\mathbf{k}s} Q_{\mathbf{k}s}^*, \frac{1}{\sqrt{V}} \sum_{\mathbf{k}'s'} \varepsilon_{\mathbf{k}'s'} \dot{Q}_{\mathbf{k}'s'} e^{i\mathbf{k}'\cdot x} \right\}$$

$$= \frac{1}{8\pi c^2 \sqrt{V}} \sum_{\mathbf{k}s} \sum_{\mathbf{k}'s'} \omega_{\mathbf{k}}^2 \varepsilon_{\mathbf{k}'s'} e^{i\mathbf{k}'\cdot x} \{Q_{\mathbf{k}s} Q_{\mathbf{k}s}^*, \dot{Q}_{\mathbf{k}'s'}\}$$

$$= \frac{2}{8\pi c^2 \sqrt{V}} \sum_{\mathbf{k}s} \sum_{\mathbf{k}'s'} \omega_{\mathbf{k}}^2 \varepsilon_{\mathbf{k}'s'} e^{i\mathbf{k}'\cdot x} Q_{\mathbf{k}s} \{Q_{\mathbf{k}s}^*, \dot{Q}_{\mathbf{k}'s'}\}$$

$$= \frac{1}{4\pi c^2 \sqrt{V}} \sum_{\mathbf{k}s} \sum_{\mathbf{k}'s'} \omega_{\mathbf{k}}^2 \varepsilon_{\mathbf{k}'s'} e^{i\mathbf{k}'\cdot x} Q_{\mathbf{k}s} (-4\pi c^2 \delta_{\mathbf{k}\mathbf{k}'} \delta_{ss'})$$

$$= \frac{1}{\sqrt{V}} \sum_{\mathbf{k}s} \omega_{\mathbf{k}}^2 \varepsilon_{\mathbf{k}s} e^{i\mathbf{k}\cdot\mathbf{x}} Q_{\mathbf{k}s} \quad (11.7.30)$$

and

$$\left\{ \frac{1}{8\pi c^2} \sum_{ks} \omega_{\mathbf{k}}^2 Q_{\mathbf{k}s} Q_{\mathbf{k}s}^*, \mathbf{E} \right\} = \left\{ \frac{1}{8\pi c^2} \sum_{ks} \omega_{\mathbf{k}}^2 Q_{\mathbf{k}s} Q_{\mathbf{k}s}^*, -\frac{1}{c} \frac{\partial \mathbf{A}}{\partial t} \right\}$$

$$= \frac{1}{c} \frac{1}{\sqrt{V}} \sum_{ks} \omega_{\mathbf{k}}^2 \varepsilon_{\mathbf{k}s} e^{i\mathbf{k}\cdot\mathbf{x}} Q_{\mathbf{k}s}$$

$$= -c\nabla^2 \mathbf{A}(\mathbf{x}, t) \qquad (11.7.31)$$

Note that

$$\{f(\mathbf{A}), \mathbf{E}\} = \sum_j \frac{\partial f}{\partial A_j} \{A_j, e\} \qquad (11.7.32)$$

Then

$$\sum_s \{ \sqrt{(m_s c^2)^2 + [c\mathbf{p}_s - e_s \mathbf{A}(\mathbf{x}_s, t)]^2}, E_i(\mathbf{x}, t) \}$$

$$= \sum_s \sum_j \frac{\partial \sqrt{(m_s c^2)^2 + [c\mathbf{p}_s - e_s \mathbf{A}(\mathbf{x}_s, t)]^2}}{\partial A_j(\mathbf{x}_s, t)}$$

$$\times \{ A_j(\mathbf{x}_s, t), E_i(\mathbf{x}, t) \}$$

$$= \sum_s \sum_j \frac{c p_{sj} - e_s A_j(\mathbf{x}_s, t)}{\sqrt{(m_s c^2)^2 + [c\mathbf{p}_s - e_s \mathbf{A}(\mathbf{x}_s, t)]^2}}$$

$$\times (-e_s) \{ A_j(\mathbf{x}_s, t), E_i(\mathbf{x}, t) \}$$

But, according to Eq. (11.7.17),

$$\frac{c p_{sj} - e_s A_j(\mathbf{x}_s, t)}{\sqrt{(m_s c^2)^2 + [c\mathbf{p}_s - e_s \mathbf{A}(\mathbf{x}_s, t)]^2}} = \frac{v_{sj}}{c}$$

Then

$$\sum_s \{ \sqrt{(m_s c^2)^2 + [c\mathbf{p}_s - e_s \mathbf{A}(\mathbf{x}_s, t)]^2}, E_i(\mathbf{x}, t) \}$$

$$= \sum_s \sum_j \frac{v_{sj}}{c} (-e_s) 4\pi c \delta_{ij}^{\text{tr}}(\mathbf{x}_s - \mathbf{x})$$

$$= 4\pi \sum_s \sum_j e_s v_{sj} \delta_{ij}^{\text{tr}}(\mathbf{x}_s - \mathbf{x}) = -4\pi j_{ti}(\mathbf{x}, t)$$

and

$$\sum_s \{\sqrt{(m_s c^2)^2 + [c\mathbf{p}_s - e_s \mathbf{A}(\mathbf{x}_s, t)]^2}, E(\mathbf{x}, t)\}$$

$$= -4\pi \mathbf{j}_t(\mathbf{x}, t) \tag{11.7.33}$$

Finally, using results (11.7.31) and (11.7.33), we write Eq. (11.7.27) as follows:

$$-\frac{1}{c}\frac{\partial^2 \mathbf{A}(\mathbf{x}, t)}{\partial t^2} = -c\nabla^2 \mathbf{A}(\mathbf{x}, t) - 4\pi \mathbf{j}_t(\mathbf{x}, t)$$

or

$$\nabla^2 \mathbf{A}(\mathbf{x}, t) - \frac{1}{c^2}\frac{\partial^2 \mathbf{A}(\mathbf{x}, t)}{\partial t^2} = -\frac{4\pi}{c}\mathbf{j}_t(\mathbf{x}, t) \tag{11.7.34}$$

This result together with the previous ones, confirms the fact that the Hamiltonian H in Eq. (11.7.1) is indeed the Hamiltonian of a closed system.

CHAPTER 11 EXERCISES

11.1. (a) A particle of mass m and charge q travels in a region of space where an electromagnetic field with potentials $\mathbf{A}$ and ϕ is present. Prove that the expression

$$L = \frac{mv^2}{2} - q\phi + \frac{q}{c}\mathbf{v} \cdot \mathbf{A}$$

represents the Lagrangian of the particle when $v \ll c$.

(b) Prove that under the same conditions the Hamiltonian of the particle is given by

$$H = \frac{1}{2}\frac{[p - (e/c)\mathbf{A}]^2}{m} + q\phi$$

Note: Do *not* derive L and H as nonrelativistic limits of the relativistic expressions.

11.2. The inhomogeneous Maxwell equations and the use of the relations

$$\mathbf{B} = \nabla \times \mathbf{A}$$

$$\mathbf{E} = -\nabla\phi - \frac{1}{c}\frac{\partial \mathbf{A}}{\partial t}$$

give

$$\nabla^2 \phi + \frac{1}{c} \frac{\partial}{\partial t} \nabla \cdot \mathbf{A} = -4\pi\rho$$

$$\Box^2 \mathbf{A} - \nabla \left(\nabla \cdot \mathbf{A} + \frac{1}{c} \frac{\partial \phi}{\partial t} \right) = -\frac{4\pi}{c} \mathbf{j}$$

(a) Show that in the case when $\nabla \times \mathbf{j} = 0$ (irrotational current pattern) it is possible to choose a gauge with $\mathbf{A} = 0$. Find the equation for ϕ.

(b) Use a new gauge $\mathbf{A}', \phi'$ so that $\phi' = 0$. Find the equation for $\mathbf{A}'$. Verify that for a stationary charge distribution we find

$$\mathbf{E} = -\frac{1}{c} \dot{\mathbf{A}}' = -\nabla \int d^3x' \frac{\rho(\mathbf{x}')}{|\mathbf{x} - \mathbf{x}'|}$$

12

Electromagnetic Properties of Matter

12.1. Normal and Anomalous Dispersion

The aim of this chapter is to examine the problem of the interaction of radiation with matter. A thorough treatment of such interaction requires the use of quantum mechanics, that is, the quantum treatment of atoms and molecules, or a more complete theory that includes the quantization of the electromagnetic field. However, a classical treatment is not only didactically valuable by presenting some useful models that may sharpen our understanding of the basic problems, but, in many instances, it may provide useful approaches to the quantitative evaluation of radiative phenomena.

We shall now consider the response of matter to the application of an external field. The dielectric constant is the physical parameter that characterizes this response. We shall assume that the applied field **E** is time independent, and that the medium on which it is acting is isotropic and has no polarization in the absence of the applied field. The dielectric constant is given by

$$K = 1 + 4\pi\chi = 1 + 4\pi\frac{\mathbf{P}}{\mathbf{E}} \qquad (12.1.1)$$

where

$$\chi = \frac{\mathbf{P}}{\mathbf{E}} = \frac{\text{dipole per unit volume}}{\text{applied field}} \qquad (12.1.2)$$

The "applied" field is an "average" field.

The dipole associated with a molecule of polarizability α is

$$\mathbf{p}_s = \alpha\mathbf{E}(\mathbf{x}_s) \qquad (12.1.3)$$

where $\mathbf{E}(\mathbf{x}_s) =$ local field at the position $\mathbf{x}_s$ of the molecule. A molecule is not in an "average" position, but in a definite position.

We make the assumption

$$\text{local field } \mathbf{E}(\mathbf{x}_s) = \text{const} \times \langle\mathbf{E}\rangle \qquad (12.1.4)$$

where $\langle\mathbf{E}\rangle$ is the "average" applied field. The local field at $\mathbf{x}_s$ is due to the applied field and to the contributions of the molecules other than the one at $\mathbf{x}_s$. In what follows the applied field will be simply indicated by $\mathbf{E}$. The local field coincides with the applied field in gases at low pressures; in gases at high pressures, in liquids, and in solids, the two fields may differ.

Let us consider a molecule of a dense system that resides between the plates of a capacitor as in Fig. 12.1. Let the molecule be surrounded by an imaginary sphere of such an extent that beyond it the dielectric can be

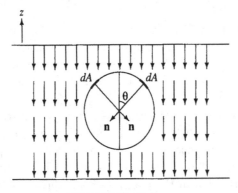

FIGURE 12.1

treated as a continuum. If the molecules inside the sphere were removed while the polarization outside the sphere remains frozen, the field acting on A would stem from two sources:

(1) Free charges at the electrodes of the capacitor's plates $\mathbf{E}$
(2) Bound charges from the ends of the dipole chains $\mathbf{E}_2$

According to Mosotti's hypothesis,[1] the field due to the molecules inside the sphere is set equal to zero.

To calculate $\mathbf{E}_2$, we start by considering the charge on an element dA of the surface of the interior of the sphere. This charge, according to Eq. (2.14.15), is given by

$$\mathbf{P} \cdot \mathbf{n}\, dA = P \cos\theta\, dA \qquad (12.1.5)$$

where $\mathbf{n} =$ inward pointing unit vector, normal to the surface. Each surface element of the sphere contributes a radial field at the center of the sphere given by

$$d\mathbf{E}_2 = \frac{P\cos\theta}{r^2}\, dA\,\mathbf{n} \qquad (12.1.6)$$

For each surface element dA there exists a counterpart that produces the same vertical component, but an equal and opposite horizontal component. Hence only the vertical components add up to create a field intensity:

$$\int_{\substack{\text{overall}\\\text{sphere}}} \frac{P\cos^2\theta}{r^2}\, dA \qquad (12.1.7)$$

The field is oriented parallel to the applied field and strengthens it. Now

$$dA = rd\theta 2\pi r\sin\theta = 2\pi r^2 \sin\theta d\theta \qquad (12.1.8)$$

$$\mathbf{E}_2 = \int_0^\pi \frac{\mathbf{P}\cos^2\theta}{r^2} 2\pi r^2 \sin\theta d\theta$$

$$= 2\pi\mathbf{P} \int_0^\pi \sin\theta \cos^2\theta d\theta$$

$$= 2\pi\mathbf{P} \left[-\frac{\cos^3\theta}{3}\right]_0^\pi = 2\pi\mathbf{P}\frac{2}{3} = \frac{4\pi\mathbf{P}}{3} \qquad (12.1.9)$$

---

[1] A. von Hippel, *Dielectrics and Waves*, John Wiley & Sons, Inc., New York, 1954, p. 97.

Therefore, we have

$$\mathbf{E}_{\text{local}} = \mathbf{E} + \frac{4}{3}\pi\mathbf{P} \tag{12.1.10}$$

Also, for the molecule s,

$$\mathbf{P}_s = \alpha\left(\mathbf{E} + \frac{4}{3}\pi\mathbf{P}\right) \tag{12.1.11}$$

Assuming all the molecules are equal,

$$\mathbf{P} = n_0\mathbf{P}_s = n_0\alpha\left(\mathbf{E} + \frac{4}{3}\pi\mathbf{P}\right) \tag{12.1.12}$$

where

$$n_0 = \text{number of molecules per unit volume}$$

Since $\mathbf{P} = \chi\mathbf{E}$,

$$\chi\mathbf{E} = n_0\alpha\left(\mathbf{E} + \frac{4}{3}\pi\chi\mathbf{E}\right) \tag{12.1.13}$$

or

$$\chi = n_0\alpha\left(1 + \frac{4}{3}\pi\chi\right) \tag{12.1.14}$$

and

$$\chi = \frac{n_0\alpha}{1 - \frac{4}{3}\pi n_0\alpha} \tag{12.1.15}$$

Also,

$$\mathbf{D} = \mathbf{E} + 4\pi\mathbf{P} = \mathbf{E} + 4\pi\chi\mathbf{E} = \mathbf{E}(1 + 4\pi\chi) = K\mathbf{E} \tag{12.1.16}$$

where

$$K = 1 + 4\pi\chi \tag{12.1.17}$$

and

$$\chi = \frac{K - 1}{4\pi} \tag{12.1.18}$$

Then

$$K - 1 = 4\pi\chi$$

$$K + 2 = 4\pi\chi + 3$$

$$\frac{K-1}{K+2} = \frac{4\pi\chi}{4\pi\chi + 3} = \frac{1}{1 + \dfrac{3}{4\pi\chi}}$$

$$= \frac{1}{1 + \dfrac{3}{4\pi}\left[\dfrac{1 - (4/3)\pi n_0\alpha}{n_0\alpha}\right]}$$

or

$$\frac{K-1}{K+2} = \frac{4\pi}{3}n_0\alpha \qquad (12.1.19)$$

This relation expresses the *Clausius–Mosotti law.*

For gases at low pressure, $K - 1 \ll 1$ and $K + 2 \simeq 3$, and

$$\frac{K-1}{4\pi} = n_0\alpha = \chi \qquad (12.1.20)$$

At this point, in order to calculate a general index of refraction the static polarization a is replaced by $\alpha(\omega)$. We have already found that if an electromagnetic monochromatic wave impinges on a molecule it produces an oscillating dipole $\mathbf{p}(t) = e\mathbf{z}(t)$ such that [see Eq. (9.4.3)]

$$\ddot{\mathbf{p}} + \omega_0^2\mathbf{p} + \gamma\,\dot{\mathbf{p}} = \frac{e^2}{m}\mathbf{E}_0 e^{i\omega t} \qquad (12.1.21)$$

where $\gamma + (2e^2\omega^2)/(3mc^3)$. In this case

$$\mathbf{p}(t) = \mathbf{p}_0 e^{i\omega t} = \alpha(\omega)\mathbf{E}_0 e^{i\omega t} \qquad (12.1.22)$$

and

$$\alpha(\omega) = \frac{e^2/m}{\omega_0^2 - \omega^2 + i\gamma\omega} \qquad (12.1.23)$$

If we set $K = n^2(\omega)$, where $n(w) = $ index of refraction at the frequency ω, and replace α with $\alpha(\omega)$ in the expression for the Clausius–Mosotti law, we obtain

$$\frac{n^2(\omega) - 1}{n^2(\omega) + 2} = \frac{4\pi}{3}n_0\alpha(\omega) \qquad (12.1.24)$$

which is called the *Lorentz–Lorenz formula.*

From relation (12.1.24), we can derive

$$n^2 - 1 = \frac{4\pi n_0 \alpha(\omega)}{1 - (4\pi/3)n_0\alpha(\omega)}$$

$$= 4\pi n_0 \frac{(e^2/m)/(\omega_0^2 - \omega^2 + i\gamma\omega)}{1 - (4\pi/3)n_0(e^2/m)/(\omega_0^2 - \omega^2 + i\gamma\omega)}$$

$$= 4\pi n_0 \frac{e^2}{m} \frac{1}{(\omega_0^2 - \omega^2 + i\gamma\omega) - (4\pi/3)n_0(e^2/m)}$$

$$= 4\pi n_0 \frac{e^2}{m} \frac{1}{\omega_1^2 - \omega^2 + i\gamma\omega} \qquad (12.1.25)$$

where

$$\omega_1^2 = \omega_0^2 - \frac{4\pi}{3}n_0\frac{e^2}{m} \qquad (12.1.26)$$

There is a shift in the frequency of resonance.

In the more general case of a system containing oscillators of different types without mutual coupling, Eq. (12.1.25) can be generalized as follows:

$$n^2(\omega) - 1 = \sum_s \frac{(4\pi n_{0s}e^2)/m_s}{\omega_s^2 - \omega^2 + i\gamma_s\omega} \qquad (12.1.27)$$

where n_{0s} = number of dispersion electrons for the oscillators of type s per unit volume and $\gamma_s = (2e^2\omega_s^2)/(3m_sc^3)$. Formula (12.1.27) is the *dispersion formula* of classical physics. Far below its resonance frequency ($\omega \ll \omega_s$), each oscillator type adds a constant contribution:

$$\frac{(4\pi n_{0s}e^2)/m_s}{\omega_s^2} \qquad (12.1.28)$$

to the dielectric constant.

Let us now consider the behavior of oscillators of type r, which have the lowest resonance frequency. In the vicinity of ω_r, setting $m_r = m$ and $\gamma_r = \gamma$, we obtain

$$n^2(\omega) = 1 + \sum_s \frac{(4\pi n_{0s}e^2)/m_s}{\omega_s^2 - \omega^2 + i\gamma_s\omega}$$

$$\simeq 1 + \sum_{s \neq r} \frac{(4\pi n_{0s}e^2)/m_s}{\omega_s^2} + \frac{(4\pi n_{0r}e^2)/m}{\omega_r^2 - \omega^2 + i\gamma\omega}$$

$$= A + \frac{(4\pi n_{0r}e^2)/m}{\omega_r^2 - \omega^2 + i\gamma\omega} = K_r + iK_i \qquad (12.1.29)$$

where K_r and K_i are the real and imaginary parts of the dielectric constant, and

$$A = 1 + \sum_{s \neq r} \frac{(4\pi n_{0s}e^2)/m_s}{\omega_s^2} \qquad (12.1.30)$$

But

$$\omega_r^2 - \omega^2 = (\omega_r + \omega)(\omega_r - \omega) \simeq 2\omega_r \Delta\omega \qquad (12.1.31)$$

where

$$\Delta\omega = \omega_r - \omega \qquad (12.1.32)$$

Therefore,

$$n^2 \simeq A + \frac{(4\pi n_{0r}e^2)/m}{2\omega_r \Delta\omega + i\gamma\omega_r} = A + \frac{(4\pi n_{0r}e^2)/m}{2\omega_r[\Delta\omega + i(\gamma/2)]}$$

$$= A + \frac{B}{\Delta\omega + i(\gamma/2)} \qquad (12.1.33)$$

where

$$B = \frac{(4\pi n_{0r}e^2)/m}{2\omega_r} \qquad (12.1.34)$$

Now we have

$$n^2 = A + \frac{B}{\Delta\omega + i(\gamma/2)} = A + \frac{B[\Delta\omega - i(\gamma/2)]}{(\Delta\omega)^2 + (\gamma/2)^2}$$

$$= A + \frac{B\Delta\omega}{(\Delta\omega)^2 + (\gamma/2)^2} - i\frac{B(\gamma/2)}{(\Delta\omega)^2 + (\gamma/2)^2}$$

$$= K_r + iK_i \qquad (12.1.35)$$

where

$$K_r = A + \frac{B\Delta\omega}{(\Delta\omega)^2 + (\gamma/2)^2} \qquad (12.1.36)$$

$$K_i = -\frac{B(\gamma/2)}{(\Delta\omega)^2 + (\gamma/2)^2} \qquad (12.1.37)$$

The real and imaginary parts of the dielectric constant are plotted in Fig. 12.2. The real part of the dielectric rises with increasing ω over the major part of the ω spectrum; this corresponds to *normal dispersion*. In the spectral region corresponding to the half width of $|K_i|$, K_r falls with increasing ω; this region corresponds to the *anomalous dispersion*. Note also

$$K_{r\,\text{max}} = A + \frac{B}{\gamma} \quad \text{at} \quad \omega = \omega_r - \frac{\gamma}{2}$$

$$K_{r\,\text{min}} = A + \frac{B}{\gamma} \quad \text{at} \quad \omega = \omega_r + \frac{\gamma}{2} \qquad (12.1.38)$$

$$K_r = A \quad \text{at} \quad \omega = \omega_r$$

On the other hand,

$$|K_i|_{\text{max}} = \frac{2B}{\gamma}, \quad \text{at} \quad \omega = \omega_r \qquad (12.1.39)$$

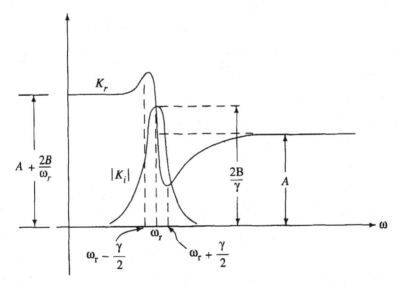

FIGURE 12.2

The slopes of the K_r and $|K_i|$ curves at $\omega = \omega_r$ are $-B/(\gamma/2)^2$ and zero, respectively. Let us follow now the general behavior of K_r and K_i.

K_r rises hyperbolically from the low frequency value $[A + (2B/\omega_r)]$ to a maximum $[A + (B/\gamma)]$, which it reaches at $\omega = \omega_r - (\gamma/2)$; it then falls with a slope $-B/(\gamma/2)^2$, going through the value A at resonance, reaches the minimum value $[A - (B/\gamma)]$ at $\omega = \omega_r + (\gamma/2)$, and rises again asymptotically to the constant value A, which it reaches at $\omega \gg \omega_r$.

$|K_i|$ starts from zero at low frequencies, reaches its maximum value $2B/\gamma$ at resonance, and falls symmetrically to zero at high frequencies. The half-values of the $|K_i|$ bell-shaped curve are reached at values of ω that differ from the resonance value ω_r by $\pm \gamma/2$.

Expressions for K_r and K_i that are valid in a larger frequency region can be derived from relation (12.1.29):

$$
\begin{aligned}
n^2 &= A + \frac{(4\pi n_{0r} e^2)/m}{\omega_r^2 - \omega^2 + i\gamma\omega} \\
&= A + \frac{4\pi n_{0r} e^2}{m} \frac{\omega_r^2 - \omega^2 - i\gamma\omega}{(\omega_r^2 - \omega^2)^2 + (\gamma\omega)^2} \\
&= K_r + iK_i
\end{aligned}
\tag{12.1.40}
$$

where

$$
K_r = A + \frac{4\pi n_{0r} e^2}{m} \frac{\omega_r^2 - \omega^2}{(\omega_r^2 - \omega^2)^2 + (\gamma\omega)^2}
\tag{12.1.41}
$$

$$
K_i = -\frac{4\pi n_{0r} e^2}{m} \frac{\gamma\omega}{(\omega_r^2 - \omega^2)^2 + (\gamma\omega)^2}
\tag{12.1.42}
$$

These expressions for K_r and K_i are more accurate than Eqs. (12.1.35) and (12.1.36), respectively.

Finally, the local field $\mathbf{E}(x_s)$ can be expressed in terms of the index of refraction n and the applied field $\mathbf{E}$:

$$
\begin{aligned}
\mathbf{E}(\mathbf{x}_s) &= \mathbf{E} + \frac{4}{3}\pi \mathbf{P} = \mathbf{E} + \frac{4}{3}\pi\chi\mathbf{E} \\
&= \left(1 + \frac{4}{3}\pi\chi\right)\mathbf{E} = \left(1 + \frac{4}{3}\pi\frac{K-1}{4\pi}\right)\mathbf{E} \\
&= \left(1 + \frac{K-1}{3}\right)\mathbf{E} = \frac{K+2}{3}\mathbf{E} = \frac{n^2+2}{3}\mathbf{E}
\end{aligned}
\tag{12.1.43}
$$

12.2. Multiple Scattering Theory of the Index of Refraction

The present approach to the calculation of the index of refraction differs from the one presented in the previous section, which started with the consideration of a time-independent field. In this section we shall begin by considering a plane wave of frequency ω and wave number k_0, propagating in a direction perpendicular to the surface of a certain dielectric medium (ooo Fig. 12.3).

The wave will come inside the medium and will produce oscillating dipoles. The oscillating dipoles will in turn emit radiation with the same frequency ω. The problem consists in trying to find the wavelength λ, or the wave number $k = 2\pi/\lambda$, with which the wave will propagate inside the medium. Once the value of k is found, the index of refraction is easily derived:

$$n = \frac{k}{\omega/c} \qquad (12.2.1)$$

We shall use the Coulomb gauge $\nabla \cdot \mathbf{A} = 0$ and characterize the field by means of a vector potential $\mathbf{A}$. There will be no need for a scalar potential. The vector $\mathbf{A}$ will be a solution of the inhomogeneous equation

$$\nabla^2 \mathbf{A} - \frac{1}{c^2}\frac{\partial^2 \mathbf{A}}{\partial t^2} = -\frac{4\pi}{c}\mathbf{j}_t \qquad (12.2.2)$$

The current density $\mathbf{j}_t$ will be expressed in terms of the polarization currents related to the oscillating dipoles. The presence of the medium will be taken into account by means of these polarization currents.

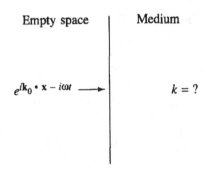

FIGURE 12.3

If the medium is a crystal and the dipoles are localized at the lattice sites, the propagating field will be undulatory only on the average, because it will present spikelike deviations at the lattice sites.

We want to find the solution that is characterized by

$$\mathbf{A}(\mathbf{x}, t) = \mathbf{A}(\mathbf{x})e^{-i\omega t}$$
$$\mathbf{j}(\mathbf{x}, t) = \mathbf{j}(\mathbf{x})e^{-i\omega t}$$
(12.2.3)

We have

$$\frac{\omega}{c} = k_0$$
(12.2.4)

Then, since

$$\Box^2 \mathbf{A}(\mathbf{x}, t) = -\frac{4\pi}{c}\mathbf{j}_t(\mathbf{x}, t)$$
(12.2.5)

we write

$$(\nabla^2 + k_0^2)\mathbf{A}(\mathbf{x}) = -\frac{4\pi}{c}\mathbf{j}_t(\mathbf{x})$$
(12.2.6)

CLAIM:

$$\mathbf{A}(\mathbf{x}) = \mathbf{A}_0 e^{i\mathbf{k}_0 \cdot \mathbf{x}} + \frac{1}{c}\int d^3\mathbf{x}' \frac{e^{ik_0|\mathbf{x}-\mathbf{x}'|}}{|\mathbf{x}-\mathbf{x}'|}\mathbf{j}_t(\mathbf{x}')$$
(12.2.7)

is the solution of the equation

$$(\nabla^2 + k_0^2)\mathbf{A}(\mathbf{x}) = -\frac{4\pi}{c}\mathbf{j}_t(\mathbf{x})$$
(12.2.8)

Proof: The first term in Eq. (12.2.7) gives

$$\nabla^2 \mathbf{A}_0 e^{i\mathbf{k}_0 \cdot \mathbf{x}} + k_0^2 \mathbf{A}_0 e^{i\mathbf{k}_0 \cdot \mathbf{x}} = 0$$
(12.2.9)

As for the second term,

$$\nabla^2 \frac{1}{c}\int d^3\mathbf{x}' \frac{e^{ik_0|\mathbf{x}-\mathbf{x}'|}}{|\mathbf{x}-\mathbf{x}'|}\mathbf{j}_t(\mathbf{x}')$$
$$= \frac{1}{c}\int d^3\mathbf{x}' \frac{\nabla^2 e^{ik_0|\mathbf{x}-\mathbf{x}'|}}{|\mathbf{x}-\mathbf{x}'|}\mathbf{j}_t(\mathbf{x}')$$

$$+\frac{1}{c}\int d^3\mathbf{x}' e^{ik_0|\mathbf{x}-\mathbf{x}'|}\nabla^2\frac{1}{|\mathbf{x}-\mathbf{x}'|}\mathbf{j}_t(\mathbf{x}')$$

$$+\frac{2}{c}\int d^3\mathbf{x}'\left[\nabla e^{ik_0|\mathbf{x}-\mathbf{x}'|}\cdot\nabla\frac{1}{|\mathbf{x}-\mathbf{x}'|}\right]\mathbf{j}_t(\mathbf{x}')\qquad(12.2.10)$$

But

$$\nabla e^{ik_0|\mathbf{x}-\mathbf{x}'|}=ik_0 e^{ik_0|\mathbf{x}-\mathbf{x}'|}\frac{\mathbf{x}-\mathbf{x}'}{|\mathbf{x}-\mathbf{x}'|}\qquad(12.2.11)$$

$$\nabla^2 e^{ik_0|\mathbf{x}-\mathbf{x}'|}=\frac{2ik_0}{|\mathbf{x}-\mathbf{x}'|}e^{ik_0|\mathbf{x}-\mathbf{x}'|}-k_0^2 e^{ik_0|\mathbf{x}-\mathbf{x}'|}\qquad(12.2.12)$$

$$\nabla^2\frac{1}{|\mathbf{x}-\mathbf{x}'|}=-4\pi\delta(\mathbf{x}-\mathbf{x}')\qquad(12.2.13)$$

$$\nabla\frac{1}{|\mathbf{x}-\mathbf{x}'|}=-\frac{\mathbf{x}-\mathbf{x}'}{|\mathbf{x}-\mathbf{x}'|^3}\qquad(12.2.14)$$

The first term in Eq. (12.2.12) and the third integral in Eq. (12.2.10), both leading to a $|x|^{-2}$ dependence, can be dropped:

$$\nabla^2\frac{1}{c}\int d^3\mathbf{x}'\frac{e^{ik_0|\mathbf{x}-\mathbf{x}'|}}{|\mathbf{x}-\mathbf{x}'|}\mathbf{j}_t(\mathbf{x}')$$

$$=-k_0^2\frac{1}{c}\int d^3\mathbf{x}'\frac{e^{ik_0|\mathbf{x}-\mathbf{x}'|}}{|\mathbf{x}-\mathbf{x}'|}\mathbf{j}_t(\mathbf{x}')$$

$$-\frac{1}{c}\int d^3\mathbf{x}' e^{ik_0|\mathbf{x}-\mathbf{x}'|}4\pi\delta(\mathbf{x}-\mathbf{x}')\mathbf{j}_t(\mathbf{x}')$$

$$=-k_0^2\frac{1}{c}\int d^3\mathbf{x}'\frac{e^{ik_0|\mathbf{x}-\mathbf{x}'|}}{|\mathbf{x}-\mathbf{x}'|}\mathbf{j}_t(\mathbf{x}')-\frac{4\pi}{c}\mathbf{j}_t(\mathbf{x})\qquad\text{Q.E.D.}$$

The solution we want to consider is the one in which the first term in Eq. (12.2.7) is zero. If we consider the medium infinitely extended, the wave propagates with a wave number k, which is different from k_0, the wave number with which the wave propagates in vacuum.

The current $\mathbf{j}_t$ is associated with the oscillators that are excited and is linear in $\mathbf{A}$, deriving from the wave that goes inside the medium. We have a linear equation in $\mathbf{A}$ that can be solved only in certain conditions; these conditions for the existence of the solution give the wave number k.

The process inside the medium is the following: All the dipoles are excited and every dipole emits spherical waves that have the effect of producing a wave-front. All the waves of wave number ko are canceled. This is in accordance with the *extinction theorem*,[2] which can be stated as follows: *If an incident electromagnetic wave traveling with speed c and wave number ko in vacuum enters a dispersive medium, its fields are canceled by part of the fields of the induced dipoles and a wave characteristic of the medium with a wave number* $k \neq k_0$ *is produced.* In effect,

(1) The extinction of the incident wave and its replacement by a wave characteristic of the medium occurs over a finite distance Δ.

(2) Δ is on the order of the distance required for the "vacuum" wave and the "medium" wave to dephase themselves significantly as a result of their different phase velocities.

Let us consider a slab of material of thickness $\Delta < \lambda$. The wave that we have at $z = \Delta$ is shifted in phase by δ (see Fig. 12.4):

$$e^{ik_0 n\Delta} = e^{ik_0 n\Delta + i\delta} \qquad (12.2.15)$$

or

$$k_0 n\Delta = k_0\Delta + \delta$$

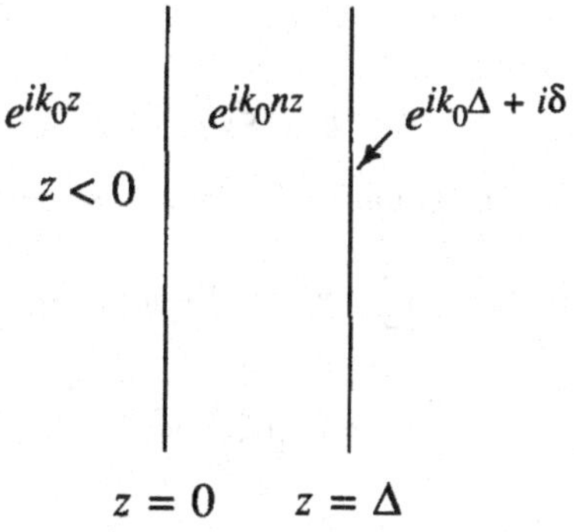

$$e^{ik_0 z} \qquad e^{ik_0 nz} \qquad e^{ik_0\Delta + i\delta}$$

$$z < 0$$

$$z = 0 \qquad z = \Delta$$

FIGURE 12.4

[2]M. Born and E. Wolf, *Principles of Optics,* 4th ed., Pergamon, Elmsford, N.Y., 1970, p. 100.

or

$$\delta = k_0 \Delta (n - 1) \tag{12.2.16}$$

The effect on the wave at $z = \Delta$ is the addition of a phase shift proportional to Δ and to k_0. Setting $\delta = 1$ in Eq. (12.2.16), we obtain, for the extinction theorem distance,

$$\Delta = \frac{1}{k_0(n-1)} = \frac{c}{\omega(n-1)} = \frac{\lambda}{2\pi(n-1)} \tag{12.2.17}$$

EXAMPLES

$\lambda = 6000 \text{ Å}$

$n = 1.5$ (glass) $\Delta = \dfrac{6000 \times 10^{-8}}{2\pi \times 0.5} = 1.91 \times 10^{-5}\,\text{cm}$

$n = 1 + 2.8 \times 10^{-4}$ $\Delta = \dfrac{6000 \times 10^{-8}}{2\pi \times 2.8 \times 10^{-4}} = 0.034\,\text{cm}$

(air at atmospheric
pressure and room
temperature)

Let us now consider $\mathbf{A}$ again. We have an equation for $\mathbf{A}$ inside the medium and $\mathbf{A}$ must have a certain k in order to exist. Assume that the medium is a crystal lattice, and that at each point in the lattice we have a vibrating dipole

$$e\boldsymbol{\xi}_s(t) = e\boldsymbol{\xi}_s e^{-i\omega t} \tag{12.2.18}$$

The current that enters the equation for $\mathbf{A}$ is associated with these dipoles:

$$
\begin{aligned}
\mathbf{j}(\mathbf{x}, t) &= \sum_s \delta(\mathbf{x} - \mathbf{x}_s) \frac{d}{dt}[e\boldsymbol{\xi}_s(t)] \\
&= \sum_s (-i\omega)e\boldsymbol{\xi}_s(t)\delta(\mathbf{x} - \mathbf{x}_s) \\
&= \sum_s (-i\omega)e\boldsymbol{\xi}_s e^{-i\omega t}\delta(\mathbf{x} - \mathbf{x}_s) \tag{12.2.19}
\end{aligned}
$$

The dipole $e\boldsymbol{\xi}_s$ is proportional to the polarizability and to the local field

$$e\boldsymbol{\xi}_s = \alpha \mathbf{E}^{\text{loc}}(\mathbf{x}_s) = \frac{i\omega}{c}\alpha \mathbf{A}^{\text{loc}}(\mathbf{x}_s) \tag{12.2.20}$$

where loc stands for local and $\alpha = \alpha(\omega)$. An electric static field may be present, but does not have to be included, because only $\mathbf{j}_t$ appears in Eq. (12.2.7). We have

$$\mathbf{j}(\mathbf{x}) = \sum_s (-i\omega)e\boldsymbol{\xi}_s\delta(\mathbf{x} - \mathbf{x}_s)$$

$$= \sum_s (-i\omega)\alpha\mathbf{E}^{\text{loc}}(\mathbf{x}_s)\delta(\mathbf{x} - \mathbf{x}_s)$$

$$= \sum_s (-i\omega)\alpha\frac{i\omega}{c}\mathbf{A}^{\text{loc}}(\mathbf{x}_s)\delta(\mathbf{x} - \mathbf{x}_s)$$

$$= \frac{\omega^2}{c}\alpha\sum_s \mathbf{A}^{\text{loc}}(\mathbf{x}_s)\delta(\mathbf{x} - \mathbf{x}_s) \qquad (12.2.21)$$

Then

$$\mathbf{A}(\mathbf{x}) = \frac{1}{c}\int d^3\mathbf{x}'\frac{e^{ik_0|\mathbf{x}-\mathbf{x}'|}}{|\mathbf{x} - \mathbf{x}'|}\mathbf{j}_t(\mathbf{x}')$$

$$= \frac{1}{c}\int d^3\mathbf{x}'\frac{e^{ik_0|\mathbf{x}-\mathbf{x}'|}}{|\mathbf{x} - \mathbf{x}'|}\sum_s \frac{\omega^2}{c}\alpha\mathbf{A}_t^{\text{loc}}(\mathbf{x}_s)\delta(\mathbf{x}' - \mathbf{x}_s)$$

$$= \frac{\omega^2}{c^2}\alpha\sum_s \frac{e^{ik_0|\mathbf{x}-\mathbf{x}_s|}}{|\mathbf{x} - \mathbf{x}_s|}\mathbf{A}_t^{\text{loc}}(\mathbf{x}_s) \qquad (12.2.22)$$

where the subscript t stands for transverse. Dropping the superfluous superscript loc,

$$\mathbf{A}(\mathbf{x}) = \alpha\frac{\omega^2}{c^2}\sum_s \frac{e^{ik_0|\mathbf{x}-\mathbf{x}_s|}}{|\mathbf{x} - \mathbf{x}_s|}\mathbf{A}_t(\mathbf{x}_s) \qquad (12.2.23)$$

To find out if this equation is satisfied for a certain wave number, let us now use Fourier transforms:

$$A_i(\mathbf{x}) = \sum_{\mathbf{k}} A_i(\mathbf{k})e^{i\mathbf{k}\cdot\mathbf{x}} \qquad (12.2.24)$$

where

$$A_i(\mathbf{k}) = \frac{1}{V}\int d^3\mathbf{x}A_i(\mathbf{x})e^{-i\mathbf{k}\cdot\mathbf{x}} \qquad (12.2.25)$$

Also,

$$j_i(\mathbf{x}) = \sum_{\mathbf{k}} j_i(\mathbf{k})e^{i\mathbf{k}\cdot\mathbf{x}} \qquad (12.2.26)$$

where

$$j_i(\mathbf{K}) = \frac{1}{V} \int d^3\mathbf{x} j_i(\mathbf{x}) e^{-i\mathbf{k}\cdot\mathbf{x}} \qquad (12.2.27)$$

On the other hand,

$$(\nabla^2 + k_0^2)\mathbf{A}(\mathbf{x}) = -\frac{4\pi}{c}\mathbf{j}_t(\mathbf{x}) \qquad (12.2.28)$$

Then

$$(\nabla^2 + k_0^2)\sum_{\mathbf{k}} A_i(\mathbf{k})e^{i\mathbf{k}\cdot\mathbf{x}}$$

$$= -\frac{4\pi}{c}\sum_{\mathbf{k}} j_{ti}(\mathbf{k})e^{i\mathbf{k}\cdot\mathbf{x}}$$

$$(\nabla^2 + k_0^2)\sum_{\mathbf{k}} A_i(\mathbf{k})e^{i\mathbf{k}\cdot\mathbf{x}}$$

$$= -\frac{4\pi}{c}\sum_{\mathbf{k}} \left[\frac{1}{V}\int d^3\mathbf{x}' e^{-i\mathbf{k}\cdot\mathbf{x}'} j_{ti}(\mathbf{x})\right] e^{i\mathbf{k}\cdot\mathbf{x}}$$

$$(k^2 - k_0^2)A_i(\mathbf{k})$$

$$= \frac{4\pi}{c}\frac{1}{V}\int d^3\mathbf{x} e^{-i\mathbf{k}\cdot\mathbf{x}} \sum_{j}\left(\delta_{ij} - \frac{k_i k_j}{k^2}\right)j_j(\mathbf{x})$$

$$= \frac{4\pi}{c}\sum_{j}\left(\delta_{ij} - \frac{k_i k_j}{k^2}\right)\frac{1}{V}\int d^3\mathbf{x} e^{-i\mathbf{k}\cdot\mathbf{x}} j_j(\mathbf{x})$$

But, according to Eq. (12.2.21),

$$j_j(\mathbf{x}) = \frac{\omega^2}{c}\alpha\sum_{s}\delta(\mathbf{x} - \mathbf{x}_s)A_j(\mathbf{x}_s)$$

Then

$$(k^2 - k_0^2)A_i(\mathbf{k}) = 4\pi\alpha\left(\frac{\omega}{c}\right)^2\sum_{j}\left(\delta_{ij} - \frac{k_i k_j}{k^2}\right)\frac{1}{V}$$

$$\times \int d^3\mathbf{x} e^{-i\mathbf{k}\cdot\mathbf{x}}\sum_{s}\delta(\mathbf{x} - \mathbf{x}_s)A_j(\mathbf{x}_s)$$

$$= 4\pi\alpha\left(\frac{\omega}{c}\right)^2\sum_{j}\left(\delta_{ij} - \frac{k_i k_j}{k^2}\right)\sum_{s}\frac{1}{V}A_j(\mathbf{x}_s)e^{-i\mathbf{k}\cdot\mathbf{x}_s}$$

But

$$A_j(\mathbf{x}_s) = \sum_{\mathbf{k}'} A_j(\mathbf{k}')e^{i\mathbf{k}'\cdot\mathbf{x}_s}$$

Then

$$(k^2 - k_0^2)A_i(\mathbf{k}) = 4\pi\alpha\left(\frac{\omega}{c}\right)^2\sum_j\left(\delta_{ij} - \frac{k_ik_j}{k^2}\right)$$

$$\times \sum_{\mathbf{k}'} A_j(\mathbf{k}')\left[\frac{1}{V}\sum_s e^{i(\mathbf{k}'-\mathbf{k})\cdot\mathbf{x}_s}\right]$$

or

$$A_i(\mathbf{k}) = 4\pi\alpha\left(\frac{\omega}{c}\right)^2\sum_j\frac{\delta_{ij} - \frac{k_ik_j}{k^2}}{k^2 - k_0^2}$$

$$\times \sum_{\mathbf{k}'} A_i(\mathbf{k}')\left[\frac{1}{V}\sum_s e^{i(\mathbf{k}'-\mathbf{k})\cdot\mathbf{x}_s}\right] \qquad (12.2.29)$$

We have now the task of calculating the sum in brackets in relation (12.2.29).

If we consider for simplicity a cubic lattice of lattice constant a and use periodic boundary conditions, the vector $\mathbf{k}' - \mathbf{k}$ must be restricted to the following values:

$$\mathbf{k}' - \mathbf{k} = 0, \quad \frac{2\pi}{a}\mathbf{n} \qquad (12.2.30)$$

where $\mathbf{n}$ = vector with integer numbers as components. We have then

$$\frac{1}{V}\sum_s e^{i(\mathbf{k}'-\mathbf{k})\cdot\mathbf{x}_s} = n_0\left[\delta_{\mathbf{k}\mathbf{k}'} + \sum_{\mathbf{n}}\delta_{\mathbf{k}', \mathbf{k} + \mathbf{n}(2\pi/a)}\right] \qquad (12.2.31)$$

where n_0 = number of lattice points per unit volume. Then

$$A_i(\mathbf{k}) = 4\pi n_0\alpha\left(\frac{\omega}{c}\right)^2\sum_j\frac{P_{ij}}{k^2 - k_0^2}$$

$$\times \left[A_j(\mathbf{k}) + \sum_{\mathbf{n}\neq 0}A_j\left(\mathbf{k} + \frac{2\pi}{a}\mathbf{n}\right)\right] \qquad (12.2.32)$$

where

$$P_{ij} = \delta_{ij} - \frac{k_ik_j}{k^2} \qquad (12.2.33)$$

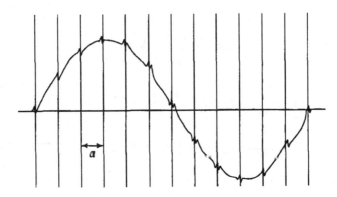

FIGURE 12.5

The wave inside the medium is modulated by the granularity of the lattice (see Fig. 12.5). The irregularities at the lattice points make the higher components of the spectrum relevant.

We want to find a wave number $k_1 \ll (2\pi/a)(\lambda \gg a = \text{lattice constant})$; then we consider Eq. (12.2.32) to its lowest approximation:

$$A_j\left(\mathbf{k} + \frac{2\pi}{a}\mathbf{n}\right) \simeq 0 \tag{12.2.34}$$

In Eq. (12.2.32), $\mathbf{A}$ is automatically transverse, because P_{ij} cancels any longitudinal part. We can then write

$$A_i(\mathbf{k}) = 4\pi n_0 \alpha \left(\frac{\omega}{c}\right)^2 \sum_j \frac{\delta_{ij} - [(k_i k_j)/k^2]}{k^2 - k_0^2} A_j(\mathbf{k})$$

$$= 4\pi n_0 \alpha \left(\frac{\omega}{c}\right)^2 \frac{1}{k^2 - k_0^2} A_j(\mathbf{k}) \tag{12.2.35}$$

$$1 = 4\pi n_0 \alpha \frac{k_0^2}{k^2 - k_0^2} \tag{12.2.36}$$

Let us define

$$\frac{k}{k_0} = n = \text{index of refraction} \tag{12.2.37}$$

$$k = n\frac{\omega}{c} = \frac{\omega}{c/n}$$

Then

$$1 = \frac{4\pi n_0 \alpha}{(k^2/k_0^2) - 1} = \frac{4\pi n_0 \alpha}{n^2 - 1}$$

and

$$n^2(\omega) - 1 = 4\pi n_0 \alpha(\omega) \qquad (12.2.38)$$

which is called the *Sellmeier formula*.

On the other hand, if we replace the static polarization α with the dynamic polarization $\alpha(\omega)$ in the Clausius–Mosotti formula (12.1.19), we obtain the Lorentz–Lorenz formula:

$$\frac{K - 1}{K + 2} = \frac{n^2(\omega) - 1}{n^2(\omega) + 2} = \frac{4\pi}{3} n_0 \alpha(\omega) \qquad (12.2.39)$$

The reason for the difference in the two results is that we have obtained the Lorentz–Lorenz formula by considering the polarizability of a free molecule and the local field. In the present formulation, we have not used the local field, and a is not the polarizability of the free molecule, but the polarizability of the molecule in the medium.

12.3. Kramers–Kronig Relations

Let us consider an ensemble of molecules under the action of a "real" electric field, and let $E(t)$ be the value of the field at a certain point. We can write

$$E(t) = \int_{-\infty}^{+\infty} E(\omega) e^{-i\omega t} d\omega \qquad (12.3.1)$$

where

$$E(\omega) = \frac{1}{2\pi} \int_{-\infty}^{+\infty} E(t) e^{i\omega t} dt \qquad (12.3.2)$$

Since $E(t)$ is real,

$$E(-\omega) = E^*(\omega) \qquad (12.3.3)$$

The field $E(t)$ can be considered a "stimulus" applied to the molecules which in turn "respond" by becoming polarized. If $P(t)$ is the polarization at time t,

$$P(t) = \int_{-\infty}^{+\infty} P(\omega) e^{-i\omega t} d\omega \qquad (12.3.4)$$

where

$$P(\omega) = \chi(\omega)E(\omega) \tag{12.3.5}$$

Then

$$P(t) = \int_{-\infty}^{+\infty} \chi(\omega)E(\omega)e^{-i\omega t}d\omega \tag{12.3.6}$$

A real field must produce a real polarization. For $P(t)$ to be real, we must have

$$\chi(-\omega) = \chi^*(\omega) \tag{12.3.7}$$

If we set

$$\chi(\omega) = \chi_r(\omega) + i\chi_i(\omega) \tag{12.3.8}$$

relation (12.3.7) is written

$$\chi_r(-\omega) + i\chi_i(-\omega) = \chi_r(\omega) - i\chi_i(\omega) \tag{12.3.9}$$

or

$$\begin{aligned} \chi_r(\omega) &= \chi_r(-\omega) \\ \chi_i(\omega) &= -\chi_i(-\omega) \end{aligned} \tag{12.3.10}$$

These relations, called *crossing relations for the susceptibility*, are derived from the condition that the response to the real field must be a real polarization.

To proceed further with this treatment, it is necessary to specify the model we are using. We shall assume that each molecule can be represented by a particle of mass m and charge e undergoing possible natural oscillations of angular frequency

$$\omega_0 = \sqrt{\frac{K}{m}} \tag{12.3.11}$$

about an equilibrium position where a charge of equal value and opposite sign resides. The equation of motion of the individual molecule under the action of a field component $E_0(\mathbf{x})e^{-i\omega t}$, which we shall assume is polarized in the z-direction, is

$$\begin{aligned} \ddot{z} &= -\frac{K}{m}z + \frac{e}{m}E_0(\mathbf{x})e^{-i\omega t} - \gamma\,\dot{z} \\ &= -\omega_0^2 z + \frac{e}{m}E_0(\mathbf{x})e^{-i\omega t} - \gamma\,\dot{z} \end{aligned} \tag{12.3.12}$$

where $-\gamma \dot{z}$ is the *damping* term. We shall make the assumption that the z displacements are always much smaller than the wavelength of the radiation and that γ is independent of ω. The steady-state solution is

$$z(\mathbf{x}, t) = z_0(\mathbf{x}) e^{-i\omega t} \tag{12.3.13}$$

where

$$z_0(\mathbf{x}) = \frac{e/m}{\omega_0^2 - \omega^2 - i\gamma\omega} E_0(\mathbf{x}) \tag{12.3.14}$$

Then induced dipole moment is

$$ez(\mathbf{x}, t) = \frac{e^2/m}{\omega_0^2 - \omega^2 - i\gamma\omega} E_0(\mathbf{x}) e^{-i\omega t} = \alpha(\omega) E_0(\mathbf{x}) e^{-i\omega t} \tag{12.3.15}$$

where the polarizability $\alpha(\omega)$ is given by

$$\alpha(\omega) = \frac{e^2/m}{\omega_0^2 - \omega^2 - i\gamma\omega} \tag{12.3.16}$$

This expression for $\alpha(\omega)$ is the same as that obtained in Eq. (9.4.12), apart from the fact that here γ is considered independent of ω and there is a minus sign preceding $i\gamma\omega$. This minus sign is due to the choice made here for the field component of frequency ω of a time dependence $e^{-i\omega t}$, rather than of a time dependence $e^{i\omega t}$, as in Sec. 9.4. The susceptibility is given by

$$\chi(\omega) = n_0 \alpha(\omega) = \frac{(n_0 e^2)/m}{\omega_0^2 - \omega^2 - i\gamma\omega} \tag{12.3.17}$$

where $n_0 = $ density of molecules. We can also write

$$
\begin{aligned}
\chi(\omega) &= \frac{(n_0 e^2)/m}{\omega_0^2 - \omega^2 - i\gamma\omega} \\
&= -\frac{(n_0 e^2)/m}{\omega_1 - \omega_2} \left[\frac{1}{\omega - \omega_1} - \frac{1}{\omega - \omega_2} \right] \tag{12.3.18}
\end{aligned}
$$

where

$$
\begin{aligned}
\omega_1 &= -\frac{1}{2} i \left[\gamma - \sqrt{\gamma^2 - 4\omega_0^2} \right] \\
\omega_2 &= -\frac{1}{2} i \left[\gamma + \sqrt{\gamma^2 - 4\omega_0^2} \right] \tag{12.3.19}
\end{aligned}
$$

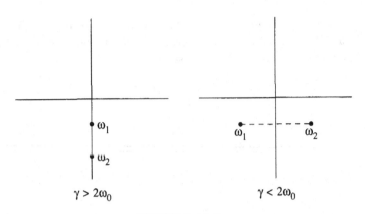

$\gamma > 2\omega_0$ $\gamma < 2\omega_0$

FIGURE 12.6

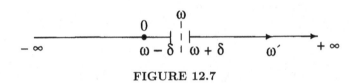

FIGURE 12.7

The positions of ω_1 and ω_2 in the complex plane are depicted in Fig. 12.6. Consider the integral[3]

$$\mathcal{I} = \mathcal{P} \int_{-\infty}^{+\infty} \frac{\chi(\omega')}{\omega' - \omega} d\omega' = \lim_{\delta \to 0} \int_{-\infty}^{\omega-\delta} \frac{\chi(\omega')}{\omega' - \omega} d\omega'$$

$$+ \lim_{\delta \to 0} \int_{\omega+\delta}^{+\infty} \frac{\chi(\omega')}{\omega' - \omega} d\omega' \qquad (12.3.20)$$

where $\mathcal{P}$ indicates the *principal value*. The path of this integral is represented in Fig. 12.7. The integral over this path is shown in Fig. 12.8 to be equal to the integral over the path A minus the two integrals over paths B and C.

The integrand in I has three poles:

$$\omega' = \omega$$
$$\omega' = \omega_1 \qquad (12.3.21)$$
$$\omega' = \omega_2$$

[3]R. Loudon, *The Quantum Theory of Light*, Oxford University Press, New York, 1973, p. 64.

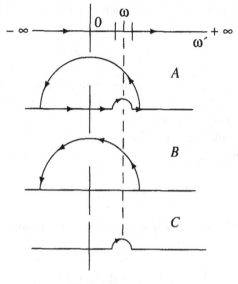

FIGURE 12.8

We now calculate the various parts of I.

Part A. This path for this integral is a closed contour. Given a certain function of a complex variable $z, f(z)$, if L is the boundary of a region in which $f(z)$ is analytic except at a finite number n of poles, a_k, then

$$\oint_L f(z)\, dz = 2\pi i \sum_{k=1}^{n} \text{Res}(a_k) \qquad (12.3.22)$$

where

$$\text{Res}(a_k) = \frac{1}{(m-1)!} \left[\frac{d^{m-1}}{dz^{m-1}} (z - a_k)^m f(z) \right]_{z=a_k} \qquad (12.3.23)$$

This is known as *Cauchy's Residue Theorem.*[4] In the present case the part A is zero, because there are no poles in the upper plane.

Part B. The contour for this integral involves very large values of ω'

$$\chi(\omega') \sim \left(\frac{1}{\omega' - \omega_1} - \frac{1}{\omega' - \omega_2} \right) \qquad (12.3.24)$$

---

[4]F. B. Hildebrand, *Advanced Calculus for Engineers*, Prentice-Hall, Inc., Englewood Cliffs, N.J., 1949, p. 523.

The integrand has an $(\omega')^{-2}$ dependence for large ω'. The length of the contour is $\pi\omega'$. The integral has then an

$$(\omega')^{-2} \times \pi\omega' \simeq (\omega')^{-1} \tag{12.3.25}$$

dependence for large ω'. For an infinite contour, the part B of the integral is zero.

Part C. If $f(z)$ has a simple pole at $z = a$ with the residue $\text{Res}(a)$, and if c_p is a circular arc of radius ρ and center at $z = a$ intercepting an angle α at $z = a$, then

$$\lim_{p \to 0} \int_{c_p} f(z)dz = \alpha i \, \text{Res}(a) \tag{12.3.26}$$

α is positive if the integration is carried out in the counterclockwise way. In our case,

$$\lim_{p \to 0} \int_{c_p} \frac{\chi(\omega')}{\omega' - \omega}d\omega = -i\pi \, \text{Res}(\omega) \tag{12.3.27}$$

$$\text{Res}(\omega) = \left[\frac{\chi(\omega')}{\omega' - \omega}(\omega' - \omega)\right]_{\omega' - \omega} = \chi(\omega) \tag{12.3.28}$$

and the part C is

$$-i\pi\chi(\omega)$$

On collecting terms,

$$i\pi\chi(\omega) = \mathcal{P}\int_{-\infty}^{+\infty} \frac{\chi(\omega')}{\omega' - \omega}d\omega' \tag{12.3.29}$$

$$i\pi(\chi_r + i\chi_i) = \mathcal{P}\int_{-\infty}^{+\infty} \frac{\chi_r(\omega')}{\omega' - \omega}d\omega'$$

$$+ i\mathcal{P}\int_{-\infty}^{+\infty} \frac{\chi_i(\omega')}{\omega' - \omega}d\omega' \tag{12.3.30}$$

$$\chi_r(\omega) = \frac{1}{\pi}\mathcal{P}\int_{-\infty}^{+\infty} \frac{\chi_i(\omega')}{\omega' - \omega}d\omega'$$

$$\chi_i(\omega) = -\frac{1}{\pi}\mathcal{P}\int_{-\infty}^{+\infty} \frac{\chi_r(\omega')}{\omega' - \omega}d\omega' \tag{12.3.31}$$

These relations must be in accord with the crossing relations:

$$\chi_r(\omega) = \frac{1}{\pi}\mathcal{P}\int_{-\infty}^{+\infty}\frac{\chi_i(\omega')}{\omega'-\omega}d\omega'$$

$$= \frac{1}{\pi}\mathcal{P}\left[\int_{-\infty}^{0}\frac{\chi_i(\omega')}{\omega'-\omega}d\omega' + \int_{0}^{\infty}\frac{\chi_i(\omega')}{\omega'-\omega}d\omega'\right]$$

$$= \frac{1}{\pi}\mathcal{P}\left[\int_{0}^{\infty}\frac{\chi_i(-\omega')}{-\omega'-\omega}d\omega' + \int_{0}^{\infty}\frac{\chi_i(\omega')}{\omega'-\omega}d\omega'\right]$$

$$= \frac{1}{\pi}\mathcal{P}\left[\int_{0}^{\infty}\frac{\chi_i(\omega')}{\omega'+\omega}d\omega' + \int_{0}^{\infty}\frac{\chi_i(\omega')}{\omega'-\omega}d\omega'\right]$$

$$= \frac{1}{\pi}\mathcal{P}\int_{0}^{\infty}\chi_i(\omega')\left[\frac{1}{\omega'+\omega}+\frac{1}{\omega'-\omega}\right] = \frac{2}{\pi}\mathcal{P}\int_{0}^{\infty}\frac{\chi_i(\omega')\omega'}{\omega'^2-\omega^2}$$

Also,

$$\chi_i(\omega) = -\frac{1}{\pi}\mathcal{P}\int_{-\infty}^{+\infty}\frac{\chi_r(\omega')}{\omega'-\omega}d\omega'$$

$$= -\frac{1}{\pi}\mathcal{P}\left[\int_{-\infty}^{0}\frac{\chi_r(\omega')}{\omega'-\omega}d\omega' + \int_{0}^{\infty}\frac{\chi_r(\omega')}{\omega'-\omega}d\omega'\right]$$

$$= -\frac{1}{\pi}\mathcal{P}\left[\int_{0}^{\infty}\frac{\chi_r(\omega')}{-\omega'-\omega}d\omega' + \int_{0}^{\infty}\frac{\chi_r(\omega')}{\omega'-\omega}d\omega'\right]$$

$$= -\frac{1}{\pi}\mathcal{P}\int_{0}^{\infty}\chi_r(\omega')\left[\frac{1}{-\omega'-\omega}+\frac{1}{\omega'-\omega}\right]d\omega'$$

$$= -\frac{2\omega}{\pi}\mathcal{P}\int_{0}^{\infty}\frac{\chi_r(\omega')d\omega'}{\omega'^2-\omega^2}$$

We can now write

$$\chi_r(\omega) = \frac{2}{\pi}\mathcal{P}\int_{0}^{\infty}\frac{\omega'\chi_i(\omega')}{\omega'^2-\omega^2}d\omega'$$

$$\chi_i(\omega) = -\frac{2\omega}{\pi}\mathcal{P}\int_{0}^{\infty}\frac{\chi_r(\omega')}{\omega'^2-\omega^2}d\omega'$$

(12.3.32)

These relations are known as *Kramers–Kronig relations* (and also as *dispersion relations*).

On the other hand, the complex index of refraction is given by

$$n(\omega) = \sqrt{K(\omega)} = \sqrt{1 + 4\pi\chi(\omega)}$$
$$\simeq 1 + 2\pi\chi(\omega) = n_r(\omega) + in_i(\omega) \tag{12.3.33}$$

$$n_r(\omega) = 1 + 2\pi\chi_r(\omega)$$
$$n_i(\omega) = 2\pi\chi_i(\omega) \tag{12.3.34}$$

Then the Kramers–Kronig relations give us

$$n_r(\omega) - 1 = \frac{2}{\pi}\mathcal{P}\int_0^\infty \frac{\omega' n_i(\omega')}{\omega'^2 - \omega^2}d\omega'$$
$$n_i(\omega) = -\frac{2\omega}{\pi}\mathcal{P}\int_0^\infty \frac{n_r(\omega')}{\omega'^2 - \omega^2}d\omega' \tag{12.3.35}$$

since

$$\mathcal{P}\int_0^\infty \frac{1}{\omega'^2 - \omega^2}d\omega' = 0 \tag{12.3.36}$$

The proof of Eq. (12.3.36) is left as a problem for the reader.

12.4. General Observations on the Kramers–Kronig Relations

The Kramers–Kronig relations have been derived by using a specific expression for the susceptibility:

$$\chi(\omega) = \frac{(n_0 e^2)/m}{\omega_0^2 - \omega^2 - i\gamma\omega} \tag{12.4.1}$$

This expression is based on a particular model of the "molecule".

However, only two properties of $\chi(\omega)$ were used:

(1) The positions of the poles ω_1 and ω_2 *below* the real axis
(2) The decrease of $\chi(\omega')/(\omega' - \omega)$ faster than $(\omega')^{-1}$ for large ω'

In general, the poles of a response function must lie in the lower half of the complex plane. The principle of causality (see Fig. 12.9) introduces an asymmetry between $t > t_0$ and $t < t_0$, which mathematically leads to the asymmetry in the frequency domain, that is, *poles* below the real axis.[5]

---

[5] J. S. Toll, *Phys. Rev. 104*, 1760 (1956); J. Hamilton, in *Progress in Nuclear Physics*, Vol. 8, edited by O. R. Frisch, Pergamon, Elmsford, N.Y., 1960, p. 145.

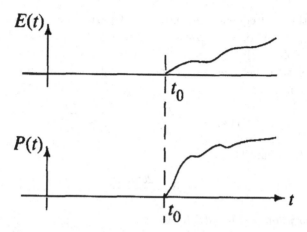

FIGURE 12.9 The principle of causality implies that the response cannot precede the stimulus.

For the classical oscillator model, the causality principle is related to the condition that γ in Eq. (12.3.17) must be positive. The real part of the induced dipole is given by

$$\text{Re}[ez(\mathbf{x}, t)] = \frac{e^2}{m} \frac{E_0(\mathbf{x})}{\sqrt{(\omega_0^2 - \omega^2)^2 + \gamma^2 \omega^2}} \cos(\omega t - \phi) \qquad (12.4.2)$$

where

$$\tan \phi = \frac{\gamma \omega}{\omega_0^2 - \omega^2} \qquad (12.4.3)$$

and

$\omega \simeq 0, \quad \tan \phi = 0, \quad \phi = 0, \quad$ the dipole follows the field

$\omega = \omega_0, \quad \tan \phi = \infty, \quad \phi = \pi/2, \quad$ the dipole moment lags by $90°$

$\omega = \infty, \quad \tan \phi = 0_-, \quad \phi = \pi, \quad$ the dipole moment lags by $180°$

$$(12.4.4)$$

The positive sign of γ places ω_1 and ω_2 below the real axis.

The Kramers–Kronig relations show that the real and imaginary parts of the susceptibility are very intimately connected. Indeed, if we know one part at all the frequencies, we know by means of the integrals the other part at *all* the frequencies. This may be very useful when it is easier to measure one part of $\chi(\omega)$ rather than the other.

Note that a frequency-dependent real part $\chi_r(\omega)$ implies a nonzero imaginary part. Also, a sharp maximum in $\chi_i(\omega)$ produces a sharp change in the slope of $\chi_r(\omega)$.

The relations between $n_r(\omega)$ and $n_i(\omega)$ also contain interesting information. We have

$$n_r(\omega) = 1 + \frac{2}{\pi}\mathcal{P}\int_0^\infty d\omega' \frac{\omega' n_i(\omega')}{\omega'^2 - \omega^2} \qquad (12.4.5)$$

We define the quantity

$$\mu(\omega) = \frac{2\omega n_i(\omega)}{c} \qquad (12.4.6)$$

as the *absorption coefficient*, and write

$$\omega' n_i(\omega') = \frac{c\mu(\omega')}{2} \qquad (12.4.7)$$

Then

$$n_r(\omega) = 1 + \frac{c}{\pi}\mathcal{P}\int_0^\infty d\omega' \frac{\mu(\omega')}{\omega'^2 - \omega^2} \qquad (12.4.8)$$

and

$$n_r(0) = 1 + \frac{c}{\pi}\mathcal{P}\int_0^\infty d\omega \frac{\mu(\omega)}{\omega^2} = \sqrt{K(0)} \qquad (12.4.9)$$

where $K(0)$ is the dielectric constant for stationary fields. Then

$$K(0) = 1 + \frac{2c}{\pi}\mathcal{P}\int_0^\infty d\omega \frac{\mu(\omega)}{\omega^2} \qquad (12.4.10)$$

But

$$\mu(\omega) = \sigma(\omega) n_0 \qquad (12.4.11)$$

where

$$\sigma(\omega) = \text{absorption cross section}$$

$$n_0 = \text{number of molecules per unit volume}$$

Therefore,

$$K(0) - 1 = \frac{2cn_0}{\pi}\mathcal{P}\int_0^\infty d\omega \frac{\sigma(\omega)}{\omega^2} \qquad (12.4.12)$$

and

$$n_r(0) = \sqrt{K(0)} = 1 + \frac{c}{\pi} \mathcal{P} \int_0^\infty d\omega \frac{\mu(\omega)}{\omega^2}$$

$$= 1 + \frac{cn_0}{\pi} \mathcal{P} \int_0^\infty d\omega \frac{\sigma(\omega)}{\omega^2} \tag{12.4.13}$$

These two relations indicate that, if $n_r(0)$ and consequently $K(0)$ are different from 1, the system absorbs in some frequency region. We can also write

$$n_r(\omega) - n_r(0) = \frac{c}{\pi} \mathcal{P} \int_0^\infty \mu(\omega') \left[\frac{1}{\omega'^2 - \omega^2} - \frac{1}{\omega'^2} \right] d\omega'$$

$$= \frac{c\omega^2}{\pi} \mathcal{P} \int_0^\infty \frac{\mu(\omega')}{\omega'^2(\omega'^2 - \omega^2)} d\omega' \tag{12.4.14}$$

The index of refraction $n_r(\omega)$ increases with the frequency ω as long as ω does not fall in the region where a resonance occurs.

The general behavior of the index of refraction n_r of a system with several resonances is depicted in Fig. 12.10. Going up in frequency, if, *after* a resonance, the index of refraction increases again and exceeds the value 1, at least another resonance is present at a higher frequency. If, after a resonance, the index of refraction goes to 1, no resonance occurs at greater frequencies.

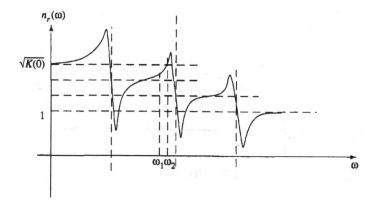

FIGURE 12.10

We note that in the frequency regions where n_r changes, n_i also changes, so we can say with Debye, *Every dispersion has to be accompanied by absorption.*[6]

We note also that all the properties considered above are derived simply by the application of the causality argument.

12.5. Relaxation

The mechanisms that affect the index of refraction may be of various kinds. We can have, for example, radiation damping. If we have a liquid with molecules that have dipole moments (for instance, water or organic substances of the benzene group) at low frequencies, the greatest contribution derives from the orientation of the elementary dipole moments; the main damping in this case is due to the *viscosity* of the medium.

Assume that we have a number of n_0 molecules per unit volume, each molecule having an electric dipole moment. Let P_s be the value that the polarization ultimately takes after the application of a constant E field. If the E field is applied at time $t = 0$, the polarization will follow the law

$$P(t) = P_s(l - e^{-t/\tau}) \tag{12.5.1}$$

where $\tau = $ *relaxation time*. We can also write

$$\frac{dP(t)}{dt} = \frac{1}{\tau}P_s e^{-t/\tau} = \frac{1}{\tau}[P_s - P(t)] \tag{12.5.2}$$

The evolution of $P(t)$ is represented in Fig. 12.11.

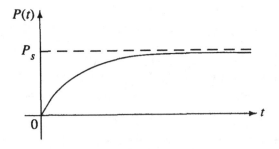

FIGURE 12.11

---

[6]P. Debye, *Polar Molecules*, Dover Publications, Inc., New York, 1929, p. 107.

For a time-dependent field, $E_0 e^{-i\omega t}$,

$$\frac{dP(t)}{dt} = \frac{1}{\tau}[n_0 \alpha_0 E_0 e^{-i\omega t} - P(t)] \tag{12.5.3}$$

where α_0 = orientational polarizability and $n_0 \alpha_0 E_0 e^{-i\omega t}$ is the saturation value of the polarization for a stationary electric field equal in value to $E_0 e^{-i\omega t}$. The solution of Eq. (12.5.3), apart from a transient $Ce^{-t/\tau}$ in which we are not interested here, is given by

$$P(t) = \frac{n_0 \alpha_0}{1 - i\omega \tau} E_0 e^{-i\omega t} = \chi(\omega) E_0 e^{-i\omega t} \tag{12.5.4}$$

Therefore,

$$\chi(\omega) = \frac{n_0 \alpha_0}{1 - i\omega \tau} = \frac{n_0 \alpha_0}{1 + \omega^2 \tau^2}(1 + i\omega \tau) \tag{12.5.5}$$

and

$$\chi_r(\omega) = \frac{n_0 \alpha_0}{1 + \omega^2 \tau^2}$$
$$\chi_i(\omega) = \frac{n_0 \alpha_0 \omega \tau}{1 + \omega^2 \tau^2} \tag{12.5.6}$$

These two functions are depicted in Fig. 12.12. An angle ϕ, a measure of the phase lag of $P(t)$ with respect to $E(t)$, is defined by the relation

$$\tan \phi = \omega \tau \tag{12.5.7}$$

and

$$\phi = 0, \quad \text{for} \quad \omega = 0$$
$$\phi = \frac{\pi}{4}, \quad \text{for} \quad \omega = \frac{1}{\tau} \tag{12.5.8}$$
$$\phi = \frac{\pi}{2}, \quad \text{for} \quad \omega = \infty$$

In liquids,[7] the relaxation time is related to the viscosity η by the approximate relation

$$\tau = \frac{4\pi \eta a^3}{k_B T} \tag{12.5.9}$$

---

[7]P. Debye, *Polar Molecules*, Dover Publications, Inc., New York, 1929, pp. 84–85.

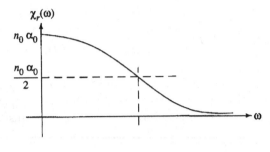

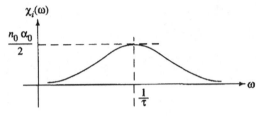

FIGURE 12.12

where

$$k_B = \text{Boltzmann's constant}$$
$$a = \text{``radius'' of the molecule}$$
$$\eta = \text{inner friction constant of the liquid}$$

For water at $T = 300\,\text{K}$

$$\eta = 0.01\,\text{gm cm}^{-1}\text{s}^{-1}$$

$$a = 2 \times 10^{-8}\,\text{cm}$$

$$\tau = \frac{4\pi \times 0.01 \times (2 \times 10^{-8})^3}{1.38 \times 10^{-16} \times 300} \simeq 1.9 \times 10^{-11}\,\text{s}$$

Relaxation times in solids are much longer than in liquids, because a solid presents a more rigid barrier to the internal motion of molecules, leading to their orientation in the direction of an applied field.

12.6. Plasma Frequency

At frequencies far above the highest resonant frequency, the dielectric constant takes a simple form. We can write, using the Eq. (12.3.17) result,

$$K(\omega) = 1 + 4\pi\chi(\omega)$$

$$= 1 + \frac{(4\pi n_0 e^2)/m}{\omega_0^2 - \omega^2 - i\gamma\omega} \xrightarrow{\omega \gg \omega_0} 1 - \frac{4\pi n_0 e^2}{m\omega^2}$$

$$= 1 - \frac{\omega_p^2}{\omega^2} \qquad (12.6.1)$$

where

$$\omega_p^2 = \frac{4\pi n_0 e^2}{m} \qquad (12.6.2)$$

n_0 = total number of free electrons per unit volume

ω_p is called the *plasma frequency*. The wave number is given by

$$k = \frac{\omega}{c/n} = \frac{\omega}{c}n = \frac{\omega}{c}\sqrt{1 - \frac{\omega_p^2}{\omega^2}} = \frac{1}{c}\sqrt{\omega^2 - \omega_p^2} \qquad (12.6.3)$$

or

$$ck\sqrt{\omega^2 - \omega_p^2}$$

or

$$\omega^2 = \omega_p^2 + c^2 k^2 \qquad (12.6.4)$$

which is the *plasma dispersion relation*.

In dielectric media,

$$K(\omega) = 1 - \frac{\omega_p^2}{\omega^2} \qquad (12.6.5)$$

applies only for $\omega \gg \omega_p$. The dielectric constant is slightly less than unity and increases with frequency. The wave number k is real and varies with frequency; as for an electromagnetic mode in a waveguide, there is a cutoff frequency ω_p.

In general, for frequencies ω much smaller than ω_p, but still well above the highest resonance,

$$K(\omega) = \frac{\omega^2 - \omega_p^2}{\omega^2} \simeq -\frac{\omega_p^2}{\omega^2} \tag{12.6.6}$$

$$k = \frac{\omega}{c}\sqrt{K(\omega)} \simeq \frac{\omega}{c}\sqrt{-\frac{\omega_p^2}{\omega^2}} = i\frac{\omega_p}{c} \tag{12.6.7}$$

$$e^{ikz} = e^{i|i(\omega_p/c)|z} = e^{-(\omega_p/c)z} \tag{12.6.8}$$

The intensity of the wave decreases as

$$e^{-(2\omega_p/c)z} = e^{-\mu z} \tag{12.6.9}$$

where

$$\mu = \frac{2\omega_p}{c} \tag{12.6.10}$$

In the ionosphere or in a tenuous electronic plasma in the laboratory,

$$n_0 = 10^{12} - 10^{16} \text{ electrons/cm}^3$$

$$\omega_p^2 = \frac{4\pi n_0 e^2}{m} = \frac{4 \times \pi \times (4.8 \times 10^{-10})^2}{9.1 \times 10^{-28}} n_0$$

$$= 3.18 \times 10^9 n_0 = 3.18 \times (10^{21} - 10^{25})\text{s}^{-2}$$

$$\omega_p \simeq 5.7 \times (10^{10} - 10^{12})\text{s}^{-1}$$

The absorption coefficient is given by

$$\mu = \frac{2\omega_p}{c} = \frac{2 \times (6 \times 10^{10} - 6 \times 10^{12})}{3 \times 10^{10}} = 4 - 400 \text{ cm}^{-1}$$

The reflectivity of metals at optical and higher frequencies is caused by essentially the same behavior as for the plasma. Metals become transparent at very high ultraviolet frequencies.

CHAPTER 12 EXERCISES

12.1. It is often said that the fact that the phase velocity c of the is electromagnetic radiation is independent from the frequency implies that photons have zero rest mass. How would you expect c to vary with frequency if the photon had a very small mass m_0?

12.2. Determine the polarizability and the dielectric constant of a gas of hydrogen atoms, assuming that each atom can be represented by an isotropic harmonic oscillator.

12.3. A dielectric contains n_0 molecules/unit volume. Every molecule has a total scattering cross section $\sigma(\omega)$ for radiation of frequency ω. The Kramers–Kronig dispersion relation states that the static $(\omega = 0)$ dielectric constant K is related to $\sigma(\omega)$ by

$$K - 1 = \frac{2}{\pi} n_0 c \mathcal{P} \int_0^\infty \frac{d\omega}{\omega^2} \sigma(\omega)$$

in the approximation $k + 1 \simeq 2$. Check this relation for a damped oscillator model of scattering. To perform this task, take advantage of reasonable approximations and note that

$$\int_0^\infty \frac{d\omega}{(\omega - \omega_0)^2 + a^2} = \frac{\pi}{a}, \quad \text{for } \omega_0 \gg a$$

12.4. Calculate the dielectric constant $K(\omega)$ of a gas of n_0 free electrons/unit volume as a function of the frequency ω. Determine the critical frequency ω_c below which the gas becomes absorptive for radiation.

12.5. Estimate the critical wavelength λ_c in A at which a typical metal becomes transparent.

12.6. Show that, if the real part of a susceptibility is independent of frequency, the imaginary part vanishes.

12.7. High energy X-rays are falling on a metal plate in which there are n_0 free electrons/cm^3. Calculate the critical angle for total external reflection.

12.8. As the last exercise of this book, comment on the quotation by Galileo Ferraris, which was reported at the very beginning:

Electricity is not only the powerful agent that, breaking through the atmosphere, frightens us with the flash of the lightning and the roar of the thunder, but also the life-giving agent that brings from the Sun to the Earth with light and heat the magic of colors and the breath of life, makes our eyes and heart participate in the beauty of the external world and transmits to the soul the charm of a glance and the enchantment of a smile.

Appendix A

How to Convert a Given Amount of a Quantity from SI Units to Gaussian Units

Quantity	Symbol	SI units	Equivalent in Gaussian units
Distance	L	1 m (meter)	10^2 cm (centimeters)
Mass	m, M	1 kg (Kilogram)	10^3 gm (grams)
Force	$\mathbf{F}$	1 N (Newton)	10^5 dynes
Work, Energy	W, U, E	1 J (Joule)	10^7 ergs
Power	P	1 W (Watt)	10^7 ergs s^{-1}
Charges	q	1 C (Coulomb)	3×10^9 statcoulombs

(*Continued*)

517

(*Continued*)

Quantity	Symbol	SI units	Equivalent in Gaussian units
Current	J, I	1 A (Ampere)	3×10^9 statcoulombs s^{-1}
Electric Potential	Φ, V	1 V (Volt)	$1/300$ statvolt
Electric Field	$\mathbf{E}$	1 V m^{-1}	$(3 \times 10^4)^{-1}$ statvolt cm^{-1}
Polarization	$\mathbf{P}$	1 C m^{-2}	3×10^5 statvolts cm^{-1}
Displacement	$\mathbf{D}$	1 C m^{-2}	$127\pi \times 10^5$ statvolts cm^{-1}
Resistance	R	1 Ω (Ohm)	$(9 \times 10^{11})^{-1}$ s cm^{-1}
Capacitance	C	1 F (Farad)	9×10^{11} cm
Magnetic Induction	$\mathbf{B}$	1 T (Tesla)	10^4 G (Gauss)
Magnetic Flux	$\mathbf{\Phi}$	1 Weber	10^8 G cm^2 (Maxwell)
Magnetization	$\mathbf{M}$	1 A m^{-1}	10^{-3} Oe (Oersted)
Magnet Field	$\mathbf{H}$	1 A m^{-1}	$4\pi \times 10^{-3}$ Oe
Inductance	L	1 H (Henry)	$(9 \times 10^{11})^{-1}$ esu units (When using esu units for currents)

Appendix B

How to Convert an Equation from SI Units to Gaussian Units

Quantity	SI units	Conversion factor for Gaussian units
Charge	q	$\sqrt{4\pi\varepsilon_0}$
Current	J, I	$\sqrt{4\pi\varepsilon_0}$
Electric Potential	Φ, V	$1/\sqrt{4\pi\varepsilon_0}$
Electric Field	$\mathbf{E}$	$1/\sqrt{4\pi\varepsilon_0}$
Polarization	$\mathbf{P}$	$\sqrt{4\pi\varepsilon_0}$
Displacement	$\mathbf{D}$	$\sqrt{\varepsilon_0/4\pi}$

(*Continued*)

519

(*Continued*)

Quantity	SI units	Conversion factor for Gaussian units
Resistance	R	$1/(4\pi\varepsilon_0)$
Capacitance	C	$4\pi\varepsilon_0$
Magnetic Induction	$\mathbf{B}$	$\sqrt{\mu_0/4\pi}$
Vector Potential	$\mathbf{A}$	$\sqrt{\mu_0/4\pi}$
Magnetic Flux	$\mathbf{\Phi}$	$\sqrt{\mu_0/4\pi}$
Magnetization	$\mathbf{M}$	$\sqrt{4\pi/\mu_0}$
Magnetic Field	$\mathbf{H}$	$1/\sqrt{4\pi\mu_0}$
Inductance	L	$1/(4\pi\varepsilon_0)$
Dielectric Constant	K	ε_0
Conductivity	σ	$4\pi\varepsilon_0$
Permeability	μ	μ_0

Bibliography

Becker, R., *Electromagnetic Fields and Interaction*, Dover Publications, Inc., New York, 1964.

Bergmann, P. G., *Introduction to the Theory of Relativity*, Dover Publications, Inc., New York, 1976.

Cheng, D. K., *Field and Wave Electromagnetics*, Addison-Wesley Publishing Company, Reading, Mass., 1989.

Debye, P., *Polar Molecules*, Dover Publications, Inc., New York, 1929.

Eyges, L., *The Classical Electromagnetic Field*, Addison-Wesley Publishing Company, Reading, Mass., 1972.

Frank, N. H., *Introduction to Electricity and Optics*, McGraw-Hill Book Co., New York, 1950.

Frankl, D. R., *Electromagnetic Theory*, Prentice Hall, Englewood Cliffs, N.J., 1989.

Griffiths, D. J., *Introduction to Electrodynamics*, Prentice Hall, Englewood Ciffs, N.J., 1989.

Jackson, J. D., *Classical Electrodynamics*, John Wiley & Sons, Inc., New York, 1975.

Lorrain, P., D. R. Corson, and F. Lorrain, *Electromagnetic Fields and Waves*, W. H. Freeman and Co., San Francisco, 1987.

Marion, J. B. and M. A. Heald, *Classical Electromagnetic Radiation*, Academic Press, Inc., New York, 1980.

Panofsky, K. H. and M. Phillips, *Classical Electricity and Magnetism*, Addison-Wesley Publishing Company, Reading, Mass., 1989.

Purcell, E. M., *Electricity and Magnetism*, Berkeley Physics Course, Volume 2, McGraw-Hill Book Co., New York, 1965.

Rojansky, V., *Electromagnetic Fields and Waves*, Dover Publications, Inc., New York, 1979.

Schwarz, W. M., *Intermediate Electromagnetic Theory*, John Wiley & Sons, Inc., New York, 1964.

Shadowitz, A., *The Electromagnetic Field*, McGraw-Hill Book Co., New York, 1975.

Smythe, W. R., *Static and Dynamic Electricity*, Hemisphere Publishing Corporation, New York, 1989.

Taylor, E. F. and J. A. Wheeler, *Spacetime Physics*, W.H. Freeman Co., San Francisco, 1966.

Index

523

Solutions Manual

Preface

This manual contains the solutions of the odd-numbered exercises that appear in my book "Classical Theory of Electromagnetism." The manual contains not only the answers, but also the procedures that can be used solve the problems.

I am indebted to my students for their contributions to this manual and especially to one of them, Brian Walsh.

Baldassare Di Bartolo

1

Exercises

1.1.

$$\vec{\nabla}(\vec{c} \cdot \vec{r}) = \left(\hat{i}\frac{\partial}{\partial x} + \hat{j}\frac{\partial}{\partial y} + \hat{k}\frac{\partial}{\partial z} \right) (c_x x + c_y y + c_z z)$$

$$= \hat{i}c_x + \hat{j}c_y + \hat{k}c_z$$

$$\vec{\nabla} \cdot \vec{r} = \left(\hat{i}\frac{\partial}{\partial x} + \hat{j}\frac{\partial}{\partial y} + \hat{k}\frac{\partial}{\partial z} \right) \cdot (\hat{i}x + \hat{j}y + \hat{k}z)$$

$$= \frac{\partial x}{\partial x} + \frac{\partial y}{\partial y} + \frac{\partial z}{\partial z} = 1 + 1 + 1 = 3$$

$$\vec{\nabla} \cdot (\vec{c} \times \vec{r}) = \begin{vmatrix} \partial/\partial x & \partial/\partial y & \partial/\partial z \\ c_x & c_y & c_z \\ x & y & z \end{vmatrix}$$

$$= \frac{\partial}{\partial x}(c_y z - c_z y) + \frac{\partial}{\partial y}(c_z x - c_x z) + \frac{\partial}{\partial z}(c_x y - c_y x)$$

$$= 0$$

$$\vec{\nabla} \times \vec{r} = \begin{vmatrix} \hat{i} & \hat{j} & \hat{k} \\ \partial/\partial x & \partial/\partial y & \partial/\partial z \\ x & y & z \end{vmatrix} = 0$$

$$\vec{\nabla} \times (\vec{c} \times \vec{r}) = \begin{vmatrix} \hat{i} & \hat{j} & \hat{k} \\ \partial/\partial x & \partial/\partial y & \partial/\partial z \\ c_y z - c_z y & c_z x - c_x z & c_x y - c_y x \end{vmatrix}$$

$$= \hat{i} 2c_x + \hat{j} 2c_y + \hat{k} 2c_z = 2\vec{c}$$

1.3.

(a) Let $\vec{k}$ be some arbitrary constant vector. By the divergence theorem:

$$\int_S \vec{k} f \cdot \hat{n}\, dS = \int_V (\vec{\nabla} \cdot \vec{k} f) d\tau$$

But,

$$\vec{\nabla} \cdot \vec{k} f = f \vec{\nabla} \cdot \vec{k} + \vec{k} \cdot \vec{\nabla} f = \vec{k} \cdot \vec{\nabla} f$$

Then

$$\int_S \vec{k} f \cdot \hat{n}\, dS = \int_V \vec{k} \cdot \vec{\nabla} f\, d\tau$$

Since $\vec{k}$ is arbitrary we can rewrite this as

$$\vec{k} \cdot \int_S f \hat{n}\, dS = \vec{k} \cdot \int_V \vec{\nabla} f\, d\tau$$

Thus,

$$\int_S f \hat{n}\, dS = \int_V \vec{\nabla} f\, d\tau$$

(b) Let $\vec{k}$ be some arbitrary constant vector. By the divergence theorem

$$\int_S (\vec{k} \cdot \vec{A})(\vec{B} \cdot \hat{n})\, dS = \int_V \vec{\nabla} \cdot [(\vec{k} \cdot \vec{A})\vec{B}] d\tau$$

$$= \int_V \left[\vec{B} \cdot \vec{\nabla}(\vec{k} \cdot \vec{A}) + (\vec{k} \cdot \vec{A})\vec{\nabla} \cdot \vec{B} \right] d\tau$$

But,

$$\vec{B} \cdot \vec{\nabla}(\vec{k} \cdot \vec{A}) = \sum_{i=1}^{3} \vec{B} \cdot \vec{\nabla}(k_i A_i)$$

$$= \sum_{i=1}^{3} k_i(\vec{B} \cdot \vec{\nabla})A_i$$

$$= \vec{k} \cdot (\vec{B} \cdot \vec{\nabla})\vec{A}$$

Then

$$\vec{k} \cdot \int_S \vec{A}(\vec{B} \cdot \hat{n})dS = \vec{k} \cdot \int_V [(\vec{B} \cdot \vec{\nabla})\vec{A} + \vec{A}(\vec{\nabla} \cdot \vec{B})]d\tau$$

and thus

$$\int_S \vec{A}(\vec{B} \cdot \hat{n})dS = \int_V \vec{A}(\vec{\nabla} \cdot \vec{B})d\tau + \int_V (\vec{B} \cdot \vec{\nabla})\vec{A}d\tau$$

1.5.

Let

$$\vec{A} = \varphi\vec{\nabla}\psi$$

Then,

$$\vec{\nabla} \times \vec{A} = \vec{\nabla} \times (\varphi\vec{\nabla}\psi) = \varphi\vec{\nabla} \times \vec{\nabla}\psi + \vec{\nabla}\varphi \times \vec{\nabla}\psi$$

$$= \vec{\nabla}\varphi \times \vec{\nabla}\psi$$

Since

$$\varphi\vec{\nabla} \times \vec{\nabla}\psi = 0$$

Now, using stokes theorem:

$$\int_S (\vec{\nabla} \times \vec{A}) \cdot d\vec{S} = \oint_l \vec{A} \cdot d\vec{l}$$

We find that

$$\int_S (\vec{\nabla}\varphi \times \vec{\nabla}\psi) \cdot d\vec{S} = \int_l \varphi\vec{\nabla}\psi \cdot d\vec{l}$$

But,

$$\vec{\nabla}\psi \cdot d\vec{l} = \frac{\partial\psi}{\partial x}dx + \frac{\partial\psi}{\partial y}dy + \frac{\partial\psi}{\partial z}dz = d\psi$$

So,

$$\int_S (\vec{\nabla}\varphi \times \vec{\nabla}\psi) \cdot d\vec{S} = \int_l \varphi \, d\psi$$

1.7.

Using the vector relation:

$$\vec{A} \wedge (\vec{D} \wedge \vec{C}) - \vec{D}(\vec{A} \ \vec{C}) \quad \vec{C}(\vec{A} \ \vec{B})$$

with $\vec{A} = \vec{C} = \hat{u}$ and $\vec{B} = \vec{a}$

$$\hat{u} \times (\vec{a} \times \hat{u}) = \vec{a}(\hat{u} \cdot \hat{u}) - \hat{u}(\vec{a} \cdot \hat{u})$$
$$= \vec{a} - \hat{u}(\vec{a} \cdot \hat{u})$$

or

$$\vec{a} = \hat{u}(\vec{a} \cdot \hat{u}) + \hat{u} \times (\vec{a} \times \hat{u})$$

Also, $\vec{a} \cdot \hat{u}$ is the component of $\vec{a}$ along $\hat{u}$. The vector $\hat{u} \times (\vec{a} \times \hat{u})$ is perpendicular to $\hat{u}$ and is then the component of $\vec{a}$ a perpendicular to $\hat{u}$.

1.9.

Assume $f(r) = \dfrac{a}{r^n}$ where n = integer

Then

$$\vec{F} = \vec{r} f(r) = a\frac{\vec{r}}{r^n} = \frac{a}{r^{n-1}}\frac{\vec{r}}{r} = F_r \hat{r}$$

Where $F_r = \dfrac{a}{r^{n-1}}$ and $\hat{r} = \dfrac{\vec{r}}{r}$ = unit vector in r direction.

$$\vec{\nabla} \times \vec{F} = \frac{1}{r}\begin{vmatrix} \hat{r} & r\hat{\theta} & (r\sin\theta)\hat{\varphi} \\ \partial/\partial r & \partial/\partial\theta & \partial/\partial\varphi \\ F_r & 0 & 0 \end{vmatrix}$$

$$= -\sin\theta \frac{\partial F_r}{\partial\theta}\hat{\varphi} = 0$$

Now, since F does not depend on θ or φ, then $\vec{\nabla} \cdot \vec{F}$ in spherical coordinates is

$$\vec{\nabla} \cdot \vec{F} = \frac{1}{r^2} \frac{\partial}{\partial r}(r^2 F_r) = \frac{1}{r^2} \frac{\partial}{\partial r}\left(\frac{a}{r^{n-3}}\right)$$

$$= \frac{1}{r^2} \frac{3-n}{r^{n-2}} = \frac{3-n}{r^n}$$

The condition that $\vec{\nabla} \cdot \vec{F} = 0$ requires

$$\frac{3-n}{r^n} = 0 \quad \text{or} \quad n = 3$$

Thus,

$$f(r) = \frac{a}{r^n} = \frac{a}{r^3}$$

1.11.

(a) $\vec{\nabla} \times \vec{A} = \vec{\nabla} \times a\frac{\vec{r}}{r} = a\vec{\nabla} \times \frac{\vec{r}}{r} = a\vec{\nabla} \times \hat{n} = 0$

Then, because of stokes' theorem

$$\int_S \vec{\nabla} \times \vec{A}\, dS = \oint \vec{A} \cdot d\vec{l} = 0$$

This means that the integral from r_1 to r_2 is independent of the path taken:

$$\int_{r_1}^{r_2} \vec{A} \cdot d\vec{r} = a\int_{r_1}^{r_2} \frac{\vec{r}}{r} \cdot d\vec{r} = a\int_{r_1}^{r_2} dr$$

$$= a(r_2 - r_1)$$

(b)

$$\int_S \vec{A} \cdot \hat{n}\, dS = \int_S a\frac{\vec{r}}{r} \cdot \hat{n}\, dS$$

$$= \int_S a\hat{n} \cdot \hat{n}\, dS = a\int_S dS$$

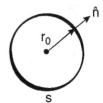

$$\int_S dS = \text{Surface area of sphere of}$$

$$\text{radius } r_0$$

$$= 4\pi r_0^2$$

So,

$$\int \vec{A} \cdot \hat{n} \, dS = 4\pi a r_0^2$$

1.13.

If $\vec{\nabla} \times \vec{A} = 0$ then $\vec{A}$ can be written as the gradient of a scalar.

$$\vec{\nabla} \times \vec{A} = 0 \longrightarrow \vec{A} = \vec{\nabla}\varphi$$

If $\vec{\nabla} \cdot \vec{A} = 0$ then $\vec{\nabla} \cdot \vec{\nabla}\varphi = \nabla^2\varphi = 0$

Let $\vec{h} = \varphi\vec{\nabla}\varphi$

Then

$$\vec{\nabla} \cdot \vec{h} = \vec{\nabla} \cdot (\varphi\vec{\nabla}\varphi) = \varphi\nabla^2\varphi + (\vec{\nabla}\varphi)^2 = (\vec{\nabla}\varphi)^2$$

because $\nabla^2\varphi = 0$

Now

$$\int_V \vec{\nabla} \cdot \vec{h} \, d\tau = \int_V \vec{\nabla} \cdot (\varphi\vec{\nabla}\varphi) d\tau$$

$$= \int_S (\varphi\vec{\nabla}\varphi \cdot \hat{n}) \, dS \quad \begin{array}{l}\text{By the divergence} \\ \text{theorem}\end{array}$$

$$\int_V \vec{\nabla} \cdot \vec{h} \, d\tau = \int_V (\vec{\nabla}\varphi)^2 \, d\tau \quad \text{By above} \quad \vec{\nabla} \cdot \vec{h} = (\vec{\nabla}\varphi)^2$$

But

$$\int_S (\varphi \vec{\nabla}\varphi \cdot \hat{n})dS = \int \varphi \vec{A} \cdot \hat{n}\, dS = 0$$

because $\vec{A} = 0$ on the surface S

Thus,

$$\int_V (\vec{\nabla}\varphi)^2 \, d\tau = 0 \quad \text{implies} \quad \vec{\nabla}\varphi = 0 \quad \text{in} \quad V$$

Since $\vec{A} = \vec{\nabla}\varphi$ then $\vec{A} = 0$ in V.

1.15.

$$\oint_C \vec{v} \cdot d\vec{l} = \oint_C (\vec{w} \times \vec{r}) \cdot d\vec{l}$$

$$= \int_S [\vec{\nabla} \times (\vec{w} \times \vec{r})] \cdot d\vec{S} \quad \text{By stokes theorem.}$$

$$\vec{\nabla} \times (\vec{\omega} \times \vec{r}) = 2\vec{\omega}$$

So,

$$\oint_c \vec{v} \cdot d\vec{l} = \int_S 2\vec{\omega} \cdot d\vec{S} = 2\omega \int_S dS$$

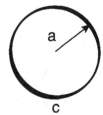

$$\int_S dS = \pi a^2$$

Thus,

$$\oint_C \vec{v} \cdot d\vec{l} = 2\omega \pi a^2$$

1.17.

Consider a vector field defined by: $\vec{A}(\vec{x}_p) = \oint d\vec{l} \times \vec{\nabla}_x \dfrac{1}{r}$, where $\vec{r} = \vec{x} - \vec{x}_p$ and $r = |\vec{x} - \vec{x}_p|$

We can write, because of relation (b) of the preceeding problem

$$\oint d\vec{l} \times \vec{\nabla}_x \frac{1}{r} = \int dS(\vec{p} \times \vec{\nabla}_x) \times \vec{\nabla}_x \frac{1}{r}$$

But

$$(\vec{A} \times \vec{B}) \times \vec{C} = \vec{B}(\vec{A} \cdot \vec{C}) - \vec{A}(\vec{B} \cdot \vec{C})$$

$$(\vec{p} \wedge \vec{\nabla}) \wedge \vec{\nabla} - \vec{\nabla}(\vec{p} \quad \vec{\nabla}) \quad \vec{P}(\vec{\nabla} \quad \vec{\nabla})$$

$$= \vec{\nabla}(\vec{p} \cdot \vec{\nabla}) - \vec{P}\nabla^2$$

Therefore

$$\oint d\vec{l} \times \vec{\nabla}_x \frac{1}{r} = \int dS\, \vec{\nabla}_x(\vec{p} \cdot \vec{\nabla}_x)\frac{1}{r} - \int \vec{P}\nabla_x^2 \frac{1}{r} dS$$

$$= -\vec{\nabla}_{x_p} \int dS(\vec{p} \cdot \vec{\nabla}_x)\frac{1}{r}$$

because $\nabla_x^2 \frac{1}{r} = 0$ for $\vec{x} \neq \vec{x}_p$

So,

$$\int d\vec{l} \times \vec{\nabla}_x \frac{1}{r} = -\vec{\nabla}_{x_p} \phi(\vec{x}_p)$$

1.19.

Let $\vec{C}$ be a constant vector. We can write the following:

$$\vec{\nabla} \cdot (\vec{A} \times \vec{C}) = \vec{C}(\vec{\nabla} \times \vec{A}) - \vec{A}(\vec{\nabla} \times \vec{C})$$

$$= \vec{C} \cdot (\vec{\nabla} \times \vec{A})$$

Also,

$$\int_V \vec{\nabla} \cdot (\vec{A} \times \vec{C})d\tau = \vec{C} \cdot \int_V (\vec{\nabla} \times \vec{A})d\tau$$

and by the divergence theorem

$$\int_V \vec{\nabla} \cdot (\vec{A} \times \vec{C}) d\tau = \int_S (\vec{A} \times \vec{C}) \cdot d\vec{S} = -\vec{C} \cdot \int \vec{A} \times d\vec{S}$$

$$= -\vec{C} \cdot \int_S (\vec{A} \times \hat{n}) dS$$

So,

$$\vec{C} \cdot \int_V (\vec{\nabla} \times \vec{A}) d\tau = -\vec{C} \cdot \int_S (\vec{A} \times \hat{n}) dS$$

or,

$$\int_V (\vec{\nabla} \times \vec{A}) d\tau = -\int_S (\vec{A} \times \hat{n}) dS$$

2

Exercises

2.1.

a) The Laplacian in spherical coordinates is expressed as follows

$$\nabla^2 = \frac{1}{r}\frac{\partial^2}{\partial r^2}r + \frac{\Omega}{r^2}$$

where,

$$\Omega = \frac{1}{\sin\theta}\frac{\partial}{\partial\theta}\left(\sin\theta\frac{\partial}{\partial\theta}\right) + \frac{1}{\sin^2\theta}\frac{\partial^2}{\partial\varphi^2}$$

The Laplace equation $\nabla^2\phi = 0$ is then written as follows in spherical coordinates:

$$\frac{1}{r}\frac{\partial^2}{\partial r^2}(r\phi) + \frac{1}{r^2\sin\theta}\frac{\partial}{\partial\theta}\left(\sin\theta\frac{\partial\phi}{\partial\theta}\right) + \frac{1}{r^2\sin^2\theta}\frac{\partial^2\phi}{\partial\varphi^2} = 0$$

(b) Setting $\phi = R(r)Y(\theta, \varphi)$, Laplace equation is

$$\frac{Y}{r}\frac{\partial^2}{\partial r^2}(Rr) + \frac{R}{r^2 \sin\theta}\frac{\partial}{\partial\theta}\left(\sin\theta\frac{\partial Y}{\partial\theta}\right) + \frac{R}{r^2 \sin^2\theta}\frac{\partial^2 Y}{\partial\varphi^2} = 0$$

With $R(r) = \dfrac{u(r)}{r}$ the above equation becomes

$$\frac{Y}{r}\frac{d^2 u}{dr^2} + \frac{u}{r}\frac{1}{r^2 \sin\theta}\frac{\partial}{\partial\theta}\left(\sin\theta\frac{\partial Y}{\partial\theta}\right) + \frac{u}{r}\frac{1}{r^2 \sin^2\theta}\frac{\partial^2 Y}{\partial\varphi^2} = 0$$

or, multiplying through by r

$$Y\frac{d^2 u}{dr^2} + \frac{u}{r^2 \sin\theta}\frac{\partial}{\partial\theta}\left(\sin\theta\frac{\partial Y}{\partial\theta}\right) + \frac{u}{r^2 \sin^2\theta}\frac{\partial^2 Y}{\partial\varphi^2} = 0$$

With $Y(\theta, \varphi) = P(\theta)\,Q(\varphi)$ the Laplace equation becomes:

$$PQ\frac{d^2 u}{dr^2} + \frac{uQ}{r^2 \sin\theta}\frac{d}{d\theta}\left(\sin\theta\frac{dP}{d\theta}\right) + \frac{uP}{r^2 \sin^2\theta}\frac{d^2 Q}{d\varphi^2} = 0$$

Multiplying through by $r^2 \sin^2\theta/(uPQ)$ gives:

$$r^2 \sin^2\theta\left[\frac{1}{u}\frac{d^2 u}{dr^2} + \frac{1}{(r^2 \sin\theta)P}\frac{d}{d\theta}\left(\sin\theta\frac{dP}{d\theta}\right)\right] + \frac{1}{Q}\frac{d^2 Q}{d\varphi^2} = 0$$

The φ dependence has been isolated in the last term since Q depends only on φ. Taking the last term to the other side of the equation gives an equation with a φ dependence on one side and an (r, θ) dependence on the other side. This can only be true if both sides are equal to a constant. So, we can write the following:

$$\frac{1}{Q}\frac{d^2 Q}{d\varphi^2} = -m^2 \tag{1}$$

$$r^2 \sin^2\theta\left[\frac{1}{u}\frac{d^2 u}{dr^2} + \frac{1}{(r^2 \sin\theta)P}\frac{d}{d\theta}\left(\sin\theta\frac{dP}{d\theta}\right)\right] = m^2$$

The first equation above gives us our differential equation for Q. The second differential equation can be rewritten as:

$$\frac{r^2}{u}\frac{d^2 u}{dr^2} + \frac{1}{P\sin\theta}\frac{d}{d\theta}\left(\sin\theta\frac{dP}{d\theta}\right) - \frac{m^2}{\sin^2\theta} = 0$$

We have isolated the r dependence in the first term and the θ dependence in the second term. Just as before, they will be equal if:

$$\frac{r^2}{u}\frac{d^2u}{dr^2} = \alpha$$

$$\frac{1}{P\sin\theta}\frac{d}{d\theta}\left(\sin\theta\frac{dP}{d\theta}\right) - \frac{m^2}{\sin^2\theta} = -\alpha$$

These two equations give us our differential equations for u and P. Thus

$$\frac{d^2u}{dr^2} = \frac{\alpha u}{r^2} \tag{2}$$

$$-\frac{1}{\sin\theta}\frac{d}{d\theta}\left(\sin\theta\frac{dP}{d\theta}\right) + \frac{m^2 P}{\sin^2\theta} = \alpha P \tag{3}$$

Equations (1), (2) and (3) determine the functions Q, u, and P.

(c) From Eq. (1) the solution is easily seen to be

$$Q(\varphi) = e^{im\varphi}$$

We require that the above solution be single valued, meaning that $Q(\varphi = 0) = Q(\varphi = 2\pi)$
That is: $e^{im(0)} = e^{im(2\pi)}$
or,

$$1 = \cos 2\pi m + i\sin 2\pi m$$

Equating real and imaginary parts, we must demand that:

$$|m| = 0, 1, 2, 3, \ldots$$

For Eq. (3), finite solutions exist everywhere for $\alpha = l(l+1)$ where

$$l = |m|, \quad |m| + 1, \quad |m| + 2, \ldots$$

They are called Associated Legendre functions and are given by:

$$P_l^m(\cos\theta) = (1 - \cos^2\theta)^{m/2}\left(\frac{d}{d(\cos\theta)}\right)P_l(\cos\theta)$$

where

$$P_l(\cos\theta) = \frac{1}{2^l l!}\left(\frac{d}{d(\cos\theta)}\right)^l(\cos^2\theta - 1)^l$$

These are the θ dependent part of the spherical harmonics. Finally, the solution to Eq. (2) is given by, for $\alpha = l(l+1)$

$$U(r) = Ar^{l+1} + Br^{-l}$$

(d) We wish to show that the spherical harmonics are formed from the product of P and Q

$$Y(\theta, \varphi) = P(\theta)Q(\varphi)$$

If $Y(\theta, \varphi)$ obeys the equation

$$\Omega Y + l(l+1)Y = 0$$

then Y is a spherical harmonic.

$$\frac{1}{\sin\theta}\frac{\partial}{\partial\theta}\left(\sin\theta\frac{\partial}{\partial\theta}\right)PQ + \frac{1}{\sin^2\theta}\frac{\partial^2(PQ)}{\partial\varphi^2} + l(l+1)PQ = 0$$

The second term can be written as

$$\frac{1}{\sin^2\theta}\frac{\partial^2(PQ)}{\partial\varphi^2} = \frac{P}{\sin^2\theta}\frac{\partial^2 Q}{\partial\varphi^2} = -\frac{m^2}{\sin^2\theta}PQ$$

So,

$$\frac{1}{\sin\theta}\frac{\partial}{\partial\theta}\left(\sin\theta\frac{\partial}{\partial\theta}\right)PQ - \frac{m^2}{\sin^2\theta}PQ + l(l+1)PQ = 0$$

With $\alpha = l(l+1)$

$$-\frac{1}{\sin\theta}\frac{\partial}{\partial\theta}\left(\sin\theta\frac{\partial P}{\partial\theta}\right) + \frac{m^2 P}{\sin^2\theta} = \alpha P$$

which is Eq. (3). So, we see that $Y(\theta, \varphi) = P(\theta)Q(\varphi)$ is a spherical harmonic.

2.3

Nothing that:

$$P_l(\cos\theta) = \sum_m (-1)^m \alpha_m Y_{lm}^*(\theta_0, \varphi_0)Y_{lm}(\theta, \varphi)$$

If $l = 1$, using the values (2.6.21) for α_m

$$P_1(\cos\theta) = \frac{4\pi}{3}\left[Y_{11}^*(\theta_0, \varphi_0)Y_{11}(\theta, \varphi)\right.$$
$$\left. + Y_{10}(\theta_0, \varphi_0)Y_{10}(\theta, \varphi)\right.$$

$$+Y_{1-1}(\theta_0, \varphi_0)Y_{1-1}(\theta, \varphi)]$$

$$= \frac{4\pi}{3}\left[\frac{3}{8\pi}\sin\theta_0 e^{-i\varphi_0}\sin\theta e^{i\varphi}\right.$$

$$+\frac{3}{4\pi}\cos\theta_0\cos\theta$$

$$\left.+\frac{3}{8\pi}\sin\theta_0 e^{i\varphi_0}\sin\theta e^{-i\varphi}\right]$$

$$= \frac{4\pi}{3}\left\{\frac{3}{8\pi}\sin\theta_0\sin\theta[e^{i(\varphi-\varphi_0)}+e^{-i(\varphi-\varphi_0)}]\right.$$

$$\left.+\frac{3}{4\pi}\cos\theta_0\cos\theta\right\}$$

$$= \frac{4\pi}{3}\left\{\frac{3}{4\pi}\sin\theta_0\sin\theta\cos(\varphi-\varphi_0)+\frac{3}{4\pi}\cos\theta_0\cos\theta\right\}$$

$$= \sin\theta_0\sin\theta\cos(\varphi_0-\varphi)+\cos\theta_0\cos\theta$$

$$= \cos\theta$$

If all the coefficients $(-1)^m\alpha_m$ were not equal, we would not be able to get the dependence on

$$\sin\theta_0\sin\theta\cos(\varphi_0-\varphi)+\cos\theta_0\cos\theta$$

2.5

(a) Let us use Gauss' law

$$\int_S \vec{E}\cdot\hat{n}dS = 4\pi\int_V \rho d\tau$$

with $\vec{E} = -\vec{\nabla}\phi$, we can write $\vec{E}\cdot\hat{n} = -\dfrac{\partial\phi}{\partial n}$
So,

$$\int_S \frac{\partial\phi}{\partial n}dS = -4\pi\int_V \rho d\tau$$

Now consider a surface S of radius r with $\partial\phi/\partial n = \partial\phi/\partial r$ and

$$\int_S \frac{\partial\phi}{\partial n}dS = \int_0^{2\pi}\int_0^\pi \frac{\partial\phi}{\partial r}r^2\sin\theta\,d\theta\,d\varphi = 4\pi r^2\frac{\partial\phi}{\partial r}$$

On the other hand, with $\rho(r) = \rho_0 e^{-\alpha r}$

$$\int_V \rho d\tau = \int_0^r \int_0^{2\pi} \int_0^\pi \rho_0 e^{-\alpha r} r^2 \sin\theta\, d\theta\, d\varphi\, dr$$

$$= 4\pi\rho_0 \int_0^r e^{-\alpha r} r^2 dr$$

$$= -4\pi\rho_0 \frac{e^{-\alpha r}}{\alpha^3} [\alpha^2 r^2 + 2\alpha r + 2]_0^r$$

$$= -4\pi\rho_0 \left[\frac{e^{-\alpha r}}{\alpha^3}(\alpha^2 r^2 + 2\alpha r + 2) - \frac{2}{\alpha^3} \right]$$

Note that this integral goes to zero as $r \to 0$.
Since $\int_S \frac{\partial \phi}{\partial n} dS = -4\pi \int_V \rho d\tau$ we find:

$$4\pi r^2 \frac{d\phi}{dr} = 16\pi^2 \rho_0 \left[\frac{e^{-\alpha r}}{\alpha^3}(\alpha^2 r^2 + 2\alpha r + 2) - \frac{2}{\alpha^3} \right]$$

or,

$$\frac{d\phi}{dr} = \frac{4\pi\rho_0}{\alpha^3} \left[\alpha^2 e^{-\alpha r} + \frac{2\alpha e^{-\alpha r}}{r} + \frac{2e^{-\alpha r}}{r^2} - \frac{2}{r^2} \right]$$

Noting that

$$\frac{d}{dr}\left(\frac{e^{-\alpha r}}{r} \right) = -\frac{\alpha e^{-\alpha r}}{r} - \frac{e^{-\alpha r}}{r^2}$$

We can write the right hand side of the equation as:

$$\frac{4\pi\rho_0}{\alpha^3} \frac{d}{dr}\left[-\alpha e^{-\alpha r} - \frac{2e^{-\alpha r}}{r} + \frac{2}{r} \right]$$

$\phi(r)$ follows immediately

$$\phi(r) = \frac{4\pi\rho_0}{\alpha^3} \left[-\alpha e^{-\alpha r} - \frac{2e^{-\alpha r}}{r} + \frac{2}{r} \right]$$

Note that $\phi(\infty) = 0$ as would be expected.

(b) Let us use the formal solution

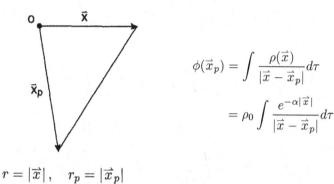

$$\phi(\vec{x}_p) = \int \frac{\rho(\vec{x})}{|\vec{x} - \vec{x}_p|} d\tau$$

$$= \rho_0 \int \frac{e^{-\alpha|\vec{x}|}}{|\vec{x} - \vec{x}_p|} d\tau$$

$$r = |\vec{x}|, \quad r_p = |\vec{x}_p|$$

Let us expand $\dfrac{1}{|\vec{x} - \vec{x}_p|}$ in terms of Legendre Polynomials

$$\frac{|\vec{x}|}{|\vec{x}_p|} < 1 : \quad \frac{1}{|\vec{x} - \vec{x}_p|} = \sum_l \frac{|\vec{x}|^l}{|\vec{x}_p|^{l+1}} P_l(\cos\theta)$$

$$\frac{|\vec{x}|}{|\vec{x}_p|} > 1 : \quad \frac{1}{|\vec{x} - \vec{x}_p|} = \sum_l \frac{|\vec{x}|^l}{|\vec{x}_p|^{l+1}} P_l(\cos\theta)$$

$$\phi(\vec{x}_p) = \rho_0 \int \frac{e^{-\alpha r}}{|\vec{x} - \vec{x}_p|} r^2 \sin\theta \, dr \, d\theta \, d\varphi$$

$$= 2\pi\rho_0 \int_0^{r_p} dr \, e^{-\alpha r} r^2 \sum_l \frac{r^l}{r_p^{l+1}} \int_0^\pi P_l(\cos\theta) \sin\theta \, d\theta$$

$$+ 2\pi\rho_0 \int_{r_p}^\infty dr \, e^{-\alpha r} r^2 \sum_l \frac{r_p^l}{r^{l+1}} \int_0^\pi P_l(\cos\theta) \sin\theta \, d\theta$$

$$= 2\pi\rho_0 \sum_l \frac{1}{r_p^{l+1}} \int_0^{r_p} dr \, e^{-\alpha r} r^{l+2} \int_0^\pi \sin\theta P_l(\cos\theta) d\theta$$

$$+ 2\pi\rho_0 \sum_l r_p^l \int_{r_p}^\infty dr \, e^{-\alpha r} (r^{l-1})^{-1} \int_0^\pi \sin\theta P_l(\cos\theta) d\theta$$

But

$$\int_0^\pi \sin\theta P_l(\cos\theta) \, d\theta = \begin{cases} 0 & \text{for} \quad l > 0 \\ 2 & \text{for} \quad l = 0 \end{cases}$$

Using this in the above expression for $\phi(\vec{x}_p)$

We find

$$\phi(\vec{x}_p) = 4\pi\rho_0 \left\{ \frac{1}{r_p} \int_0^{r_p} r^2 e^{-\alpha r}\,dr + \int_{r_p}^{\infty} r e^{-\alpha r}\,dr \right\}$$

But,

$$\int_0^{r_p} r^2 e^{-\alpha r}\,dr = \frac{d^2}{d\alpha^2}\int_0^{r_p} e^{-\alpha r}\,dr = \frac{d^2}{d\alpha^2}\left(-\frac{1}{\alpha}e^{-\alpha r}\right)\Big|_0^{r_p}$$

$$= -\frac{e^{-\alpha r_p}}{\alpha^3}(\alpha^2 r_p^2 + 2\alpha r_p + 2) + \frac{2}{\alpha^3}$$

$$\int_{r_p}^{\infty} r e^{-\alpha r}\,dr = -\frac{d}{d\alpha}\int_{r_p}^{\infty} e^{-\alpha r}\,dr = \frac{d}{d\alpha}\left(-\frac{1}{\alpha}e^{-\alpha r}\right)\Big|_{r_p}^{\infty}$$

$$= \frac{e^{-\alpha r_p}}{\alpha^2}(\alpha r_p + 1)$$

$$\phi(\vec{x}_p) = 4\pi\rho_0 \left\{ -\frac{e^{-\alpha r_p}}{\alpha}r_p - \frac{2e^{-\alpha r_p}}{\alpha^2} - \frac{2e^{-\alpha r_p}}{\alpha^3 r_p} + \frac{2}{\alpha^3 r_p} \right.$$

$$\left. + \frac{e^{-\alpha r_p}}{\alpha}r_p + \frac{e^{-\alpha r_p}}{\alpha^2} \right\}$$

$$= 4\pi\rho_0 \left\{ -\frac{e^{-\alpha r_p}}{\alpha^2} - \frac{2e^{-\alpha r_p}}{\alpha^3 r_p} + \frac{2}{\alpha^3 r_p} \right\}$$

$$= \frac{4\pi\rho_0}{\alpha^3}\left[-\alpha e^{-\alpha r_p} - \frac{2e^{-\alpha r_p}}{r_p} + \frac{2}{r_p} \right]$$

2.7.

$$\phi(r) = q\frac{e^{-\alpha r}}{r}$$

(a) Let us first find the $\vec{E}$-field

$$\vec{E} = -\vec{\nabla}\phi(r) = -q\vec{\nabla}\frac{e^{-\alpha r}}{r} = -q\frac{\vec{r}}{r}\frac{\partial}{\partial r}\left(\frac{e^{-\alpha r}}{r}\right)$$

$$= -q\frac{\vec{r}}{r}\left(\frac{-\alpha r\,e^{-\alpha r} - e^{-\alpha r}}{r^2}\right)$$

$$= q\frac{\vec{r}}{r}e^{-\alpha r}\left(\frac{1}{r^2} + \frac{\alpha}{r}\right) = qe^{-\alpha r}\left(\frac{1}{r^2} + \frac{\alpha}{r}\right)\hat{r}$$

We now apply Gauss' theorem

$$\int \vec{E} \cdot d\vec{S} = 4\pi Q$$

$$\int \vec{E} \cdot d\vec{S} = q \int_0^{2\pi} \int_0^{\pi} e^{-\alpha r} \left(\frac{1}{r^2} + \frac{\alpha}{r} \right) \hat{r} \cdot \hat{r} r^2 \sin\theta \, d\theta \, d\varphi$$

$$= 4\pi q e^{-\alpha r} (1 + \alpha r)$$

So, the total charge on the sphere is:

$$Q = q e^{-\alpha r} (1 + \alpha r)$$

Note that as $r \to \infty, Q = 0$ as expected.

(b) The charge distribution can be found using Poisson's equation

$$\nabla^2 \phi = -4\pi \rho$$

or,

$$\rho = -\frac{1}{4\pi} \nabla^2 \phi = -\frac{q}{4\pi} \nabla^2 \frac{e^{-\alpha r}}{r}$$

But,

$$\nabla^2 fg = g\nabla^2 f + f\nabla^2 g + 2\vec{\nabla} f \cdot \vec{\nabla} g$$

Therefore, with $f = \dfrac{1}{r}$ and $g = e^{-\alpha r}$

$$\nabla^2 \phi = q \left[e^{-\alpha r} \nabla^2 \frac{1}{r} + \frac{1}{r} \nabla^2 e^{-\alpha r} + 2\vec{\nabla} \frac{1}{r} \cdot \vec{\nabla} e^{-\alpha r} \right]$$

In spherical-polar coordinates

$$\nabla^2 = \frac{1}{r^2} \frac{\partial}{\partial r} \left(r^2 \frac{\partial}{\partial r} \right)$$

$$\nabla^2 e^{-\alpha r} = \frac{1}{r^2} \frac{\partial}{\partial r} \left(r^2 \frac{\partial e^{-\alpha r}}{\partial r} \right) = \frac{1}{r^2} \frac{\partial}{\partial r} (-\alpha r^2 e^{-\alpha r})$$

$$= \frac{1}{r^2} [\alpha^2 r^2 e^{-\alpha r} - 2r\alpha e^{-\alpha r}]$$

$$= \alpha^2 e^{-\alpha r} - \frac{2\alpha}{r} e^{-\alpha r}$$

$$\vec{\nabla} \frac{1}{r} \cdot \vec{\nabla} e^{-\alpha r} = \left(-\frac{1}{r^2} \right) (-\alpha e^{-\alpha r}) = \frac{\alpha e^{-\alpha r}}{r^2}$$

$$\nabla^2 \frac{1}{r} = \nabla_x^2 \frac{1}{|\vec{x} - \vec{x}'|} = -4\pi\delta(\vec{x} - \vec{x}')$$

$$= -4\pi\delta(\vec{x}) \quad \text{for} \quad \vec{x}' = 0$$

Thus,

$$\nabla^2\phi = q\left[-4\pi e^{-\alpha r}\delta(\vec{x}) + \frac{\alpha^2 e^{-\alpha r}}{r} - \frac{2\alpha e^{-\alpha r}}{r^2} + \frac{2\alpha e^{-\alpha r}}{r^2}\right]$$

$$= q\left[\frac{\alpha^2 e^{-\alpha r}}{r} - 4\pi e^{-\alpha r}\delta(\vec{x})\right]$$

The charge density is then given by

$$\rho(\vec{x}) = -\frac{q}{4\pi}\left[\frac{\alpha^2 e^{-\alpha r}}{r} - 4\pi e^{-\alpha r}\delta(\vec{x})\right]$$

(c)

$$Q = \int d\tau \rho(\vec{x})$$

$$= -\frac{q}{4\pi}\left\{\int d\tau\left[\frac{\alpha^2 e^{-\alpha r}}{r} - 4\pi e^{-\alpha r}\delta(\vec{x})\right]\right\}$$

$$= -\frac{q}{4\pi} \cdot 4\pi \int_0^\infty \frac{\alpha^2 e^{-\alpha r}}{r} r^2 dr + q\int_0^\infty e^{-\alpha r}\delta(\vec{x})d\tau$$

$$= -q\int_0^\infty \alpha^2 e^{-\alpha r} r\, dr + q$$

To find the charge at a distance r

$$\int_0^r \alpha^2 e^{-\alpha r} r\, dr = [e^{-\alpha r}(-\alpha r - 1)]_0^r$$

$$= -\alpha r e^{-\alpha r} - e^{-\alpha r} + 1$$

Thus,

$$Q = -q(-\alpha r e^{-\alpha r} - e^{-\alpha r} + 1) + q$$

$$= qe^{-\alpha r}(1 + \alpha r)$$

which is the same as in part (a). Also, $Q \to 0$ as $r \to \infty$ which can be seen from the above expression or by calculating $\int_0^\infty \ldots$ instead of $\int_0^r \ldots$

2.9.

For a spherically symmetric charge distribution Gauss' law gives

$$\int_S \vec{E} \cdot d\vec{S} = 4\pi \int_V \rho(r) d\tau = 16\pi^2 \int_0^r \rho(r) r^2 dr$$

$\vec{E}$ will be a constant normal to the surface at a radius r, so

$$\int_S \vec{E} \cdot d\vec{S} = 4\pi r^2 E$$

Thus,

$$\vec{E} = 4\pi \frac{\hat{r}}{r^2} \int_0^r \rho(r) r^2 dr$$

(a) $\underline{r > R}$:

$$\vec{E} = 4\pi \frac{\hat{r}}{r^2} \int_0^R \frac{A}{r} r^2 dr = \frac{2\pi A R^2}{r^2} \hat{r}$$

with $\phi(\infty) = 0$

$$\phi(r) = \int_r^\infty \vec{E} \cdot d\vec{r} = 2\pi A R^2 \int_r^\infty \frac{dr}{r^2}$$

$$= -2\pi A R^2 \frac{1}{r} \Big|_r^\infty = \frac{2\pi A R^2}{r}$$

$\underline{r < R}$:

$$\vec{E} = 4\pi \frac{\hat{r}}{r^2} \int_0^r \frac{A}{r} r^2 dr = 2\pi A \hat{r}$$

$$\phi(R) - \phi(r) = -\int_r^R \vec{E} \cdot d\vec{r} = -\int_r^R 2\pi A dr = -2\pi A(R - r)$$

or,

$$\phi(r) = \phi(R) + 2\pi A(R - r)$$

but from the $r > R$ case we know that

$$\phi(r) = \frac{2\pi A R^2}{r} \rightarrow \phi(R) = 2\pi A R$$

So,

$$\phi(r) = 2\pi A R + 2\pi A(R - r) = 2\pi A(2R - r)$$

We can sketch $\vec{E}$ and ϕ as follows

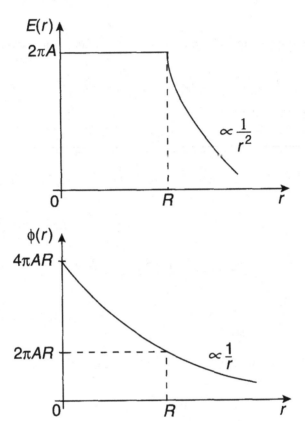

(b) $\underline{r > R}$:

$$\vec{E} = 4\pi \frac{\hat{r}}{r^2} \int_0^R \rho_0 r^2 dr = \frac{4\pi R^3}{3r^2} \rho_0 \hat{r}$$

$$\phi(r) = \frac{4\pi R^3}{3r} \rho_0, \quad \phi(R) \frac{4\pi R^2}{3} \rho_0$$

$\underline{r < R}$:

$$\vec{E} = 4\pi \frac{\hat{r}}{r^2} \int_0^r \rho_0 r^2 dr = \frac{4\pi r}{3} \rho_0 \hat{r}$$

$$\phi(r) - \phi(R) = \int_r^R \vec{E} \cdot d\vec{r} = \int_r^R \frac{4\pi r}{3} \rho_0 dr$$

$$= \frac{4\pi\rho_0}{3}\frac{1}{2}(R^2 - r^2)$$

$$\phi(r) = \frac{4\pi R^2}{3}\rho_0 + \frac{4\pi\rho_0}{3}\cdot\frac{1}{2}(R^2 - r^2) = \frac{2}{3}\pi\rho_0(3R^2 - r^2)$$

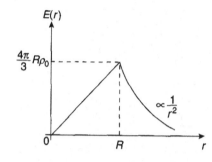

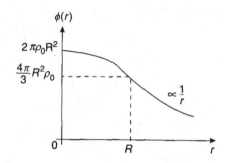

2.11.

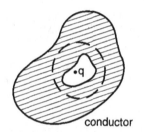

The field inside any conductor is zero

$$\int_S \vec{E}\cdot\hat{n}\,dS = 4\pi Q = 0$$

Thus, if the surface S is chosen to be the dashed curve above, then the charge contained within is $Q = 0$. Since there is a charge q inside the cavity, then there must be a charge $-q$ induced on the surface of the cavity.

2.13.

For a spherically symmetric charge distribution the electric field is given by (see exercise 2.9)

$$\vec{E} = 4\pi\frac{\hat{r}}{r^2}\int_0^r \rho(r)r^2\,dr$$

(a) $\underline{r \leq R}$:

$$\vec{E} = 4\pi\frac{\hat{r}}{r^2}\int_0^r (ar^2)r^2\,dr = \frac{4\pi ar^3}{5}\hat{r}$$

$\underline{r > R}$:

$$\vec{E} = 4\pi\frac{\hat{r}}{r^2}\int_0^R (ar^2)r^2 dr = \frac{4\pi aR^5}{5r^2}\hat{r}$$

(b) $\underline{r > R}$:

$$\phi(r) - \phi(\infty) = \int_r^\infty \vec{E}\cdot d\vec{r} = \frac{4\pi aR^5}{5}\int_r^\infty \frac{dr}{r^2}$$

with $\phi(\infty) = 0$ and $\displaystyle\int_r^\infty \frac{dr}{r^2} = -\frac{1}{r}\Big|_r^\infty = \frac{1}{r}$

$$\phi(r) = \frac{4\pi aR^5}{5r}$$

$\underline{r < R}$:

$$\phi(r) - \phi(R) = \int_r^R \frac{4\pi a}{5}r^3 dr = \left(\frac{4\pi a}{5}\frac{r^4}{4}\right)\Big|_r^R = \frac{\pi a}{5}(R^4 - r^4)$$

$$\phi(r) = \phi(R) + \frac{\pi a}{5}(R^4 - r^4) = \frac{4\pi aR^4}{5} + \frac{\pi a}{5}(R^4 - r^4)$$

$$\phi(r) = \frac{\pi a}{5}(5R^4 - r^4)$$

(c)

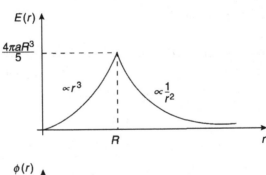

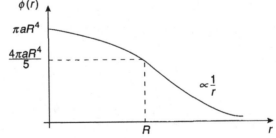

2.15.

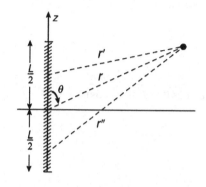

$$r'^2 = r^2 + z^2 - 2zr\cos\theta$$

$$r''^2 = r^2 + z^2 - 2zr\cos\theta$$

For a charge per length

$$\phi = \int \frac{\lambda}{r} dl$$

In this case, with the charge distribution aligned along the z-axis, $dl = dz$ and we have two integrals, one for the range $0 \to L/2$ with $r = r'$ and the other for the range $-L/2 \to 0$ with $r = r''$. So,

$$\phi = \int \frac{\lambda}{r} dl = \lambda \int_0^{L/2} \frac{dz}{r'} + \lambda \int_{-L/2}^0 \frac{dz}{r''}$$

$$= \lambda \int_0^{L/2} \frac{dz}{(r^2 + z^2 - 2zr\cos\theta)^{1/2}}$$

$$+ \lambda \int_{-L/2}^0 \frac{dz}{(r^2 + z^2 + 2zr\cos\theta)}$$

$$= \lambda \left\{ \left[\ln\left(2\sqrt{r^2 + z^2 - 2zr\cos\theta}\right) + 2z - 2r\cos\theta \right]_0^{L/2} \right.$$

$$+ \left[\ln(2\sqrt{r^2 + z^2 + 2zr\cos\theta}) \right.$$

$$\left. +2z + 2r\cos\theta \right]_{-L/2}^0 \Bigg\}$$

$$= \lambda \left\{ \ln\left(2\sqrt{r^2 + \frac{L^2}{4} - Lr\cos\theta} + L - 2r\cos\theta\right) \right.$$

$$-\ln(2r - 2r\cos\theta) + \ln(2r - 2r\cos\theta)$$

$$-\ln\left(2\sqrt{r^2 + \frac{L^2}{4} + Lr\cos\theta} - L - 2r\cos\theta\right)\Bigg\}$$

or,

$$\phi = \lambda\ln\frac{\sqrt{4r^2 + L^2 - 2Lr\cos\theta} + L - 2r\cos\theta}{\sqrt{4r^2 + L^2 + 2Lr\cos\theta} - L - 2r\cos\theta}$$

2.17.

We begin with equation (2.13.32)

$$\sigma(\xi) = -\frac{\phi}{4\pi}\sum_{l=1}^{\infty}(-1)^l(2l+1)\frac{R_0^{l-1}}{R^{l+1}}P_l(\xi)$$

If we move the charge Q to very great distances, but at the same time increase its value so that the field produced by it, $E_0 = \dfrac{Q}{R^2}$, will remain constant. When we do this all the terms with $l > 1$ will become very small due to the fact that at large $R, 1/R^{l+1}$ becomes small. So, we are left with

$$\sigma(\xi) = -\frac{Q}{4\pi}(-1)\frac{3}{R^2}P_1(\xi)$$

with $P_1 = \cos\xi$

$$\sigma(\xi) = \frac{3}{4\pi}\frac{Q}{R^2}\cos\xi$$

The induced charge is found from

$$Q_{\text{ind}} = \int\sigma(\xi)dS = \frac{3Q}{4\pi}\int_0^{2\pi}\int_0^\pi\frac{\cos\xi}{R^2}\cdot R^2\sin\xi\,d\xi\,d\varphi$$

$$= \frac{3Q}{4\pi}\int_0^{2\pi}\int_0^\pi\cos\xi\,\sin\xi\,d\xi\,d\varphi = 0$$

As one might have expected since $Q_{\text{initial}} = 0$ and no charge was added to the sphere. It was simply polarized by the E-field.

2.19.

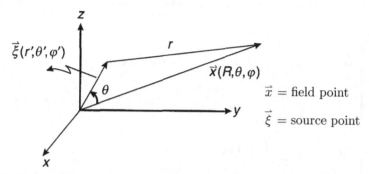

$\vec{x}$ = field point

$\vec{\xi}$ = source point

(a) We can express r in the following way

$$\frac{1}{r} = \sum_{l=0}^{\infty} \frac{r'^l}{R^{l+1}} P_l(\cos\theta)$$

$$= \sum_{l=0}^{\infty} \frac{r'^l}{R^{l+1}} \frac{4\pi}{2l+1} \sum_m (-1)^m Y_{l,-m}(\theta',\varphi') Y_{lm}(\theta,\varphi)$$

Therefore

$$\phi(\vec{x}) = \int \frac{\rho(\vec{\xi})}{r} d\tau'$$

$$\phi(\vec{x}) = \sum_l \int \rho(\vec{\xi}) \frac{r'^l}{R^{l+1}} \frac{4\pi}{2l+1} \sum_m (-1)^m Y_{l,-m}(\theta',\varphi') Y_{lm}(\theta,\varphi)$$

$$= \sum_{lm} \sqrt{\frac{4\pi}{2l+1}} (-1)^m$$

$$\times \left[\sqrt{\frac{4\pi}{2l+1}} \int d\tau' \rho(\vec{\xi}) r'^l Y_{l,-m}(\theta',\varphi') \right] \frac{Y_{lm}l(\theta,\varphi)}{R^{l+1}}$$

$$= \sum_{lm} \sqrt{\frac{4\pi}{2l+1}} (-1)^m D_{l,-m} \frac{Y_{lm}(\theta,\varphi)}{R^{l+1}}$$

where,

$$D_{lm} = \sqrt{\frac{4\pi}{2l+1}} \int d\tau' \rho(\vec{\xi}) r'^l Y_{lm}(\theta',\varphi')$$

(b) We have

$$D_{00} = \sqrt{4\pi} \int d\tau' \rho(\vec{\xi}) Y_{00} = \int d\tau' \rho(\vec{\xi})$$

$$
\begin{cases}
D_{10} = \sqrt{\dfrac{4\pi}{3}} \int d\tau' \rho(\vec{\xi}) r' Y_{10} = \int d\tau' \rho(\vec{\xi}) r' \cos\theta' \\[3ex]
D_{11} = \sqrt{\dfrac{4\pi}{3}} \int d\tau' \rho(\vec{\xi}) r' Y_{11} = -\dfrac{1}{\sqrt{2}} \int d\tau' \rho(\vec{\xi}) r' \sin\theta' e^{i\varphi'} \\[3ex]
D_{1-1} = \sqrt{\dfrac{4\pi}{3}} \int d\tau' \rho(\vec{\xi}) r' Y_{1-1} \\[3ex]
\quad\quad = \dfrac{1}{\sqrt{2}} \int d\tau' \rho(\vec{\xi}) r' \sin\theta' e^{-i\varphi'}
\end{cases}
$$

$$
\begin{cases}
D_{2,-2} = \sqrt{\dfrac{4\pi}{5}} \int d\tau' \rho(\vec{\xi}) r'^{2} Y_{2-2} \\[3ex]
\quad\quad = \sqrt{\dfrac{3}{8}} \int d\tau' \rho(\vec{\xi}) r'^{2} \sin^2\theta' e^{-2i\varphi'} \\[3ex]
D_{2,-1} = \sqrt{\dfrac{4\pi}{5}} \int d\tau' \rho(\vec{\xi}) r'^{2} Y_{2-1} \\[3ex]
\quad\quad = \sqrt{\dfrac{3}{2}} \int d\tau' \rho(\vec{\xi}) r'^{2} \sin\theta' \cos\theta' e^{-i\varphi'} \\[3ex]
D_{20} = \sqrt{\dfrac{4\pi}{5}} \int d\tau' \rho(\vec{\xi}) r'^{2} Y_{20} \\[3ex]
\quad\quad = \dfrac{1}{2} \int d\tau' \rho(\vec{\xi}) r'^{2} (3\cos^2\theta' - 1) \\[3ex]
D_{21} = \sqrt{\dfrac{4\pi}{5}} \int d\tau' \rho(\vec{\xi}) r'^{2} Y_{21} \\[3ex]
\quad\quad = -\sqrt{\dfrac{3}{2}} \int d\tau' \rho(\vec{\xi}) r'^{2} \sin\theta' \cos\theta' e^{i\varphi'} \\[3ex]
D_{22} = \sqrt{\dfrac{4\pi}{5}} \int d\tau' \rho(\vec{\xi}) r'^{2} Y_{22} \\[3ex]
\quad\quad = \sqrt{\dfrac{3}{8}} \int d\tau' \rho(\vec{\xi}) r'^{2} \sin^2\theta' e^{2i\varphi'}
\end{cases}
$$

Using

$$x = r \sin \theta \cos \varphi$$

$$y = r \sin \theta \sin \varphi$$

$$z = r \cos \theta$$

and the equation $D_i = \int d\tau \rho(\vec{x}) x_i$ we find that

$$D_z = \int \rho(\vec{x}) z d\tau = \int \rho(\vec{x}) r \cos \theta d\tau = D_{10}$$

$$D_x = \int \rho(\vec{x}) x d\tau = \int \rho(\vec{x}) r \sin \theta \cos \varphi d\tau$$

$$D_y = \int \rho(\vec{x}) y d\tau = \int \rho(\vec{x}) r \sin \theta \sin \varphi d\tau$$

Noting that

$$\sqrt{2} D_x = -Re[D_{11}] = Re[D_{1-1}]$$

$$\sqrt{2} D_y = -Im[D_{11}] = -Im[D_{1-1}]$$

Thus

$$D_{11} = -\frac{1}{\sqrt{2}} (D_x + i D_y)$$

$$D_{1-1} = \frac{1}{\sqrt{2}} (D_x - i D_y)$$

$$D_{10} = D_z \text{ as found above}$$

To find the next set $\{D_2, m\}$ in terms of D_x, D_y, D_z we first note the following

$$r^2 (3 \cos^2 \theta - 1) = 3 r^2 \cos^2 \theta - r^2 = 3 z^2 - r^2$$

$$r^2 \sin \theta \cos \theta e^{\pm i \varphi} = (x \pm iy) z$$

$$r^2 \sin^2 \theta e^{\pm i 2 \varphi} = (x \pm iy)(x \pm iy)$$

$$\text{and use } Q_{ik} = \int d\tau \rho(\vec{x}) x_i x_k$$

Therefore

$$D_{20} = \frac{1}{2} \int d\tau \rho(\vec{x}) r^2 (3\cos^2\theta - 1)$$

$$= \frac{1}{2} \int d\tau \rho(\vec{x})(3z^2 - r^2)$$

$$= \frac{1}{2} \int d\tau \rho(\vec{x})(3z^2 - x^2 - y^2 - z^2)$$

$$= \frac{1}{2} \int d\tau \rho(\vec{x}) 2z^2 - \frac{1}{2} \int d\tau \rho(\vec{x}) x^2 - \frac{1}{2} \int d\tau \rho(\vec{x}) y^2$$

$$= Q_{zz} - \frac{1}{2} Q_{xx} - \frac{1}{2} Q_{yy}$$

$$D_{20} = -\sqrt{\frac{3}{2}} \int d\tau \rho(\vec{x}) r^2 \sin\theta \cos\theta e^{i\varphi}$$

$$= -\sqrt{\frac{3}{2}} \int d\tau \rho(\vec{x})(xz - iyz)$$

$$= -\sqrt{\frac{3}{2}} Q_{xz} - i\sqrt{\frac{3}{2}} Q_{yz}$$

$$D_{2-1} = \sqrt{\frac{3}{2}} Q_{xz} - i\sqrt{\frac{3}{2}} Q_{yz}$$

$$D_{22} = \sqrt{\frac{3}{8}} \int d\tau \rho(\vec{x}) r^2 \sin^2\theta e^{2i\varphi}$$

$$= \sqrt{\frac{3}{8}} \int d\tau \rho(\vec{x})(x^2 - y^2 + 2ixy)$$

$$= \sqrt{\frac{3}{8}}(Q_{xx} - Q_{yy} + 2iQ_{xy})$$

$$D_{2-2} = \sqrt{\frac{3}{8}}(Q_{xx} - Q_{yy} - 2iQ_{xy})$$

(c) Assume

$$Q = D_{00} = \int \rho(\vec{x}) d\tau = 0$$

If we were to displace the origin by $-a$ then

$$D_{x'} = \int d\tau \rho(\vec{x})(x + a)$$

$$= \int d\tau \rho(\vec{x})x + a \int d\tau \rho(\vec{x})$$

$$= D_x + aQ$$

So, D_x will remain constant with a change of origin only if $Q = 0$. Similarly if $D_x = D_y = D_z = 0$ then a shift of origin $-a$ in $x, -b$ in y

$$D_{x'y'} = \int d\tau (\rho\vec{x})(x + a)(y + b)$$

$$= aD_y + bD_x + abQ + D_{xy}$$

Thus the multipole moments remain constant with a shift of origin if $D_x = D_y = D_z = D_\infty = 0$.

Generalization:

If all the electric multipole moments of order 0 to m are equal to zero, then the multipole moments of order $m + 1$ are independent of the choice of origin.

2.21.

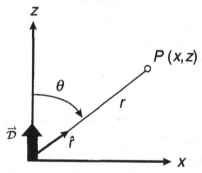

The potential at a point $P(x, z)$ due to a dipole at the origin is:

$$\phi = \frac{\vec{\mathcal{D}} \cdot \vec{r}}{r^3} = \frac{\mathcal{D}r\hat{z} \cdot \hat{r}}{r^3} = \frac{\mathcal{D}\cos\theta}{r^2}$$

Working in the xz-plane

$$\cos\theta = \frac{z}{r} = \frac{z}{(x^2 + z^2)^{1/2}} \quad \text{and} \quad r^2 = x^2 + z^2$$

So,

$$\phi(x, z) = \mathcal{D}\frac{z/(x^2 + z^2)^{1/2}}{x^2 + z^2} = \frac{\mathcal{D}z}{(x^2 + z^2)^{3/2}}$$

To find the E-field we use $\vec{E} = -\vec{\nabla}\phi$

$$E_x = -\frac{\partial\phi}{\partial x} = \frac{\mathcal{D}z \cdot \frac{3}{2}(x^2+z^2)^{1/2} \cdot 2x}{(x^2+z^2)^3} = \frac{3\mathcal{D}xz}{(x^2+z^2)^{5/2}}$$

$$E_z = -\frac{\partial\phi}{\partial z} = -\mathcal{D}\frac{(x^2+z^2)^{3/2} - z \cdot \frac{3}{2}(x^2+z^2)^{1/2} \cdot 2x}{(x^2+z^2)^3}$$

$$= \mathcal{D}\left\{-\frac{1}{(x^2+z^2)^{3/2}} + \frac{3z^2}{(x^2+z^2)^{5/2}}\right\} = \mathcal{D}\frac{(-x^2-z^2+3z^2)}{(x^2+z^2)^{5/2}}$$

$$= \frac{\mathcal{D}(2z^2-x^2)}{(x^2+z^2)^{5/2}}$$

2.23.

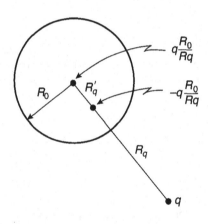

R_q = distance of charge q from center of sphere.

R_q' = distance of mirror charge from center of sphere.

$$R_q' = \frac{R_0^2}{R_q}$$

(a) Take charge q and bring it to an infinite distance from the sphere in such a way that the field produced by it is:

$$|\vec{E}_0| = q/R_q^2 = \text{constant}$$

The image charge will go to the origin in such a way that

$$q\frac{R_0}{R_q}R_q' = q\frac{R_0}{R_q}\frac{R_0^2}{R_q} = \frac{q}{R_q^2}R_0^3 = E_0 R_0^3$$

remains constant. The equivalent dipole is

$$\vec{\mathcal{D}} = \vec{E}_0 R_0^{\ 3}$$

(b)

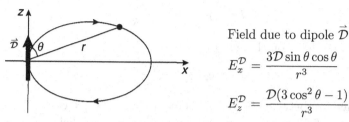

Field due to dipole $\vec{\mathcal{D}}$

$$E_x^{\mathcal{D}} = \frac{3\mathcal{D}\sin\theta\cos\theta}{r^3}$$

$$E_z^{\mathcal{D}} = \frac{\mathcal{D}(3\cos^2\theta - 1)}{r^3}$$

At the surface of the sphere we have a field given by

$$E_x = E_x^{\mathcal{D}} = \frac{3\mathcal{D}\sin\theta\cos\theta}{R_0^3} = 3E_0\sin\theta\cos\theta$$

$$E_z = E_z^{\mathcal{D}} = \mathcal{D}\frac{(3\cos^2\theta - 1)}{R_0^3} + E_0 = 3E_0\cos^2\theta$$

The unit vector $\hat{n}$ perpendicular to the surface of the sphere has the two components

$$n_x = \sin\theta$$
$$n_y = \cos\theta$$

The ratio E_x/E_z is equal to n_x/n_z, therefore $\vec{E}_{\text{tot}}$ and $\hat{n}$ are parallel at the surface,

(c) The total field at the surface is

$$E_{\text{tot}} = \sqrt{E_x^2 + E_y^2} = 3E_0(\sin^2\theta\cos^2\theta + \cos^4\theta)^{1/2} = 3E_0\cos\theta$$

The charge density is thus given by

$$\sigma = \frac{1}{4\pi}E_{\text{tot}} = \frac{3E_0\cos\theta}{4\pi}$$

2.25.

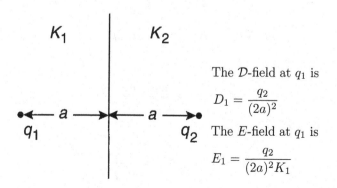

K_1　　K_2

The $\mathcal{D}$-field at q_1 is

$$D_1 = \frac{q_2}{(2a)^2}$$

The E-field at q_1 is

$$E_1 = \frac{q_2}{(2a)^2 K_1}$$

The force on charge q_1 is:

$$F_1 = q_1 E_1 = \frac{q_1 q_2}{(2a)^2 K_1}$$

Likewise, the force on charge q_2 is:

$$F_2 = q_2 E_2 = \frac{q_1 q_2}{(2a)^2 K_2}$$

The difference between F_1 and F_2 is due to the presence of the dielectric. The electric field polarizes the dielectric; the polarized dielectric exercises some additional local force on each charge, so $F_1 \neq F_2$ so long as $K_1 \neq K_2$.

2.27.

(a)

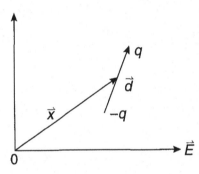

Let the dipole be placed at a position $\vec{x}$. Let $\phi(\vec{x})$ be the electrical potential at $\vec{x}$. The interaction energy is

$$U = \sum_i q_i \phi(\vec{x}_i)$$

$$U = -q\phi\left(\vec{x} - \frac{\vec{d}}{2}\right) + q\phi\left(\vec{x} + \frac{\vec{d}}{2}\right)$$

$$= -q\left[\phi(\vec{x}) - \frac{\vec{d}}{2} \cdot \vec{\nabla}\phi(\vec{x})\right] + q\left[\phi(\vec{x}) + \frac{\vec{d}}{2} \cdot \vec{\nabla}\phi(\vec{x})\right]$$

$$= q\vec{d} \cdot \vec{\nabla}\phi = -\vec{E} \cdot (q\vec{d}) = -\vec{E} \cdot \vec{p}$$

(b) The electric field due to $\vec{p}_1$ is

$$\vec{E}_1(\vec{x}) = \frac{3(\vec{p}_1 \cdot \hat{n})\hat{n} - \vec{p}_1}{r^3}$$

if $\vec{p}_1$ is at the origin.

$$\hat{n} = \frac{\vec{x}}{|\vec{x}|} = \frac{\vec{r}_{12}}{|\vec{r}_{12}|}$$

The interaction of $\vec{p}_2$ with this field gives the interaction between the two dipoles.

$$U_{12} = -\vec{E}_1 \cdot \vec{p}_2 = \frac{\vec{p}_1 \cdot \vec{p}_2}{r_{12}{}^3} - \frac{3(\vec{p}_1 \cdot \vec{r}_{12})(\vec{p}_2 \cdot \vec{r}_{12})}{r_{12}{}^5}$$

2.29.

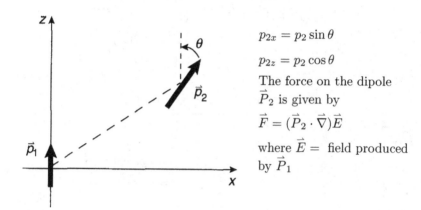

$p_{2x} = p_2 \sin \theta$

$p_{2z} = p_2 \cos \theta$

The force on the dipole $\vec{P}_2$ is given by

$$\vec{F} = (\vec{P}_2 \cdot \vec{\nabla})\vec{E}$$

where $\vec{E} = $ field produced by $\vec{P}_1$

Let E_x and E_z be the components of $\vec{E}$.
We can write

$$\vec{F} = \left(p_2 \sin \theta \frac{\partial}{\partial x} + p_2 \cos \theta \frac{\partial}{\partial z} \right) \vec{E} = \left(p_{2x} \frac{\partial}{\partial x} + p_{2z} \frac{\partial}{\partial z} \right) \vec{E}$$

and

$$F_x = \left(p_{2x} \frac{\partial}{\partial x} + p_{2z} \frac{\partial}{\partial z} \right) E_x = p_{2x} \frac{\partial E_x}{\partial x} + p_{2z} \frac{\partial E_x}{\partial z}$$

$$F_z = \left(p_{2x} \frac{\partial}{\partial x} + p_{2z} \frac{\partial}{\partial z} \right) E_z = p_{2x} \frac{\partial E_z}{\partial x} + p_{2z} \frac{\partial E_z}{\partial z}$$

The fields E_x and E_z due to the dipole $\vec{p}_1$ are given by (see exercise 2.21)

$$E_x = \frac{3p_1 xz}{(x^2 + z^2)^{5/2}}$$

$$E_z = \frac{p_1(2z^2 - x^2)}{(x^2 + z^2)^{5/2}}$$

We find

$$\frac{\partial E_x}{\partial x} = \frac{3z(z^2 - 4x^2)}{(x^2 + z^2)^{7/2}}p_1; \qquad \frac{\partial E_z}{\partial z} = \frac{2z(5x^2 - 3z^2)}{(x^2 + z^2)^{7/2}}p_1$$

$$\frac{\partial E_x}{\partial z} = \frac{3x(x^2 - 4z^2)}{(x^2 + z^2)^{7/2}}p_1; \qquad \frac{\partial E_z}{\partial x} = \frac{x(8z^2 - 7x^2)}{(x^2 + z^2)^{7/2}}p_1$$

Therefore

$$F_x = \frac{3z(z^2 - 4x^2)}{(x^2 + z^2)^{7/2}}p_{2x}p_1 + \frac{3x(x^2 - 4z^2)}{(x^2 + z^2)^{7/2}}p_{2z}p_1$$

$$F_z = \frac{x(8z^2 - 7x^2)}{(x^2 + z^2)^{7/2}}p_{2x}p_1 + \frac{2z(5x^2 - 3z^2)}{(x^2 + z^2)^{7/2}}p_{2z}p_1$$

(b)

$$x = 0, \quad z = a, \quad \theta = 0$$

$$p_{2z} = p_2 \cos 0 = p_2$$
$$p_{2x} = p_2 \sin 0 = 0$$

$$F_x = 0; \quad F_z = -\frac{\sigma}{a^4}p_1p_2$$

$$x = 0, \quad z = a, \quad \theta = 90°$$

$$p_{2z} = p_2 \cos 90 = 0$$
$$p_{2x} = p_2 \sin 90 = p_2$$

$$F_x = \frac{3}{a^4}p_1p_2; \quad F_z = 0$$

$$x = a, \quad z = 0, \quad \theta = 0$$

$$p_{2x} = p_2 \sin 0$$
$$p_{2z} = p_2 \cos 0 = p_2$$

$$F_x = \frac{3}{a^4} p_1 p_2; \quad F_z = 0$$

$$x = a, \quad z = 0, \quad \theta = 90$$

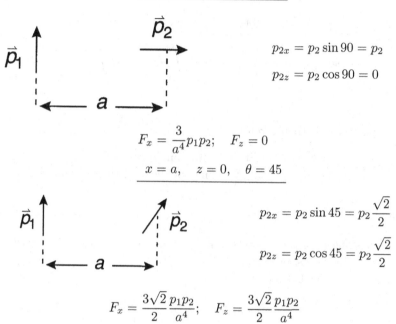

$$p_{2x} = p_2 \sin 90 = p_2$$

$$p_{2z} = p_2 \cos 90 = 0$$

$$F_x = \frac{3}{a^4} p_1 p_2; \quad F_z = 0$$

$$x = a, \quad z = 0, \quad \theta = 45$$

$$p_{2x} = p_2 \sin 45 = p_2 \frac{\sqrt{2}}{2}$$

$$p_{2z} = p_2 \cos 45 = p_2 \frac{\sqrt{2}}{2}$$

$$F_x = \frac{3\sqrt{2}}{2} \frac{p_1 p_2}{a^4}; \quad F_z = \frac{3\sqrt{2}}{2} \frac{p_1 p_2}{a^4}$$

2.31.

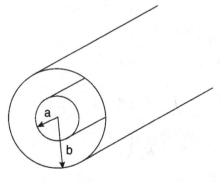

Assume that the charge q is spread over the unit length of internal conductor.

Using Gauss' law

$$\int_S \vec{E} \cdot \hat{n} dS = 4\pi q$$

$$E \cdot 2\pi r = 4\pi q \Rightarrow E = \frac{2q}{r}$$

$$\phi(a) - \phi(b) = -\int_b^a E dr = \int_a^b E dr = 2q \int_a^b \frac{dr}{r}$$

$$= 2q \ln \frac{b}{a}$$

The capacitance is then given by

$$C = \frac{q}{\phi(a) - \phi(b)} = \frac{1}{2 \ln \frac{b}{a}} = \frac{\text{pure number}}{\text{(independent of the units used)}}$$

2.33

Let us assume that the charges q and $-q$ are the charges of the internal and external spheres, respectively.

At every point between the spheres

$$\oint \vec{D} \cdot \hat{n} dS = 4\pi q \Rightarrow D = \frac{q}{r^2}$$

The E-field can be found from $D = KE$

$$E = \frac{D}{K} = \begin{cases} \dfrac{q}{K_1 r^2} a < r < R \\[4mm] \qquad\qquad\qquad \text{RADIALLY} \\ \qquad\qquad\qquad \text{DIRECTED} \\[2mm] \dfrac{q}{K_2 r^2} R < r < b \end{cases}$$

The potential difference between a and b is given by

$$\phi(a) - \phi(b) = \int_a^b E dr = \int_b^a \frac{q}{K_1 r^2} dr + \int_R^b \frac{q}{K_2 r^2} dr$$

$$= \frac{q}{K_1} \left(-\frac{1}{r}\right)_a^R + \frac{q}{K_2} \left(-\frac{1}{r}\right)_R^b$$

$$= \frac{q}{K_1} \left(\frac{1}{a} - \frac{1}{R}\right) + \frac{q}{K_2} \left(\frac{1}{R} - \frac{1}{a}\right)$$

The capacitance is now easily found

$$\frac{1}{c} = \frac{\phi(a) - \phi(b)}{q} = \frac{1}{K_1}\left(\frac{1}{a} - \frac{1}{R}\right) + \frac{1}{K_2}\left(\frac{1}{R} - \frac{1}{b}\right)$$

$$= \frac{1}{K_1 a}\frac{1}{K_2 b} + \frac{1}{R}\left(\frac{1}{K_2} - \frac{1}{K_1}\right)$$

(b) Let $b \to \infty, K_2 = 1, K_1 = K$

Then

$$\frac{1}{c} = \frac{1}{Ka} + \frac{1}{R}\left(1 - \frac{1}{K}\right) = \frac{1}{Ka}\left[1 + \frac{a(K-1)}{R}\right]$$

$$= \frac{(K-1)a + R}{KaR}$$

When $R \gg a(K-1)$ then it is as if the sphere is in an infinite medium of dieletric constant K

$$\frac{1}{c} = \frac{R}{KaR} = \frac{1}{Ka}$$

or

$$c = Ka$$

2.35.

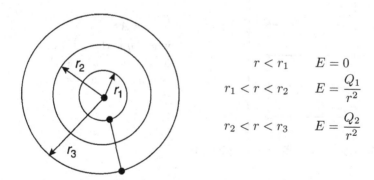

$$r < r_1 \qquad E = 0$$
$$r_1 < r < r_2 \qquad E = \frac{Q_1}{r^2}$$
$$r_2 < r < r_3 \qquad E = \frac{Q_2}{r^2}$$

$$\phi(r_1) - \phi(r_2) = \phi_{12} = \int_{r_1}^{r_2} E\,dr = \int_{r_1}^{r_2} Q_1\frac{dr}{r^2} = \frac{r_2 - r_1}{r_1 r_2}Q_1$$

$$\phi(r_2) - \phi(r_3) = \phi_{23} = \frac{r_3 - r_2}{r_2 r_3}Q_2$$

$$c_{12} = \frac{Q_1}{\phi_{12}} = \frac{r_2 r_1}{r_2 - r_1}; \quad c_{23} = \frac{Q_2}{\phi_{23}} = \frac{r_3 r_2}{r_3 - r_2}$$

$$C_{\text{TOTAL}} = \frac{Q_{\text{TOTAL}}}{\phi_{\text{TOTAL}}} \left.\begin{cases} Q_{\text{TOTAL}} = Q_1 + Q_2 \\ \phi_{\text{TOTAL}} = \phi_{12} = \phi_{23} = \phi \end{cases}\right.$$

$$C_{\text{TOTAL}} = \frac{Q_1 + Q_2}{\phi} = \frac{c_{12}\phi_{12} + c_{23}\phi_{23}}{\phi} = c_{12} + c_{23}$$

Therefore

$$C_{\text{TOTAL}} = \frac{r_2 r_1}{r_2 - r_1} + \frac{r_3 r_2}{r_3 - r_2} = \frac{r_2^2(r_3 - r_1)}{(r_2 - r_1)(r_3 - r_2)}$$

2.37.

(a)

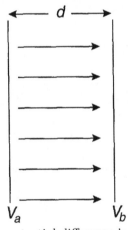

The electric field between the plates is given by

$$E = \frac{4\pi q}{KA}$$

where q is the total charge on the plate a, and A is the area of the plate

The potential difference is given by

$$V_a - V_b = \int_0^d \vec{E} \cdot d\vec{l} \, \frac{4\pi q}{KA} d$$

Therefore

$$C = \frac{q}{V_a - V_b} = \frac{KA}{4\pi d}$$

(b) A constant voltage battery attached to the capacitor would provide a current until the charge reaches the value

$$\frac{AK(V_a - V_b)}{4\pi d}$$

But, since the metal short-circuits the capacitor, the charging process would go on indefinitely as if $c = \infty$, and therefore $K = \infty$.

3

Exercises

3.1.

The electric field at either side of the charge sheet will be pointing perpendicularly to the sheet. We can apply Gauss' theorem to the volume in the figure below.

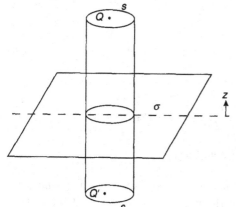

Let points Q and Q' be at some distance from the sheet. We can write:

$$SE(Q) + SE(Q')$$

$$= 2SE(Q)$$

$$= 4\pi\sigma S$$

where,

$$E(Q) = \text{Electric field at point } Q$$
$$E(Q') = \text{Electric field at point } Q'$$

Therefore,

$$E = 2\pi\sigma, \quad z > 0$$
$$E = -2\pi\sigma, \quad z < 0$$

3.3.

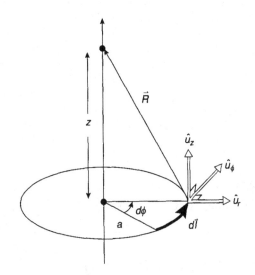

$$d\vec{l} = \hat{u}_\phi a d\phi$$
$$\vec{R} = \hat{u}_z z - \hat{u}_r a$$
$$R = \sqrt{z^2 + a^2}$$

where $\hat{u}_r, \hat{u}_\phi, \hat{u}_z$ are unit vectors in the r, ϕ and z directions respectively.

The Biot and Savart law gives:

$$d\vec{B} = \frac{I d\vec{l} \times \vec{R}}{CR^3} = \frac{I}{CR^3}[\hat{u}_\phi a d\phi \times (\hat{u}_z z - \hat{u}_r a)]$$

$$= \frac{I}{CR^3}[\hat{u}_r a z d\phi + \hat{u}_z a^2 d\phi]$$

If we integrate from $\phi = 0 \to 2\pi$ we see that we have an equal number of oppositely directed unit vectors in the r direction. So, the first term above integrates to zero.

$$\vec{B} = I \int \frac{\vec{dl} \times \vec{R}}{R^3} = \hat{u}_z I \int_0^{2\pi} \frac{a^2 d\phi}{c(z^2 + a^2)^{3/2}}$$

$$= \hat{u}_z \frac{Ia^2 \cdot 2\pi}{c(z^2 + a^2)^{3/2}} = \hat{u}_z \frac{\mu_0 Ia^2}{2(z^2 + a^2)^{3/2}}$$

(in gaussian units)   (in SI units)

3.5.

(a) The magnetic dipole moment is given by:

$$\vec{m} = \frac{J}{c} \times (\text{Area of Loop}) = \frac{\pi R^2 J}{c} \hat{z} \begin{pmatrix} \text{directed in} \\ +z \text{ direction} \end{pmatrix}$$

(b) The magnetic dipole potential is given by:

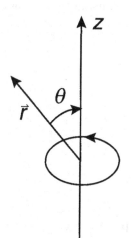

$$\phi_m = -\vec{m} \cdot \vec{\nabla} \frac{1}{r} = -\vec{m} \cdot \left(-\frac{1}{r^2}\right) \frac{\vec{r}}{r}$$

$$= \frac{m \cos\theta}{r^2}$$

where θ = angle between $\vec{r}$ and $\vec{m}$.

Now, $\nabla = \hat{r}\frac{\partial}{\partial r} + \frac{\hat{\theta}}{r}\frac{\partial}{\partial \theta} + \frac{\hat{\varphi}}{r\sin\theta}\frac{\partial}{\partial \varphi}$ in spherical cord. Using this with $\vec{B} = -\nabla\phi$ and noting that ϕ_m does not have a dependence on φ (only on r, θ)

$$\vec{B} = -\left(\hat{r}\frac{\partial}{\partial r} + \frac{\hat{\theta}}{r}\frac{\partial}{\partial \theta}\right)\frac{m\cos\theta}{r}$$

$$= \frac{2m\cos\theta}{r^3}\hat{r} + \frac{m\sin\theta}{r^3}\hat{\theta}$$

(c) Equation of motion is $\vec{F} = \frac{q}{c}\vec{v} \times \vec{B}$ where $\vec{v}$ is the velocity vector of charge q and $\vec{B}$ is the asymptotic field found in part (b)

Now, in the xy-plane $(\theta = \frac{\pi}{2})$

$$\vec{B} = \frac{2m\cos\frac{\pi}{2}}{r^3}\hat{r} + \frac{m\sin\frac{\pi}{2}}{r^3}\hat{\theta} = -\frac{m}{r^3}\hat{z}$$

because $\sin\frac{\pi}{2} = 1$ and $\hat{\theta} - \hat{z}$ in the xy-plane. Nothing that $\vec{v} = v_x\hat{x} + v_y\hat{y}$ initially, then the equation of motion is

$$\vec{F} = M\vec{r} = \frac{q}{c}(v_x\hat{x} + v_y\hat{y})x\left(-\frac{m}{r^3}\hat{z}\right)$$

$$= \frac{qm}{cr^3}(v_x\hat{y} - v_y\hat{x}) = \frac{qm(v_x\hat{y} - v_y\hat{x})}{c(x^2 + y^2)^{3/2}}$$

The force is initially in the xy-plane, and since the magnetic field $\vec{B}$ is the same everywhere ($r =$ const.) then the force remains in the xy plane and so the particle stays in the xy plane.

3.7.

The force acting on a proton is:

$$F = qvB = \frac{mv^2}{r}\text{(SI)}$$

where,

$$m = \text{mass of proton}$$
$$q = \text{charge of proton}$$
$$r = \text{radius of circular path}$$

Solving for B in the above equation

$$B = \frac{mv}{qr} = \frac{P}{qr}$$

The kinetic energy for a proton is

$$T = \frac{P^2}{2m}$$

Combining this with the equation for B we can write

$$B^2 = \frac{p^2}{q^2 r^2} = \frac{2mT}{q^2 r^2} \quad \text{(for protons)}$$

$$B_\alpha^2 = \frac{2m_\alpha T_\alpha}{q_\alpha^2 r_\alpha^2} \quad \text{(for alpha particles)}$$

$$\frac{B_\alpha^2}{B^2} = \frac{m_\alpha q^2 r^2}{m q_\alpha^2 r_\alpha^2} \cdot \frac{T_\alpha}{T} = \frac{m_\alpha q^2 r^2}{m q_\alpha^2 r_\alpha^2} \cdot \frac{q_\alpha v_\alpha}{q V}$$

with $m_\alpha = 4m$, $q_\alpha = 2q$, $V_\alpha = 2 \times 10^6$, $V = 2.5 \times 10^6$ and $r = r_\alpha$ (Requirement that path is same.)

$$B_\alpha = \sqrt{1.6} B = 1.265 \times 8000 = 10119 \text{ Gauss.}$$

3.9.

A particle of mass m and charge q in a cyclotron is subjected to a force

$$Bqv = \frac{mv^2}{r}$$

where v = tangential velocity, r = radius of the orbit.

The period of the circular motion is

$$T = \frac{2\pi r}{v} = \frac{2\pi r}{(Bqr/m)} = \frac{2\pi}{B(q/m)}$$

This result is independent of r and states that for any given value of (q/m) the time required for the charge q to transverse a circumference (complete an orbit) is determined only by the intensity of B.

If R is the outside radius of the cyclotron

$$v_{\max} = \frac{BqR}{m}$$

and the corresponding kinetic energy is

$$\frac{1}{2} m v_{\max}^2 = \frac{1}{2} \left(\frac{q^2}{m} \right) B^2 R^2$$

The potential difference required to produce the above kinetic energy is easily found.

$$qV = \frac{1}{2}mv_{max}^2 = \frac{1}{2}\left(\frac{q^2}{m}\right)B^2R^2$$

Thus,

$$V = \frac{1}{2}\left(\frac{q^2}{m}\right)B^2R^2 \left.\right\}$$
Potential difference required to make particles in cyclotron to orbit at radius R

The frequency of the orbiting particles is

$$\nu = \frac{1}{T} = \frac{B}{2\pi}\left(\frac{q}{m}\right)$$

For 12 MeV deuterons to be produced we need a voltage of 12×10^6 volts. Now,

$$\frac{q}{m_{electron}} = 1.758 \times 10^{11} \frac{coul.}{kg}$$

$$\frac{q}{m_{proton}} = \frac{q}{1836.12\, m_{electron}} = 9.57 \times 10^7 \frac{coul.}{kg}$$

$$\frac{q}{m_{deuteron}} = \frac{q}{2\, m_{proton}} = 4.787 \times 10^7 \frac{coul.}{kg}$$

From the equation $V = \frac{1}{2}(\frac{q}{m})B^2R^2$ found earlier, we calculate the radius R of the cyclotron using 12 MeV deuterons in a B-field of 15000 Gauss $= 1.5\,weber/m^2$:

$$R = \left[\frac{2V}{(q/m_d)}\right]^{1/2} \cdot \frac{1}{B} = \left[\frac{2 \times 12 \times 10^6}{4.787 \times 10^7}\right]^{1/2} \times \frac{1}{1.5} = 0.47\,m$$

(a) To find B we use

$$B = \left[\frac{2V}{(q/m)}\right]^{1/2} \times \frac{1}{R}$$

where $R = 0.47\,m$

$V = 16 \times 10^6$ volts to produce $16\,\text{MeV}$ deuterons

$$q/m = 4.787 \times 10^7 \, \frac{\text{coul.}}{\text{kg}}$$

plugging in the numbers we find the required B-field:

$$B = 1.74 \, \frac{\text{weber}}{\text{m}^2} = 17400\,\text{Gauss.}$$

The frequency follows from this result

$$\nu = \frac{B}{2\pi} \left(\frac{q}{m} \right) = 13.26\,\text{Mc/sec.}$$

(b) For $8\,\text{MeV}$ protons:

$$V = 8 \times 10^6\,\text{volts}$$

$$\frac{q}{m} = 9.57 \times 10^7 \, \frac{\text{coul}}{\text{kg}}$$

$$R = 0.47\,\text{m}$$

We find, plugging the numbers into the same equations we used for pt. a

$$B = 0.87 \, \frac{\text{weber}}{\text{m}^2} = 8717\,\text{Gauss.}$$

$$\nu = 13.25\,\text{Mc/sec.}$$

3.11.

ELECTROSTATICS	MAGNETOSTATICS
$\vec{\nabla} \times \vec{E} = 0$	$\vec{\nabla} \times \vec{B} = \dfrac{4\pi}{c}(\vec{J}_{\text{true}} + \vec{J}_m)$
$\vec{\nabla} \cdot \vec{E} = 4\pi(\rho_{\text{true}} + \rho_{\text{pol}})$	where $\vec{J}_m = c\,\vec{\nabla} \times \vec{M}$
where $\rho_{\text{pol}} = -\vec{\nabla} \cdot \vec{P}$	$\vec{\nabla} \cdot \vec{B} = 0$
$\vec{\nabla} \times \vec{D} = 4\pi\,\vec{\nabla} \times \vec{P}$	$\vec{\nabla} \times \vec{H} = \frac{4\pi}{c}\,\vec{J}_{\text{true}}$
$\vec{\nabla} \cdot \vec{D} = 4\pi\rho_{\text{true}}$	$\vec{\nabla} \cdot \vec{H} = -4\pi\,\vec{\nabla} \cdot \vec{M}$
$\vec{D} = \vec{E} + 4\pi\,\vec{P}$	$\vec{H} = \vec{B} - 4\pi\,\vec{M}$

FERROELECTRICS

$(\rho_{\text{true}} = 0)$

$\vec{\nabla} \times \vec{E} = 0$

$\vec{\nabla} \cdot \vec{E} = 4\pi\rho_{\text{pol.}} = -4\pi \, \vec{\nabla} \cdot \vec{P}$

$\vec{\nabla} \times \vec{D} = 4\pi \, \vec{\nabla} \times \vec{P}$

$\vec{\nabla} \cdot \vec{D} = 0$

$\vec{E} = - \vec{\nabla}_\phi$

$\nabla^2\phi = - \vec{\nabla} \cdot \vec{E} = 4\pi \, \vec{\nabla} \cdot \vec{P}$

$$= -4\pi\rho_{\text{pol.}}$$

$\rho_{\text{pol.}} = - \vec{\nabla} \cdot \vec{P}$

$\sigma_{\text{pol.}} = \hat{n} \cdot \vec{P}$

FERROMAGNETICS

$(\vec{J}_{\text{true}} = 0)$

$\vec{\nabla} + \vec{B} = 4\pi \, \vec{\nabla} \times \vec{M}$

$\vec{\nabla} \cdot \vec{B} = 0$

$\vec{\nabla} \times \vec{H} = 0$

$\vec{\nabla} \cdot \vec{H} = -4\pi \, \vec{\nabla} \cdot \vec{M}$

$\vec{H} = - \vec{\nabla}_{\phi H}$

By analogy with

electrostatic case

$\nabla^2\phi_H = - \vec{\nabla} \cdot \vec{H}$

$$= 4\pi \, \vec{\nabla} \cdot \vec{M}$$

So, we postulate by analogy:

$$\rho_M = - \vec{\nabla} \cdot \vec{M}$$

$$\sigma_M = \hat{n} \cdot \vec{M}$$

3.13.

(a) The ring of charge q rotates about it own axis, so it acts like a current loop. The magnetic moment is:

$$m = \frac{I}{C} \times (\text{Area of Loop}) = \frac{I}{C}\pi R^2$$

But $I = q/T$ where $T = \text{Period} = \frac{2\pi}{\omega}$

So, we can write the magnetic moment as

$$m = \frac{q}{CT}\pi R^2 = \frac{q\omega R^2}{2C}$$

The angular momentum of the rotating ring is

$$L = I\omega = MR^2\omega$$

So, the gyromagnetic ratio is given by:

$$\frac{m}{L} = \frac{q\omega R^2/2C}{MR^2\omega} = \frac{q}{2CM}$$

(b)

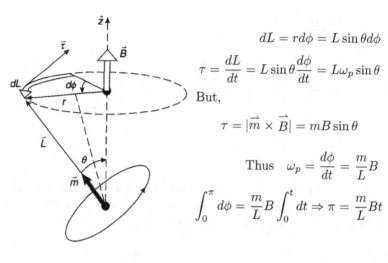

$$dL = r d\phi = L \sin\theta \, d\phi$$

$$\tau = \frac{dL}{dt} = L \sin\theta \frac{d\phi}{dt} = L\omega_p \sin\theta$$

But,

$$\tau = |\vec{m} \times \vec{B}| = mB \sin\theta$$

Thus $\omega_p = \dfrac{d\phi}{dt} = \dfrac{m}{L} B$

$$\int_0^\pi d\phi = \frac{m}{L} B \int_0^t dt \Rightarrow \pi = \frac{m}{L} Bt$$

$$t = \frac{\pi}{B(m/L)} = \frac{2\pi cM}{Bq} = \text{Time to process through } \phi = \pi$$

3.15.

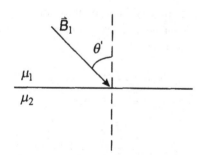

Assume that at the interface between two linear magnetic media of permeabilities μ_1 and μ_2 the field $\vec{B}_1$ makes an angle θ_1 with the normal to the interface.

Application of the boundary conditions for B and H are

$$B_{1n} = B_{2n}$$

$$H_{1t} = H_{2t} \rightarrow \frac{B_{1t}}{\mu_1} = \frac{B_{2t}}{\mu_2}$$

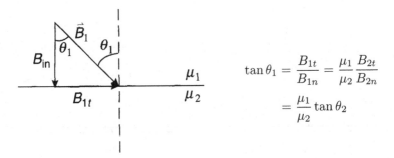

$$\tan\theta_1 = \frac{B_{1t}}{B_{1n}} = \frac{\mu_1}{\mu_2}\frac{B_{2t}}{B_{2n}}$$

$$= \frac{\mu_1}{\mu_2}\tan\theta_2$$

If μ_2 is very large then $\tan\theta_1$ is very small, and $\vec{B}_1$ will be very nearly perpendicular to the boundary between the magnetic mediums. This is analogous to the electric field just outside a good conductor.

3.17.

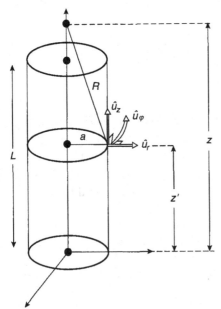

We can solve this problem by considering the fact that the magnet is like a cylindrical sheet with a surface current density

$$\vec{K}_m = \vec{M} \times \hat{n}$$

$$= M\hat{u}_z \times \hat{u}_r$$

$$= M\hat{u}_\varphi$$

$$d\vec{B} = \frac{\mu_0}{4\pi}\frac{\vec{K}_m \times \vec{R}}{R^3}dS$$

For points along the axis:

$$\vec{R} = (z - z')\hat{u}_z - a\hat{u}_r, \quad R^3 = [(z - z')^2 + a^2]^{3/2}$$

$$dS = ad\varphi dz' \begin{pmatrix} \text{We don't need to consider} \\ \text{the top and bottom surface} \\ \text{areas because } K_m = 0 \text{ there} \end{pmatrix}$$

$$d\vec{B} = \frac{\mu_0}{4\pi} \frac{M\hat{u}_\varphi \times [(z - z')\hat{u}_z - a\hat{u}_r]}{[(z - z')^2 + a^2]^{3/2}} ad\varphi dz'$$

$$dB_z = u_z \cdot d\vec{B} = \frac{\mu_0}{4\pi} \frac{Ma^2 d\varphi dz'}{[(z - z')^2 + a^2]^{3/2}}$$

$$B_z = \int dB_z = \int_0^{2\pi} d\varphi \int_0^L \frac{\mu_0}{4\pi} \frac{Ma^2 dz'}{[(z - z')^2 + a^2]^{3/2}}$$

Doing the integration gives:

$$B_z = \frac{\mu_0 M}{2} \left[\frac{z}{\sqrt{z^2 + a^2}} - \frac{(z - L)}{\sqrt{(z - L)^2 + a^2}} \right] \text{(SI)}$$

$$\vec{H} = \frac{\vec{B}}{\mu_0} - M$$

$$H_z = \frac{B_z}{\mu_0} - M_z = \frac{B_z}{\mu_0} - M \quad \text{since } M = M_z$$

Thus,

$$H_z = \frac{M}{2} \left[\frac{z}{\sqrt{z^2 + a^2}} - \frac{(z - L)}{\sqrt{(z - L)^2 + a^2}} - 2 \right] \text{(SI)}$$

Note the following:
At $z = L$

$$B_z = \frac{\mu_0 M}{2} \frac{L}{\sqrt{L^2 + a^2}}, \quad H_z = \frac{M}{2} \left[\frac{L}{\sqrt{L^2 + a^2}} - 2 \right]$$

As $L \to 0$

$$B_z \to 0, \quad H_z \to -M$$

As $L \to \infty; L \gg a$

$$B_z \to \mu_0 M, \quad H_z \to 0$$

3.19.

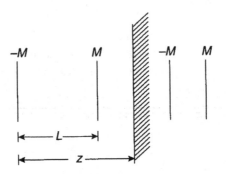

Because of the medium of infinite permeability on the right, we have "images" as in the figure above. The effect is to double the intensity of the field at the surface of the medium:

$$B = H = \mu_0 M \left(\frac{z}{\sqrt{z^2 + a^2}} - \frac{z - L}{\sqrt{(z - L)^2 + a^2}} \right) \text{(SI)}$$

or,

$$B = H = 4\pi M \left(\frac{z}{\sqrt{z^2 + a^2}} - \frac{z - L}{\sqrt{(z - L)^2 + a^2}} \right) \text{(gaussian)}$$

The answers were obtained by doubling the results in exercise 3.17

3.21.

Let us consider a rectangular hole in a permanent homogeneous magnet of magnetization M. We can write

$$\vec{\nabla} \cdot \vec{H} = -4\pi \vec{\nabla} \cdot \vec{M} = 4\pi \rho_M \quad \text{where,} \quad \rho_M = -\vec{\nabla} \cdot \vec{M}$$

Therefore,

$$\vec{H}(\vec{x}) = \vec{\nabla}_x \int d\tau' \frac{\vec{\nabla}' \cdot \vec{M}(\vec{x}')}{|\vec{x} - \vec{x}'|}$$

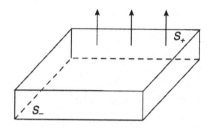

There is a surface charge $-M$ on S_+ and M on S_-. So, $\vec{\nabla} \cdot \vec{M} \neq 0$ only on S_+ and S_-

Then,

$$H(\vec{x}) = \vec{\nabla}_x \left[\iint dS_+ \frac{M}{r} - \iint dS_- \frac{M}{r} \right]$$

This holds for all space. We also have

$$\vec{B} = \vec{H} \qquad \text{inside hole}$$

$$\vec{B} = \vec{H} + 4\pi\vec{M} \qquad \text{outside hole}$$

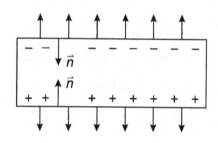

$\sigma_m = \hat{n} \cdot \vec{M}$

(a)

Flat Cavity: $H_{\text{inside}} = 4\pi\vec{M}$

(b)

Thin Cavity: $H_{\text{inside}} = 0$

3.23.

$$\vec{P} = \text{constant} \quad \text{inside}$$
$$\vec{P} = 0 \qquad \text{outside}$$

(a) The surface charge density is given by

$$\sigma_{\text{pol.}} = \hat{n} \cdot \vec{P}(\vec{x}) = \hat{n} \cdot \vec{P}$$

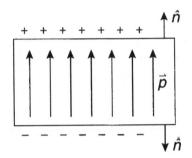

Sides : $\sigma_{\text{pol.}} = 0$
Top : $\sigma_{\text{pol.}} = P$
Bottom : $\sigma_{\text{pol.}} = -P$

(b) Since $\vec{\nabla} \cdot \vec{D} = 0$ then from the boundary conditions between macroscopic media

$$D_n(\text{inside}) = D_n(\text{outside})$$

Thus, $D_z = $ constant through the crystal

(c)

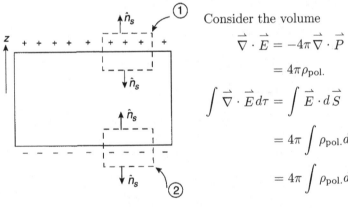

Consider the volume

$$\vec{\nabla} \cdot \vec{E} = -4\pi \vec{\nabla} \cdot \vec{P}$$
$$= 4\pi \rho_{\text{pol.}}$$
$$\int \vec{\nabla} \cdot \vec{E}\, d\tau = \int \vec{E} \cdot d\vec{S}$$
$$= 4\pi \int \rho_{\text{pol.}} d\tau$$
$$= 4\pi \int \rho_{\text{pol.}} dS$$

Therefore,

$$E_z(\text{outside}) - E_z(\text{inside})$$
$$= 4\pi \sigma_{\text{pol.}} = 4\pi P$$

Now consider the volume ②

$$\int \vec{E} \cdot d\vec{S} = 4\pi \int \sigma_{\text{pol}}.dS$$

$$= -4\pi \int \hat{n} \cdot \vec{P} dS$$

$$= -4\pi \int \vec{P} \cdot d\vec{S}$$

$$E_z(\text{inside}) - E_z(\text{outside}) = -4\pi P$$
$$E_z(\text{outside}) - E_z(\text{inside}) = 4\pi P$$

The following sketch illustrates the answer

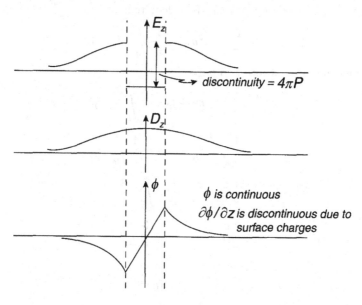

3.25.

$$\pi^+ \text{ meson: } mc^2 = 139.6 \,\text{MeV}$$

$$\pi^0 \text{ meson: } mc^2 = 135.0 \,\text{MeV}$$

$$\text{Energy} = 139.6 - 135.0 = 4.6 \,\text{MeV}$$

$$= 7.36 \times 10^{-6} \,\text{ergs}.$$

Let q be the charge of the π^+ meson. The charge density is given by:

$$\rho = \frac{q}{\frac{4}{3}\pi R^3} = \frac{3q}{4\pi R^3} \quad \begin{pmatrix} \text{where } R = \text{radius} \\ \text{of the } \pi^+ \text{ meson} \end{pmatrix}$$

For $\underline{0 < r < R}$

$$E \cdot 4\pi r^2 = 4\pi \int \rho d\tau = 4\pi q \frac{r^3}{R^3}$$

$$E = q\frac{r}{R^3}$$

For $\underline{r \geq R}$

$$E = \frac{q}{r^2}$$

The electrostatic energy can now be calculated.

$$\text{Energy} = \frac{1}{8\pi} \int E^2 d\tau = \frac{1}{8\pi} \int_0^\infty E^2 \cdot 4\pi r^2 dr$$

Breaking the integral into two parts

$$\text{Energy} = \frac{1}{2} \int_0^R E^2 r^2 dr + \frac{1}{2} \int_R^\infty E^2 r^2 dr$$

$$= \frac{1}{2} \int_0^R \frac{q^2 r^2}{R^6} r^2 dr + \frac{1}{2} \int_R^\infty \frac{q^2}{r^4} r^2 dr$$

$$= \frac{1}{2} \frac{q^2}{R^6} \frac{R^5}{5} + \frac{1}{2} q^2 \int_R^\infty \frac{1}{r^2} dr$$

$$= \frac{1}{2} \frac{q^2}{R} \frac{1}{5} + \frac{1}{2} \frac{q^2}{R} = \frac{3}{5} \frac{q^2}{R}$$

Now,

$$q = 4.8 \times 10^{-10} \text{e.s.u.}$$

$$\text{Energy} = 7.36 \times 10^{-6} \text{ergs.}$$

So,

$$7.36 \times 10^{-6} = \frac{3}{5} \frac{(4.8 \times 10^{-10})^2}{R}$$

$$R = \frac{3 \times (4.8 \times 10^{-10})^2}{5 \times 7.36 \times 10^{-6}} = 1.88 \times 10^{-14} \text{cm}$$

4

Exercises

4.1.

(a) The displacement current is given by

$$\vec{J}_D = \frac{1}{4\pi} \frac{\partial \vec{D}}{\partial t}$$

The conduction current is

$$\vec{J} = \sigma \vec{E}$$

where E = electric field

Assume that $J = j \sin 2\pi\nu t$

In the medium of conductivity σ and dielectric constant K:

$$J_D = \frac{1}{4\pi} \frac{\partial D}{\partial t} = \frac{K}{4\pi} \frac{\partial E}{\partial t} = \frac{K}{4\pi\sigma} \frac{\partial J}{\partial t}$$

$$= \frac{K}{4\pi\sigma} 2\pi\nu j \cos 2\pi\nu t = \frac{K\nu}{2\sigma} j \cos 2\pi\nu t$$

$\frac{J_D}{J}$ is on the order of $\frac{\nu}{\sigma}$

If $\sigma \gg \nu$ (in good conductors $\sigma = 10^{17}$) we can disregard the displacement current,

(b) If the displacement current is negligible

$$\vec{\nabla} \times \vec{H} = \frac{4\pi}{c}\vec{J}, \quad \vec{\nabla} \times \vec{E} = -\frac{\mu}{c}\frac{\partial \vec{H}}{\partial t}$$

Then,

$$\vec{\nabla} \times (\vec{\nabla} \times \vec{E}) = \vec{\nabla}(\vec{\nabla} \cdot \vec{E}) - \nabla^2 \vec{E} = -\frac{\mu}{c}\frac{\partial}{\partial t}(\vec{\nabla} \times \vec{H})$$

$$= -\frac{\mu}{c}\frac{\partial}{\partial t}\left(\frac{4\pi}{c}\vec{J}\right)$$

$$\nabla^2\vec{E} - \vec{\nabla}(\vec{\nabla} \cdot \vec{E}) = \frac{\mu}{c}\frac{\partial}{\partial t}\left(\frac{4\pi}{c}\vec{J}\right)$$

But $\vec{E} = \frac{\vec{J}}{\sigma}$

$$\nabla^2\vec{J} - \vec{\nabla}(\vec{\nabla} \cdot \vec{J}) = \frac{4\pi\mu\sigma}{c^2}\frac{\partial \vec{J}}{\partial t}$$

If $\rho = 0$, then

$$\nabla^2\vec{J} - \frac{4\pi\mu\sigma}{c^2}\frac{\partial \vec{J}}{\partial t} = 0$$

4.3.

(a) The motion of the metal stick causes a change in the magnetic flux through the circuit, which in turn induces an εmf given by:

$$V_A - V_B = \frac{1}{q}\oint_C \vec{F_q} \cdot d\vec{l} = \frac{1}{c}\oint d\vec{l} \cdot (v \times \vec{B})$$

$$= \frac{1}{c}\oint dl\hat{y} \cdot (v\hat{x} \times B\hat{z}) = -\frac{vBl}{c}$$

(b) The current which flows from terminal B to A is given by Ohm's law

$$I = \frac{V}{R} = \frac{V_B - V_A}{R} = \frac{vBl}{cR}$$

The electric power dissipated in R

$$P_e = I^2 R = \frac{v^2 B^2 l^2}{c^2 R^2} R = \frac{v^2 B^2 l^2}{c^2 R}$$

The mechanical power necessary to move the stick is given by

$$P_m = \vec{F} \cdot \vec{v}$$

In order for the stick to move $\vec{F}$ must be equal and opposite to the magnetic force on the stick

$$\vec{F} = -\vec{F}_{\text{mag}} = -\frac{I}{c} \int_{B'}^{A'} d\,\vec{l} \times \vec{B} = \frac{v^2 B^2 l^2}{c^2 R}\hat{x}$$

So,

$$P_m = \vec{F} \cdot \vec{v} = \frac{v^2 B^2 l^2}{c^2 R} = P_e$$

4.5.

$$\vec{v} = wr\hat{u}_{\varphi}, \quad \vec{B} = B\hat{k}, \quad d\,\vec{l} = \hat{u}_r dr$$

The εmf between the axis and the rim can be found from

$$V_0 = \int \left(\frac{\vec{v}}{c} \times \vec{B}\right) \cdot d\,\vec{l} = \frac{1}{c} \int (wr\hat{u}_{\varphi} \times B\hat{k}) \cdot \hat{u}_r dr$$

$$= -\frac{wB}{c} \int_0^R r\,dr = -\frac{wBR^2}{2c}$$

We can also use the formula (Faraday's law)

$$V_0 = -\frac{1}{c}\frac{d\Phi}{dt}$$

where $\Phi = $ magnetic flux

$$\Phi = \int_S \vec{B} \cdot d\,\vec{S} = B \int_0^R \int_0^{\omega t} r\,d\varphi\,dr$$

$$= B\omega t \frac{R^2}{2} = \frac{wBR^2}{2}t$$

Then,

$$V_0 = -\frac{1}{c}\frac{d\Phi}{dt} = -\frac{wBR^2}{2c}$$

Both methods give the same result.

4.7.

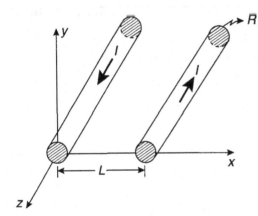

The self inductance per unit length of each wire is, $\frac{1}{2c^2}$. Therefore, the self inductance of the two wires is $\frac{1}{c^2}$.

Between the two wires we have the 5-fields

$$B_{y1} = \frac{2I}{cx}, \quad B_{y2} = \frac{2I}{c(L-x)}$$

The flux of a unit length of transmission line

$$\Phi = \int_R^{L-R}(B_{y1}+B_{y2})dx = \frac{2I}{c}\int_R^{L-R}\left(\frac{1}{x}-\frac{1}{L-x}\right)dx$$

$$= \frac{2I}{c}[\ln x - \ln(L-x)]_R^{L-R}$$

$$= \frac{2I}{c}[\ln(L-R) - \ln R - \ln R + \ln(L-R)]$$

$$= 4\frac{I}{c}\ln\frac{L-R}{R} \simeq 4\frac{I}{c}\ln\frac{L}{R} \quad \text{since } L \gg R$$

Then,

$$L = \frac{\Phi}{cI} = \frac{4}{c^2}\ln\frac{L}{R}$$

and,

$$L_{\text{tot}} = \frac{1}{c^2} + \frac{4}{c^2}\ln\frac{L}{R} = \frac{1}{c^2}\left(1 + 4\ln\frac{L}{R}\right) = \frac{\mu_0}{4\pi}\left(1 + 4\ln\frac{L}{R}\right)$$

(gaussian)   (SI)

4.9.

(a)

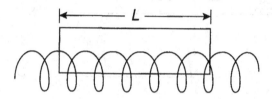

Using Ampere's circuital law

$$\oint \vec{B}\cdot d\vec{l} = \frac{4\pi}{c}I_{\text{enclosed}} = \frac{4\pi}{c}nLI$$

which gives

$$B = \frac{4\pi}{c}nI \quad \text{since} \quad \oint \vec{B}\cdot d\vec{l} = BL$$

(b) The flux of B through a single loop is

$$\Phi = BS = \frac{4\pi}{c}nIS$$

For n turns per unit length the total flux is

$$n\Phi = \frac{4\pi}{c}n^2 IS$$

The inductance per unit length is then given by

$$L = \frac{n\Phi}{CI} = \frac{4\pi n^2 S}{C^2} \quad \text{(gaussian)} = \mu_0 n^2 S \quad \text{(SI)}$$

4.11.

Lenz's law can be stated as follows:

"The sense of an induced εmf is such that if the εmf were allowed
to drive a current, the field produced by that current would tend
to cancel out the charge which produced the εmf."

The flux through coil b is downward. When the current in coil a is interrupted this downward flux is diminished. An εmf is induced in coil b which would tend to increase the downward flux: therefore the induced εmf in coil b is clockwise.

4.13.

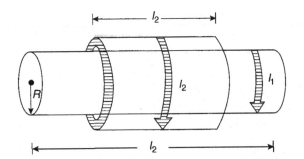

The B-field due to I_1 in the inner coil is given by

$$B = \frac{4\pi}{c} \frac{N_1}{l_1} I_1$$

The flux Φ_{12} in the core that links with the outer coil is given by

$$\Phi_{12} = \pi R^2 B = \frac{4\pi^2 N_1 I_1 R^2}{c l_1}$$

The linkage with the outer coil is

$$N_2 \Phi_{12} = \frac{4\pi^2 N_1 N_2 R^2 I_1}{c l_1}$$

Then the mutual inductance is

$$L_{12} = \frac{N_2 \Phi_{12}}{c I_1} = \frac{4\pi^2 N_1 N_2 R^2}{c^2 l_1}$$

4.15.

The value of the magnetic field for $a < r < b$ can be obtained from

$$\vec{\nabla} \times \vec{H} = \frac{4\pi}{c} \vec{J} \Rightarrow \int \vec{\nabla} \times \vec{H} dS = \frac{4\pi}{c} I$$

where $I = J\pi a^2$

$$\oint \vec{H} \cdot d\vec{l} = 2\pi r H = \frac{4\pi}{c} I$$

$$H = \frac{2I}{cr}(a \le r \le b), \quad H(a) = \frac{2I}{ca}$$

The energy stored in 1 cm of coaxial cable is given by

$$\frac{1}{8\pi} \int H^2 d\tau = \frac{1}{8\pi} \int_a^b \frac{4I^2}{c^2 r^2} 2\pi r \, dr = \frac{I^2}{c^2} \int_a^b \frac{dr}{r}$$

$$= \frac{I^2}{c^2} \ln \frac{b}{a} = \frac{1}{2} L I^2$$

Then,

$$L_{\text{ext}} = \frac{2}{c^2} \ln \frac{b}{a}$$

Now, for $r < a$

$$\vec{\nabla} \times \vec{H} = \frac{4\pi}{c} \vec{J} \Rightarrow \int \vec{\nabla} \times \vec{H} dS = \frac{4\pi}{c} J \pi r^2$$

$$2\pi r H = \frac{4\pi}{c} \frac{I}{\pi a^2} \pi r^2$$

$$H = \frac{4\pi}{c} \frac{I r^2}{a^2} \frac{1}{2\pi r} = \frac{2 I r}{c a^2} \quad (r \leq a), \quad H(a) = \frac{2I}{ca}$$

The energy stored in 1 cm of the internal conductor is

$$\frac{1}{8\pi} \int H^2 d\tau = \frac{1}{8\pi} \int_0^a \frac{4I^2 r^2}{c^2 a^4} 2\pi r \, dr$$

$$= \frac{I^2}{c^2 a^4} \int_0^a r^3 dr = \frac{I^2}{c^2 a^4} \frac{a^4}{4}$$

$$= \frac{I^2}{4c^2} = \frac{1}{2} L I^2$$

Then,

$$L_{\text{int}} = \frac{1}{2c^2}$$

The total inductance per unit length of coaxial cable is given by

$$L = L_{\text{ext}} + L_{\text{int}}$$

$$= \frac{2}{c^2} \ln \frac{b}{a} + \frac{1}{2c^2}$$

$$= \frac{1}{2c^2} \left(1 + 4 \ln \frac{b}{a} \right)$$

4.17.

(a)

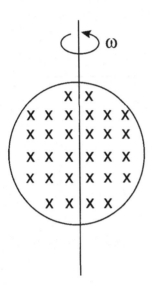

When H is perpendicular to area of loop the flux is:

$$\Phi_0 = \pi a^2 H$$

As the loop rotates it forms an angle $\theta = wt$ with H after a time t.

$$\Phi = \Phi_0 \cos wt$$

For n turns we have, by Faraday's law, an induced εmf given by

$$\varepsilon = -n \frac{d\Phi}{dt} \quad \text{(SI)}$$

$$= -n \frac{d\Phi}{dt}(\Phi_0 \cos wt) = nw\Phi_0 \sin wt$$

This situation is like an LR circuit obeying the differential equation:

$$L\frac{dI}{dt} + RI = nw\Phi_0 \sin wt$$

which has the solution

$$I(t) = Ae^{-\frac{R}{L}t} + \alpha \sin wt + \beta \cos wt$$

$$= Ae^{-\frac{R}{L}t} + \frac{nw\Phi_0}{R^2 + \omega^2 L^2}(R \sin wt - \omega L \cos wt)$$

where α, β are found by substituting $I(t)$ into the differential equation and equating coefficients of $\sin wt, \cos wt$. This can be done because $\alpha \sin wt + \beta \cos wt$ is the particular solution to the DE.

The result obtained can be written as

$$I(t) = Ae^{-\frac{R}{L}t} + \frac{n\omega\Phi_0}{\sqrt{R^2 + \omega^2 L^2}} \left\{ \frac{R}{\sqrt{R^2 + \omega^2 L^2}} \sin\omega t \right.$$

$$\left. - \frac{\omega L}{\sqrt{R^2 + \omega^2 L^2}} \cos\omega t \right\}$$

$$= Ae^{-\frac{R}{L}t} + \frac{n\omega\Phi_0}{\sqrt{R^2 + \omega^2 L^2}} \sin(\omega t - \varepsilon)$$

where

$$\cos\varepsilon = \frac{R}{\sqrt{R^2 + \omega^2 L^2}}, \quad \sin\varepsilon = \frac{\omega L}{\sqrt{R^2 + \omega^2 L^2}}$$

So, in a steady state situation, recalling that $\Phi_0 = \pi a^2 H$

$$I = \frac{n\pi a^2 H\omega}{\sqrt{R^2 + L^2\omega^2}} \sin(\omega t - \varepsilon)$$

(b)

$$P = \varepsilon I = \frac{n^2\omega^2(\pi a^2)^2 H^2}{\sqrt{R^2 + L^2\omega^2}} \sin(\omega t - \varepsilon)\sin\omega t$$

$$\bar{P} = \frac{1}{2} \frac{n^2\omega^2(\pi a^2)H^2}{\sqrt{R^2 + L^2\omega^2}} \cos\varepsilon$$

$$= \frac{n^2\pi^2 a^4 H^2\omega^2}{2(R^2 + \omega^2 L^2)} R$$

4.19.

We can think of the circular trajectory as a circuit with a fixed radius R. The εmf induced in this circuit by the varying magnetic flux is given by

$$V = -\frac{d\Phi}{dt}(\text{SI}) \tag{1}$$

The work done on an electron in one revolution is: $W = eV$. The tangential force acting on an electron over a distance ds does an amount of work

$$dW = Fds \Rightarrow F = \frac{dW}{ds} \tag{2}$$

considering one revolution

$$F = \frac{eV}{2\pi R} = \frac{e}{2\pi R}\frac{d\Phi}{dt} \tag{3}$$

From Newton's second law

$$F = \frac{d}{dt}(mv) = \frac{e}{2\pi r}\frac{d\Phi}{dt}$$

$$d(mv) = \frac{e}{2\pi r}d\Phi \tag{4}$$

The electron also experiences a radial force pointing toward the center of the orbit

$$Bev = \frac{mv^2}{R} \tag{5}$$

Then

$$BR = \frac{mv}{e}$$

where B = value of the field at the electron's orbit. We can write

$$mv = BeR$$

If R = constant then

$$d(mv) = eRdB \tag{6}$$

From Eqs. (4) and (6)

$$\frac{e}{2\pi R}d\Phi = eRdB \Rightarrow d\Phi = 2\pi R^2 dB$$

Integrating between 0 and Φ gives

$$\Phi = 2\pi R^2 B \tag{7}$$

(a)

$$Period = T = \frac{2\pi R}{v} = \frac{2\pi Rm}{mv} = \frac{2\pi Rm}{BeR} = \frac{2\pi m}{Be}$$

The momentum increase of the electron during one revolution is the (*Force*) × (*period*)

$$FT = \frac{e}{2\pi R}\frac{d\Phi}{dt}\frac{2\pi m}{eB} = \frac{m}{BR}\frac{d\Phi}{dt} = \frac{m}{mv/e}\frac{d\Phi}{dt}$$

$$= \frac{e}{v}\frac{d\Phi}{dt} \tag{8}$$

(b)

$$\Phi = 2\pi R^2 B \tag{9}$$

Note that the magnetic flux within the orbit is twice what it would have been if B were uniform throughout the orbit at the value B.

4.21.

a)

$$\vec{F} = q\left(\vec{E} + \frac{1}{c}\,\vec{v} \times \vec{B}\right)$$

But, from Maxwell equations

$$\oint \vec{E} \cdot d\vec{l} = -\frac{1}{c}\int \frac{\partial B}{\partial t}\,dA$$

$$2\pi r E_\theta = -\frac{1}{c}\pi r^2 \frac{\partial B}{\partial t}$$

then

$$E_\theta = -\frac{r}{2c}\frac{\partial \vec{B}}{\partial t}$$

We can now write the equation of motion:

$$\begin{cases} ma_r = m\dfrac{d^2 r}{dt^2} - mr\omega^2 = \dfrac{q}{c}Br\omega & (1) \\[2ex] ma_\theta = m\dfrac{1}{r}\dfrac{d}{dt}(r^2\omega) = \dfrac{1}{r}\dfrac{d}{dt}(J) & \\[2ex] \qquad = -\dfrac{qr}{2c}\dfrac{\partial B}{\partial t} - \dfrac{qB}{c}\dfrac{dr}{dt} & (2) \\[2ex] ma_z = 0 & (3) \end{cases}$$

(b) Assume

$$r = \frac{A}{B^{1/2}} = K(t + t_0)^{1/2} \tag{4}$$

Then

$$\frac{dr}{dt} = \frac{K}{2}(t + t_0)^{-1/2}$$

$$\frac{d^2 r}{dt^2} = -\frac{K}{4}(t + t_0)^{-3/2} \tag{5}$$

From Eq. (2)

$$\frac{dJ}{dt} = -\frac{qr^2}{2c}\frac{\partial B}{\partial t} - \frac{qBr}{c}\frac{dr}{dt} = -\frac{d}{dt}\left(\frac{qr^2 B}{2c}\right)$$

and, setting $q = -e$

$$J = J_0 + \frac{er^2 B}{2c} = mr^2 \omega$$

Then,

$$\omega = \frac{J_0}{mr^2} + \frac{eB}{2mc} \tag{6}$$

$$\omega^2 = \frac{J_0^2}{m^2 r^4} + \frac{e^2 B^2}{4m^2 c^2} + \frac{J_0 eB}{m^2 cr^2} \tag{7}$$

From Eq. (1), using Eqs. (5), (6) and (7)

$$-\frac{e}{c} Be\omega = m\frac{d^2 r}{dt^2} - mr\omega^2$$

$$-\frac{e}{c}\frac{BJ_0}{mr} - \frac{eBr}{c} \cdot \frac{eB}{2mc} = -\frac{Km(t-t_0)^{-3/2}}{4}$$

$$+ \frac{J_0^2}{mr^3} + \frac{e^2 B^2 r}{4mc^2} + \frac{J_0 eB}{mcr}$$

Rearranging and canceling terms

$$\frac{Km}{4}(t+t_0)^{-3/2} + \frac{J_0^2}{mr^3} + \frac{1}{4}\frac{e^2 B^2 r}{mc^2} = 0$$

$$\frac{Km}{4}(t+t_0)^{-3/2} + \frac{J_0^2}{mK^3}(t+t_0)^{-3/2}$$

$$+ \frac{1}{4}\frac{e^2 (B_0 t_0)^2}{mc^2} K(t+t_0)^{-3/2} = 0$$

Since the time dependence cancels out

$$\frac{Km}{4} + \frac{J_0^2}{mK^3} + \frac{1}{4}\frac{e^2 B_0^2 t_0^2}{mc^2} K = 0$$

then statement (4) is true.

(c) Since the angular momentum is proportional to $r^2 B$, we find:

$$r^2 B = K^2 (t+t_0)\frac{B_0 t_0}{t+t_0} = K^2 B_0 t_0 = \text{constant.}$$

Thus $J = \text{constant}$

5

Exercises

5.1.

The "relaxation time" (see Chapter 2 exercises) is given by

$$\tau = \frac{4\pi\sigma}{k}$$

5.3.

The magnetic field produced by the motion of q_1 is zero at the position of q_2, and therefore q_1 exercises no force on q_2. On the other hand the field of q_2 does not vanish at q_1, and it exerts a (transversal) force on q_1.

Action and reaction forces are not equal and opposite in this system: the total momentum of the system is not conserved. The law of conservation of linear momentum can be retained only by including the momentum of the electromagnetic field.

5.5.

The outdated view: "Define your terms before you proceed with the formulation of the laws" must be replaced by the more modern point of view that in physics, defining concepts, physical laws and methods of measurements are to be introduced at the same time.

The conservation of momentum is not just a matter of definition. In analyzing an initial set of collisions, we may use the conservation law to find out or define the momenta of several objects. After some collisions the momenta of the objects are known, and the validity of the conservation laws is verified; it is no longer a matter of definition, but of the internal ways in which the collisional process enfolds.

5.7.

The two antennas should be located on the south end of the island a distance $\lambda/4$ apart (see figure below), where $\lambda = c/v$ and $v =$ the broadcasting frequency. The signal from antenna 1 should be anticipated in phase by $90°$

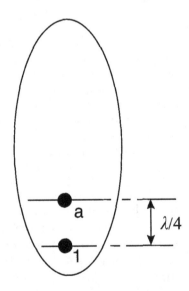

5.9.

We consider a cubic cavity of sides L_1, L_2 and L_3 and in the x, y and z directions, respectively. We impose periodic boundary conditions by accepting only the $\vec{k}$ vector with components

$$\begin{cases} k_x = n_x \dfrac{2\pi}{L_1} \\[2mm] k_y = n_y \dfrac{2\pi}{L_2} \\[2mm] k_z = n_z \dfrac{2\pi}{L_3} \end{cases}$$

where $n_x, n_y, n_z = 0, \pm1, \pm2, \pm3, \ldots$

We can also write

$$\begin{cases} dn_x = \dfrac{L_1}{2\pi} dk_x \\[2mm] dn_y = \dfrac{L_2}{2\pi} dk_y \\[2mm] dn_z = \dfrac{L_3}{2\pi} dk_z \end{cases}$$

The number of cavity modes with k_x in $(k_x, k_x + dk_x)$, k_y in $(k_y, k_y + dk_y)$, and k_z in $(k_z, k_z + dk_z)$ is

$$dn_x dn_y dn_z = \frac{L_1 L_2 L_3}{8\pi^3} dk_x dk_y dk_z$$

The number of cavity modes per unit volume with k in $(k, k + dk)$ is given by

$$2 \cdot \frac{4\pi k^2 dk}{8\pi^3} = \frac{k^2}{\pi^2} dk$$

where the extra factor 2 is due to the two independent polarizations corresponding to each $\vec{k}$.

The number of cavity modes with frequency v in $(v, v + dv)$ is given by

$$n(v)dv = \frac{k^2}{\pi^2} \frac{dk}{dv} dv$$

But,

$$k = \frac{2\pi v}{c}, \quad dk = \frac{2\pi}{c} dv$$

Then

$$n(v)dv = \frac{1}{\pi^2} \frac{4\pi^2 v^2}{c^2} \frac{2\pi}{c} dv = \frac{8\pi^2 v^2}{c^3} dv$$

Each mode has energy $K_B T$, so

$$u(v) = \frac{8\pi v^2}{c^3} K_B T$$

where $K_B =$ Boltzmann's constant and thus

$$c = \frac{8\pi v^2 K_B}{c^3}$$

5.11.

We know that $p = \dfrac{u}{3}$

where $u = aT^4 \left(a = 4\dfrac{\sigma}{c}\right)$

For an adiabatic compression we must have

$$dS = \left(\frac{\partial S}{\partial T}\right)_V dT + \left(\frac{\partial S}{\partial V}\right)_T dV = 0$$

But one of the Maxwell relations gives us

$$\left(\frac{\partial S}{\partial V}\right)_T = \left(\frac{\partial P}{\partial T}\right)_V$$

Then

$$dS = \frac{C_v}{T} dT + \left(\frac{\partial P}{\partial T}\right)_V dV$$

$$= \frac{V}{T} \frac{\partial U}{\partial T} dT + \left(\frac{\partial P}{\partial T}\right)_V dv$$

$$= \frac{V}{T} 4aT^3 dT + \frac{4}{3} aT^3 dV = 0$$

or

$$\frac{4aT^2}{\frac{4}{3}aT^3} dT + \frac{dV}{V} = 0 \Rightarrow 3\frac{dT}{T} = -\frac{dV}{V}$$

But $dP = \dfrac{4}{3} aT^3 dT \Rightarrow \dfrac{dP}{P} = \dfrac{\frac{4}{3}aT^3 dT}{\frac{a}{3}T^4} = 4\dfrac{dT}{T}$

Then

$$3\frac{dT}{T} = \frac{3}{4}\frac{dP}{P}$$

$$\frac{dP}{P} = -\frac{4}{3}\frac{dV}{V}$$

or

$$PV^{4/3} = \text{constant}.$$

5.13.

(a) Incident beam

$$\vec{E} = \hat{i}E_x + \hat{j}E_y \tag{1}$$

where

$$\begin{cases} E_x = E_0 \cos\theta \sin 2\pi v \left(t + \dfrac{z}{c} \right) \\[2mm] E_y = E_0 \sin\theta \sin 2\pi v \left(t + \dfrac{z}{c} \right) \end{cases}$$

Inside the crystal

$$\begin{cases} E_x = E_0 \cos\theta \sin 2\pi v \left(t + \dfrac{z}{v_x} \right) \\[2mm] E_y = E_0 \sin\theta \sin 2\pi v \left(t + \dfrac{z}{v_y} \right) \end{cases} \tag{2}$$

where

$$v_x = \frac{c}{\sqrt{K_x}}, \quad v_y = \frac{c}{\sqrt{K_y}}$$

The emergent beam at $z = -l$ is made of

$$\begin{cases} E_x = E_0 \cos\theta \sin 2\pi v \left(t - \dfrac{l}{v_x} \right) \\[2mm] E_y = E_0 \sin\theta \sin 2\pi v \left(t - \dfrac{z}{v_y} \right) \end{cases} \tag{3}$$

We can write Eq. (3) as follows

$$\begin{cases} E_x = E_0 \cos\theta \sin(\omega t - \varphi_1) \\ E_y = E_0 \sin\theta \cos(\omega t - \varphi_2) \end{cases}$$

where

$$\omega = 2\pi v, \quad \varphi_1 = 2\pi v \frac{l}{v_x}, \quad \varphi_2 = 2\pi v \frac{l}{v_y}$$

Also,

$$\begin{cases} E_x = E_0 \cos\theta \sin(\omega t - \varphi_2 + \varphi) \\ E_y = E_0 \sin\theta \cos(\omega t - \varphi_2) \end{cases}$$

where

$$\varphi = \varphi_2 - \varphi_1$$

Now, if $\sin\theta = \cos\theta$, i.e., $\theta = \dfrac{\pi}{4}, \dfrac{3\pi}{4}, \ldots$

and $\varphi = \pi/2$,

$$\begin{cases} E_x = E_0 \dfrac{\sqrt{2}}{2} \sin\left(\omega t - \varphi_2 + \dfrac{\pi}{2}\right) = -\dfrac{\sqrt{2}}{2} E_0 \cos(\omega t - \varphi_2) \\[3mm] E_y = E_0 \dfrac{\sqrt{2}}{2} \sin(\omega t - \varphi_2) \end{cases}$$

and

$$E_x^2 + E_y^2 = \text{constant}$$

Thus, for circular polarization the incident plane of polarization must be at 45° to the x and y axes and l must be such that E_x and E_y, initially in phase, must emerge 90° out of phase.

5.15.

(a)

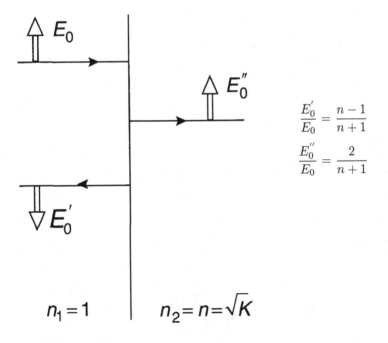

$$\frac{E_0'}{E_0} = \frac{n-1}{n+1}$$

$$\frac{E_0''}{E_0} = \frac{2}{n+1}$$

Therefore, the ratio of incident energy flux to reflected energy flux is

$$\frac{|E_0'|^2}{|E_0|^2} = \left(\frac{n-1}{n+1}\right)^2 = \left(\frac{\sqrt{K}-1}{\sqrt{K}+1}\right)$$

(b)

Light reflected back: R

$$\frac{n-1}{n+1} + \frac{4n}{(1+n)^2}\frac{1-n}{1+n} + \frac{4n}{(1+n)^2}\left(\frac{1-n}{1+n}\right)^3 + \cdots$$

$$= \frac{1-n}{1+n} + \frac{4n}{(1+n)^2}\frac{1-n}{1+n}\left[1 + \left(\frac{1-n}{1+n}\right)^2 + \left(\frac{1-n}{1+n}\right)^4 + \cdots\right]$$

$$= \frac{1-n}{1+n} + \frac{4n}{(1+n)^2}\frac{1-n}{1+n}\left[\frac{1}{1-\left(\frac{1-n}{1+n}\right)^2}\right]$$

$$= \frac{1-n}{1+n} + \frac{4n}{(1+n)^2}\frac{1-n}{1+n}\left[\frac{(1+n)^2}{1+2n+n^2-1-n^2+2n}\right]$$

$$= \frac{1-n}{n+1} + \frac{1-n}{1+n} = 0$$

Light Transmitted: T

$$\frac{2}{1+n}\frac{2n}{1+n}+\frac{2}{1+n}\left(\frac{1-n}{1+n}\right)^2\frac{2n}{1+n}+\cdots$$

$$=\frac{4n}{(1+n)^2}\left[1+\left(\frac{1-n}{1+n}\right)^2+\left(\frac{1-n}{1+n}\right)^4+\cdots\right]$$

$$=\frac{4n}{(1+n)^2}\left[\frac{1}{1-\left(\frac{1-n}{1+n}\right)^2}\right]$$

$$=\frac{4n}{(1+n)^2}\frac{(1+n)^2}{(1+n)^2-(1-n)^2}=\frac{4n}{1+n^2+2n-1-n^2+2n}$$

$$=\frac{4n}{4n}=1$$

Note that $R+T=1$ as it should.

5.17.

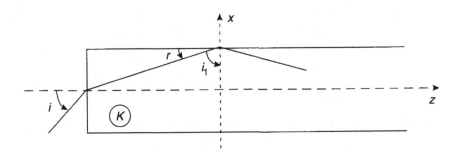

In order to have internal reflection we need to have

$$\sin i_1 \geq \sin i_c$$

where i_c is the critical angle for which no reflection results. It is determined by setting the z component of the K vector of the reflected wave to zero. The result is

$$\sin i_c = \frac{1}{\sqrt{K}}$$

Now, from the diagram $i_1 = \frac{\pi}{2} - r$

Then, $\sin(\frac{\pi}{2} - r) \geq \sin i_c$

Since $\sin(\frac{\pi}{2} - r) = \cos r$ we have

$$\cos r \geq \frac{1}{\sqrt{K}}$$

From Snell's law $\frac{1}{\sqrt{K}} \sin i = \sin r$

$$\cos r = \sqrt{1 - \sin^2 r} = \sqrt{1 - \frac{1}{K} \sin i} \geq \frac{1}{\sqrt{K}}$$

$$1 - \frac{\sin^2 i}{K} \geq \frac{1}{K} \Rightarrow K \geq 1 + \sin^2 i$$

Max value for i is $\frac{\pi}{2}, K \geq 2$ satisfied by glass and quartz.

5.19.

For the conducting surface at $x = o$

$$E_y = A \sin \omega t$$
$$H_z = A \sin \omega t$$

So, the radiation pressure is given by

$$u = \frac{1}{8\pi}(E^2 + B^2) = \frac{1}{8\pi}(E^2 + H^2)$$

$$= \frac{1}{8\pi} \cdot 2E^2 = \frac{A^2}{4\pi} \sin^2 \omega t$$

But this is an instantaneous quantity. To find an average pressure per unit time we can do a time average. $\sin^2 \omega t$ ranges between 0 and 1, so it has a time averaged value of $1/2$. Thus, the time averaged radiation pressure is

$$\langle u \rangle = \frac{A^2}{4\pi} \langle \sin^2 \omega t \rangle \frac{A^2}{8\pi}$$

5.21.

From Maxwell's equations

$$\vec{\nabla} \times \vec{E} = -\frac{1}{c} \frac{\partial \vec{H}}{\partial t}$$

or,

$$\vec{\nabla} \times \vec{E} = \begin{pmatrix} \hat{x} & \hat{y} & \hat{z} \\ \partial/\partial x & \partial/\partial y & \partial/\partial z \\ 0 & E_y & 0 \end{pmatrix} = \frac{\partial E_y}{\partial x} \hat{z}$$

Then
$$\frac{\partial E_y}{\partial x} = -\frac{1}{c}\frac{\partial H_z}{\partial t}$$

Also

$$\vec{\nabla} \times \vec{H} = \frac{1}{c}\frac{\partial \vec{E}}{\partial t} + \frac{4\pi}{c}\vec{J}$$

$$\vec{\nabla} \times \vec{H} = \begin{pmatrix} \hat{x} & \hat{y} & \hat{z} \\ \partial/\partial x & \partial/\partial y & \partial/\partial z \\ 0 & 0 & H_z \end{pmatrix} = \frac{\partial H_z}{\partial y}\hat{x} + \left(-\frac{\partial H_z}{\partial x}\right)\hat{y}$$

$$= \frac{1}{c}\frac{\partial E_y}{\partial t}\hat{y} + \frac{4\pi}{c}J\hat{y}$$

Then
$$\frac{\partial H_z}{\partial x} = -\frac{1}{c}\frac{\partial E_y}{\partial t} - \frac{4\pi}{c}J$$

and
$$J = -\frac{1}{4\pi}\left(\frac{\partial E_y}{\partial t} + c\frac{\partial H_z}{\partial x}\right)$$

Now, on the other hand $\vec{F} = F_x\hat{x}$

where

$$F_x = \frac{1}{c}JH_z = \frac{1}{c}H_z\left[-\frac{1}{4\pi}\left(\frac{\partial E_y}{\partial t} + c\frac{\partial H_z}{\partial x}\right)\right]$$

$$= -\frac{1}{4\pi}\left(H_z\frac{\partial H_z}{\partial x} + \frac{1}{c}H_z\frac{\partial E_y}{\partial t}\right)$$

$$= -\frac{1}{4\pi}\left[\frac{1}{2}\frac{\partial H_z^2}{\partial x} - \frac{1}{c}E_y\frac{\partial H_z}{\partial t} + \frac{1}{c}\frac{\partial}{\partial t}(H_z E_y)\right]$$

$$= -\frac{1}{4\pi}\left[\frac{1}{2}\frac{\partial H_z^2}{\partial x} + E_y\frac{\partial E_y}{\partial x} + \frac{1}{c}\frac{\partial}{\partial t}(H_z E_y)\right]$$

$$= -\frac{1}{4\pi}\left[\frac{1}{2}\frac{\partial H_z^2}{\partial x} + \frac{1}{2}\frac{\partial E_y^2}{\partial x} + \frac{1}{c}\frac{\partial}{\partial t}(H_z E_y)\right]$$

$$= -\frac{1}{8\pi}\frac{\partial}{\partial x}(H_z^2 + E_y^2) - \frac{1}{8\pi c}\frac{\partial}{\partial t}(H_z E_y)$$

Then we obtain for the time average

$$F_x = -\frac{1}{8\pi}\frac{\partial}{\partial x}(H_z^2 + E_y^2)$$

and

$$\int_0^\infty F_x dx = \int \left[-\frac{1}{8\pi}\overline{\frac{\partial}{\partial x}(H_z^2 + E_y^2)} \right] dx$$

$$= \left[\frac{1}{8\pi}\overline{(E_y^2 + H_z^2)} \right]_{x=0}$$

$$= \frac{A^2}{8\pi}\left(\frac{1}{2} + \frac{1}{2} \right) = \frac{A^2}{8\pi}$$

5.23.

(a)

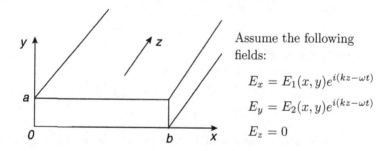

Assume the following fields:

$$E_x = E_1(x,y)e^{i(kz-\omega t)}$$
$$E_y = E_2(x,y)e^{i(kz-\omega t)}$$
$$E_z = 0$$

These fields are solutions to the wave equation

$$\nabla^2 \vec{E} - \frac{1}{c^2}\frac{\partial^2 \vec{E}}{\partial t^2}$$

This implies that

$$\begin{cases} \left(\dfrac{\partial^2}{\partial x^2} + \dfrac{\partial^2}{\partial y^2} \right) E_1 - \left(k^2 - \dfrac{\omega^2}{c^2} \right) E_1 = 0 \\[2mm] \left(\dfrac{\partial^2}{\partial x^2} + \dfrac{\partial^2}{\partial y^2} \right) E_2 - \left(k^2 - \dfrac{\omega^2}{c^2} \right) E_2 = 0 \end{cases} \qquad (3)$$

The solution to Eq. (3) are compatible with the boundary conditions

$$\begin{cases} E_1 = E_0^1 \cos\left(\frac{m\pi x}{a}\right) \sin\left(\frac{n\pi y}{b}\right) \\[2mm] E_2 = E_0^2 \sin\left(\frac{m\pi x}{a}\right) \cos\left(\frac{n\pi y}{b}\right) \end{cases}$$

provided that

$$-\frac{m^2\pi^2}{a^2} - \frac{n^2\pi^2}{b^2} - k^2 + \frac{\omega^2}{c^2} = 0$$

or,

$$\omega = c\left[k^2 + \pi\left(\frac{m^2}{a^2} + \frac{n^2}{b^2}\right)\right]^{1/2}$$

(b) The phase velocity is given by

$$v_p = \frac{\omega}{k} = c\left[1 + \frac{\pi^2}{k^2}\left(\frac{m^2}{a^2} + \frac{n^2}{b^2}\right)\right]^{1/2}$$

and the group velocity is given by

$$v_g = \frac{d\omega}{dk} = \frac{c}{\left[1 + \frac{\pi^2}{k^2}\left(\frac{m^2}{a^2} + \frac{n^2}{b^2}\right)\right]^{1/2}}$$

Note that $v_p v_g = c^2$

(c) Given m and n the cutoff frequency is

$$(v_c)_{mn} = \frac{c}{2}\left(\frac{m^2}{a^2} + \frac{n^2}{b^2}\right)^{1/2}$$

A certain v_c corresponds to a certain (m, n) "mode" of propagation. The smallest of these v_c values corresponds to $n = 0, m = 1$ and is

$$v_c = \frac{c}{2a}$$

Other modes are possible in which the magnetic field is transverse; these modes have a higher cutoff frequency.

6

Exercises

6.1.

In the quoted poem there are two important points:

(i) Travels in time are without return, contrary to travels in space.
(ii) It is not possible "to let the anchor down", i.e., to stop time. Regardless of the conditions in which it is measured (Newton), or affected by such conditions (Einstein), time is always moving forward.

6.3.

We could prove that S' is also an inertial frame in two simple ways:

(i) By verifying that the speed of the test particle is constant in S'.
(ii) By using a second particle, running crosswise. If both particles move in straight lines in both frames, then S' is an inertial frame.

6.5.

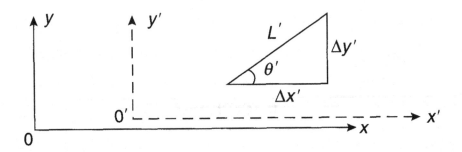

(a)

$$\Delta x' = L' \cos \theta'$$

$$\Delta y' = L' \sin \theta'$$

$$\tan \theta' = \frac{\Delta y'}{\Delta x'} \quad \text{in } S'$$

Now, in the lab frame S:

$$\Delta y = \Delta y'$$

$$\Delta x = \Delta x' \sqrt{1 - \beta^2}$$

Then

$$\tan \theta = \frac{\Delta y}{\Delta x} = \frac{\Delta y'}{\Delta x' \sqrt{1 - \beta^2}} = \frac{\tan \theta'}{\sqrt{1 - \beta^2}}$$

(b)

$$L = \sqrt{(\Delta x)^2 + (\Delta y)^2} = \sqrt{\Delta x'^2(1 - \beta^2) + \Delta y'^2}$$

$$= L' \sqrt{(1 - \beta^2) \cos^2 \theta' + \sin^2 \theta'}$$

$$= L' \sqrt{1 - \beta^2 \cos^2 \theta'}$$

6.7.

(a)

Observer in S' "sees"

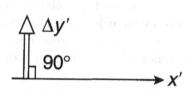

Observer in S "sees"

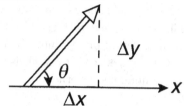

$$\Delta y' = c\Delta t'$$

$$\Delta x' = 0$$

$$\Delta y = \Delta y' = c\Delta t'$$

$$\Delta x = v\Delta t$$

Using the relation for time dilation we can rewrite Δx as:

$$\Delta x = v\Delta t = v\gamma\Delta t'$$

Then

$$\tan\theta = \frac{\Delta y}{\Delta x} = \frac{c\Delta t'}{v\gamma\Delta t'} = \frac{1}{\beta\gamma}$$

(b) for an initial angle of θ' in S' we have

$$\Delta y' = c\Delta t' \sin\theta'$$

$$\Delta x' = c\Delta t' \cos\theta'$$

In the S-frame, the Lorentz equations give

$$\Delta x = \gamma\Delta x' + \gamma v\Delta t', \quad \Delta y = \Delta y'$$

Then,

$$\tan\theta = \frac{\Delta y}{\Delta x} = \frac{c\Delta t' \sin\theta'}{\gamma c\Delta t' \cos\theta' + \gamma v\Delta t'} = \frac{\sin\theta'}{\gamma(\cos\theta' + \beta)}$$

Note that for $\theta' = 90°$, $\tan\theta = \frac{1}{\beta\gamma}$ as in part a.

6.9.

Let Q be a particle that has speed c in a certain frame. A race between particle Q and a photon P would be a tie in such a frame.

It must, therefore, be a tie in any other inertial frame; otherwise the outcome could allow us to "label" frames according to the result of the race. This is consistent with the idea that there are no preferred inertial frames. Such preferred or absolute frames have not been found to exist in nature (i.e., the experiment by Michelson and Morley failed to detect the presence of the luminiferous ether).

6.11.

$\underline{v \ll c}$

The number of particles which have not decayed in time t is

$$N = N_0 e^{-t/t_0}$$

The probability that a particle survives for a time t is

$$\frac{N}{N_0} = e^{-t/t_0}$$

If the particle moves a distance $L = vt$, the probability of moving a distance L without decaying is given by

$$e^{-t/t_0} = e^{-L/vt_0}$$

$\underline{v \sim c}$

The time t_0 relates a frame of reference in which the particle is at rest. In the laboratory the time t_0 has to be replaced by the time

$$t_0' = \frac{t_0}{\sqrt{1 - v^2/c^2}}$$

The probability that the particle moves a distance L without decaying is now

$$e^{-t/t_0'} = \exp\left(-\frac{L\sqrt{1 - v^2/c^2}}{vt_0}\right)$$

6.13.

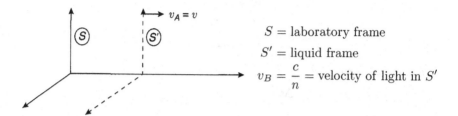

S = laboratory frame

S' = liquid frame

$v_B = \dfrac{c}{n}$ = velocity of light in S'

The velocity of light in the laboratory frame is given by the relativistic formula for the addition of velocities

$$W = \frac{v_A + v_B}{1 + \dfrac{v_A v_B}{c^2}} = \frac{v + \dfrac{c}{n}}{1 + \dfrac{vc/n}{c^2}} = \frac{v + \dfrac{c}{n}}{1 + \dfrac{v}{nc}}$$

$$\simeq \left(v + \frac{c}{n}\right)\left(1 - \frac{v}{nc}\right) = v + \frac{c}{n} - \frac{v^2}{nc} - \frac{v}{n^2}$$

$$= v + \frac{c}{n} - \frac{v}{n}\frac{v}{c} - \frac{v}{n^2} \simeq \frac{c}{n} + v - \frac{v}{n^2}$$

$$= \frac{c}{n} + \left(1 - \frac{1}{n^2}\right)v \qquad \begin{array}{l}\text{Velocity of light}\\ \text{in liquid as observed}\\ \text{from laboratory frame.}\end{array}$$

6.15.

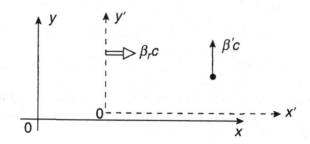

If we assume that $\Delta x' = 0$

$$\Delta y = \Delta y'$$
$$\Delta x = \beta_r \gamma_r \Delta t'$$
$$\Delta t = \gamma_r \Delta t'$$

We can then write

$$\beta_y = \frac{\Delta y}{\Delta t} = \frac{\Delta y'}{\gamma_r \Delta t'} = \frac{1}{\gamma_r} \beta_y' = \frac{1}{\gamma_r} \beta'$$

$$\beta_x = \frac{\Delta x}{\Delta t} = \frac{\beta_r \gamma_r \Delta t'}{\gamma_r \Delta t'} = \beta_r$$

6.17.

(a) We know that

$$\begin{cases} x' = \gamma x - \beta \gamma t \\ t' = -\beta \gamma x + \gamma t \end{cases}$$

Set $t = 0$; then

$$\begin{cases} x' = \gamma x \\ t' = -\beta \gamma x \end{cases}$$

If $x > 0, t' < 0$; if $x < 0, t' > 0$

(b) The inverse Lorentz transformation gives us

$$\begin{cases} x = \gamma x' + \beta \gamma t' \\ t = \beta \gamma x' + \gamma t' \end{cases}$$

Set $t' = 0$; then

$$\begin{cases} x = \gamma x' \\ t = \beta \gamma x' \end{cases}$$

If $x' > 0, t' > 0$; if $x' < 0, t' < 0$

(c) The apparent asymmetry derives from the fact that while the $+x$-axis of S is in the direction of the S' motion, the x'-axis of S' is not in the direction of the S motion.

If the same choice is made for each frame (+x in the direction of the other frame's motion) then the paradox is eliminated.

6.19.

There is no question about what happens in S; the ruler is Lorentz-contracted and goes through the hole.

In S' the plate, while moving upward, is tilted, with its right end being the highest point. The ruler, longer than the hole, will get through (see figure)

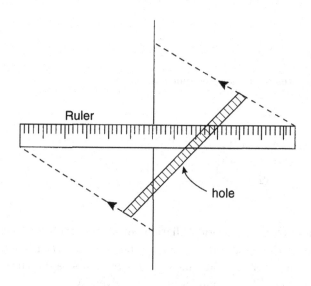

6.21.

$$\begin{cases} \Box^2 \phi = -4\pi\rho \\ \Box^2 \vec{A} = -\dfrac{4\pi}{c}\,\vec{J} \end{cases}$$

Lorentz gauge: $\vec{\nabla} \cdot \vec{A} + \frac{1}{c}\frac{\partial \phi}{\partial t} = 0$

$$\Box^2 \frac{1}{c}\frac{\partial \phi}{\partial t} = -4\pi\frac{1}{c}\frac{\partial \rho}{\partial t}$$

$$\Box^2 \vec{\nabla} \cdot \vec{A} = -\frac{4\pi}{c}\,\vec{\nabla} \cdot \vec{J}$$

Then

$$\Box^2 \left(\vec{\nabla} \cdot \vec{A} + \frac{1}{c} \frac{\partial \phi}{\partial t} \right) = -\frac{4\pi}{c} \left(\vec{\nabla} \cdot \vec{J} + \frac{\partial \rho}{\partial t} \right)$$

Since

$$\vec{\nabla} \cdot \vec{A} + \frac{1}{c} \frac{\partial \phi}{\partial t} = 0$$

We have also

$$\vec{\nabla} \cdot \vec{J} + \frac{\partial \rho}{\partial t} = 0$$

6.23.

Frequency

The mirror sees a smaller frequency given by

$$v' = v_1 \frac{1 - \beta}{\sqrt{1 - \beta^2}}$$

The observer sees a still smaller frequency

$$v_2 = v' \frac{1 - \beta}{\sqrt{1 - \beta^2}} = v_1 \frac{(1 - \beta)^2}{1 - \beta^2} = v_1 \frac{1 - \beta}{1 + \beta}$$

Intensity

The beam of electromagnetic radiation consists of photons which we shall assume are at the distance x_0 in the x direction from each other. Consider a photon that is reflected at time $t = 0$. A photon at a distance x_0 behind it will be reflected by the mirror at a time given by

$$ct = x_0 + vt \Rightarrow t = \frac{x_0}{c - v}$$

The distance separating the two photons, after reflection, is

$$ct + vt = (c + v)t = x_0 \frac{c + v}{c - v} = x_0 \frac{1 + \beta}{1 - \beta}$$

The rate at which the photons will be arriving at the observer is reduced by a factor $(1 - \beta)/(1 + \beta)$.

The frequency is reduced the same factor. Then

$$I_2 = I_1 \left(\frac{1 - \beta}{1 + \beta} \right)^2$$

6.25.

(a) $\dfrac{11.3}{0.5} = 22.6$ years

(b) $22.6 + 5 = 27.6$ years

(c) $27.6 + 22.6 = 50.2$ years

(d) $\gamma = \dfrac{1}{\sqrt{1 - \beta^2}} = \dfrac{1}{\sqrt{1 - 0.5^2}} = 1.1547$

$\dfrac{27.6}{1.1547} = 19.57$ years

(e) $19.57 + 5 = 24.57$ years

(f) $24.57 + 19.57 = 44.14$ years

(g) $\dfrac{11.3}{1.1547} = 9.79$ light-years

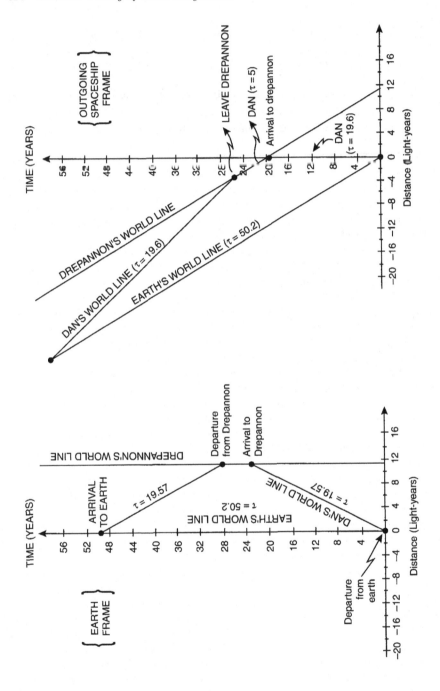

6.27.

(a) John's proper time is 50.2 years. Daniel boards the outgoing frame and travels 11.3 light-years of space in 22.6 years of time, stays 5 years in Drepannon, and then goes back to earth in 22.6 years.

(b) Daniel travels 19.6 years in the outgoing frame. He then disembarks and spends 5 years in the earth frame. How long is this period of time in the outgoing frame?

$$5 \times 1.1547 = 5.77 \text{ years}$$

Then we have

$$19.6 + 5.77 = 25.37 \text{ years}$$

In the 5 years that Daniel spent in the earth frame, how far did the outgoing frame travel

$$5.77 \times 0.5 = 2.885 \text{ light-years}$$

How far did the earth go from the outgoing frame? We know that the time elapsed was 50.2 years on earth. Then the relevant time is

$$50.2 \times 1.1547 = 57.96 \text{ years}$$

6.29.

Daniel has aged by

$$\Delta \tau_D = \sqrt{20^2 - 12^2} = \sqrt{256} = 16 \text{ years}$$

John has aged by 4 years plus

$$\Delta \tau = \sqrt{(20-4)^2 - 12^2} = \sqrt{256 - 144}$$
$$= 10.58 \text{ years}$$

or

$$10.58 + 4 = 14.58 \text{ years}$$

7

Exercises

7.1.

(a) Let us express the invariants

$$\sum_{\mu v}(F_{\mu v})^2 \quad \text{and} \quad \sum_{\mu v}F_{\mu v}\hat{F}_{\mu v}$$

in terms of $\vec{E}$ and $\vec{B}$

We have:

$$F_{12} = B_3 \qquad F_{14} = -iE_1$$
$$F_{13} = -B_2 \qquad F_{24} = -iE_2$$
$$F_{23} = B_1 \qquad F_{34} = -iE_3$$
$$\hat{F}_{12} = -iE_3 \qquad \hat{F}_{14} = B_1$$
$$\hat{F}_{13} = iE_2 \qquad \hat{F}_{24} = B_2$$
$$\hat{F}_{23} = -iE_1 \qquad \hat{F}_{34} = B_3$$

Then

$$\sum_{\mu v} F_{\mu v}^2 = F_{12}^2 + F_{13}^2 + F_{14}^2 + F_{21}^2 + F_{23}^2 + F_{24}^2$$

$$+ F_{31}^2 + F_{32}^2 + F_{34}^2 + F_{41}^2 + F_{42}^2 + F_{43}^2$$

$$= B_3^2 + B_2^2 - E_1^2 + B_3^2 + B_1^2 - E_2^2$$

$$+ B_2^2 + B_1^2 - E_3^2 - E_1^2 - E_2^2 - E_3^2$$

$$= 2(B^2 - E^2)$$

Also

$$\sum_{\mu v} F_{\mu v} \hat{F}_{\mu v} = F_{12}\hat{F}_{12} + F_{13}\hat{F}_{13} + F_{14}\hat{F}_{14}$$

$$+ F_{21}\hat{F}_{21} + F_{23}\hat{F}_{23} + F_{24}\hat{F}_{24}$$

$$+ F_{31}\hat{F}_{31} + F_{32}\hat{F}_{32} + F_{34}\hat{F}_{34}$$

$$+ F_{41}\hat{F}_{41} + F_{42}\hat{F}_{42} + F_{43}\hat{F}_{43}$$

$$= B_3(-iE_3) + (-B_2)(iE_2) + (-iE_1)B_1$$

$$+ (-B_3)(iE_3) + B_1(-iE_1) + (-iE_2)B_2$$

$$+ B_2(-iE_2) + (-B_1)(iE_1) + (-iE_3)B_3$$

$$+ (iE_1)(-B_1) + (iE_2)(-B_2) + (iE_3)(-B_3)$$

$$= -2i(B_3E_3 + B_1E_1 + B_2E_2)$$

$$- 2i(B_2E_2 + B_1E_1 + B_3E_3)$$

$$= -4i(\vec{E} \cdot \vec{B})$$

(b) Let us show the invariance of the property $|\vec{E}| = |\vec{B}|$ of our wave.

Define,

$\vec{E}, \vec{B}$ = fields in rest frames

$\vec{E}', \vec{B}'$ = fields in frame S' moving with respect to S with velocity $\vec{v}$

We have, if we take $\vec{v}$ along the z-axis

$$
\begin{cases}
E'_z = E_z \qquad\qquad B'_z = B_z \\[2mm]
E'_x = \dfrac{E_x - \dfrac{v}{c}B_y}{\sqrt{1-\beta^2}} \quad B'_x = \dfrac{B_x + \dfrac{v}{c}E_y}{\sqrt{1-\beta^2}} \\[4mm]
E'_y = \dfrac{E_y + \dfrac{v}{c}B_x}{\sqrt{1-\beta^2}} \quad B'_y = \dfrac{B_y - \dfrac{v}{c}E_x}{\sqrt{1-\beta^2}}
\end{cases}
$$

We shall assume

$$
E_x^2 + E_y^2 + E_z^2 = B_x^2 + B_y^2 + B_z^2
$$

Now let us calculate

$$
E_x'^2 + E_y'^2 + E_z'^2
$$

$$
= E_z^2 + \frac{\left(E_x - \dfrac{v}{c}B_y\right)^2 + \left(E_y + \dfrac{v}{c}B_x\right)^2}{1-\beta^2}
$$

$$
= E_z^2 + \frac{E_x^2 + \beta^2 B_y - 2\beta E_x B_y + E_y^2 + \beta^2 B_x^2 + 2\beta E_y B_x}{1-\beta^2}
$$

$$
B_x'^2 + B_y'^2 + B_z'^2
$$

$$
= B_z^2 + \frac{\left(B_x + \dfrac{v}{c}E_y\right)^2 + \left(B_y - \dfrac{v}{c}E_x\right)^2}{1-\beta^2}
$$

$$
= B_z^2 + \frac{B_x^2 + \beta^2 E_y + 2\beta B_x E_y + B_y^2 + \beta^2 E_x^2 - 2\beta B_y E_x}{1-\beta^2}
$$

If $E_x'^2 + E_y'^2 + E_z'^2 = B_x'^2 + B_y'^2 + B_z'^2$
we must have

$$
E_z^2(1-\beta^2) + E_x^2 + E_y^2 + \beta^2(B_x^2 + B_y^2)
$$
$$
= B_z^2(1-\beta^2) + B_x^2 + B_y^2 + \beta^2(E_x^2 + E_y^2)
$$

or

$$
E_x^2 + E_y^2 + E_z^2 - \beta^2 E_z^2 + \beta^2(B_x^2 + B_y^2)
$$
$$
= B_x^2 + B_y^2 + B_z^2 - \beta^2 B_z^2 + \beta^2(E_x^2 + E_y^2)
$$

or

$$-\beta^2 E_z^2 + \beta^2(B_x^2 + B_y^2) = -\beta^2 B_z^2 + \beta^2(E_x^2 + E_y^2)$$
$$B_x^2 + B_y^2 + B_z^2 = E_x^2 + E_y^2 + E_z^2$$

(c) Let us show the invariance of the circular polarization. To show this
we have to prove two things

(i) The 90^δ "time" angle is preserved in the transformation.

(ii) The $90°$ "space" angle is preserved in the transformation.

The circularly polarized wave consists of the superposition of two
linearly polarized waves whose $\vec{E}$ fields are at $90°$ and lag with respect
to each other by $90°$

Let $\vec{E}_1$, $\vec{B}_1$ and $\vec{E}_2$, $\vec{B}_2$ define the two waves

$$\vec{E}_1 \equiv (E_1^x, E_1^y, E_1^z); \quad E_2 \equiv (E_2^x, E_2^y, E_2^z)$$
$$\vec{B}_1 \equiv (B_1^x, B_1^y, B_1^z); \quad B_2 \equiv (B_2^x, B_2^y, B_2^z)$$

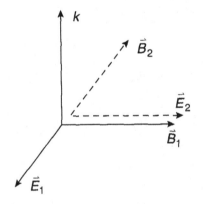

The conditions that
assume circular
polarization are

$$E_2^x = iB_1^x \quad B_2^x = -iE_1^x$$
$$E_2^y = iB_1^y \quad B_2^y = -iE_1^y$$
$$E_2^z = iB_1^z \quad B_2^z = -iE_1^z$$

Now

$$\begin{cases} E_1^{z'} = E_1^z & B_1^{z'} = B_1^z \\[2mm] E_1^{x'} = \dfrac{E_1^x - \beta B_1^y}{\sqrt{1-\beta^2}} & B_1^{x'} = \dfrac{B_1^x + \beta E_1^y}{\sqrt{1-\beta^2}} \\[2mm] E_1^{y'} = \dfrac{E_1^y + \beta B_1^x}{\sqrt{1-\beta^2}} & B_1^{y'} = \dfrac{B_1^y - \beta E_1^x}{\sqrt{1-\beta^2}} \end{cases}$$

We also have equations for E_2 and B_2.

$$\begin{cases} E_2^{z'} = E_2^z = iB_1^z = iB_1^{z'} \\[2mm] E_2^{x'} = \dfrac{E_2^x - \beta B_2^y}{\sqrt{1-\beta^2}} = \dfrac{iB_1^x - \beta(-iE_1^y)}{\sqrt{1-\beta^2}} = iB_1^{x'} \\[2mm] E_2^{y'} = \dfrac{E_2^y + \beta B_2^x}{\sqrt{1-\beta^2}} = \dfrac{iB_1^y + \beta(-iE_1^x)}{\sqrt{1-\beta^2}} = iB_1^{y'} \end{cases}$$

and

$$\begin{cases} B_2^{z'} = B_2^z = -iE_1^z = -iE_1^{z'} \\[2mm] B_2^{x'} = \dfrac{B_2^x + \beta E_2^y}{\sqrt{1-\beta^2}} = -\dfrac{iE_1^x + \beta(iB_1^y)}{\sqrt{1-\beta^2}} = -iE_1^{x'} \\[2mm] B_2^{y'} = \dfrac{B_2^y - \beta E_2^x}{\sqrt{1-\beta^2}} = \dfrac{-iE_1^y - \beta(iB_1^x)}{\sqrt{1-\beta^2}} = -iE_1^{y'} \end{cases}$$

7.3.

Consider a plane wave expressed by the four-potential

$$A_\mu = \alpha_\mu e^{i\sum_\lambda k_\lambda x_\lambda}$$

where α_μ is a constant four-vector
Also,

$$\sum_\lambda k_\lambda^2 = 0 \quad \text{because}$$

$$k_\mu \equiv \left(\vec{k}, i\frac{\omega}{c}\right) \quad \text{and} \quad k^2 - \frac{\omega^2}{c^2} = 0$$

(a) Let us express the Lorentz condition on α_μ

$$\sum_\mu \frac{\partial A_\mu}{\partial x_\mu} = 0$$

$$\frac{\partial A_\mu}{\partial x_\mu} = ik_\mu \alpha_\mu e^{i\sum_\lambda k_\lambda x_\lambda}$$

Therefore the Lorentz condition gives us

$$\sum_\mu \alpha_\mu k_\mu = 0$$

(b) Let us find the fields $F_{\mu v}$

$$F_{\mu v} = \frac{\partial A_v}{\partial X_\mu} - \frac{\partial A_\mu}{\partial x_v}$$

$$= ik_\mu \alpha_v e^{i \sum_\lambda k_\lambda x_\lambda} - ik_v \alpha_\mu e^{i \sum_\lambda k_\lambda x_\lambda}$$

$$= ie^{i \sum_\lambda k_\lambda x_\lambda} (k_\mu \alpha_v - k_v \alpha_\mu)$$

(c) Let us show that

$$\sum_v F_{\mu v} k_v = 0$$

$$\sum_v F_{\mu v} k_v = ie^{i \sum_\lambda k_\lambda x_\lambda} \sum_v (k_\mu \alpha_v - k_v \alpha_\mu) k_v$$

$$= ie^{i \sum_\lambda k_\lambda x_\lambda} \left[\sum_v k_\mu \alpha_v k_v - \sum_v k_v \alpha_\mu k_v \right]$$

$$= ie^{i \sum_\lambda k_\lambda x_\lambda} \left[k_\mu \sum_v \alpha_v k_v - \alpha_\mu \sum_v k_v^2 \right]$$

$$= 0$$

because $\sum_v \alpha_v k_v = 0$ and $\sum_v k_v^2 = 0$
To show that $\sum_v \hat{F}_{\mu v} k_v = 0$, we have

$$\hat{F}_{\mu v} = \frac{1}{2} \sum_{\sigma,\tau} \epsilon_{\mu v \sigma \tau} F_{\sigma \tau}$$

Now

$$\sum_v \hat{F}_{\mu v} k_v = \frac{1}{2} \sum_v \sum_{\sigma,\tau} \epsilon_{\mu v \sigma \tau} F_{\sigma \tau} k_v$$

$$= \frac{1}{2} \sum_v \sum_{\sigma,\tau} \epsilon_{\mu v \sigma \tau} (ie^{i \sum_\lambda k_\lambda x_\lambda})(k_\sigma \alpha_\tau - k_\tau \alpha_\sigma) k_v$$

It is enough to know that

$$\sum_v \sum_{\sigma,\tau} \epsilon_{\mu v \sigma \tau} (k_\sigma \alpha_\tau - k_\tau \alpha_\sigma) k_v = 0$$

Take for example $\mu = 1$:

$$\sum_v \hat{F}_{\mu v} k_v = \hat{F}_{12} k_2 + \hat{F}_{13} k_3 + \hat{F}_{14} k_4$$

$$= F_{34} k_2 + F_{42} k_3 + F_{23} k_4$$

$$= (i e^{i \sum_\lambda k_\lambda X_\lambda}) \{ (k_3 \alpha_4 - k_4 \alpha_3) k_2$$

$$+ (k_4 \alpha_2 - k_2 \alpha_4) k_3$$

$$+ (k_2 \alpha_3 - k_3 \alpha_2) k_4 \}$$

$$= (i e^{i \sum_\lambda k_\lambda X_\lambda}) \{ k_2 k_3 \alpha_4 - k_2 k_4 \alpha_3$$

$$+ k_3 k_4 \alpha_2 - k_2 k_3 \alpha_4$$

$$+ k_2 k_4 \alpha_3 - k_3 k_4 \alpha_2 \}$$

$$= 0$$

(d) $\sum_v F_{\mu v} k_v = 0$

$$\begin{cases} F_{12} k_2 + F_{13} k_3 + F_{14} k_4 = 0 \rightarrow B_3 k_2 - B_2 k_3 + \left(-i E_1 \dfrac{i\omega}{c} \right) = 0 \\[2mm] F_{21} k_1 + F_{23} k_3 + F_{34} k_4 = 0 \rightarrow -B_3 k_1 + B_1 k_3 + \left(-i E_2 \dfrac{i\omega}{c} \right) = 0 \\[2mm] F_{31} k_1 + F_{32} k_2 + F_{34} k_4 = 0 \rightarrow B_2 k_1 - B_1 k_2 + \left(-i E_3 \dfrac{i\omega}{c} \right) = 0 \\[2mm] F_{41} k_1 + F_{42} k_2 + F_{43} k_3 = 0 \rightarrow -i E_1 k_1 - i E_2 k_2 - i E_3 k_3 = 0 \end{cases}$$

The fourth relation gives

$$E_1 k_1 + E_2 k_2 + E_3 k_3 = \vec{E} \cdot \vec{k} = 0$$

The first three relations give

$$\vec{B} \times \vec{k} = \frac{\omega}{c} \vec{E}$$

$$\vec{B} \times \vec{k} = \begin{pmatrix} \hat{i} & \hat{j} & \hat{k} \\ B_1 & B_2 & B_3 \\ k_1 & k_2 & k_3 \end{pmatrix}$$

$$= \hat{i}(B_2 k_3 - B_3 k_2) + \hat{j}(B_3 k_1 - B_1 k_3)$$

$$+ \hat{k}(B_1 k_2 - B_2 k_1)$$

$$
\begin{cases}
\hat{F}_{12}k_2 + \hat{F}_{13}k_3 + \hat{F}_{14}k_4 = 0 \rightarrow (-iE_3)k_2 + (iE_2)k_3 + B_1\left(\dfrac{i\omega}{c}\right) = 0 \\[2ex]
\hat{F}_{21}k_1 + \hat{F}_{23}k_3 + \hat{F}_{24}k_4 = 0 \rightarrow (iE_3)k_1 + (-iE_1)k_3 + B_2\left(\dfrac{i\omega}{c}\right) = 0 \\[2ex]
\hat{F}_{31}k_1 + \hat{F}_{32}k_2 + \hat{F}_{34}k_4 = 0 \rightarrow (-iE_2)k_1 + (iE_1)k_2 + B_3\left(\dfrac{i\omega}{c}\right) = 0 \\[2ex]
\hat{F}_{41}k_1 + \hat{F}_{42}k_2 + \hat{F}_{43}k_3 = 0 \rightarrow (-B_1)k_1 + (-B_2)k_2 + (-B_3)k_3 = 0
\end{cases}
$$

The fourth relation gives

$$
B_1 k_1 + B_2 k_2 + B_3 k_3 = \vec{B} \cdot \vec{k} = 0
$$

The other three relations are

$$
\begin{cases}
E_2 k_3 - E_3 k_2 = -B_1 \dfrac{\omega}{c} \\[2ex]
E_3 k_1 - E_1 k_3 = -B_2 \dfrac{\omega}{c} \\[2ex]
E_1 k_2 - E_2 k_1 = -B_3 \dfrac{\omega}{c}
\end{cases}
$$

or

$$
\begin{cases}
k_2 E_3 - k_3 E_2 = B_1 \dfrac{\omega}{c} \\[2ex]
k_3 E_1 - k_1 E_3 = B_2 \dfrac{\omega}{c} \\[2ex]
k_1 E_2 - k_2 E_1 = B_3 \dfrac{\omega}{c}
\end{cases}
$$

or

$$
\vec{k} \times \vec{E} = \frac{\omega}{c}\,\vec{B}
$$

7.5.

(a)

$$
F_\mu = m\frac{du_\mu}{d\tau}
$$

$$
\sum_\mu F_\mu u_\mu = m \sum_\mu \frac{du_\mu}{d\tau} u_\mu
$$

$$
= \frac{1}{2} m \frac{d}{d\tau} \sum_\mu u_\mu^2 = \frac{1}{2} m \frac{d}{d\tau}(-c^2) = 0
$$

(b) We know that

$$u_\mu \equiv \left(\frac{\vec{v}}{\sqrt{1-\beta^2}}, \frac{ic}{\sqrt{1-\beta^2}} \right)$$

$$d\tau = dt\sqrt{1-\beta^2}$$

Then

$$\dot{u}_\mu = \frac{du_\mu}{d\tau}$$

$$\dot{u}_1 = \left(\frac{d}{dt} \frac{v_1}{\sqrt{1-\beta^2}} \right) \frac{dt}{d\tau} = \left[\frac{d}{dt} \left(\frac{v_1}{\sqrt{1-\beta^2}} \right) \right] \frac{1}{\sqrt{1-\beta^2}}$$

$$= \frac{\dot{v}_1}{\sqrt{1-\beta^2}} \frac{1}{\sqrt{1-\beta^2}} + \frac{v_1}{\sqrt{1-\beta^2}} \frac{1}{1-\beta^2} \frac{2\dfrac{\vec{v} \cdot \dot{\vec{v}}}{c^2}}{2\sqrt{1-\beta^2}}$$

$$= \frac{a_1}{1-\beta^2} + \frac{v_1(\vec{v} \cdot \vec{a}\,/c^2)}{(1-\beta^2)^2}$$

$$\dot{u}_4 = \left(\frac{d}{dt} \frac{ic}{\sqrt{1-\beta^2}} \right) \frac{dt}{d\tau} = \frac{1}{\sqrt{1-\beta^2}} \left(ic \frac{1}{1-\beta^2} \frac{\vec{v} \cdot \dot{\vec{v}}/c^2}{\sqrt{1-\beta^2}} \right)$$

$$= -\frac{ic}{1-\beta^2} \frac{\vec{v} \cdot \vec{a}/c^2}{1-\beta^2}$$

$$m\dot{\vec{u}} = \frac{m\vec{a}}{1-\beta^2} + m\vec{v} \frac{\vec{v} \cdot \vec{a}/c^2}{(1-\beta^2)^2}$$

On the other hand

$$\vec{p} = \frac{m\vec{v}}{\sqrt{1-\beta^2}}$$

and

$$\vec{F} = \frac{d\vec{p}}{dt} = \frac{m\vec{a}}{\sqrt{1-\beta^2}} + m\vec{v} \frac{\vec{v} \cdot \vec{a}/c^2}{1-\beta^2} \frac{1}{\sqrt{1-\beta^2}}$$

$$\frac{\vec{F}}{\sqrt{1-\beta^2}} = \frac{m\vec{a}}{1-\beta^2} + m\vec{v} \frac{\vec{v} \cdot \vec{a}/c^2}{(1-\beta^2)^2} = m\dot{\vec{u}}$$

also

$$\frac{\vec{F} \cdot \vec{v}}{\sqrt{1-\beta^2}} = \frac{m}{\sqrt{1-\beta^2}} \left(\vec{a} \cdot \vec{v} + \beta^2 \frac{\vec{v} \cdot \vec{a}}{1-\beta^2} \right)$$

$$= \frac{m}{1-\beta^2} \left[\frac{(1-\beta^2)(\vec{a} \cdot \vec{v}) + \beta^2 \, \vec{v} \cdot \vec{a}}{1-\beta^2} \right]$$

$$= \frac{m(\vec{v} \cdot \vec{a})}{(1-\beta^2)^2}$$

and

$$\frac{i \vec{F} \cdot \vec{v}}{c\sqrt{1-\beta^2}} = m \frac{ic}{1-\beta^2} \frac{\vec{v} \cdot \vec{a} / c^2}{1-\beta^2} = m \dot{u}_4$$

Then

$$F_\mu \equiv \left(\frac{\vec{F}}{\sqrt{1-\beta^2}}, \frac{i \vec{F} \cdot \vec{v}}{c\sqrt{1-\beta^2}} \right)$$

7.7.

Let

S = frame of the observer

S' = frame instantaneously at rest relative to the particle.

and

$$S = S' \quad \text{at} \quad t = t' = 0$$

At every time the acceleration in S' is $\vec{a}' = a\hat{z}'$ where $\hat{z}'$ = unit vector in the z' direction in the frame S'. The problem can be solved in two steps:

(1) First we find the acceleration in S in terms of a, and then
(2) we integrate to get $\vec{v}$

$$\vec{v} \int_0^T \vec{a} \, dt$$

The following observations can be made:

(i) S' is always chosen so that $\vec{v}'$ is equal to the velocity in $S' = 0$

(ii) $\hat{z}'$ will always be in the same direction and parallel to $\hat{z}$. The particle will start in a certain direction and will continue to go in the same direction in S and S'.

The problem is <u>one-dimensional</u>. Since the particle is always at rest in S', the problem consists of evaluating d^2v/dt^2, the acceleration of the particle <u>and</u> S', as measured in S, in terms of a, the acceleration of S', as measured in S'. The basic question is: what does a constant acceleration in one frame appear to be in another frame?

$$dx = \sqrt{1 - v^2/c^2}dx' \quad \text{Lorentz contraction}$$

$$dt = \sqrt{1 - v^2/c^2}dt' \quad \text{time dilation}$$

Therefore

$$d^2x = \sqrt{1 - v^2/c^2}dx' \quad (v \text{ is a function of time, not } x)$$

$$dt^2 = (1 - v^2/c^2)dt^2$$

and the acceleration of S' as measured by an observer in S is

$$\frac{d^2x}{dt^2} = \left(1 - \frac{v^2}{c^2}\right)^{3/2} \frac{d^2x'}{dt'^2} = \left(1 - \frac{v^2}{c^2}\right)^{3/2} a$$

We now have

$$\frac{dv}{dt} = \left(1 - \frac{v^2}{c^2}\right)^{3/2} a \Rightarrow \frac{dv}{(1 - v^2/c^2)^{3/2}} = adt$$

$$\int_0^v \frac{dv}{(1 - v^2/c^2)^{3/2}} = \int_0^T adt \Rightarrow \frac{v}{\sqrt{1 - v^2/c^2}} = aT$$

$$v^2 = a^2T^2\left(1 - \frac{v^2}{c^2}\right) \Rightarrow v = \frac{aT}{\sqrt{1 + \left(\dfrac{aT}{c}\right)^2}}$$

7.9.

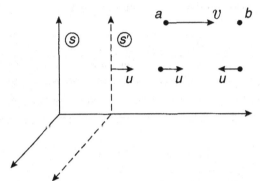

v is the velocity of the incident electron a in the frame S in which the electron b is at rest. u is the velocity of b in frame S' that moves with respect to S with velocity u.

The velocity of electron a in S' must be u. The law of addition of velocities gives

$$v = \frac{2u}{1 + u^2/c^2}$$

Then

$$\frac{1}{\sqrt{1 - v^2/c^2}} = \frac{1}{\sqrt{1 - \dfrac{1}{c^2}\dfrac{4u^2}{\left(1 + \dfrac{u^2}{c^2}\right)}}} = \frac{1 + \dfrac{u^2}{c^2}}{\sqrt{\left(1 + \dfrac{u^2}{c^2}\right)^2 - \dfrac{4u^2}{c^2}}}$$

$$= \frac{1 + \dfrac{u^2}{c^2}}{\sqrt{1 + \dfrac{u^4}{c^4} + \dfrac{2u^2}{c^2} - \dfrac{4u^2}{c^2}}} = \frac{1 + \dfrac{u^2}{c^2}}{1 - \dfrac{u^2}{c^2}} = \frac{c^2 + u^2}{c^2 - u^2}$$

Therefore

$$E = \frac{mc^2}{\sqrt{1 + \dfrac{v^2}{c^2}}} = mc^2 \left(\frac{1 + \dfrac{u^2}{c^2}}{1 - \dfrac{u^2}{c^2}}\right) = mc^2 \frac{c^2 + u^2}{c^2 - u^2}$$

$$\frac{E}{mc^2}\left(1 - \frac{u^2}{c^2}\right) = 1 + \frac{u^2}{c^2} \Rightarrow \frac{u^2}{c^2}\left(1 + \frac{E}{mc^2}\right) + 1 = \frac{E}{mc^2}$$

or,

$$\frac{u^2}{c^2} = \frac{\dfrac{E}{mc^2} - 1}{1 + \dfrac{E}{mc^2}}$$

$$1 - \frac{u^2}{c^2} = 1 - \frac{\dfrac{E}{mc^2} - 1}{1 + \dfrac{E}{mc^2}} = \frac{2}{1 + \dfrac{E}{mc^2}} = \frac{2mc^2}{E + mc^2}$$

Then

$$E_u = \frac{mc^2}{\sqrt{1 - \dfrac{u^2}{c^2}}} = \frac{mc^2}{\sqrt{\dfrac{2mc^2}{E + mc^2}}} = \sqrt{E + mc^2} \cdot \sqrt{\frac{1}{2}mc^2}$$

$$= \sqrt{\frac{1}{2}mc^2(E + mc^2)}$$

7.11.

In the entire region occupied by the field there is vacuum, i.e., $p = \vec{J} = 0$ and

$$f_\mu = \frac{1}{c} \sum_v F_{\mu v} J_v = 0 \quad [J_v \equiv (\vec{J}, ic\rho)]$$

But

$$f_\mu = \sum_\sigma \frac{\partial T_{\mu\sigma}}{\partial x_\sigma} = \sum_\sigma \frac{\partial T_{\mu\sigma}}{\partial x_\sigma} = 0$$

where

$$T_{\mu\sigma} = \frac{1}{4\pi} \left[\sum_v F_{\mu v} F_{v\sigma} + \frac{1}{4}\delta_{\mu\sigma} \left(\sum_{\lambda\mu} F_{\lambda\mu}^2 \right) \right]$$

the tensor T is given by

$$T \equiv \begin{pmatrix} T_{11} & T_{12} & T_{13} & -icG_1 \\ T_{21} & T_{22} & T_{23} & -icG_2 \\ T_{31} & T_{32} & T_{33} & -icG_3 \\ -icG_1 & -icG_2 & -icG_3 & W \end{pmatrix}$$

where

$$\vec{G} = \frac{c}{4\pi}(\vec{E} \times \vec{B}) = \text{momentum density}$$

$$W = \frac{1}{8\pi}(E^2 + B^2) = \text{energy density}$$

We know from Sec. 7.3 of the text that if

$$\sum_\sigma \frac{\partial T_{\sigma\mu}}{\partial x_\sigma} = 0$$

then $\int d^3\vec{x} T_{4\mu}$ is a four-vector:

$$\int d^3\vec{x}\, T_{4\mu} \equiv \left(-ci \int \vec{G} d^3\vec{x},\, \int W d^3\vec{x} \right)$$

$$= (-ic\,\vec{P}, E)$$

where

$$\vec{P} = \int \vec{G} d^3\vec{x} = \text{total electromagnetic momentum.}$$

$$E = \int W d^3\vec{x} = \text{total energy of the field.}$$

7.13.

The general expression for the energy is given by

$$E^2 = p^2 c^2 + (m_0 c^2)^2$$

The interaction momentum and the total energy are conserved. Let us use subscripts $i(f)$ for quantities before (after) collision.

$$\text{Initial momentum} \quad P_i = \frac{1}{c}E_\gamma$$

$$\text{Initial energy} \quad E_i = E_\gamma + m_0 c^2$$

After the collision we have

$$E_f^2 = P_f^2 c^2 + (m_0' c^2)^2$$

where

$$P_f = P_i = \frac{1}{c}E_\gamma \quad \text{(conservation of momentum)}$$

$$m_0' = m_0 + \frac{\Delta E}{c^2} \quad \text{(conservation of mass-energy)}$$

But $E_f = E_i$, so

$$(E_\gamma + m_0c^2)^2 = \frac{E_\gamma^2}{c^2}c^2 + \left[\left(m_0 + \frac{\Delta E}{c^2}\right)c^2\right]^2$$

$$E_\gamma^2 = (E_\gamma + m_0c^2)^2 - \left[\left(m_0 + \frac{\Delta E}{c^2}\right)c^2\right]^2$$

$\underbrace{\qquad\qquad}\qquad\underbrace{\qquad\qquad}$

$\Downarrow\qquad\qquad\qquad\Downarrow$

initial energy   rest energy of
                   excited particle

$$= (E_\gamma + m_0c^2)^2 - (mc^2 + \Delta E)^2$$

$$= E_\gamma^2 + m^2c^4 + 2mc^2E_\gamma - m^2c^4$$

$$- (\Delta E)^2 - 2mc^2\Delta E$$

$$2mc^2E_\gamma = (\Delta E)^2 + 2mc^2\Delta E$$

$$E_\gamma = \Delta E + \frac{(\Delta E)^2}{2mc^2}$$

$$= \Delta E\left(1 + \frac{\Delta E}{2mc^2}\right)$$

m_0 = rest mass of particle in ground state
m_0' = rest mass of excited particle = $m_0 + \dfrac{\Delta E}{c^2}$

7.15.

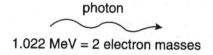

photon

1.022 MeV = 2 electron masses   electron at rest

We can represent the situation in the following diagram where masses are encircled:

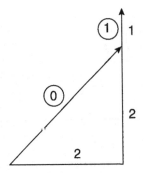 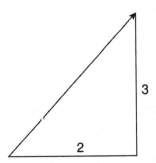

We have to use the following equations where the subscript 1 (2) indicates the electron (photon)

$$
\begin{cases}
E_1^2 - p_1^2 = 1 \\
E_2^2 - p_2^2 = 0 \\
E_1 + E_2 = 3 \\
p_1 + p_2 = 2
\end{cases}
$$

where we have set for simplicity $c = 1$.

Some calculations follow

$$
\begin{cases}
E_1^2 = 1 + p_1^2 \\
E_2^2 = p_2^2 \\
E_2 = 3 - E_1 \\
p_2 = 2 - p_1
\end{cases}
$$

$$(3 - E_1) = (2 - p_1)^2$$

$$9 + E_1^2 - 6E_1 = 4 + p_1^2 - 4p_1$$

$$5 + 1 + p_1^2 - 6\sqrt{1 + p_1^2} = p_1^2 - 4p_1$$

$$3 + 2p_1 = 3\sqrt{1 + p_1^2}$$

$$9 + 12p_1 + 4p_1 = 9 + 9p_1^2$$

$$5p_1^2 - 12p_1 = 0$$

$$p_1 = \begin{cases} 0 \\ \dfrac{12}{5} = 2.4 \end{cases}$$

Then

$p_1 = 0$	$p_1 = 2.4$
$p_2 = 2$	$p_2 = 2 - 2.4 = -0.4$
$E_2 = 2$	$E_2^2 = p_2^2 \rightarrow E_2 = 0.4$
$E_1 = 1$	$E_1 = 3 - E_2 = 3 - 0.4 = 2.6$

Electron is kicked forward with a momentum 2.4 times the electron mass, much less than the $2m_e$ of energy and momentum with which it approached:

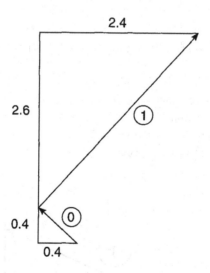

7.17.

Conservation of momentum:

$$\begin{cases} p_0 = p_1 \cos\theta + p_2 \cos\varphi \\ p_1 \sin\theta = p_2 \sin\varphi \end{cases}$$

$$\begin{cases} (p_0 - p_1 \cos\theta)^2 = p_2^2 \cos^2\varphi \\ p_1^2 \sin^2\theta = p_2^2 \sin^2\varphi \end{cases}$$

Adding the above two equations

$$p_0^2 + p_1^2 \cos^2\theta - 2p_0 p_1 \cos\theta + p_1^2 \sin^2\theta = p_2^2$$

$$p_o^2 + p_1^2 - 2p_0 p_1 \cos\theta \tag{1}$$

Conservation of energy, setting $c = 1$:

Energy of positron $= 3m_0$

Rest mass $= m_0$, which we may set $= 1$

momentum $= \sqrt{9-1} = \sqrt{8} = 2\sqrt{2} = p_0$

Total energy $= 4m_0 = p_1 + P_2$

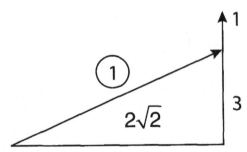

Going back to Eq. (1)

$$p_0^2 + p_1^2 - 2p_0p_1 \cos\theta = p_2^2$$

Now

$$p_0 = 2\sqrt{2}, \quad \theta = 30°, \quad \cos\theta = 0.866$$

Then

$$8 + p_1^2 - 2 \times 2\sqrt{2} \times 0.866 p_1 = p_2^2$$
$$p_1^2 - 4.899p_1 + 8 = p_2^2 = (4 - p_1)^2 = 16 + p_1^2 - 8p_1$$
$$(8 - 4.899)p_1 = 8 \Rightarrow p_1 = \frac{8}{3.101} = 2.5798$$

Then

$$\begin{cases} p_1 = 2.5798 = E_1 \\ p_2 = 4 - 2.5798 = 1.4202 = E_2 \end{cases}$$

Also,

$$p_1 \sin 30 = p_2 \sin\varphi$$
$$\sin\varphi = \frac{p_1 \sin 30}{p_2} = \frac{2.5798 \times 0.5}{1.4202} = 0.908$$
$$\varphi = 65.3°$$

Therefore

$$E_1 = 2.5798 \times 0.511\,\text{MeV} = 1.318\,\text{MeV}$$
$$E_2 = 1.4202 \times 0.511\,\text{MeV} = 0.726\,\text{MeV}$$

7.19.

(a) We know that

$$\gamma = \frac{\text{total energy}}{\text{rest energy}} = \frac{\text{rest energy} + \text{kinetic energy}}{\text{rest energy}}$$

In the first case, kinetic energy $\equiv 0.7\,\text{MeV}$

$$\gamma = \frac{140 + 0.7}{140} = \frac{140.7}{140} = 1.005$$

and

$$\beta = \sqrt{1 - \frac{1}{\gamma^2}} = 0.1$$

The time dilation is negligible. The π^+ will travel, before disappearing, a distance

$$0.1 \times 3 \times 10^{10} \times 2.6 \times 10^{-8} = 78 \,\text{cm}$$

(b) Kinetic energy of $\pi^+ = 30 \,\text{MeV}$

$$\gamma = \frac{170}{140} = 1.21$$

The speed is given by

$$\beta_c = c\sqrt{1 - \frac{1}{\gamma^2}} = 0.57 \, c$$

The "stretched" lifetime (i.e., the lifetime in the lab) will be

$$2.6 \times 10^{-8} \times 1.21 = 3.15 \times 10^{-8} \,\text{sec.}$$

The distance traveled in the time available is

$$0.57 \times 3 \times 10^8 \times 3.15 \times 10^{-8} = 5.4 \,\text{m}$$

(c) Kinetic energy $= 700 \,\text{MeV}$

$$\gamma = \frac{840}{140} = 6$$

The "stretched" lifetime is:

$$6 \times 2.6 \times 10^{-8} = 15.6 \times 10^{-8} \,\text{sec.}$$

The speed is:

$$\beta = \sqrt{1 - \frac{1}{\gamma^2}} = \sqrt{1 - \frac{1}{36}} = 0.986$$

The distance traveled is:

$$15.6 \times 10^{-8} \times 0.986 \times 3 \times 10^8 = 46.1 \,\text{m}$$

7.21.

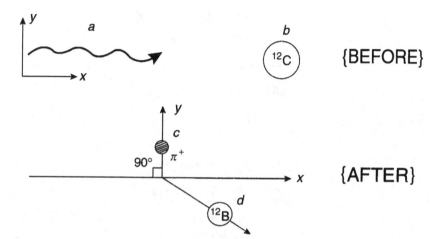

We set again $c = 1$.

$$E_a = p_{xa} = 800\,\text{MeV}$$

$$m_b = E_b = 11\,175\,\text{MeV}$$

$$m_c = 139.57\,\text{MeV};\ p_{xc} = 0$$

$$m_d = 11\,188\,\text{MeV}$$

Conservation of momentum:

$$p_{xa} = p_{xd} = 800\,\text{MeV} \tag{1}$$

$$p_{yc} = -p_{yd} \tag{2}$$

Conservation of energy

$$E_a + m_b = E_c + E_d \tag{3}$$

or

$$800 + 11\,1755 = 11\,975 = E_c + E_d \tag{3'}$$

In addition

$$m_c^2 = E_c^2 - p_{yc}^2 - p_{xc}^2$$
$$139.57^2 = 19\,479.785 = E_c^2 - p_{yc}^2 \tag{4}$$

and

$$m_d^2 = E_d^2 - p_{yd}^2 - p_{xd}^2$$
$$m_d^2 + P_{yd}^2 = E_d^2 - p_{yd}^2$$
$$11\,188^2 + 800^2 = E_d^2 - p_{yd}^2$$
$$125\,811\,344 = E_d^2 - p_{yd}^2$$

(5)

In the same way

$$E_c^2 - p_{yc}^2 = 19\,479\,785 \tag{4}$$

$$E_d^2 - p_{yd}^2 = 125\,811\,344 \tag{5}$$

$$E_c + E_d = 11\,975 \tag{3'}$$
$$p_{yc} = -p_{yd} \tag{2}$$

From Eqs. (4) and (5), taking into account Eqs. (3′) and (2) we obtain

$$E_d^2 - E_c^2 = -E_c^2 - 2 + (11\,975 - E_c)^2 = 125\,791\,864.2$$
$$E_c = 735.23\,\text{MeV}$$

The various quantities can be set in a table.

	m	E	p_x	p_y	
					B
(a) γ-ray	0	800	800	0	E
					F
(b) ^{12}C	11 175	11 175	0	0	O
					R
System	11 948.25	11 975	800	0	E
System	11 948.25	11 975	800	0	A
					F
(d) ^{12}B	11 188	11 239.77	800	−721.86	T
					E
(e) π^+	139.57	735.23	0	−721.86	R

7.23.

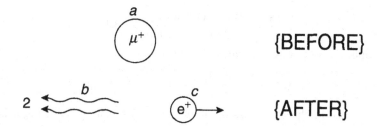

The basic data are

$$m_a = 105.66 \quad E_a = 105.66 \quad p_a = 0$$
$$m_b = 0 \qquad\qquad E_b = p_b$$
$$m_c = 0.511$$

Then, setting $c = 1$

$$E_a = E_b + E_c \rightarrow E_c = E_a - E_b$$
$$p_a = 0 = p_b + p_c \rightarrow p_c = -p_b$$

and

$$m_c^2 = E_c^2 - p_c^2 = (E_a - E_b)^2 - p_b^2$$
$$= E_a^2 + E_b^2 - 2E_aE_b - p_b^2 = E_a^2 - 2E_aE_b$$
$$= 105.66^2 - 2 \times 105.66\, E_b = 0.511^2$$
$$E_b = 52.829$$
$$E_c = 105.66 - 52.829 = 52.831$$

	m	E	p
π^+	105.66	105.66	0
System	105.66	105.66	0
e^+	0.511	52.831	52.829
v	0	52.829	-52.829

7.25.

$$130 \, \frac{\text{miles}}{\text{hour}} = \frac{130 \times 1609}{3600} \, \frac{\text{meter}}{\text{second}} = 58.1 \frac{m}{s} \ll c$$

The kinetic energy of each train is given by

$$\frac{1}{2} m v^2 = \frac{10^6 \times 58.1^2}{2} = 1.688 \times 10^9 \, \text{joules}$$

The increase in the rest mass of the track trains system following a head on collision is

$$\frac{2 \times 1.687 \times 10^9}{9 \times 10^{16}} = 3.7 \times 10^{-8} \, \text{kg}$$

7.27.

(a) $2\sqrt{2}m$

(b) $2\sqrt{6}m$

(c) $\sqrt{7}m$

(d) $2\sqrt{3}E$

(e) 0

(f) $\sqrt{10}m$

(g) $\sqrt{6}E$

8

Exercises

8.1.

Let us assume that a particle at rest at time $t = 0$ is subjected to a force $\vec{F}$ constant in the x direction. We have

$$\vec{p} = \frac{m\vec{v}}{\sqrt{1 - v^2/c^2}}, \quad \vec{F} = \frac{d\vec{p}}{dt}$$

Then

$$\frac{d\vec{p}}{dt} = \frac{d}{dt} \frac{m\vec{v}}{\sqrt{1 - v^2/c^2}} = \vec{F}$$

and assuming the particle is at rest at time $t = 0$

$$\frac{m\vec{v}}{\sqrt{1 - v^2/c^2}} = \vec{F}t$$

We have to consider only the x direction

$$\frac{mv}{\sqrt{1 - v^2/c^2}} = Ft$$

$$m^2v^2 = F^2t^2(1 - v^2/c^2)$$

$$m^2v^2 = F^2t^2 - F^2t^2\frac{v^2}{c^2}$$

$$\left(m^2 + \frac{F^2t^2}{c^2}\right)v^2 = F^2t^2$$

$$v = \frac{Ft}{\sqrt{m^2 + \frac{F^2t^2}{c^2}}} = \frac{dx}{dt}$$

At time $t = 0, x = 0$. Integrating the above equation

$$x = \int_0^t \frac{Ft}{\sqrt{m^2 + \frac{(Ft)^2}{c^2}}}dt = c\int_0^t \frac{Ft}{\sqrt{m^2c^2 + (Ft)^2}}dt$$

Let

$$Ft = z \rightarrow dt = dz/F$$

$$mc = a$$

$$x = \frac{c}{F}\int_0^{Ft} \frac{z}{\sqrt{a^2 + z^2}} = \frac{c}{F}[\sqrt{a^2 + z^2}]_0^{Ft}$$

$$= \frac{c}{F}[\sqrt{a^2 + (Ft)^2} - a] = \frac{c}{F}\sqrt{(mc)^2 + (Ft)^2} - \frac{mc^2}{F}$$

or

$$x + \frac{mc^2}{F} = \frac{c}{F}\sqrt{m^2c^2 + F^2t^2}$$

$$\left(x + \frac{mc^2}{F}\right)^2 = \frac{c^2}{F^2}(m^2c^2 + F^2t^2) = \left(\frac{mc^2}{F}\right)^2 + c^2t^2$$

or

$$\left(x + \frac{mc^2}{F}\right) - (ct)^2 = \left(\frac{mc^2}{F}\right)^2$$

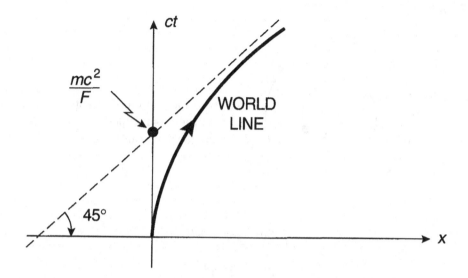

8.3.

The formula for the transformation of velocities is given by

$$vx' = \frac{v_x - u}{1 - \frac{v_x u}{c^2}} = \text{velocity of electron moving towards } x \text{ measured in } S'$$

where

$$v_x = \text{velocity of electron moving towards } x$$
$$\text{measured in } S \text{ (laboratory frame)}$$
$$= 0.6c$$
$$u = \text{velocity of frame } S' = 0.6c$$

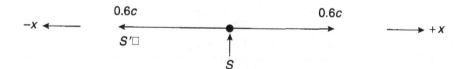

Then

$$v'_x = \frac{(0.6 + 0.6)c}{1 + \frac{0.6 \times 0.6c^2}{c^2}} = \frac{1.2c}{1 + (0.6)^2} = 0.88c$$

8.5.

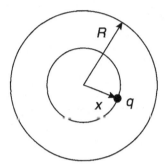

A point charge q, a distance x from the center of the atom, experiences a force which can be found from

$$\int E ds = 4\pi \int \rho dV$$

$$4\pi x^2 E = 4\pi Q \frac{x^3}{R^3}$$

$$F = qE = \frac{qQ}{R^3} x$$

The equation of motion of the charge q is

$$m \frac{d^2 x}{dt^2} = \frac{qQ}{R^3} x$$

Q is positive, q is negative and thus the equation above is the equation of a simple, linear harmonic oscillator. Considering the H atom: $Q = e, q = -e$

Then,

$$\omega = \sqrt{\frac{qQ}{R^3 m}} = \sqrt{\frac{e^2}{mR^3}}$$

where

$$e = 4.8 \times 10^{-10} \text{ esu}$$
$$m = 9.1 \times 10^{-28} \text{ gm}$$
$$R = 10^{-8} \text{ cm}$$

We find

$$v = \frac{1}{2\pi} \sqrt{\frac{e^2}{mR^3}} = 2.53 \times 10^{15}; \quad \lambda = \frac{c}{v} = 1185 \,\text{Å}$$

8.7.

$$\rho(\vec{x},t) = e\delta^{(3)}(\vec{x} - \vec{x}_p(t))$$

$$\vec{J}(\vec{x},t) = e\vec{v}\delta^{(3)}(\vec{x} - x_p(t))$$

Consider the expression

$$u_\mu \int d\tau\, \delta^{(3)}(\vec{x} - x_p(t))\delta(t - t_p(\tau))$$

(a) The expression above is a four-vector because u_μ is a four-vector

$$\int d\tau\, \delta^{(3)}(x - x_p(t))\delta(t - t_p(\tau))$$

$$= c \int \delta^{(3)}(x - x_p(t))\delta[c(t - t_p(\tau))]d\tau$$

$$= c \int \delta^{(4)}(x - x_p(\tau))d(\tau)$$

$$= c \int \delta^{(4)}(|x - x_p(\tau)|)d\tau$$

$$= \text{a scalar}$$

Note that the delta function is an even function and that $|x - x_p(\tau)|$ is invariant.

(b)

$$u_\mu \int d\tau\, \delta^{(3)}(x - x_p(\tau))\delta(t - t_p(\tau))$$

$$= \begin{cases} \dfrac{\vec{v}}{\sqrt{1 - \beta^2}} \displaystyle\int dt\sqrt{1 - \beta^2}\delta^{(3)}(x - x_p(\tau)) \\ \quad \times\, \delta(t - t_p(\tau)) = \vec{v}\delta^{(3)}(x - x_p(\tau)) \\ \dfrac{ic}{\sqrt{1 - \beta^2}} \displaystyle\int dt\sqrt{1 - \beta^2}\delta^{(3)}(x - x_p(\tau)) \\ \quad \times\, \delta(t - t_p(\tau)) = ic\delta^{(3)}(x - x_p(\tau)) \end{cases}$$

Therefore

$$\vec{J}(x,t) = e\vec{u} \int d\tau\, \delta^{(3)}(x - x_p(\tau))\delta(t - t_p(\tau))$$

$$\rho(x,t) = eu_4 \int d\tau\, \delta^{(3)}(x - x_p(\tau))\delta(t - t_p(\tau))$$

8.9.

Let S be the laboratory frame and S' a frame that moves with the electron with velocity v in the x direction:

$$
\begin{cases}
x = \dfrac{x' + vt'}{\sqrt{1 - \beta^2}} \\[2mm]
y = y' \\[1mm]
z = z' \\[2mm]
ct = \gamma\left(ct - c\dfrac{v}{c^2}x\right)
\end{cases}
$$

$(\vec{x}, ict)$ forms a four-vector

$(A, i\phi)$ forms a four-vector

Then

$$
\begin{cases}
A_x = \dfrac{A_{x'} + \beta\phi'}{\sqrt{1 - \beta^2}} \\[3mm]
A_y = A_{y'} \\[1mm]
A_z = A_{z'} \\[2mm]
\phi = \dfrac{\phi' + (\beta/c)A_{x'}}{\sqrt{1 - \beta^2}}
\end{cases}
$$

On the other hand

$$
\vec{A}' = 0, \quad \phi' = \frac{e}{r'}
$$

where

$$
r' = (x'^2 + y'^2 + z'^2)^{1/2}
$$

Therefore

$$
\begin{cases}
A_x = \dfrac{\beta\phi'}{\sqrt{1 - \beta^2}} = \dfrac{\beta}{\sqrt{1 - \beta^2}}\dfrac{e}{r'} \\[3mm]
A_y = A_z = 0 \\[2mm]
\phi = \dfrac{\phi'}{\sqrt{1 - \beta^2}} = \dfrac{1}{\sqrt{1 - \beta^2}}\dfrac{e}{r'}
\end{cases}
$$

We can write

$$\begin{cases} x' = \dfrac{x - vt}{\sqrt{1 - \beta^2}} \\[2mm] y' = y \\[1mm] z' = z \end{cases}$$

$$r' = \left[\frac{(x - vt)^2}{1 - \beta^2} + y^2 + z^2 \right]^{1/2}$$

and

$$S = r'\sqrt{1 - \beta^2} = [(x - vt)^2 + (y^2 + z^2)(1 - \beta^2)]^{1/2}$$

Finally

$$\begin{cases} A_x = \dfrac{v}{c}\dfrac{e}{S} = \beta\dfrac{e}{S} \\[2mm] A_y = A_z = 0 \\[1mm] \phi = \dfrac{e}{S} \end{cases}$$

To find $\vec{E}$ and $\vec{B}$ we use

$$\vec{E} = -\vec{\nabla}\phi - \frac{1}{c}\frac{\partial \vec{A}}{\partial t}$$

$$\vec{B} = \vec{\nabla} \times \vec{A}$$

Let us make the following calculations

$$\frac{\partial \phi}{\partial x} = \frac{\partial \phi}{\partial S}\frac{\partial S}{\partial x} = -\frac{e}{S^2}\frac{1}{S}(x - vt) = -\frac{e}{S^3}(x - vt)$$

$$\frac{\partial \phi}{\partial y} = \frac{\partial \phi}{\partial S}\frac{\partial S}{\partial y} = -\frac{e}{S^2}\frac{y}{S}(1 - \beta^2) = -\frac{ey}{S^3}(1 - \beta^2)$$

$$\frac{\partial \phi}{\partial t} = -\frac{ez}{S^3}(1 - \beta^2)$$

$$\frac{\partial A_x}{\partial t} = \frac{\partial A_x}{\partial S}\frac{\partial S}{\partial t}\left(-\beta\frac{e}{S^2}\right)\left(-\frac{v(x - vt)}{S}\right)$$

$$= \beta\frac{ev}{S^3}(x - vt)$$

Then

$$\begin{cases} E_x = -\dfrac{\partial \phi}{\partial x} - \dfrac{1}{c}\dfrac{\partial A_x}{\partial t} = \dfrac{e}{S^3}(x - vt) - \beta^2 \dfrac{e}{S^3}(x - vt) \\[2mm] \qquad = \dfrac{e}{S^3}(x - vt)(1 - \beta^2) \\[2mm] E_y = -\dfrac{\partial \phi}{\partial y} = \dfrac{ey}{S^3}(1 - \beta^2) \\[2mm] F_y = -\dfrac{\partial \phi}{\partial z} = \dfrac{ez}{S^3}(1 - \beta^2) \end{cases}$$

Now

$$\vec{B} = \vec{\nabla} \times \vec{A} = \begin{pmatrix} \hat{i} & \hat{j} & \hat{k} \\ \partial/\partial x & \partial/\partial y & \partial/\partial z \\ A_x & 0 & 0 \end{pmatrix} = \hat{j}\frac{\partial A_x}{\partial z} - \hat{k}\frac{\partial A_x}{\partial y}$$

$$\frac{\partial A_x}{\partial z} = \frac{\partial}{\partial z}\left(\beta\frac{e}{S}\right) = e\beta\frac{\partial}{\partial z}\frac{1}{S} = -\frac{e\beta}{S^2}\frac{\partial S}{\partial z} = -\frac{e\beta}{S^3}(1 - \beta^2)z$$

$$\frac{\partial A_x}{\partial y} = -\frac{e\beta}{S^3}(1 - \beta^2)y$$

We can write finally

$$\begin{cases} \vec{E} = \dfrac{e}{S^3}(\vec{r} - \vec{v}t)(1 - \beta^2) \\[2mm] \vec{B} = \dfrac{\vec{v}}{c} \times \vec{E} \end{cases}$$

where

$$S = [(x - vt)^2 + (1 - v^2/c^2)(y^2 + z^2)]^{1/2}$$

8.11.

(a) To show that ϕ and $\vec{A}$ satisfy the Lorentz condition we do the following

$$\frac{\partial f(x - \vec{v}t)}{\partial t} = -(\vec{v} \cdot \vec{\nabla})f$$

$$\frac{\partial f(x, y, z - vt)}{\partial t} = -v\frac{\partial f}{\partial z}$$

In the present case

$$\frac{\partial \phi}{\partial t} = -v\frac{\partial \phi}{\partial z}, \qquad \frac{1}{c}\frac{\partial \phi}{\partial t} = -\frac{v}{c}\frac{\partial \phi}{\partial z}$$

Also

$$\vec{\nabla} \cdot \vec{A} = \frac{\partial \vec{A}}{\partial t} = \frac{\partial}{\partial z}\left(\frac{v}{c}\phi\right) = \frac{v}{c}\frac{\partial \phi}{\partial z}$$

Therefore

$$\vec{\nabla} \cdot \vec{A} + \frac{1}{c}\frac{\partial \phi}{\partial t} = 0$$

(b)

$$\vec{E} = -\vec{\nabla}\phi - \frac{1}{c}\vec{A} = -\vec{\nabla}\phi - \frac{v}{c^2}\frac{\partial \phi}{\partial t}$$

$$= -\vec{\nabla}\phi + \frac{v}{c^2}v\frac{\partial \phi}{\partial t} = -\vec{\nabla}\phi + \frac{v^2}{c^2}\frac{\partial \phi}{\partial t}\hat{k}$$

$$B = -\vec{\nabla} \times \vec{A} = \vec{\nabla}\phi \times \frac{v}{c} = -\frac{v}{c} \times \vec{\nabla}\phi$$

Therefore

$$\frac{v}{c} \times \vec{E} = -\frac{v}{c} \times \vec{\nabla}\phi = \vec{B}$$

(c) $\phi(\vec{x}, t) = \frac{e}{S}$ where $S = \sqrt{(1 - \beta^2)(x^2 + y^2) + (z - vt)^2}$

Therefore

$$\frac{\partial \phi}{\partial z} = -\frac{e(1 - \beta^2)}{S^3}$$

$$\frac{\partial \phi}{\partial y} = -\frac{e(1 - \beta^2)}{S^3}$$

$$\frac{\partial \phi}{\partial z} = -\frac{e(z - vt)}{S^3}$$

and

$$\begin{cases} E_x = -\dfrac{\partial \phi}{\partial x} = \dfrac{e(1 - \beta^2)x}{S^3} \\[2mm] E_y = -\dfrac{\partial \phi}{\partial y} = \dfrac{e(1 - \beta^2)y}{S^3} \\[2mm] E_z = -\dfrac{\partial \phi}{\partial z} + \dfrac{v^2}{c^2}\dfrac{\partial \phi}{\partial z} = (-1 + \beta^2)\dfrac{\partial \phi}{\partial z} \end{cases}$$

$$= \frac{e(1 - \beta^2)(z - vt)}{S^3}$$

It is evident that $\vec{E} \| \vec{n}$

$$\vec{B} = \frac{\vec{v}}{c} \times \vec{E} = \begin{pmatrix} \hat{i} & \hat{j} & \hat{k} \\ 0 & 0 & v/c \\ E_x & E_y & E_z \end{pmatrix} = \hat{i} \left(-\frac{v}{c} E_y \right) + \hat{j} \left(\frac{v}{c} E_x \right)$$

Therefore

$$\begin{cases} B_x = -\frac{v}{c} E_y = -\frac{ev}{cS^0}(1 - \beta^2)y \\ B_y = \frac{v}{c} E_x = \frac{ev}{cS^3}(1 - \beta^2)x \\ B_z = 0 \end{cases}$$

9

Exercises

9.1.

(a)

$$p_\mu \equiv \left(\frac{m\vec{v}}{\sqrt{1-\beta^2}}, \frac{icm}{\sqrt{1-\beta^2}} \right)$$

$$F_\mu \equiv \left(\frac{\vec{F}}{\sqrt{1-\beta^2}}, \frac{i\vec{F} \cdot \vec{v}}{c\sqrt{1-\beta^2}} \right)$$

$$\sum_\mu F_\mu p_\mu = \frac{m\vec{F} \cdot \vec{v}}{1-\beta^2} - \frac{m\vec{F} \cdot \vec{v}}{1-\beta^2} = 0$$

(b)

$$F_\mu^{\text{rad}} = \frac{2e^2}{3mc^2} \left[\frac{d^2 p_\mu}{d\tau^2} - \frac{p_\mu}{m^2 c^2} \sum_v \left(\frac{dp_v}{d\tau} \right)^2 \right]$$

or

$$F_\mu^{\text{rad}} = \frac{2e^2}{3c^3}\left[\frac{d^2u_\mu}{d\tau^2} - \frac{u_\mu}{c^2}\sum_v\left(\frac{du_v}{d\tau}\right)^2\right]$$

$$= \frac{2e^2}{3c^3}\left[\frac{d^2u_\mu}{d\tau^2} - \frac{u_\mu}{c^2}\frac{a^2 - \left(\vec{a}\times\dfrac{\vec{v}}{c}\right)^2}{(1-\beta^2)^3}\right]$$

because

$$\sum_v\left(\frac{du_v}{d\tau}\right)^2 = \frac{a^2 - \left(\vec{a}\times\dfrac{\vec{v}}{c}\right)}{(1-\beta^2)^3}$$

Let us verify that $\sum_\mu F_\mu^{\text{rad}}p_\mu = 0$

We know that

$$\sum_\mu p_\mu^2 = -m^2c^2$$

Taking the first derivative

$$2\sum_\mu p_\mu\frac{dp_\mu}{d\tau} = 0$$

Taking the second derivative

$$\sum_\mu p_\mu\frac{d^2p_\mu}{d\tau^2} + \sum_\mu\left(\frac{dp_\mu}{d\tau}\right)^2 = 0$$

Now

$$F_\mu^{\text{rad}} \sim \frac{d^2p_\mu}{d\tau^2} - \frac{p_\mu}{m^2\tau^2}\sum_v\left(\frac{dp_v}{d\tau}\right)^2$$

$$\sum_\mu F_\mu^{\text{rad}}p_\mu \propto \sum_\mu p_\mu\frac{d^2p_\mu}{d\tau^2} - \frac{1}{m^2\tau^2}\left(\sum_\mu p_\mu^2\right)\sum_v\left(\frac{dp_v}{d\tau}\right)^2$$

$$= \sum_\mu P_\mu\frac{d^2P_\mu}{d\tau^2} + \sum_\mu\left(\frac{dP_\mu}{d\tau}\right)^2 = 0$$

because $\sum_\mu P_\mu^2 = -m^2c^2$

Let us verify now that F_μ^{rad} goes to the proper limits when $\vec{v} \to 0$. We should have:

$$F_\mu^{\text{rad}} \xrightarrow[\vec{v} \to 0]{} \left(\frac{2e^2}{3c^3} \dot{\vec{a}}, 0 \right)$$

We have in general

$$F_\mu^{\text{rad}} = \frac{2e^2}{3c^3} \left[\frac{d^2 u_\mu}{d\tau^2} - \frac{u_\mu}{c^2} \frac{a^2 - \left(\vec{a} \times \dfrac{\vec{v}}{c} \right)^2}{(1 - \beta^2)^3} \right]$$

We have to show that

$$\frac{d^2 u_k}{d\tau^2} \xrightarrow[v \to 0]{} \dot{a}_k \quad (k = 1, 2, 3)$$

and

$$\frac{d^2 u_4}{d\tau^2} - \frac{u_4}{c^2} \frac{a^2 - \left(\vec{a} \times \dfrac{\vec{v}}{c} \right)^2}{(1 - \beta^2)^3} \xrightarrow[v \to 0]{} 0$$

We can write

$$u_1 = \frac{v_1}{\sqrt{1 - \beta^2}}$$

$$\frac{du_1}{d\tau} = \left(\frac{d}{dt} \frac{v_1}{\sqrt{1 - \beta^2}} \right) \frac{dt}{d\tau}$$

$$= \frac{\dot{v}_1}{\sqrt{1 - \beta^2}} \frac{1}{\sqrt{1 - \beta^2}} + \frac{v_1}{\sqrt{1 - \beta^2}} \frac{1}{\sqrt{1 - \beta^2}} \frac{2 \dfrac{\vec{v} \cdot \dot{\vec{v}}}{c^2}}{\sqrt{1 - \beta^2}}$$

$$= \frac{a_1}{1 - \beta^2} + \frac{v_1 \dfrac{\vec{v} \cdot \vec{a}}{c^2}}{(1 - \beta^2)^2} = \frac{a_1}{1 - \beta^2} + \frac{v_1 (\vec{v}_1 \cdot \vec{a})}{c^2 (1 - \beta^2)^2}$$

$$\frac{d^2 u_1}{d\tau^2} = \frac{d}{d\tau}\left(\frac{du_1}{d\tau}\right) = \frac{1}{\sqrt{1-\beta^2}}\frac{d}{d\tau}\left[\frac{a_1}{1-\beta^2} + \frac{v_1(\vec{v}\cdot\vec{a})}{c^2(1-\beta^2)^2}\right]$$

$$= \frac{1}{\sqrt{1-\beta^2}}\left\{\frac{\dot{a}_1}{1-\beta^2} + \frac{a_1 2\frac{\vec{v}\cdot\dot{\vec{v}}}{c^2}}{(1-\beta^2)^2}\right.$$

$$+ \frac{\dot{v}_1(\vec{v}\cdot\vec{a}) + v_1\frac{\partial}{\partial t}(\vec{v}\cdot\vec{a})}{c^2(1-\beta^2)^2}$$

$$\left. + \frac{v_1(\vec{v}\cdot\vec{a})2(1-\beta^2)^2\frac{\vec{v}\cdot\vec{v}}{c_2}}{c^4(1-\beta^2)^4}\right\}$$

$$\xrightarrow[v\to 0]{} \dot{a}_1$$

Also

$$u_4 = \frac{ic}{\sqrt{1-\beta^2}}$$

$$\frac{du_4}{d\tau} = \frac{1}{\sqrt{1-\beta^2}}\frac{du_4}{d\tau} = \frac{1}{\sqrt{1-\beta^2}}\frac{ic\frac{\vec{v}\cdot\dot{\vec{v}}}{c^2}}{\sqrt{1-\beta^2}}\frac{1}{\sqrt{1-\beta^2}}$$

$$= ic\frac{\vec{v}\cdot\vec{a}}{c^2(1-\beta^2)^2}$$

$$\frac{d^2 u_4}{d\tau^2} = \frac{i}{c\sqrt{1-\beta^2}}\left[\frac{a^2 + \vec{v}\cdot\dot{\vec{a}}}{(1-\beta^2)^2} + \vec{v}\cdot\vec{a}\frac{2(1-\beta^2)^2\frac{\vec{v}\cdot\dot{\vec{v}}}{c^2}}{(1-\beta^2)^4}\right]$$

$$\xrightarrow[v\to 0]{} \frac{ia^2}{c}$$

Therefore

$$\frac{d^2 u_4}{d\tau^2} - \frac{u_4}{c^2}\frac{a^2 - \left(\vec{a}\times\frac{\vec{v}}{c}\right)^2}{(1-\beta^2)^3} \xrightarrow[v\to 0]{} \frac{ia^2}{c} - \frac{ic}{c^2}a^2 = 0$$

9.3.

(a)

$$K_c = K_r + ik_i$$

$$n = n_r + n_i = \sqrt{K_r + ik_i}$$

Then

$$n^2 = n_r^2 - n_i^2 + 2n_i n_r = K_r + iK_i$$

and

$$\begin{cases} n_r^2 - n_i^2 = K_r & (1) \\ 2n_i n_r = K_i & (2) \end{cases}$$

(b) The intensity of the wave is proportional to

$$\left(e^{-i\eta x}\right)^2 = e^{-i2\eta x} = e^{-i2\frac{\omega}{c}(n_r + in_i)x}$$

$$= e^{-i2\frac{\omega}{c}n_r x} e^{2\frac{\omega}{c}n_i x}$$

and

$$\mu(\omega) = -2\frac{\omega}{c}n_i = -\frac{\omega K_i}{n_r}$$

K_i is intrinsically negative. K_r must be positive for propagation. Therefore n_i is intrinsically negative.

(c) Let us consider the expression for K_r in the limit of $A \to 0$ (no damping)

$$K_r = 1 + \frac{4\pi n_0 e^2/m}{\omega_0^2 - \omega} = 1 - \frac{4\pi n_0 e^2/m}{\omega^2 - \omega_0^2}$$

The condition for plane wave propagation is $K_r > 0$
Take $\omega > \omega_0$, then $K_r > 0$ if

$$\omega^2 - \omega_0^2 - 4\pi \frac{n_0 e^2}{m} \geq 0$$

$$\omega^2 > \omega_0^2 + \frac{4\pi n_0 e^2}{m} = \omega_0^2 \left(1 + \frac{4\pi n_0 e^2}{m\omega_0^2} \right)$$

$$\omega > \omega_0 \sqrt{1 + \frac{4\pi n_0 e^2}{m\omega^2}}$$

Take $\omega < \omega_0$, then $K_r > 0$ if

$$\omega_0^2 - \omega^2 + \frac{4\pi n_0 e^2}{m} > 0$$

$$\omega^2 < \omega_0^2 + \frac{4\pi n_0 e^2}{m}$$

which means that if $\omega < \omega_0, K_r > 0$.

9.5.

The appearance of the factor $4/3$ shows that the dynamical behavior of the electron cannot be explained by taking into account only its field.

In the dynamics of the electron other forces act besides those due to the electromagnetic field. This comes out of any attempt of building a model of the electron.

Since charges of the same sign repel each other, such a model can be stable only if other forces are present which counter balance the electrostatic repulsive forces.

Therefore we must have

$$B.E. + \frac{4}{3}u_{el} = U_{el}$$

where $B.E. = $ binding energy $= -\frac{1}{3}u_{el}$.

9.7.

(a) The homogeneous equation of motion is

$$\ddot{x} + \gamma \dot{x} + \omega_0^2 x = 0$$

We try the solution

$$x = e^{-\frac{\gamma}{2}t + i\omega_0' t}$$

Now

$$\dot{x} = \left(-\frac{\gamma}{2} + i\omega_0'\right) x$$

$$\ddot{x} = \left(-\frac{\gamma}{2} + i\omega_0'\right) \dot{x} = \left(-\frac{\gamma}{2} + i\omega_0'\right)\left(-\frac{\gamma}{2} + i\omega_0'\right) x$$

$$= \left(\frac{\gamma^2}{2} - i\omega_0'\gamma - \omega_0'^2\right) x$$

$$\gamma \dot{x} = \gamma \left(-\frac{\gamma}{2} + i\omega_0'\right) x = \left(-\frac{\gamma^2}{2} + i\omega_0'\gamma\right) x$$

$$\ddot{x} + \gamma \dot{x} + \omega_0^2 x$$

$$= \left[\left(\frac{\gamma^2}{4} - i\omega_0'\gamma - \omega_0'^2\right) + \left(-\frac{\gamma^2}{2} + i\omega_0'\gamma\right) + \omega_0'^2\right] x$$

$$= \left(-\frac{\gamma^2}{4} - \omega_0'^2 + \omega_0^2\right) x = 0$$

if

$$\omega_0^2 = \omega_0'^2 + \frac{\gamma^2}{4}$$

Now, suppose we had used an alternate trial solution:

$$x = e^{-\frac{\gamma}{2}t - i\omega_0't}$$

$$\dot{x} = \left(-\frac{\gamma}{2} - i\omega_0'\right) x$$

$$\ddot{x} = \left(-\frac{\gamma}{2} - i\omega_0'\right)^2 x = \left(\frac{\gamma^2}{2} + i\omega_0'\gamma - \omega_0'^2\right) x$$

$$\gamma \dot{x} = \left(-\frac{\gamma^2}{2} - i\omega_0'\gamma\right) x$$

and

$$\ddot{x} + \gamma \dot{x} + \omega_0^2 x$$

$$= \left(\frac{\gamma^2}{4} + i\omega_0'\gamma - \omega_0'^2 - \frac{\gamma^2}{2} - i\omega_0'\gamma + \omega_0^2\right) x$$

$$= \left(-\frac{\gamma^2}{4} - \omega_0'^2 + \omega_0^2\right) x = 0$$

if

$$\omega_0^2 = \omega_0'^2 + \frac{\gamma^2}{4}$$

The general solution of the homogeneous equation is ($\omega_0 \gg \gamma$)

$$x = A_+ e^{-\frac{\gamma}{2}t + i\omega_0 t} + A_- e^{-\frac{\gamma}{2}t - i\omega_0 t}$$

where A_+ and A_- are determined by the initial conditions.

The general solution of the inhomogeneous differential equation is given by

$$x = \frac{\dfrac{eE_0}{m}e^{i\omega t}}{\omega_0^2 - \omega^2 + i\gamma\omega} + A_+ e^{-\frac{\gamma}{2} + i\omega_0 t}$$
$$+ A_- e^{-\frac{\gamma}{2}t - i\omega_0 t}$$

Now, in a steady state

$$x = \frac{\dfrac{eE_0}{m}e^{i\omega t}}{\omega_0^2 - \omega^2 + i\gamma\omega}$$

$$d = \frac{e^2 \dfrac{E_0}{m}e^{i\omega t}}{\omega_0^2 - \omega^2 + i\gamma\omega}$$

and

$$d_0 = \frac{e^2 \dfrac{E_0}{m}}{\omega_0^2 - \omega^2 + i\gamma\omega}$$

$$d_0 d_0^* = \frac{e^4 \dfrac{E_0^2}{m^2}}{(\omega_0^2 - \omega^2)^2 + (\gamma\omega)^2}$$

Now

$$S = \frac{2}{3c^3}(\ddot{d})^2$$

$$\bar{S} = \frac{\omega^4}{3c^3}d_0 d_0^* = \frac{\omega^4}{3c^3}\frac{e^4 \dfrac{E_0^2}{m^2}}{(\omega_0^2 - \omega^2)^2 + (\gamma\omega)^2}$$

and

$$\sigma = \frac{\bar{S}}{I} = \frac{\bar{S}}{\dfrac{c}{8\pi}E_0^2} = \frac{8\pi}{3}\frac{e^4}{m^2 c^4}\frac{\omega^4}{(\omega_0^2 - \omega^2)^2 + (\gamma\omega)^2}$$

$$= \frac{8\pi}{3}r_0^2 \frac{\omega^4}{(\omega_0^2 - \omega^2)^2 + (\gamma\omega)^2}$$

At $\omega = \omega_0$

$$\sigma = \frac{8\pi}{3}r_0^2 \frac{\omega_0^2}{\gamma^2}$$

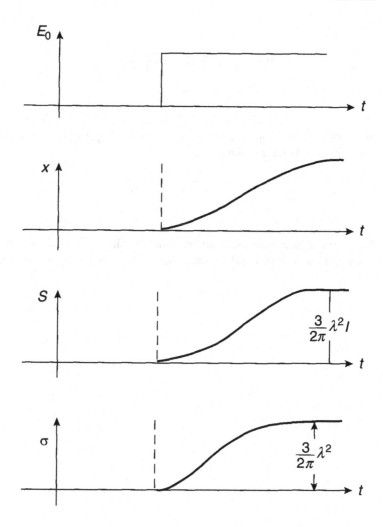

But

$$\gamma = \frac{2}{3}\frac{r_0}{c}\omega_0^2$$

Then

$$\frac{\omega_0}{\gamma} = \frac{3}{2}\frac{c}{r_0\omega_0}, \quad \frac{\omega_0^2}{\gamma^2} = \frac{9}{4}\frac{c^2}{r_0^2\omega_0^2}$$

and

$$\sigma = \frac{8\pi}{3}r_0^2\frac{\omega_0^2}{\gamma_2} = \frac{8\pi}{3}r_0^2\frac{9}{4}\frac{c^2}{r_0^2\omega_0^2}$$

$$= \frac{2\pi 3 c^2}{4\pi^2 v_0^2} = \frac{3}{2\pi}\lambda^2$$

(b) The answer to the apparent paradox is provided by the transient term. The transient behavior lasts a time

$$\gamma^{-1} = \frac{1}{\dfrac{2}{3}\dfrac{r_0}{c}\omega_0^2} = \frac{3c}{2\omega_0^2 r_0}$$

where $r_0 = e^2/mc^2$

If $e \to 0$ the transient lasts an ∞ time. At time $t = 0$, at which the field is turned on, $x = 0$; it takes some time for x to build up to its steady state value.

10

Exercises

10.1.

(a)

$$z(t') = z_0 \cos \omega t'$$

$$r(t') = r_0 - z(t') \cos \theta$$

$$t' = t - \frac{r(t')}{c} = t - \left[\frac{r_0}{c} - \frac{z(t') \cos \theta}{c} \right]$$

$$= t - \frac{r_0}{c} + \frac{z(t')}{c} \cos \theta$$

$$\cos \omega t' = \cos \omega \left[t - \frac{r(t')}{c} \right]$$

$$= \cos \left[\omega \left(t - \frac{r_0}{c} \right) + \frac{\omega z(t')}{c} \cos \theta \right]$$

$$\simeq \cos \omega \left(t - \frac{r_0}{c} \right) - \frac{\omega z(t')}{c} \cos \theta \sin \omega \left(t - \frac{r_0}{c} \right)$$

671

$$= \cos\omega\left(t - \frac{r_0}{c}\right) - \frac{\omega}{c}\cos\theta$$

$$\times \left[z_0\cos\omega\left(t - \frac{r_0}{c}\right)\right]\sin\omega\left(t - \frac{r_0}{c}\right)$$

$$= \cos\omega\left(t - \frac{r_0}{c}\right) - \frac{\omega}{c}\cos\theta\frac{z_0}{2}\sin 2\omega\left(t - \frac{r_0}{c}\right)$$

The frequency of the correction term is 2ω

Since $|\vec{B}| = |\vec{E}| = \frac{e}{c^2 r}a\sin\theta$

We have for the correction

$$|\vec{B}| = |\vec{E}| \sim \sin\theta\cos\theta \sim \sin 2\theta$$

(b)

$$a_1(t) = \ddot{z}_1(t) = -\omega^2 z_0\cos\omega t$$

$$a_2(t) = \ddot{z}_2(t) = \omega^2 z_0\cos\omega t$$

$$B_1^{\mathrm{rad}} = \left[\frac{e}{c^2 r}|\vec{a}_1 \times \hat{n}|\right]_{t'=t-\frac{r_0}{c}}$$

$$= -\frac{e}{c^2 r}\omega^2 z_0\sin\theta\cos\omega\left(t - \frac{r_0}{c}\right)$$

$$B_2^{\mathrm{rad}} = \left[\frac{e}{c^2 r}|\vec{a}_2 \times \hat{n}|\right]_{t'=t-\frac{r_0}{c}}$$

$$= \frac{e}{c^2 r}\omega^2 z_0\sin\theta\cos\omega\left(t - \frac{r_0}{c}\right)$$

Therefore the pure dipole radiation is equal to the sum of B_1^{rad} and B_2^{rad} and is thus zero.

$$z_1(t') = z_0 \cos \omega t'$$
$$z_2(t') = -z_0 \cos \omega t'$$

For the upper electron

$$r_1(t') = r_0 - z_1(t') \cos \theta$$

and

$$t' = t - \frac{r_1(t')}{c} = t - \frac{r_0}{c} + \frac{z_1(t')}{c} \cos \theta$$

$$\cos \omega t' = \cos \omega \left[t - \frac{r_0}{c} + \frac{z_1(t')}{c} \cos \theta \right]$$

$$= \cos \left[\omega \left(t - \frac{r_0}{c} \right) + \frac{\omega z_1(t')}{c} \cos \theta \right]$$

$$\simeq \cos \omega \left(t - \frac{r_0}{c} \right) - \frac{\omega z_1(t')}{c} \cos \theta \sin \omega \left(t - \frac{r_0}{c} \right)$$

$$\simeq \cos \omega \left(t - \frac{r_0}{c} \right) - \frac{\omega}{c} \cos \theta z_0 \cos \omega \left(t - \frac{r_0}{c} \right)$$

$$\times \sin \omega \left(t - \frac{r_0}{c} \right)$$

$$= \cos \omega \left(t - \frac{r_0}{c} \right) - \frac{\omega}{c} \cos \theta \frac{z_0}{2} \sin 2\omega \left(t - \frac{r_0}{c} \right)$$

For the lower electron

$$r_2(t') = r_0 - z_2(t') \cos \theta$$

$$t' = t - \frac{r_2(t')}{c} = t - \frac{r_0}{c} + \frac{z_2(t')}{c} \cos \theta$$

$$\cos \omega t' = \cos \omega \left[t - \frac{r_2(t')}{c} \right] = \cos \omega \left[t - \frac{r_0}{c} - \frac{z_2(t')}{c} \cos \theta \right]$$

$$= \cos \left[\omega \left(t - \frac{r_0}{c} \right) - \frac{\omega z_2(t')}{c} \cos \theta \right]$$

$$\simeq \cos \omega \left(t - \frac{r_0}{c} \right) + \frac{\omega}{c} z_2(t') \cos \theta \sin \omega \left(t - \frac{r_0}{c} \right)$$

$$\simeq \cos \omega \left(t - \frac{r_0}{c} \right) + \frac{\omega}{c} \cos \theta z_0 \cos \omega \left(t - \frac{r_0}{c} \right)$$

$$\times \sin \omega \left(t - \frac{r_0}{c} \right)$$

$$= \cos \omega \left(t - \frac{r_0}{c} \right) + \frac{\omega}{c} \cos \theta \frac{z_0}{2} \sin 2\omega \left(t - \frac{r_0}{c} \right)$$

Therefore dipole radiation is eliminated and only quadrupole radiation is present.

10.3.

Set

$$A = \sqrt{l(l+1) - m(m+1)}$$

$$B = \sqrt{l(l+1) - m(m-1)}$$

Now

$$L_+ = L_x + iL_y \rightarrow L_+ Y_l^m = A Y_l^{m+1}$$

$$L_- = L_x - iL_y \rightarrow L_- Y_l^m = B Y_l^{m-1}$$

Then

$$L_x = \frac{1}{2}(L_+ + L_-)$$

$$L_y = \frac{1}{2i}(L_+ - L_-)$$

Therefore

$$L_x Y_l^m = \frac{1}{2} \{ L_+ Y_l^m + L_- Y_l^m \}$$

$$= \frac{1}{2} [A Y_l^{m+1} + B Y_l^{m-1*}]$$

$$(L_x Y_l^m)^* = \frac{1}{2} [A Y_l^{m+1*} + B Y_l^{m-1*}]$$

$$(L_x Y_l^m)(L_x Y_l^m)^* = \frac{1}{4}[A^2 |Y_l^{m+1}|^2 + B^2 |Y_l^{m-1}|^2$$

$$+ ABY_l^{m+1}Y_l^{m-1*}$$

$$+ ABY_l^{m-1}Y_l^{m+1*}]$$

$$L_y Y_l^m = \frac{1}{2i}\{L_+ Y_l^m - L_- Y_l^m\}$$

$$= \frac{1}{2i}[AY_l^{m+1} - BY_l^{m-1}]$$

$$(L_y Y_l^m)^* = -\frac{1}{2i}[AY_l^{m+1*} - BY_l^{m-1*}]$$

$$(L_y Y_l^m)(L_y Y_l^m)^* = \frac{1}{4}[A^2 |Y_l^{m+1}|^2 + B^2 |Y_l^{m-1}|^2$$

$$- ABY_l^{m+1}Y_l^{m-1*} - ABY_l^{m-1}Y_l^{m+1*}]$$

Therefore

$$(L_x Y_l^m)(L_x Y_l^m)^* + (L_y Y_l^m)(L_y Y_l^m)^*$$

$$= \frac{1}{2}A^2 Y_l^{m+1}Y_l^{m+1*} + \frac{1}{2}B^2 Y_l^{m-1}Y_l^{m-1*}$$

$$= \frac{1}{2}[l(l+1) - m(m+1)]Y_l^{m+1}Y_l^{m+1*}$$

$$+ \frac{1}{2}[l(l+1) - m(m-1)]Y_l^{m-1}Y_l^{m-1*}$$

$$\sum_m [(L_x Y_l^m)(L_x Y_l^m)^* + (L_y Y_l^m)(L_y Y_l^m)^* + (L_z Y_l^m)(L_z Y_l^m)^*]$$

$$= \frac{1}{2}\sum_m \{[l(l+1) - m(m+1)]Y_l^{m+1}Y_l^{m+1*}$$

$$+ [l(l+1) - (m+1)m]Y_l^m Y_l^{m*}$$

$$+ 2m^2 Y_l^m Y_l^{m*}\}$$

$$= \sum_m \left[l(l+1) - \frac{1}{2}(m^2 - m) - \frac{1}{2}(m^2 + m) + m^2 \right] Y_l^m Y_l^{m*}$$

$$= \sum_m [l(l+1) - m^2 + m^2] Y_l^m Y_l^{m*}$$

$$= l(l+1) \sum_m Y_l^m Y_l^{m*} = \frac{2l+1}{4\pi} l(l+1)$$

$$= \text{constant.}$$

10.5.

(a) The current is given by

$$J(z,t) = J_0 \cos \frac{\pi z}{L} \cos \omega t$$

The continuity equation gives us:

$$\frac{\partial \rho(\vec{x},t)}{\partial t} + \vec{\nabla} \cdot \vec{j}\,(\vec{x},t) = 0$$

$$S \frac{\partial \rho(\vec{x},t)}{\partial t} \vec{\nabla} \cdot \vec{J}\,(z,t) = 0 \quad (\vec{J} = \vec{j}S)$$

$$S \frac{\partial \rho(\vec{x},t)}{\partial t} - J_0 \frac{\pi}{L} \sin \frac{\pi z}{L} \cos \omega t = 0$$

$$S\rho(\vec{x},t) = \text{charge per unit length}$$

$$= Q(z,t) = \int_0^t J_0 \frac{\pi}{L} \sin \frac{\pi z}{L} \cos \omega t \, dt$$

$$= \frac{\pi J_0}{\omega L} \sin \frac{\pi z}{L} \sin \omega t$$

(b) We can write

$$\vec{q}(\vec{k}) = \frac{1}{c} \int dz \, \vec{J}_0 \cos \frac{\pi z}{L} e^{-ikz \cos \theta}$$

$$= \frac{\vec{J}_0}{c} \int_{\frac{-L}{2}}^{\frac{L}{2}} \cos \frac{\pi z}{L} e^{-ikz \cos \theta} dz$$

$$= \frac{\vec{J}_0}{c} \frac{2}{\frac{\pi}{L} - \frac{L}{\pi} k^2 \cos^2 \theta} \cos\left(k \cos \theta \frac{L}{2}\right)$$

for $\frac{kL}{\pi} = 1$ we get

$$q_z(\vec{k}) = \frac{J_0}{c} \frac{2}{\dfrac{\pi}{L} - \dfrac{L}{\pi}\dfrac{\pi^2}{L^2}\cos^2\theta} \cos\left(\frac{\pi}{L}\cos\theta\frac{L}{2}\right)$$

$$= \frac{J_0}{c} \frac{2}{\dfrac{\pi}{L}(1 - \cos^2\theta)} \cos\left(\frac{\pi}{2}\cos\theta\right)$$

$$= \frac{2J_0 L}{\pi c} \frac{\cos\left(\dfrac{\pi}{2}\cos\theta\right)}{\sin^2\theta}$$

Calculation of the integral of $q(\vec{k})$

$$\int e^{ax}\cos px\,dx = \frac{e^{ax}(a\cos px + p\sin px)}{a^2 + p^2}$$

We set

$$a = -ik\cos\theta$$
$$p = \pi/L$$

and we obtain:

$$\int_{-L/2}^{L/2} e^{-ikz\cos\theta}\cos\left(\frac{\pi}{L}z\right)dz$$

$$= \left[\frac{e^{-ikz\cos\theta}\left(-ik\cos\theta\cos\dfrac{\pi}{L} + \dfrac{\pi}{L}\sin\dfrac{\pi}{L}z\right)}{-k^2\cos^2\theta + \dfrac{\pi^2}{L^2}}\right]_{-L/2}^{L/2}$$

$$= \frac{1}{\dfrac{\pi^2}{L^2} - k^2\cos^2\theta}$$

$$\times \left[e^{-ik\cos\theta\frac{L}{2}}\left(-ik\cos\theta\cos\frac{\pi}{L}\frac{L}{2} + \frac{\pi}{L}\sin\frac{\pi}{L}\frac{L}{2}\right)\right.$$

$$\left. - e^{-ik\cos\theta\frac{L}{2}}\left(-ik\cos\theta\cos\frac{\pi}{2} - \frac{\pi}{2}\sin\frac{\pi}{2}\right)\right]$$

$$= \frac{1}{\dfrac{\pi^2}{L^2} - k^2 \cos^2 \theta} \left[e^{-ik \cos \theta \frac{L}{2} \frac{\pi}{L}} + e^{ik \cos \theta \frac{L}{2} \frac{\pi}{L}} \right]$$

$$= \frac{2\dfrac{\pi}{L}}{\dfrac{\pi^2}{L^2} - k^2 \cos^2 \theta} \cos \left(k \cos \theta \frac{L}{2} \right)$$

$$= \frac{2}{\dfrac{\pi}{L} - \dfrac{L}{\pi} k^2 \cos^2 \theta} \cos \left(k \cos \theta \frac{L}{2} \right)$$

(c) Let us calculate $q_z(\vec{k})$ in the electric dipole approximation

$$q_z(\vec{k}) = \frac{J_0}{c} \int_{-L/2}^{L/2} dz \cos \frac{\pi z}{L}$$

$$= \frac{J_0}{c} \left[\frac{\sin \dfrac{\pi z}{L}}{\pi/L} \right]_{-L/2}^{L/2}$$

$$= \frac{J_0}{c} \frac{L}{\pi} \left[\sin \frac{\pi}{L} \frac{L}{2} + \sin \frac{\pi}{L} \frac{L}{2} \right]$$

$$= \frac{2J_0 L}{\pi c}$$

10.7.

The perturbation due to A can affect the field at large distances from A as follows

$$\vec{E} = \vec{E_0} - \vec{\nabla} V$$

where

$\vec{E_0}$ = constant electric field in the absence of the perturbation

V = perturbing potential

The potential can be expanded in a series of spherical harmonics

$$V = \sum_{l,m} A_{lm} \frac{Y_{lm}(\theta, \varphi)}{r^{l+1}}$$

where the origin is taken inside the region A. The first term in this expansion is

$$\frac{A_{00}Y_{00}}{r} = \frac{A_{00}}{\sqrt{4\pi}}\frac{1}{r}$$

Let us now assume that V consists solely of this term. We encircle the region A with a spherical surface and perform the following integration

$$\int_S \vec{E} \cdot d\vec{S} = \int_S (\vec{E_0} - \vec{\nabla}V) \cdot d\vec{S} = -\int \vec{\nabla}V \cdot d\vec{S}$$

$$= \frac{A_0}{\sqrt{4\pi}} \int_0^\pi \frac{1}{r^2} 2\pi r^2 \sin\theta d\theta$$

$$= \frac{A_0}{\sqrt{4\pi}} 4\pi = \text{constant}$$

If this term is different from zero, then charges are leaving the region A at the rate:

$$\int \vec{j} \cdot d\vec{S} = \sigma \int \vec{E} \cdot d\vec{S}$$

contrary to the assumption of steady state conditions. Therefore $A_{00} = 0$ and the next term is

$$V \sim \frac{1}{r^2} \sum_m A_{1m} Y_1^m(\theta, \varphi)$$

Then

$$|\vec{E} - \vec{E_0}| \propto \frac{1}{r^3}$$

11

Exercises

11.1.

(a)

$$L = \frac{1}{2}mv^2 - q\phi + \frac{q}{c}\vec{v}\cdot\vec{A}$$

$$= \sum_{i=1}^{3}\frac{1}{2}m\dot{x}_i^2 - q\phi + \frac{q}{c}\sum_{i=1}^{3}\dot{x}_i A_i$$

Consider the x-component of the particles position vector

$$\frac{d}{dt}\left(\frac{\partial L}{\partial \dot{x}}\right) = m\ddot{x} + \frac{q}{c}\frac{dA_x}{dt}$$

$$\frac{\partial L}{\partial x} = -q\frac{\partial \phi}{\partial x} + \frac{\partial}{\partial x}\left(\frac{q}{c}\vec{v}\cdot\vec{A}\right)$$

The equation of motion in the x direction is given by

$$m\ddot{x} + \frac{q}{c}\frac{dA_x}{dt} = -q\frac{\partial \phi}{\partial x} + \frac{\partial}{\partial x}\left(\frac{q}{c}\vec{v}\cdot\vec{A}\right)$$

or

$$m\ddot{x} = q\left[-\frac{\partial \phi}{\partial x} + \frac{1}{c}\frac{\partial}{\partial x}(\vec{v}\cdot\vec{A}) - \frac{1}{c}\frac{dA_x}{dt}\right]$$

Then

$$m\ddot{\vec{r}} = q\left[-\vec{\nabla}\phi + \frac{1}{c}\vec{\nabla}(\vec{v}\cdot\vec{A}) - \frac{1}{c}\frac{d\vec{A}}{dt}\right]$$

But

$$\vec{v}\times(\vec{\nabla}\times\vec{A}) = \vec{\nabla}(\vec{v}\cdot\vec{A}) - (\vec{v}\cdot\vec{\nabla})\vec{A}$$

and

$$\frac{d\vec{A}}{dt} = \frac{\partial\vec{A}}{dt} + (\vec{v}\cdot\vec{\nabla})\vec{A}$$

Therefore

$$m\ddot{\vec{r}} = q\left\{-\vec{\nabla}\phi + \frac{1}{c}[\vec{v}\times(\vec{\nabla}\times\vec{A}) + (\vec{v}\cdot\vec{\nabla})\vec{A}]\right.$$

$$\left. -\frac{1}{c}\frac{\partial\vec{A}}{\partial t} - \frac{1}{c}(\vec{v}\cdot\vec{\nabla})\vec{A}\right\}$$

$$= q\left\{-\vec{\nabla}\phi - \frac{1}{c}\frac{\partial\vec{A}}{\partial t} + \frac{1}{c}\vec{v}\times(\vec{\nabla}\times\vec{A})\right\}$$

$$= q\left\{\vec{E} + \frac{1}{c}(\vec{v}\times\vec{B})\right\}$$

where we have used the relations

$$\vec{E} = -\vec{\nabla}\phi - \frac{1}{c}\frac{\partial\vec{A}}{\partial t}$$

$$\vec{B} = \vec{\nabla}\times\vec{A}$$

Since the expression for $m\ddot{\vec{r}}$ represents the Lorentz force, the Lagrangian we started with is correct.

(b) Having found the Lagrangian, it is possible to find the Hamiltonian by first deriving the generalized momenta

$$p_i = \frac{\partial L}{\partial \dot{x}_i} = m\dot{x}_i + \frac{q}{c}A_i$$

The Hamiltonian is then given by

$$H = \sum_{i=1}^{3} p_i \dot{x}_i - L$$

$$= \sum_{i=1}^{3} \left(m\dot{x}_i + \frac{q}{c}A_i \right) \dot{x}_i - \left(\frac{1}{2}mv^2 - q\phi + \frac{q}{c}\vec{v} \cdot \vec{A} \right)$$

$$= \sum_{i=1}^{3} m\dot{x}_i^2 + \frac{q}{c}\sum_{i=1}^{3} A_i \dot{x}_i - \frac{1}{2}mv^2 + q\phi - \frac{q}{c}\vec{v} \cdot \vec{A}$$

$$= \sum_{i=1}^{3} \frac{1}{2}m\dot{x}_i^2 + q\phi = \sum_{i=1}^{3} \frac{(m\dot{x}_i)^2}{2m} + q\phi$$

But

$$m\dot{x}_i = p_i - \frac{q}{c}A_i$$

Then we can write

$$H = \frac{(\vec{p} - \frac{q}{c}\vec{A})^2}{2m} + q\phi$$

12

Exercises

12.1.

We know that

$$E^2 = p^2 c^2 + m_0^2 c^4$$
$$E = h v, \quad p = h/\lambda$$
$$v_{\text{phase}} = v \lambda$$

Then

$$h^2 v^2 = \frac{h^2}{\lambda^2} c^2 + m_0^2 c^4$$

$$\frac{h^2}{\lambda^2} c^2 = h^2 v^2 - m_0^2 c^4$$

$$\frac{1}{\lambda^2} = \frac{v^2}{c^2} - \frac{m_0^2 c^4}{h^2 c^2} = \frac{v^2}{c^2} - \frac{m_0^2 c^2}{h^2}$$

$$= \frac{v^2}{c^2} \left(1 - \frac{m_0^2 c^4}{h^2 v^2} \right)$$

$$\frac{1}{\lambda} = \frac{v}{c} \left(1 - \frac{m_0^2 c^4}{h^2 v^2} \right)^{1/2}$$

Therefore

$$v_{\text{phase}} = v\lambda = \frac{c}{\left(1 - \dfrac{m_0^2 c^4}{h^2 v^2} \right)^{1/2}} \simeq c \left(1 + \frac{m_0^2 c^4}{h^2 v^2} \right)$$

12.3.

The Kramers Kronig dispersion relation states that the static ($\omega = 0$) dielectric constant is related to $\sigma(\omega)$ by

$$K - 1 = \frac{2}{\pi} n_0 c \mathcal{P} \int_0^\infty \frac{d\omega}{\omega^2} \sigma(\omega) \tag{1}$$

Let us check this relation for a damped oscillator model of scattering. For such a model we have

$$K(\omega) = 1 + 4\pi n_0 \alpha(\omega)$$

$$= 1 + 4\pi n_0 \frac{e^2/m}{\omega_0^2 - \omega^2 + i\omega\gamma} \tag{2}$$

where $\gamma = 2e^2 \omega^2 / 3mc^3$
and

$$K(0) - 1 = \frac{4\pi n_0 e^2/m}{\omega_0^2} \tag{3}$$

But the total scattering cross section is given by

$$\sigma_{tot}(\omega) = \frac{8\pi}{3} r_e^2 f(\omega) \tag{4}$$

where

$$\frac{f(\omega)}{\omega^2} = \frac{\omega^2}{(\omega_0^2 - \omega^2) + \left(\dfrac{2e^2}{3mc^3} \omega^3 \right)^2}$$

$$\simeq \frac{\omega_0^2}{4\omega_0^2 (\omega - \omega_0)^2 + \left(\dfrac{2e^2}{3mc^3} \omega_0^3 \right)^2}$$

$$\simeq \frac{1}{4(\omega - \omega_0)^2 + \left(\dfrac{2e^2\omega_0^2}{3mc^3}\right)^2}$$

$$= \frac{1}{4} \frac{1}{(\omega - \omega_0)^2 + \left(\dfrac{e^2\omega_0^2}{3mc^3}\right)^2} \tag{5}$$

We have also

$$\int \frac{f(\omega)}{\omega^2} d\omega = \frac{1}{4} \int_0^\infty \frac{d\omega}{(\omega - \omega_0)^2 + \left(\dfrac{e^2\omega_0^2}{3mc^3}\right)^2}$$

$$= \frac{1}{4} \frac{\pi}{e^2\omega_0^2/3mc^3} = \frac{3\pi mc^3}{4e^2\omega_0^2} \tag{6}$$

and

$$\int_0^\infty \frac{\sigma(\omega)}{\omega^2} d\omega = \int_0^\infty \frac{8\pi}{3} r_e^2 \frac{f(\omega)}{\omega^2} d\omega$$

$$= \frac{8\pi}{3} \left(\frac{e^2}{mc^2}\right)^2 \int_0^\infty \frac{f(\omega)}{\omega^2} d\omega$$

$$= \frac{8\pi}{3} \frac{e^4}{m^2c^4} \frac{3\pi mc^3}{4e^2\omega_0^2} = \frac{2\pi^2 e^2}{m\omega_0^2 c} \tag{7}$$

Therefore

$$\frac{2}{\pi} n_0 c \mathcal{P} \int_0^\infty \frac{\sigma(\omega)}{\omega^2} d\omega = \frac{2}{\pi} n_0 c \int_0^\infty \frac{\sigma(\omega)}{\omega^2} d\omega$$

$$= \frac{2}{\pi} n_0 c \frac{2\pi^2 e^2}{m\omega_0^2 c}$$

$$= \frac{4\pi n_0 e^2 / m}{\omega_0^2} \tag{8}$$

which is the same result obtained in Eq. (3) for a damped oscillator model of scattering.

12.5.

$$\frac{e^2}{mc^2} = r_e = 2.8 \times 10^{-13} \, \text{cm}$$

$$\frac{\omega_c^2}{c^2} = 4\pi n_0 \frac{e^2}{mc^2} = 4\pi n_0 \times 2.8 \times 10^{-13}$$

$$= 3.52 \times 10^{-12} n_0$$

$$\lambda_c = \frac{c}{v_L} = \frac{2\pi c}{\omega_c} = 2\pi \sqrt{c^2/\omega_c^2}$$

$$= \frac{2\pi}{\sqrt{3.52 \times 10^{-12}}} \frac{1}{\sqrt{n_0}} = \frac{3.34 \times 10^6}{\sqrt{n_0}} \, \text{cm}$$

$$= \frac{3.34 \times 10^{14}}{\sqrt{n_0}} \, \text{Å}$$

If $n_0 = 10^{22} \, \text{cm}^{-3}$

Then we have

$$\lambda_c = \frac{3.34 \times 10^{14}}{\sqrt{10^{22}}} \simeq 3300 \, \text{Å}$$

12.7.

The dielectric constant of the metal is given by

$$K(\omega) = 1 - \frac{4\pi n_0 e^2}{m\omega^2} = 1 - \frac{\omega_p^2}{\omega^2}$$

where

$$\omega_p = \sqrt{\frac{4\pi n_0 e^2}{m}}$$

On the other hand

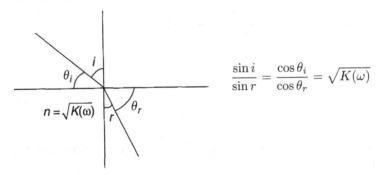

$$\frac{\sin i}{\sin r} = \frac{\cos \theta_i}{\cos \theta_r} = \sqrt{K(\omega)}$$

The optical angle for which $r = \frac{\pi}{2}$ (or $\theta_r = 0$) is given by

$$\cos\theta_{ic} = \sqrt{1 - \frac{\omega_p^2}{\omega^2}}$$

or

$$\cos^2\theta_{ic} = 1 - \frac{\omega_p^2}{\omega^2}$$

$$1 - \cos^2\theta_{ic} = \sin^2\theta_{ic} = \frac{\omega_p^2}{\omega^2}$$

$$\sin\theta_{ic} = \frac{\omega_p}{\omega}$$

For $\omega < \omega_p$ the index of refraction is purely imaginary and the radiation is totally reflected at all angles.

ELECTROSTATICS

Coulomb's law: $\mathbf{E} = -\boldsymbol{\nabla}\phi = -\boldsymbol{\nabla}\dfrac{q}{r}$

Gauss's flux theorem: $\displaystyle\int_S \mathbf{E} \cdot d\mathbf{S} = 4\pi q$

Poisson's equation: $\nabla^2 \phi = -4\pi\rho$

Laplace's equation: $\nabla^2 \phi = 0$

Capacitance: $C = \dfrac{Q}{V_a - V_b} \longrightarrow \underset{\text{plates}}{\dfrac{A}{4\pi d}} \longrightarrow \underset{\text{shells}}{\dfrac{ab}{b-a}}$

Mean value theorem: $\phi(\mathbf{x}_0) = \dfrac{1}{4\pi R^2}\displaystyle\int_S dS\,\phi(S)$

Addition theorem:

$$P_l(\cos\theta) = \frac{4\pi}{2l+1}\sum_{m=-l}^{l} Y_l^{m^*}(\zeta_0, \varphi_0) Y_l^m(\zeta, \varphi)$$

Uniqueness theorem:

$$\phi(\mathbf{x}_p) = \int_{\substack{\text{all space}\\ \text{but } V}} d\tau\,\rho(\mathbf{x})G(\mathbf{x},\mathbf{x}_p) + \frac{1}{4\pi}\int_S dS\left[G(\mathbf{x},\mathbf{x}_p)\frac{\partial\phi(S)}{\partial n} - \phi(S)\frac{\partial G(\mathbf{x},\mathbf{x}_p)}{\partial n}\right]$$

MAXWELL'S EQUATION

$\boldsymbol{\nabla}\times\mathbf{E} = 0\ \left[+\dfrac{1}{c}\dfrac{\partial\mathbf{B}}{\partial t}\right]$

$\boldsymbol{\nabla}\cdot\mathbf{E} = 4\pi\rho(\rho_{\text{true}} + \rho_{\text{pol}})$

where $\rho_{\text{pol}} = -\boldsymbol{\nabla}\cdot\mathbf{P}$

$\boldsymbol{\nabla}\cdot\mathbf{D} = 4\pi\rho_{\text{true}}$

$\boldsymbol{\nabla}\times\mathbf{D} = 4\pi\boldsymbol{\nabla}\times\mathbf{P}$

where $\mathbf{D} = \mathbf{E} + 4\pi\mathbf{P}$

Time dependent terms are in [

WAVES

EM momentum density: $\mathbf{G} = \dfrac{\mathbf{E}\times\mathbf{B}}{4\pi c} = \dfrac{\mathbf{N}}{c^2} \longrightarrow \dfrac{\mathbf{D}\times\mathbf{B}}{4\pi c} - \dfrac{\mathbf{N}}{(c/n)^2}$

Poynting vector: $\mathbf{N} = \dfrac{c}{4\pi}(\mathbf{E}\times\mathbf{B}) \longrightarrow \dfrac{c}{4\pi}(\mathbf{E}\times\mathbf{H})$

Reflection: $E_\parallel$ plane of incidence

$$\frac{E_0'}{E_0} = \frac{\tan(i-r)}{\tan(i+r)} \xrightarrow{i\to 0} \frac{n_2 - n_1}{n_2 + n_1},$$

$$\frac{E_0''}{E_0} = \frac{2\sin r\cos i}{\sin(i+r)\cos(i-r)} \xrightarrow{i\to 0} \frac{2n_1}{n_1 + n_2}$$

$E_\perp$ plane of incidence

$$\frac{E_0'}{E_0} = \frac{\sin(i-r)}{\sin(i+r)}, \qquad \frac{E_0''}{E_0} = \frac{2\cos i\sin r}{\sin(i+r)}$$

Lorenz gauge: $\boldsymbol{\nabla}\cdot\mathbf{A} + \dfrac{1}{c}\dfrac{\partial\phi}{\partial t} = 0$

$$\Box^2\phi = -4\pi\rho \longrightarrow \phi(\mathbf{x},t) = \phi_0(\mathbf{x},t) + \int\frac{\rho(\mathbf{x}',t_{\text{ret}})}{|\mathbf{x}-\mathbf{x}'|}d\tau'$$

$$\Box^2\mathbf{A} = -\frac{4\pi}{c}\mathbf{j} \longrightarrow \mathbf{A}(\mathbf{x},t) = \mathbf{A}_0(\mathbf{x},t) + \frac{1}{c}\int\frac{\mathbf{j}(\mathbf{x}',t_{\text{ret}})}{|\mathbf{x}-\mathbf{x}'|}d\tau'$$

$$\left(\Box^2 = \sum_{\mu=1}^{4}\frac{\partial^2}{\partial x_\mu^2}\right) \qquad \left(t_{\text{ret}} - \frac{|\mathbf{x}-\mathbf{x}'|}{c}\right)$$

FIELD ENERGY

in terms of

$\dfrac{1}{2}\displaystyle\iint d\tau\,d\tau'\,\dfrac{\rho(\mathbf{x})\rho(\mathbf{x}')}{\tau}$

charges

$\dfrac{1}{8\pi}\displaystyle\int d\tau\cdot\mathbf{E}\cdot\mathbf{D}$ $\dfrac{1}{2}CV_{ab}^2$

fields   capacitan

MAGNETOSTATICS

Lorentz force: $\mathbf{F} = \dfrac{q}{c}(\mathbf{v} \times \mathbf{B})$

Biot–Savart law: $\mathbf{B} = \dfrac{J}{c} \oint d\mathbf{l} \times \boldsymbol{\nabla}_x \dfrac{1}{|\mathbf{x} - \mathbf{x}_p|}$

Vector potential: $\mathbf{A} = \dfrac{1}{c} \int d\tau' \dfrac{\mathbf{j}(\mathbf{x}')}{|\mathbf{x}' - \mathbf{x}|}$

Ampere's law:

$$\int (\boldsymbol{\nabla} \times \mathbf{B}) \cdot d\mathbf{S} = \oint \mathbf{B} \cdot d\mathbf{l} = \dfrac{4\pi}{c} \int \mathbf{j} \cdot d\mathbf{S}$$

Continuity equation: $\boldsymbol{\nabla} \cdot \mathbf{j} \left[+ \dfrac{\partial \rho}{\partial t} \right] = 0$

Larmor frequency: $\omega_L = \dfrac{eB}{2m_e c}$

Force between current loops: $\mathbf{F}_{12} = \dfrac{J_1 J_2}{c^2} \oint d\mathbf{l}_2 \times \oint d\mathbf{l}_1 \times \boldsymbol{\nabla}_1 \dfrac{1}{r_{12}}$

$$(r_{12} = |\mathbf{x}_1 - \mathbf{x}_2|)$$

INDUCTION PHENOMENA

Faraday law of induction:

$$\oint \mathbf{E} \cdot d\mathbf{l} = -\dfrac{1}{c} \dfrac{d}{dt} \int dS \, \mathbf{n} \cdot \mathbf{B}$$

Self-inductance: $L_{22} = \dfrac{1}{c^2} \oint \oint \dfrac{d\mathbf{l}_1 \cdot d\mathbf{l}_2'}{r_{22}'}$

Mutual inductance:

$$L_{12} = \dfrac{1}{c^2} \oint \oint \dfrac{d\mathbf{l}_1 \cdot d\mathbf{l}_2}{r_{12}}$$

Lenz's law: "The sense of an induced emf is such that, if the emf were allowed to drive a current, the field produced by that current would tend to cancel out the change that produced the emf."

Coulomb gauge: $\boldsymbol{\nabla} \cdot \mathbf{A} = 0$

$$\nabla^2 \phi = -4\pi\rho \longrightarrow \phi(\mathbf{x}, t) = \int \dfrac{\rho(\mathbf{x}', t)}{|\mathbf{x} - \mathbf{x}'|} d\tau'$$

$$\Box^2 \mathbf{A} = -\dfrac{4\pi}{c} \mathbf{j}_{\text{tr}} \longrightarrow \mathbf{A}(\mathbf{x}_1 t)$$

$$= \mathbf{A}_0(\mathbf{x}, t) + \dfrac{1}{c} \int \dfrac{\mathbf{j}_{\text{tr}}(\mathbf{x}', t_{\text{ret}})}{|\mathbf{x} - \mathbf{x}'|} d\tau'$$

where

$$\mathbf{j}_{\text{tr}} = \mathbf{j} - \boldsymbol{\nabla} \int \dfrac{\partial \rho(\mathbf{x}', t)}{\partial t} \Big/ (4\pi|\mathbf{x} - \mathbf{x}'|) d\tau'$$

$$\boldsymbol{\nabla} \cdot \mathbf{B} = 0$$

$$\boldsymbol{\nabla} \times \mathbf{B} = \dfrac{4\pi}{c}(\mathbf{j}_{\text{true}}[+\mathbf{j}_{\text{pol}}] + \mathbf{j}_M) \left[+ \dfrac{1}{c} \dfrac{\partial \mathbf{E}}{\partial t} \right]$$

where $\mathbf{j}_{\text{pol}} = \dfrac{\partial \mathbf{P}}{\partial t}, \quad \mathbf{j}_M = c\boldsymbol{\nabla} \times \mathbf{M}$

$$\boldsymbol{\nabla} \times \mathbf{H} = \dfrac{4\pi}{c} \mathbf{j}_{\text{true}} \left[+ \dfrac{1}{c} \dfrac{\partial \mathbf{D}}{\partial t} \right]$$

$$\boldsymbol{\nabla} \cdot \mathbf{H} = -4\pi\boldsymbol{\nabla} \cdot \mathbf{M}$$

where $\mathbf{H} = \mathbf{B} - 4\pi\mathbf{M}$

$$\dfrac{1}{8\pi} \int E^2 d\tau + \dfrac{1}{8\pi} \int B^2 d\tau$$
fields   fields

$$\dfrac{1}{8\pi} \int d\tau \mathbf{B} \cdot \mathbf{H} \qquad \dfrac{1}{2} L\mathbf{J}^2$$
fields   inductance

RELATIVITY

Addition of velocities: $v_{12} = \dfrac{v_1 + v_2}{1 + (v_1 v_2/c^2)}$

Length of a 4 − vector: $l = \displaystyle\sum_{\mu=1}^{4}(y_\mu - x_\mu)^2 \begin{cases} > 0: \text{spatial} \\ = 0: \text{lightlike} \\ < 0: \text{temporal} \end{cases}$

Lorentz transformation

$$x = x' \qquad\qquad x' = x$$
$$y = y' \qquad\qquad y' = y$$
$$z = \gamma(z' + vt') \qquad z' = \gamma(z - vt)$$
$$t = \gamma\left(t' + \frac{v}{c^2}z'\right) \qquad t' = \gamma\left(t - \frac{v}{c^2}z\right)$$
$$S \longrightarrow S' \qquad\qquad S' \longrightarrow S$$

Maxwell's equation

$$\begin{cases} \displaystyle\sum_\nu \frac{\partial F_{\mu\nu}}{\partial x_\nu} = \frac{4\pi}{c}j_\mu \\ \displaystyle\sum_\nu \frac{\partial \hat{F}_{\mu\nu}}{\partial x_\nu} = 0 \end{cases}$$

$$E = \gamma mc^2 = \sqrt{p^2 c^2 + m^2 c^4}$$
$$p = m\gamma v = Ev/c^2$$

$$F_{12} = B_3 = \hat{F}_{34}; \quad F_{23} = B_1 = \hat{F}_{14}; \quad F_{31} = B_2 = \hat{F}_{24}$$
$$F_{14} = -iE_1 = \hat{F}_{23}; \quad F_{24} = -iE_2 = \hat{F}_{31}; \quad F_{34} = -iE_3 = \hat{F}_{12}$$

Minkowski force: \qquad\qquad Doppler effect:

$$F_\mu = m\frac{du_\mu}{d\tau} = \frac{1}{c}\sum_\nu F_{\mu\nu}u_\nu \qquad \nu = \nu_2'\frac{\sqrt{1-\beta^2}}{1 - \mathbf{n}\cdot(\mathbf{v}/c)}$$

4-vectors:

$$u_\mu \equiv \left(\frac{\mathbf{v}}{\sqrt{1-\beta^2}}, \frac{ic}{\sqrt{1-\beta^2}}\right), \quad p_\mu \equiv \left(\mathbf{p}, \frac{iE}{c}\right)$$

$$\frac{du_\mu}{d\tau} \equiv \left[\frac{1}{1-\beta^2}\left(\mathbf{a} + \mathbf{v}\frac{\mathbf{v}\cdot\mathbf{a}/c^2}{\sqrt{1-\beta^2}}\right), \frac{ic}{1-\beta^2}\frac{\mathbf{v}\cdot\mathbf{a}/c^2}{\sqrt{1-\beta^2}}\right]$$

$$\mathbf{j}_\mu \equiv (\mathbf{j}, ic\rho), \quad A_\mu \equiv (\mathbf{A}, i\phi)$$

Transformation of fields:

$$\mathbf{E}'_\parallel = \mathbf{E}_\parallel \qquad\qquad \mathbf{B}'_\parallel = \mathbf{B}_\parallel$$

$$\mathbf{E}'_\perp = \frac{\mathbf{E}_\perp + [(\mathbf{v}/c)\times\mathbf{B}]}{\sqrt{1-\beta^2}} \qquad \mathbf{B}'_\perp = \frac{\mathbf{B}_\perp + [(\mathbf{v}/c)\times\mathbf{E}]}{\sqrt{1-\beta^2}}$$

Liénard–Wiechert potentials: $\phi(\mathbf{x},t) = \dfrac{e}{r}\dfrac{1}{1-\mathbf{n}\cdot\dfrac{\mathbf{v}}{c}}\Bigg|_{t_p=t-(r/c)}$ $\mathbf{A}(\mathbf{x},t) = \dfrac{e}{r}\dfrac{\mathbf{v}/c}{1-\mathbf{n}\cdot\mathbf{v}/c}\Bigg|_{t_p=t-(r/c)}$

Velocity fields:
$$\begin{cases}
\mathbf{B}^{\mathrm{near}}(\mathbf{x},t) = \dfrac{e}{r^2}\left(\dfrac{\mathbf{v}}{c}\times\mathbf{n}\right)\dfrac{1-\beta^2}{\left(1-\mathbf{n}\cdot\dfrac{\mathbf{v}}{c}\right)^3}\Bigg|_{t_p=t-(r/c)} \\[3em]
\mathbf{E}^{\mathrm{near}}(\mathbf{x},t) = \dfrac{e}{r^2}\left(\mathbf{n}-\dfrac{\mathbf{v}}{c}\right)\dfrac{1-\beta^2}{\left(1-\mathbf{n}\cdot\dfrac{\mathbf{v}}{c}\right)^3}\Bigg|_{t_p=t-(r/c)}
\end{cases}$$

Acceleration fields:
$$\begin{cases}
\mathbf{B}^{\mathrm{rad}}(\mathbf{x},t) = \dfrac{e}{c^2 r}\left[\dfrac{\mathbf{a}\times\mathbf{n}}{\left(1-\mathbf{n}\cdot\dfrac{\mathbf{v}}{c}\right)^2}+\dfrac{(\mathbf{n}\cdot\mathbf{a})\left(\dfrac{\mathbf{v}}{c}\times\mathbf{n}\right)}{\left(1-\mathbf{n}\cdot\dfrac{\mathbf{v}}{c}\right)^3}\right]\Bigg|_{t_p=t-(r/c)} \\[3em]
\mathbf{E}^{\mathrm{rad}}(\mathbf{x},t) = \dfrac{e}{c^2 r}\left[\dfrac{\mathbf{a}}{\left(1-\mathbf{n}\cdot\dfrac{\mathbf{v}}{c}\right)^2}+\dfrac{(\mathbf{n}\cdot\mathbf{a})\left(\mathbf{n}-\dfrac{\mathbf{v}}{c}\right)}{\left(1-\mathbf{n}\cdot\dfrac{\mathbf{v}}{c}\right)^3}\right]\Bigg|_{t_p=t-(r/c)}
\end{cases}$$

Self-force: $\mathbf{F}_{\mathrm{self}} = \underbrace{-\dfrac{4}{3}\dfrac{d\mathbf{v}}{dt}\dfrac{U_{el}}{c^2}}_{\text{self-energy}}+\underbrace{\dfrac{2e^2}{3c^3}\dfrac{d^2\mathbf{v}}{dt^2}}_{\substack{\text{radiation}\\\text{damping}}}$

Radiative power $= \dfrac{2e^2}{3c^3}\dfrac{a^2-\left(\mathbf{a}\times\dfrac{\mathbf{v}}{c}\right)^2}{(1-\beta^2)^3} \xrightarrow[v/c\ll1]{} \dfrac{2e^2 a^2}{3c^3}$

Multipole expansion:
$$\begin{cases}
\mathbf{E}(\mathbf{x},t) = ik[\mathbf{q}-\mathbf{n}(\mathbf{n}\cdot\mathbf{q})]\dfrac{e^{ikr-i\omega t}}{r} \\[1.5em]
\mathbf{B}(\mathbf{x},t) = (i\mathbf{k}\times\mathbf{q})\dfrac{e^{ikr-i\omega t}}{r}
\end{cases}$$

where

$\mathbf{q}(\mathbf{k}) = \mathbf{q}^{el}(\mathbf{k})+\mathbf{q}^{\mathrm{magn}}(\mathbf{k})$

$\mathbf{q}^{el}(\mathbf{k}) = -\dfrac{i\omega}{c}\displaystyle\int d^3x\,\rho_0(\mathbf{x})\mathbf{x}\int_0^1 d\lambda\,e^{-i\lambda\mathbf{k}\cdot\mathbf{x}}$; $\mathbf{q}^{\mathrm{magn}}(\mathbf{k}) = i\mathbf{k}\times\displaystyle\int d^3x\,\dfrac{1}{c}(\mathbf{x}\times\mathbf{j}_0)\int_0^1 d\lambda\,\lambda e^{-i\lambda\mathbf{k}\cdot\mathbf{x}}$

Scattering cross-section:

Rayleigh $(\omega\ll\omega_0)$ $\sigma = \dfrac{8\pi}{3}r_e^2\left(\dfrac{\omega}{\omega_0}\right)^4$

Resonance $(\omega\simeq\omega_0)$ $\sigma = \dfrac{8\pi}{3}r_e^2\dfrac{\omega_0^2/4}{(\omega-\omega_0)^2+\dfrac{1}{4}\gamma^2}$

Thomson $(\omega\gg\omega_0)$ $\sigma = \dfrac{8\pi}{3}r_e^2$

$\left(r_e = \dfrac{e^2}{m_e c^2} = 2.8\times10^{-13}\,\mathrm{cm}\right)$

Printed in the United States
By Bookmasters